Springer Series in Optical Sciences   Volume 49

Edited by Arthur L. Schawlow

Springer Series in Optical Sciences

Editorial Board: J.M. Enoch D.L. MacAdam A.L. Schawlow K. Shimoda T. Tamir

Volumes 1–41 are listed on the back inside cover

Laser
Spectroscopy VII

Proceedings of the Seventh International Conference,
Hawaii, June 24–28, 1985

Editors

T. W. Hänsch and Y. R. Shen

With 298 Figures

Springer-Verlag
Berlin Heidelberg GmbH

Professor Theo W. Hänsch
Department of Physics, Stanford University,
Stanford, CA 94305, USA

Professor Yuen Ron Shen
Department of Physics, University of California,
Berkeley, CA 94720, USA

ISBN 978-3-662-15253-9   ISBN 978-3-540-39664-2 (eBook)
DOI 10.1007/978-3-540-39664-2

Library of Congress Cataloging in Publication Data. Main entry under title: Laser spectroscopy VII. (Springer series in optical sciences ; v. 49). Papers from the Seventh International Conference on Laser Spectroscopy, held at the Maui Surf Hotel, Hawaii, June 24–28, 1985. Includes index. 1. Laser spectroscopy–Congresses. I. Hänsch, T.W. (Theo W.), 1941-. II. Shen, Y.R. III. Title: Laser spectroscopy seventh. IV. Title: Laser spectroscopy 7th. V. International Conference on Laser Spectroscopy (7th : 1985 : Maui Surf Hotel, Hawaii) VI. Series. QC454.L3L345 1985 535.5'8 85-20801

Preface

The Seventh International Conference on Laser Spectroscopy or SEICOLS'85 was held at the Maui Surf Hotel, Hawaii, USA, June 24 to 28, 1985. Like its predecessors at Vail, Megève, Jackson Lake, Rottach-Egern, Jasper Park, and Interlaken, SEICOLS '85 aimed at providing an informal setting for active scientists to meet and discuss recent developments and applications in laser spectroscopy. The Conference site on the sunny sands of famed Kaanapali Beach on the Island of Maui, although perhaps not the traditional mountain resort, offered nonetheless an atmosphere most inspiring to creative discussions during the unscheduled afternoons.

The Conference was truly international: 223 scientists represented 19 countries, including Australia, Canada, People's Republic of China, Denmark, Finland, France Germany (FRG), Great Britain, Israel, Italy, Japan, South Korea, Netherlands, New Zealand, Poland, Spain, Sweden, Switzerland, and U.S.A.

The intense scientific program included 14 topical sessions with 59 invited talks. Approximately 60 additional invited papers and 16 postdeadline papers were presented during three lively evening poster sessions. The present Proceedings contain oral as well as poster and postdeadline papers.

We thank all authors for the timely preparation of their manuscirpts, now available to a wider audience. We would also like to thank the members of the International Steering Committee for their valuable suggestions and advice. Our special thanks go to the members of the Program Committee for their painstaking efforts.

SEICOLS could not have been held without the help and support from many people and organizations. We gratefully acknowledge financial support by the International Union for Pure and Applied Physics (IUPAP), the U.S. Army Research Office, the U.S. National Science Foundation, and the U.S. Office of Naval Research. We are also indebted to our 18 corporate sponsors in the U.S., Japan, and Germany. Their contributions allowed us to add some rather special touches, such as delectable tropical refreshments served during the intermissions. We are particularly grateful to Rita Jones and Fred-a Jurian for their gracious and tireless efforts during all phases of the Conference organization. Finally, we would like to thank the staff of the Maui Surf Hotel, whose efficient and courteous assistance contributed much to the success of this meeting.

July 1985   *T.W. Hänsch · Y.R. Shen*

List of Sponsors

We gratefully acknowledge support of SEICOLS '85 from the following agencies:

> IUPAP - International Union of Pure & Applied Physics
> U.S. Army Research Office
> U.S. National Science Foundation
> U.S. Office of Naval Research

Sponsorship of SEICOLS '85 has been provided by the following companies:

> Burleigh Instruments, Inc.
> Coherent, Inc.
> Cooper Laser Sonics
> Fujitsu, Ltd.
> Hamamatsu Photonics, K.K.
> Hitachi, Ltd.
> Hoya Corpation
> Lambda Physik
> Lumonics, Inc.
> NEC Corporation
> Newport Corporation
> Nippon Kogaku, K.K.
> Quanta-Ray, Inc.
> Sony Corporation
> Spectra-Physics, Inc.
> Springer-Verlag
> Sumitomo Electric Industries, Ltd.
> Toshiba Corporation

Contents

Part III   Rydberg States

Part IV   Atomic Spectroscopy

Part V   Molecular and Ion Spectroscopy

Part VI   VUV and X-Ray, Sources and Spectroscopy

Part VII Nonlinear Optics and Wave Mixing Spectroscopy

Part VIII   Quantum Optics, Squeezed States, and Chaos

Part IX   Coherent Transient Effects

Part X   Surfaces and Clusters

Part XI Ultrashort Pulses and Applications to Solids

Part XII Spectroscopic Sources and Techniques

Part XIII Miscellaneous Applications

Part I

Laser Cooling, Trapping, and Manipulation of Atoms and Ions

Single Atomic Particle at Rest in Free Space: New Value for Electron Radius

H.G. Dehmelt, G. Gabrielse, K. Helmerson, G. Janik, W.G. Nagourney, P.B. Schwinberg, and R.S. Van Dyck, Jr.

Department of Physics, University of Washington, Seattle, WA 98195, USA

Zero-point confinement in a suitable trap is briefly discussed as a quantum-mechanical equivalent of the classical single particle at rest in free space. So far, such confinement has been realized, Fig.1, only for the 150 GHz cyclotron motion in geonium, a single electron permanently confined in a Penning trap. The most important result of stored ion spectroscopy is a new, 10 000 times smaller, radius for the electron, $R < 10^{-20}$ cm [1]. This result was obtained by analyzing our g-factor data [2], now $g/2 = 1.\ 001\ 159\ 652\ 193(4)$, on the basis of a near-Dirac particle model, Fig.2. The best current theoretical value [3] is $g/2 = 1.001\ 159\ 652\ 460(145)$. RF spectroscopy in geonium relies on the Continuous Stern-Gerlach Effect [4,5], in which a spin flip is detected as a small change in the $\approx$ 60 MHz axial oscillation frequency of the electron in the trap, Fig.3. The Kaufmann- Einstein Effect, or relativistic mass shift of the electron, may become a superior alternative: operating a geonium atom as a frequency selective mini-synchro-cyclotron [6] has produced an easily detectable shift in the axial frequency, Fig.4.

Like in Habann's 1926 split-anode magnetron, energy is quickly transferred from the oscillating electron to a resonant circuit,

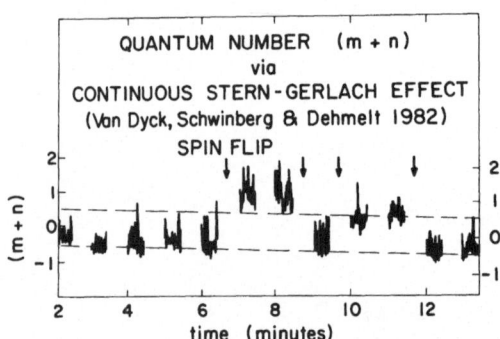

Fig.1 Demonstration of zero-point confinement for the cyclotron motion of an electron in a Penning trap by means of the Continuous Stern Gerlach Effect. The apparatus, which continuously reads out the quantum number sum n + m , is calibrated by externally induced spin flips, m = -½ → +½, occurring at random time intervals. The noise on the gated signal is in part amplifier noise, in part due to brief excursions from the n = 0 to the n = 1 cyclotron level of two Bohr magneton orbital magnetism, induced by the thermal (4 K) background radiation.

Near-Dirac Particle

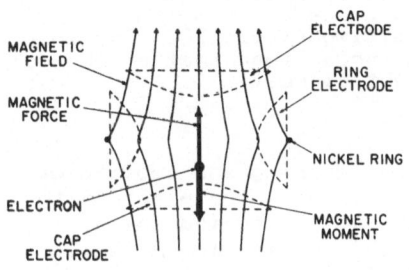

- Compton
  wave length

λ_C

⟨R⟩ particle
radius

$c \longrightarrow$

Fig.2 Near-Dirac particle (schematic). A Dirac (point) particle of mass M carries out a spontaneous quasi-circular Zitter-Bewegung (trembling motion) at the speed of light. The radius of this motion is the Compton wavelength, $\lambda_C = \hbar / M c$. For a particle charge e this motion produces the correct spin magnetism of one particle magneton, e $\hbar / 2 M c$. The less a near-Dirac particle of finite radius R looks like a point in relation to its Zitter-Bewegung radius λ_C, the more does its g-factor deviate from the Dirac value 2. We now claim that for <u>any</u> near-Dirac particle g – 2 will be ≈ 3.6 when R equals $\approx \lambda_C$, as is the case for the proton. Linear extrapolation gives R / $\lambda_C \approx |g_{exp} - g_{qed}| / 3.6 \approx 10^{-10}$ for the electron.

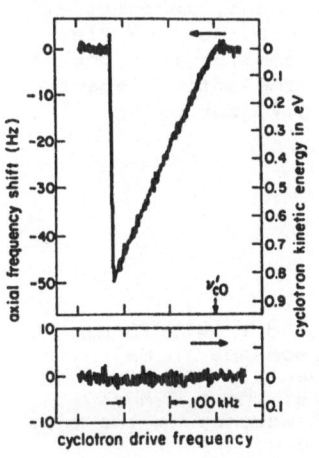

CAP
ELECTRODE

MAGNETIC
FIELD

RING
ELECTRODE

MAGNETIC
FORCE

NICKEL RING

ELECTRON

MAGNETIC
MOMENT

CAP
ELECTRODE

Fig.3 Continuous Stern-Gerlach Effect (schematic). An electron slowly moving along a field line in an inhomogeneous magnetic field with its magnetic moment anti-parallel to the field direction is driven towards weaker fields. We show here the minute magnetic force which adds to the strong axial electric restoring force and slightly deepens the net parabolic axial trapping potential for the magnetic moment direction shown. This, in turn, changes by ≈ 1 Hz the continuously monitored axial oscillation frequency of the electron when the spin is flipped.

Fig.4 Mini-Synchro-Cyclotron. As in the anharmonic oscillator, the cyclotron resonance frequency is pushed ahead when the drive frequency is slowly swept to lower values, and the kinetic energy increases. Via the Kaufmann-Einstein effect or relativistic mass shift the electron mass increases in proportion, and the cylotron resonance frequency drops correspondingly. Detection of this mass increase is via axial frequency shift.

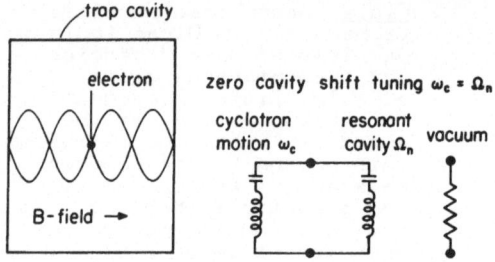

Fig.5 <u>Zero-cyclotron frequency shift tuning</u> for electron closely localized in center of low-loss cavity. One might think, that tuning ω_c to one of the cavity eigenfrequencies Ω_n of the type shown would result in zero-shift as it places an electric field node on the electron. In an equivalent lumped circuit picture valid near Ω_n the cavity looks like a simple series RLC circuit which is resonant at Ω_n. For an elongated cavity supporting only the TE11 mode, the assumed coincidence of the series resonances Ω_n' with the eigenfrequencies Ω_n is a fair approximation. According to formula 4 of reference [7] a convenient way to ascertain zero-shift tuning $\omega_c = \Omega_n'$ is by maximizing the <u>measured</u> cyclotron damping time. This procedure remains approximately valid in the general multimode case [8]. For the electron in vacuum a resistance much larger than that of the coil replaces the equivalent series resonant circuit.

but by us for the purpose of detection, damping and cooling of the oscillatory motion. Conversely, localizing the electron to ≈ 60 μm in the center of the trap cavity, and tuning the cyclotron frequency to a value where this cavity looks like a short [7,8], Fig.5, may make it possible to decrease the natural line width hundred-fold. By driving the axial frequency not on resonance but on a side band higher by the ≈ 12 kHz magnetron freqency, it has been possible to force the magnetron motion to absorb the excess in the photon energy and thereby shrink the magnetron radius to ≈ 15 μm [9].

By an analogous laser spectroscopic procedure [10] the oscillation amplitude of a Ba^+ ion in a different, zero-magnetic field trap has been reduced to ≈ 170 nm or less. As in Nuclear Magnetic Resonance, confinement is now much smaller than the wave length and <u>no motional side bands</u> appear in the optical spectrum [11], Fig.6. Such a mono-ion oscillator, Fig.7, but using Tl^+ or

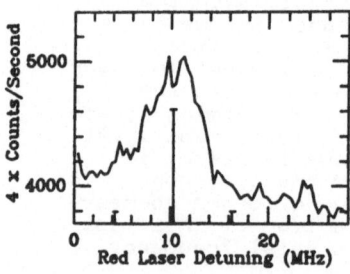

Fig.6 Side band free 2070 nm two-photon resonance in Ba^+ obtained in Seattle. No side bands spaced at the ≈ 6 MHz vibrational frequency in the trap are visible here.

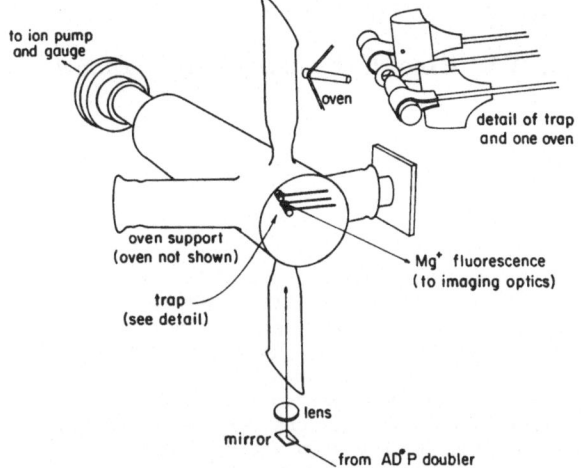

Fig.7 Simple mono-ion oscillator apparatus used in Seattle experiments. The laser beams are directed vertically through the trap and pass between the endcaps and ring. The trap axis is tilted at about 45° to the vertical to provide an unobstructed pass for the incident and fluorescent light.

to ion pump and gauge

oven

detail of trap and one oven

oven support (oven not shown)

trap (see detail)

Mg⁺ fluorescence (to imaging optics)

lens

mirror

from AD⁺P doubler

a similar ion, may make an optical frequency standard with a 1000-day reproducibility of 10^{-18} possible [4] and thereby allow one to watch the expansion of the universe [12] from one's laboratory.

The present paper is a condensed version of a more detailed publication [13]. H.D. thanks R.Beringer, R.H.Dicke, E.N.Fortson, E.L.Ginzton and D.Thouless for discussions. We gratefully acknowledge the Support of the National Science Foundation.

1. S. J. Brodsky, and S. D. Drell, Phys. Rev. D <u>22</u>, 2236 (1980)
2. R. S. Van Dyck, Jr., P. B. Schwinberg, and H. G. Dehmelt, in <u>ATOMIC PHYSICS 9</u>, Eds. R. S. Van Dyck and E. N. Fortson (World Scientific Book Publishers, New York 1985)
3. T. Kinoshita and W. B. Lindquist, ibidem.
4. Hans Dehmelt, in <u>Advances in Laser Spectroscopy</u>, Eds. F. T. Arecchi, F. Strumia & H. Walther (Plenum, New York 1983)
5. Hans Dehmelt, Phys. Rev. D, submitted.
6. G. Gabrielse, H. Dehmelt, and W. Kells, Phys. Rev. Letters <u>54</u>, 537 (1985)
7. Hans Dehmelt, Proc. Natl. Acad. Sci. USA <u>81</u>, 8037 (1984). Erratum: The equivalent transmission lines for the 10 propagating modes should be in series, not in parallel.
8. L. S. Brown, G. Gabrielse, K. Helmerson and J. Tan, to be published
9. R. S. Van Dyck, Jr., P. B. Schwinberg, and H. G. Dehmelt, in <u>New Frontiers in High Energy Physics</u>, Eds. B. Kursunoglu, A. Perlmutter, and L. Scott (Plenum, New York 1978)
10. D. Wineland, and H. Dehmelt, Bull. Am. Phys. Soc. <u>20</u>, 637 (1975)
11. G. Janik, W. Nagourney, and H. Dehmelt, J. O. S. A. B, in press
12. F. J. Dyson, Rev. Mod. Phys. <u>51</u>, 450 (1979).
13. Hans Dehmelt, Annales de Physique, in press.

Two-Photon Optical Spectroscopy of Trapped Hg II

J.C. Bergquist, D.J. Wineland, W.M. Itano, H. Hemmati*, H.-U. Daniel†, and G. Leuchs††

Time and Frequency Division, National Bureau of Standards, Boulder, CO 80303, USA

Traditionally, precision frequency measurements and lifetime measurements have provided numerous and stringent tests of physical principles. Two of the important merits of trapped ions for precision frequency spectroscopy and lifetime measurements are long confinement times and a gentle, highly non-perturbative environment. Present experiments give the first glimpse of the deep reservoir of precision spectroscopic investigations possible with stored ions. Examples include the g-2 experiment that yields the highest precision yet obtained in fundamental tests of QED [1], and the measurement of the electron to proton mass ratio to an accuracy of 0.04 ppm [2]. In other experiments, trapped ions are radiatively cooled to temperatures below 0.1K [3-5]. This directly reduces Doppler effects to all orders and thereby enhances resolution and precision in the method of trapped ion spectroscopy. Furthermore, because the radiation of the laser used for cooling is usually tuned to near resonance with a highly allowed transition in order to facilitate efficient cooling, many photons per second per ion are scattered; the high rate of scattered photons provides a powerful scheme not only to detect ions but also to detect atomic transitions to metastable states by the method of double resonance [6,7]. This method can give unit detection efficiency or, equivalently, a signal-to-noise (S/N) ratio limited only by the quantum-statistical fluctuation in the number of ions that make the transition and not by detection solid angle, detector quantum efficiency, etc. [8]. This is the maximum S/N ratio possible. We further note that this highly efficient detector of an atomic transition works even if the cycling/cooling transition and the transition to the metastable state share no common level [9]. With the continual and rapid maturation of laser cooled, stored ion spectroscopy, there has come a flurry of activity. Examples include a 300 fold improvement in the limits for spatial anisotropy by using laser cooled, electromagnetically trapped $^9Be^+$ ions [10], a laser cooled atomic frequency standard with an accuracy comparable to the best cesium beam atomic frequency standards [11], a demonstration of strong coupling in a small cloud of laser compressed and laser cooled ions [12], and a precise measurement of the electron-to-proton mass ratio [13]. In this paper we describe some of the first results obtained in our study of the Doppler-free, two-photon $5d^{10} 6s \, ^2S_{1/2} - 5d^9 6s^2 \, ^2D_{5/2}$ transition in singly ionized mercury stored in a miniature rf trap [14].

A cross-sectional view of the trap electrodes is shown in Fig. 1. In contrast to the customary hyperbolic surfaces which produce a harmonic potential nearly everywhere in the trap volume, our trap was

---

Work of the US Government, not subject to US copyright.
*Present address: Allied Bendix Aerospace Corporation, Columbia, MD
†Present address: Springer Verlag, Heidelberg, West Germany
††Heisenberg Fellow of the Deutsche Forschungsgemeinschaft. Present address: Max Planck Institute für Quantenoptik, Garching, West Germany

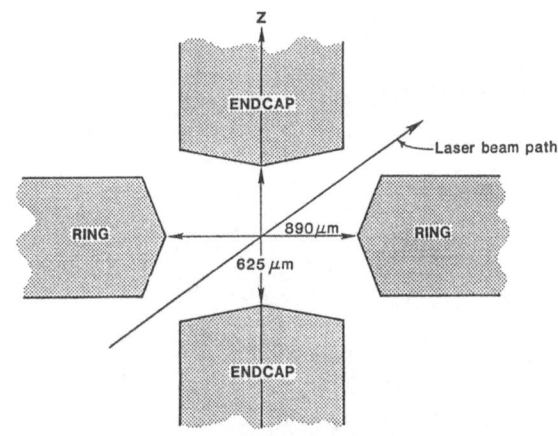

Fig. 1. Schematic showing cross-section view of trap electrodes. The electrodes are figures of revolution about the z-axis and are made from molybdenum.

machined with simple conical cuts. However, the trap dimensions and angles were chosen to make the fourth and sixth order anharmonic terms in the trapping potential vanish [15]. For practical purposes, then, the potential was harmonic in the central volume of the trap. Trapping was obtained by driving the electrodes with an oscillating electric field with a voltage amplitude $V_o \leq 1$ kV and a frequency near 21 MHz. The trap was housed in a vacuum vessel which was baked and pumped out to a partial pressure of background gas $\leq 10^{-7}$ Pa. A small amount of ^{198}Hg was admitted into the vacuum vessel and crossed by a low current electron beam near trap center in order to produce the trapped mercury ions. After loading 50-200 ions, the mercury vapor was frozen out in a liquid nitrogen cold trap and the system was back-filled with about 10^{-2} Pa of dry helium. The helium buffer gas was used to collisionally cool the mercury ions to near room temperature.

The $5d^{10} 6s\ ^2S_{1/2} - 5d^9 6s^2\ ^2D_{5/2}$ transition, driven by two photons with a wavelength near 563nm, was detected by the very sensitive optical double resonance scheme noted above. About 5 μW of narrowband, cw, sum-frequency-generated radiation near 194 nm [16] was focused through the trap between the ring electrodes and the endcaps. The beam waist, w_o, was approximately 25 μm and located near trap center. The frequency of the 194 nm radiation was tuned into coincidence with the 6s $^2S_{1/2}$ - 6p $^2P_{1/2}$ first resonance transition. The photons scattered by the ions into a small solid angle perpendicular to the 194 nm beam were detected and recorded. Rejection of stray photons scattered elsewhere in the system was achieved by means of a single spatial filter placed in the collection channel at the image of the ion cloud. The overall detection efficiency including solid angle, collection optics, filter and photomultiplier sensitivity was about 10^{-4}. The signal level was typically 2-10 x 10^3 counts/s and the signal to background level was better than 10/1. When the ions were driven out of the $^2S_{1/2}$ ground state into the metastable $^2D_{5/2}$ state, there was a decrease in the 194 nm fluorescence corresponding to the number of ions in the D state.

About 100 mW of the output from a cw ring dye laser oscillating near 563 nm was coupled through a single mode fiber and mode matched into a near concentric cavity placed around the rf trap. The relative position of the cavity and the trap was adjusted so that the cavity beam waist ($w_o \cong 30$μm) was located near the center of the cloud of trapped ions. The axes of the 563 nm beam, the 194 nm beam and the collection optics were mutually perpendicular. The cavity enhanced the one way power of the 563 nm radiation by about a factor of fifty, to give nearly 5 W of circulating power. Additionally, the high finesse of the cavity better ensured nearly equal intensity counter-propagating beams, which are necessary for high-

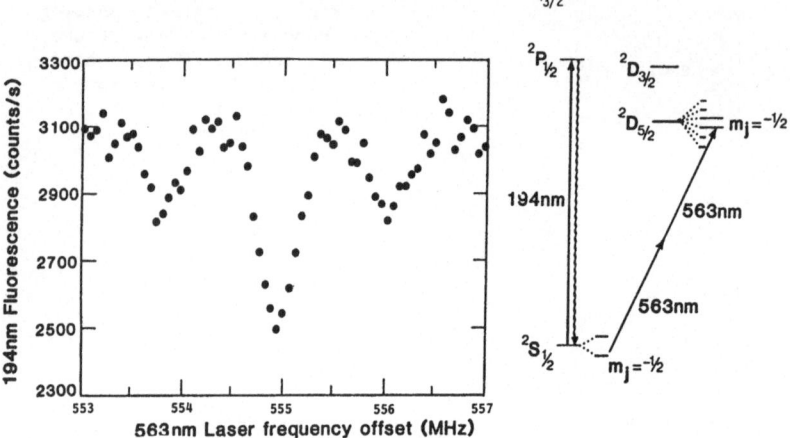

Fig. 2. Two-photon $^2S_{1/2}-^2D_{5/2}$ transition in $^{198}Hg^+$. AM sidebands caused by the harmonic secular motion of the ions are visible in this scan. The frequency scan is 4 MHz at the fundamental laser frequency ($\lambda \cong 563$ nm). The depth of the central component is about 25% of full scale. The integration time is 2 s/point. In the inset is a simplified energy-level diagram of $^{198}Hg^+$, depicting the levels of interest.

resolution Doppler-free two-photon spectroscopy. The linewidth of the ring dye laser in these preliminary experiments was of the order of 300 kHz. The frequency of the laser was offset locked and precisely scanned with respect to a second stabilized ring dye laser that was locked to a hyperfine component in the Doppler-free saturated absorption spectrum of $^{127}I_2$.

A 4 MHz scan of the frequency of the dye laser through the two-photon resonance is shown in Fig. 2. Also included in Fig. 2 is a simplified energy-level diagram showing the pertinent optical levels in Hg^+. A small magnetic field of approximately 11.6×10^{-4} T(11.6 G) which differentially Zeeman splits the ground and excited states, was applied by means of a pair of Helmholtz coils placed around the trap. The electric field vector of the linearly polarized 563 nm laser is oriented parallel to this field. The selection rule for the two-photon transition for this polarization is $\Delta m_J = 0$, and, thus, only two Zeeman components are expected, separated by approximately 13 MHz (approximately 6.5 MHz at the dye laser frequency). In Fig. 2, we scanned over only one of these components ($m_J = -1/2 \Leftrightarrow m_J = -1/2$) but saw sideband structure. This structure is due to amplitude modulation (AM) of the 563 nm laser intensity due to the harmonic secular motion of the ions in the rf trap, which carries the ions back and forth through the laser beam. Recall that the 563 nm waist is about 30 μm whereas the ion cloud volume is characterized by a linear extent of about 60μm. We note that any frequency modulation (FM) of the laser caused by the motion of the ions [17] is strongly suppressed if the cavity is well aligned, so that the k-vectors of the counter running light beams are anti-parallel. By changing the well depth of the trap, the harmonic frequency of the secular motion was changed, thereby shifting the sideband components on the two-photon signal. To our knowledge, this is the first observation of secular motion sidebands at optical frequencies. The depth of the central feature in Fig. 2 is nearly 25% of full scale, implying that we have nearly saturated the strongly forbidden two-photon transition. For the data of Fig. 2, the 194 nm laser irradiated the ions continuously. The linewidth of the two-photon resonance is about 420 kHz, and is determined in nearly equal parts by the 563 nm laser linewidth of about

320 kHz and by the nearly 270 kHz excitation rate of the $^2S_{1/2}$ state by the 194 nm radiation. When the 194 nm radiation is chopped, the two-photon linewidth drops to approximately 320 kHz.

In the near future we anticipate narrowing the 563 nm laser linewidth to the order of a few kHz and studying various systematic effects including pressure broadening and shifts, power broadening, and light shifts. Ultimately, we would like to narrow the laser linewidth to a value near that imposed by the 0.1 s natural lifetime of the D state, and to drive the two-photon (or single photon, electric quadrupole [8]) transition on a single, laser-cooled ion.

We gratefully acknowledge the support of the US Air Force Office of Scientific Research and the US Office of Naval Research. We wish to thank R. Blatt (University of Hamburg) for technical assistance in the construction of the rf trap and H. Layer (NBS) for supplying the ^{198}Hg. We take particular pleasure in acknowledging the help of W. Martin, J. Reader and C. Sansonetti (NBS) who provided the correct wavelength for the $^2D_{5/2} - {}^2S_{1/2}$ transition. We also wish to thank D. Huestis (SRI) and our colleague R. Drullinger for many useful conversations and insights.

1. P. B. Schwinberg, R. S. Van Dyck, Jr., and H. G. Dehmelt, Phys. Rev. Lett. 47, 1679 (1981).
2. R. S. Van Dyck, Jr. and P. B. Schwinberg, Phys. Rev. Lett. 47, 395 (1981)
3. D. J. Wineland, R. E. Drullinger, and F. L. Walls, Phys. Rev. Lett. 40, 1639 (1978).
4. W. Neuhauser, M. Hohenstatt, P. Toschek, and H. G. Dehmelt, Phys. Rev. Lett. 41, 233 (1978).
5. W. Nagourney, G. Janik, and H. G. Dehmelt, Proc. Natl. Acad. Sci. USA, 80, 643 (1983).
6. D. J. Wineland, W. M. Itano, J. C. Bergquist, J. J. Bollinger, and J. D. Prestage, in "Atomic Physics 9", R. S. Van Dyck, Jr. and E. N. Fortson, eds. (World Scientific Publ. Co., Singapore, 1984) p. 3.
7. H. G. Dehmelt, in "Advances in Laser Spectroscopy," F. T. Arecchi, F. Strumia, and H. Walther, eds. (Plenum, New York, 1983) p. 153.
8. D. J. Wineland, J. C. Bergquist, R. E. Drullinger, H. Hemmati, W. M. Itano, and F. L. Walls, J. Phys. (Orsay, Fr.) 42, C8-307 (1981); D. J. Wineland, W. M. Itano, J. C. Bergquist, and F. L. Walls, Proc. of 35th Annu. Symp. on Freq. Control, (1981) p. 602 (copies available from Electronic Industries Assoc., 2001 Eye St., NW, Washington, DC 200006).
9. W. M. Itano and D. J. Wineland, Phys. Rev. A 24, 1364 (1981).
10. J. D. Prestage, J. J. Bollinger, W. M. Itano, and D. J. Wineland, Phys. Rev. Let. 54, 2387 (1985).
11. J. J. Bollinger, J. D. Prestage, W. M. Itano, and D. J. Wineland, Phys. Rev. Lett. 54, 1000 (1985).
12. J. J. Bollinger and D. J. Wineland, Phys. Rev. Lett. 53, 348 (1984).
13. D. J. Wineland, J. J. Bollinger, and W. M. Itano, Phys. Rev. Lett. 50, 628; erratum: 50, 1333 (1983).
14. J. C. Bergquist, D. J. Wineland, W. M. Itano, H. Hemmati, H.-U. Daniel, and G. Leuchs, submitted for publication.
15. E. C. Beaty, to be published.
16. H. Hemmati, J. C. Bergquist, and W. M. Itano, Opt. Lett. 8, 73 (1983).
17. H. A. Schuessler, Appl. Phys. Lett. 18, 117 (1971); F. G. Major and J. L. Duchene, J. Phys. (Orsay) 36, 953 (1975); H. S. Lakkaraju and H. A. Schuessler, J. Appl. Phys. 53, 3967 (1982); M. Jardino, F. Plumelle, and M. Desaintfuscien, in "Laser Spectroscopy VI", H., P. Weber and W. Lüthy, eds. (Springer-Verlag, NY, 1983) p. 173; L. S. Cutler, R. P. Giffard and M. D. McGuire, Appl. Phys. B36, 137 (1985)

Cooling, Stopping, and Trapping Atoms

A.L. Migdall[†], T. Bergeman[†], J. Dalibard[*], H. Metcalf[†], W.D. Phillips,
J.V. Prodan[††], and I. So[†]

Electricity Division, National Bureau of Standards, MET B258,
Gaithersburg, MD 20899, USA

While it has long been possible to trap charged particles, until now no
one has conclusively demonstrated the use of non-material walls to confine
neutral atoms [1,2]. All of the traps that have been proposed suffer from
the same difficulty: they require atoms with extremely low kinetic energy
in order to work [3]. Practical atom traps are necessarily very shallow
because the electromagnetic forces which can be applied to neutral atoms
are generally quite small, since they arise from dipole or higher order
moments. The successful trapping described here occurs because we have
developed the means (via laser manipulation of an atomic beam) of producing
atoms cold enough (typically less than 1 K) to be contained in a trap. We
have demonstrated magnetic trapping and we believe our technique can load
any type of neutral atom trap, including the various laser traps proposed
[3].

Interest in trapping of atoms comes from intrinsic interest in the
dynamics and stability of the trap itself, as well as from the possible
uses of trapped atoms, such as the measurement of long, metastable-state
lifetimes, refrigeration of atoms to energies as low as a few microkelvin
[4], and observation of quantum collective effects at low temperature and
high density [5,6]. Highly refrigerated atoms could be released from a
trap for use in ultra-high-resolution spectroscopy, free of motional
Doppler or transit time effects, and in some instances spectroscopy might
be performed on atoms held in the trap.

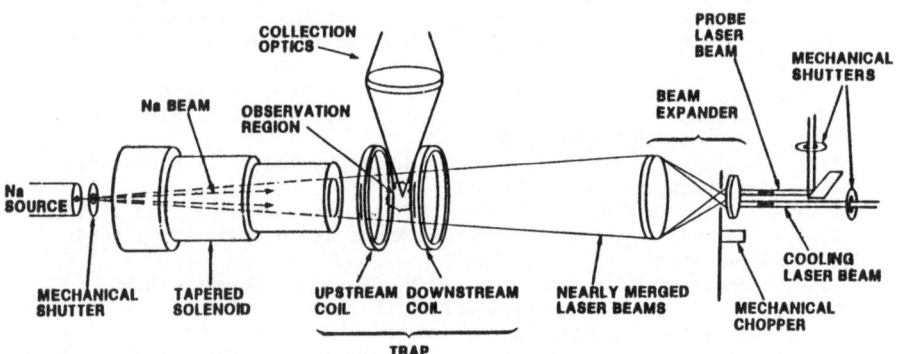

Fig 1 Schematic of the apparatus.

---

† Dept. of Physics, State University of New York at Stony Brook
* Ecole Normale Superieure, Paris
†† Lockheed Missles and Space Co., Sunnyvale, CA.

Production of a stationary gas of very cold atoms is accomplished in three steps (see Fig. 1): Sodium atoms from a thermal source are first slowed from 1000 m/s to 100 m/s in a long solenoid, thereby removing 99% of their energy. These 100 m/s atoms are brought to rest in the center of the trapping coils, 40 cm away from the solenoid; then the trap field is switched on, confining the atoms. Our slowing technique is described in detail elsewhere [7,8] and is discussed here only briefly, while our means of stopping and trapping the slow atoms is explained more fully.

Atoms in a thermal atomic beam with mean velocity of about 1000 m/s are decelerated and cooled as they scatter light from a near resonant, circularly polarized, counter-propagating laser. As the atoms in the beam slow down, their changing Doppler shift would take them out of resonance with the laser were it not for the spatially varying Zeeman shift provided by the tapered solenoid. In this way, nearly all atoms initially slower than 1000 m/s are decelerated and finally stopped near the exit end of the solenoid. Furthermore, all these atoms are optically pumped into a state which is suitably oriented for magnetic trapping.

In order to allow the slow atoms near the end of the solenoid to reach the trapping region, free of any fast atom background, we abruptly shut off both the cooling laser light and the atomic beam. In our experiments, we choose 100 m/s atoms which reach the center of the trap/observation region after 4 ms, although we can choose delays up to 75 ms for atoms as slow as 5 m/s. Once atoms with the desired velocity reach the observation/trap region, a 400 µs pulse of light from the cooling laser decelerates them to rest,while only the upstream (nearest the Na source) trap coil is on. This coil produces a spatially varying magnetic field which both shifts the atoms into resonance with the fixed-frequency cooling laser and partially compensates for the changing Doppler shift of the decelerating atoms,in the same manner as the field in the main solenoid. Because we have the flexibility to alter the delay and duration of the stopping laser pulse, the magnetic field in the observation region, and the frequency of the cooling laser, we can choose the final velocity and location of the atoms to be loaded into the trap. (Figure 3 of Ref [8] highlights the selectability of the final velocity as a function of the cooling laser frequency.)

After the atoms are brought to rest, the trap is completed by simply turning on the opposing downstream coil, and is maintained for a selected trapping time. In order to detect the trapped atoms, the trap field is turned off (about 5 ms are required for the field to fall to zero), the weak (2 mW/cm^2) probe laser is turned on for 100 µs, and fluorescence induced by the probe in a volume of about 1 cm^3 is observed. The entire cycle (slowing, stopping, trapping, and probing) is repeated as the probe laser frequency is slowly scanned. In this way we determine the number and velocity distribution of the atoms that remain after a given trapping time. Observation in zero field eliminates any Zeeman shifts or broadening.

Magnetic trapping is possible because an inhomogeneous magnetic field exerts forces on atoms with a magnetic dipole moment. Atoms in quantum states whose energy increases with increasing field are trappable,and the potential energy is equal to the effective magnetic moment µ times the magnitude of the magnetic field B. For Na, where µ is about one Bohr magneton, a maximum field of 0.025 T, as in our trap, forms a potential well only 17 mK deep, corresponding to an atomic velocity of 3.5 m/s. This small energy depth illustrates the difficulties in realizing a magnetic trap.

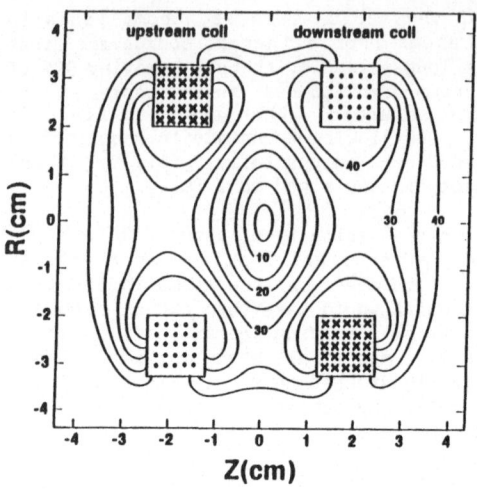

Fig. 2. Equipotentials (equivalent to equal field magnitudes shown in mT) of quadrupole trap in a plane containing the axis of symmetry (z axis).

Our trap consists of two 3.4 cm diameter coaxial loops separated by 2.7 cm and each carrying 1900 ampere-turns. This trap, quadrupolar in nature, is the simplest of the magnetic traps described by Paul [9,10]. Figure 2 shows our trap and its equipotentials and Fig. 1 shows its placement in the apparatus used to load it.

We determine the atomic density inside of the trap from the size of the probe-induced fluorescence signals. The density decreases exponentially with trapping time from an initial density of several times 10^3 with a decay time constant of 0.83(7) s. The decay time is limited mainly by collisions with background gas at a pressure of about 10^{-8} torr in the trap region.

A more fundamental limitation to the lifetime of atoms in the trap arises from non-adiabatic or Majorana transitions,which can change the orientation of the atoms' magnetic moment so that they are expelled from the trap. As long as the Larmor frequency of the atoms is greater than their instantaneous orbital frequency, the orientation of the atoms is preserved. Violation of this condition is most likely to occur for fast atoms near the center of the trap,where the field passes through zero as it reverses direction. For our atoms, the region where such transitions are significant is less than 10^{-10} of the total trap volume, so we expect for typical (unclosed) orbit times of a few tens of ms that atoms can last many seconds before suffering a Majorana transition.

In conclusion, we have demonstrated the utility of laser manipulation of atoms by successfully loading a very shallow neutral atom trap and observing those trapped atoms for over a second. The trapped atoms have energies less than 17 mK which is comparable to the lowest energies reported [11] for laser-cooled trapped ions,and is achieved not by refrigeration but by the rejection of atoms with energies higher than that which can be contained in the trap. We expect future developments to include a laser trap loaded with the same cold atoms as in the present trap, the cooling of trapped atoms, and the achievement of higher trapped atom densities.

We thank the Office of Naval Research for its support of this work. The SUNY authors thank the National Science Foundation for its support.

1. B. Martin, Bonn-IR-75-8 (1975) and N. Niehues, Bonn-IR-76-35 (1976), theses, Universitat Bonn, unpubl. Martin saw some indications of magnetic trapping.

2. A. L. Migdall, et al. Phys. Rev. Lett., 54, 2596 (1985)

3. W. Phillips, et al. J. Opt. Soc. Am. B to be published. This contains a review of trap proposals.

4. D. E. Pritchard, Phys. Rev. Lett., 51, 1336 (1983)

5. W. Stwalley and L. Nosanow, Phys. Rev. Lett., 36, 910 (1976). and W. Stwalley, Prog. Quant. Electr. 8,203 (1984).

6. H. Hess, Bull. Am. Phys. Soc. 30, 854 (1985).

7. W. Phillips et al. Phys. Rev. Lett. 48, 596 (1982), and J. Prodan et al. Phys. Rev. Lett. 49, 1149 (1981).

8. J. V. Prodan et al. Phys. Rev. Lett., 54, 992 (1985).

9. W. Paul, unpublished.

10. R. Golub and J. Pendlebury, Rep. Prog. Phys. 42, 439 (1979).

11. W. Nagourney and H. Dehmelt, Proc. Natl. Acad. Sci. USA, 80 643 (1983).

Three-Dimensional Confinement and Cooling of Atoms by Resonance Radiation Pressure

S. Chu, L. Hollberg, J.E. Bjorkholm, A. Cable, and A. Ashkin

AT&T Bell Laboratories, Holmdel, NJ 07733, USA

The scattering force due to resonance radiation pressure was first detected by Frisch in 1933.[1] Later, Ashkin[2] pointed out that laser light can exert a substantial force suitable for the optical manipulation of atoms, and numerous proposals to cool and trap neutral atoms with laser light.[3] Atoms in an atomic beam have been stopped by light,[4] in which the final velocity spread corresponds to a temperature of 50 - 100 mK. We report here the confinement and cooling of atoms with laser light, in which the atoms are localized in a 0.2 cm^3 volume for a time in excess of 0.1 second and cooled to a temperature of T = 2.4 x 10^{-4}K.[5]

The scheme we use, is briefly outlined. Consider an atom irradiated by a laser beam. For each photon absorbed, the atom receives a net change of atomic velocity $\Delta v = h/m\lambda$ directed along the laser beam. The subsequent emission of a photon has no preferred direction,so that the average force due to the re-emitted photon is zero. Hansch and Schawlow [6] noted that if counterpropagating beams were tuned to the low-frequency side of the absorption line, a moving atom will Doppler shift an opposing beam closer to resonance and shift the co-propagating beam further from resonance. Thus, an atom immersed in six counterpropagating beams along three orthogonal axes will always see a viscous damping force $\vec{F} = -\alpha\vec{v}$ opposing its motion.

In order to calculate the minimum temperature achievable by this technique, statistical fluctuations in the optical force must be considered. These fluctuations lead to heating and are balanced by the cooling effect of the damping force. In the absence of damping, the continual absorption and emission of photons will cause the atom to execute a random walk in velocity. Although $<v> = 0$, $<v^2>$ increases linearly with the total number of scattered photons. The heating effect of the scattering force was first observed by Bjorkholm, et al. [7] A detailed analysis of quantum fluxuations, including stimulated processes[8], predicts a minimum temperature based on a balance of heating and cooling of $kT_{min} = h\gamma/2$ where γ is the width (fwhm) of the absorption line. For sodium, $\gamma = 10$ MHz and $T_{min} = 240\mu K$.

An estimate of the confinement time is made by noting that the damped motion of atoms in the photon bath will be analogous to Brownian motion. The time t necessary for a displacement $<x^2>$ is given by $t = \alpha<x^2>/2kT$. This result assumes an infinite medium, but for a random walk in a medium surrounded by a spherical escape boundary [9], the storage is reduced by a factor of 3.1 for our experimental conditions.

The experiment arrangement is shown in Fig. 1, and the optical layout is given in Fig. 2. An intense pulsed atomic beam is produced by irradiating a pellet of sodium metal with a 10 nsec long, 30 mj pulse at 532 nm focused to ~ 5 x 10^{-2} cm^2. Atoms with initial velocities of ~ 2 x 10^4 cm/sec are

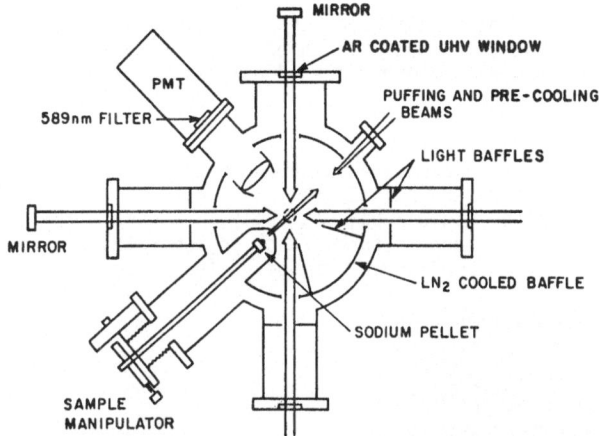

Figure 1 - Schematic of the vacuum chamber and intersecting laser beams and atomic beam. The vertical confining beam is indicated by the dotted circle. The "puffing" beam is from the pulsed YAG laser.

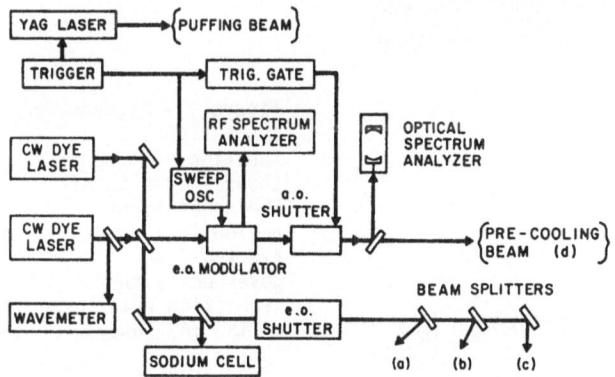

Figure 2 - Optical schematic use to produce the cooling and confining laser beams.

slowed down (pre-cooled) in less than 5 cm to velocities of ~2 x 10^3 cm/sec via a counter-propagating laser beam. The laser frequency is rapidly changed using an electro-optic modulator similar to Ertmer, et al.[4] compensate for the decreasing Doppler shift. After slowing for 0.5 ms, the pre-cooling beam is turned off, and the atoms drift into the confinement region defined by the six interacting laser beams, where they are cooled and viscously confined. The pre-cooling and confining laser beams are obtained from the combined beams of two cw stabilized dye lasers. The lasers are operated at a frequency difference of ~1.7 GHz to prevent optical pumping of the sodium ground state. The power of each confining beam was varied between 4-20 mW and the beam radius is ω_0 = 0.36 cm.

The inset to Fig. 2 shows the averaged fluorescence signal of 16 atomic beam pulses. The initial spike is due to fast atoms passing through the interaction region. The base line is the scattered light level obtained by blocking the pre-cooling beam. We estimate that densities 10^6 atoms/cm^3 are initially confined. At these densities, the cloud of atoms is clearly visible by eye. If any of the confining laser beams is blocked, the cloud vanishes. In Fig. 2, we compare the fluorescence signal decay (dots) to the predicted atom density using the model of a random-walk with a spherical boundary. The horizontal axis is in the dimensionless unit Dt/R^2 where D is the diffusion constant, t is the time, and R is an effective radius of the spherical boundary. Using a computed maximum value of damping constant $\alpha_{max} = 5.8 \times 10^{-18}$ gm/sec and R = 0.4 cm we obtain an upper limit on $T < T_{max} = D\alpha_{max}/k = 1.9$ mK.

A more direct measurement of the temperature is shown in Fig. 3. After cooling for 15 msec, all six beams are turned off in 0.1 msec. The atoms will then leave the observation region ballistically with their instantaneous velocities. When the light is turned back on, only the slower moving atoms will be observed in the confinement region. The fraction of

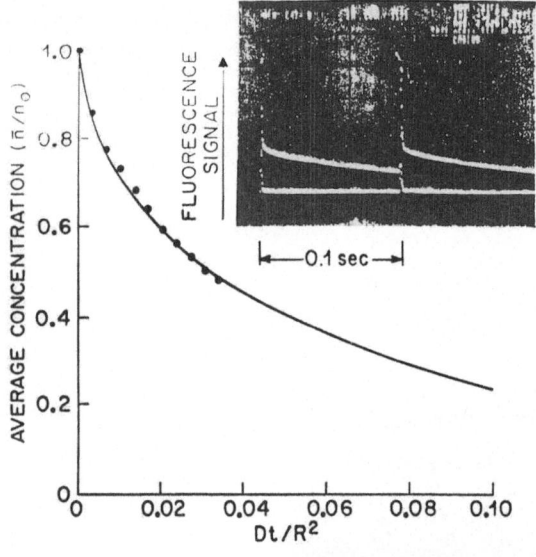

Figure 3 - Fluorescence signal as a function of time is shown in the inset. The confinement region is loaded every 0.1 sec governed by the repetition rate of the YAG laser.

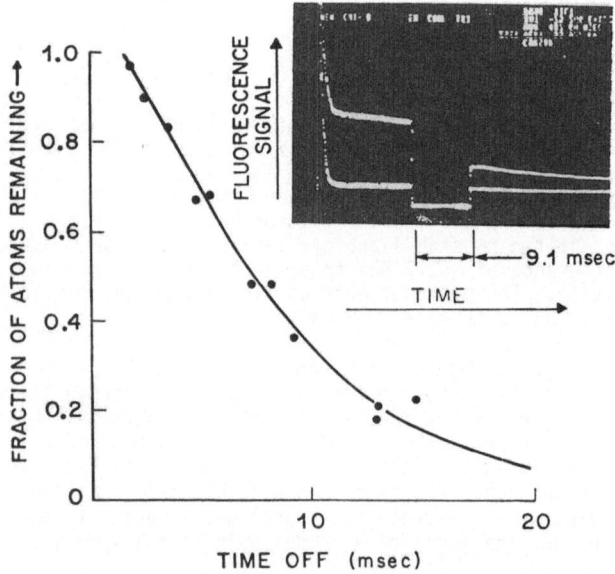

Figure 4 - Inset shows the effect of blocking the confining laser beams 15 ms after the YAG laser fires. The horizontal trace is the scattered light level. The fraction of atoms remaining is plotted a function of the time-off period.

remaining atoms as a function of the light-off time is compared to a calculation based on a initial uniform spherical distribution of atoms with a Maxwell-Boltzman distribution of velocities.[10] Given an observation radius of 0.4 cm, we obtain a measured temperature $T = 240 \ \mu K \left(^{+200 \ \mu K}_{-60 \ \mu K} \right)$

In summary, we have cooled sodium atoms to the quantum limit $T_{min} \cong h\gamma/2$ imposed by our experimental conditions. With a beam waist of $\omega_0 = 0.36$ cm, we have obtained a confinement time in excess of 0.1 sec at a density of 10^6 atoms/cm^3. Note that the confinement time can be increased by increasing ω_0. For example, if ω_0 is increased by a factor of 3, the confinement time would have a 1/e decay time of ~1.6 seconds. After a storage time of 1 second, the density would still be ~4 x 10^4 atoms/cm^3.

We acknowledge helpful discussions with J. P. Gordon and assistance from L. Buhl in the preparation of the LiTaO$_3$ crystal used in the electrooptic modulator.

REFERENCES

1. O. R. Frisch, Z. Phys. 86, 42 (1933).

2. A. Ashkin, Phys. Rev. Lett., 24, 156 (1970); Phys. Rev. Lett., 25, 1321 (1970).

3. See, for example A. Ashkin, Science, 210, 1081 (1980); Laser Cooled and Trapped Atoms, ed. W. Phillips, Prog. in Quantum Electronics, 8, 204 (1984); For recent proposals, see D. E. Pritchard, Phys. Rev. Lett., 51, 1336 (1983); A. Ashkin, Opt. Lett., 9, 454 (1984); J. Dalibard, S. Reynaud, and C. Cohen-Tannoudji, J. Phys. B, 17, 4577 (1984).

4. J. Prodan, A. Migdall, W. D. Phillips, I. So, H. Metcalf, and J. Dalibard, Phys. Rev. Lett., 54, 992 (1985); W. Ertmer, R. Blatt, J. L. Hall, and M. Zhu, Phys. Rev. Lett., 54, 996 (1985).

5. S. Chu, L. Hollberg, J. E. Bjorkholm, A. Cable, A. Ashkin, Phys. Rev. Lett., 55, 48 (1985).

6. T. W. Hansch and A. L. Schawlow, Optics Comm., 13, 68 (1975).

7. J. E. Bjorkholm, R. R. Freeman, A. Ashkin and D. B. Pearson, Optics Lett. 5, 111 (1980).

8. J. P. Gordon and A. Ashkin, Phys. Rev. A, 21, 1606 (1980); R. J. Cook, Phys. Rev. A, 20, 224 (1979); R. J. Cook, Phys. Rev. Lett., 44, 976 (1980).

9. N. A. Fuchs, The Mechanics of Aerosols, (Pergamon Press, Oxford, 1964) pp. 193-200.

10. J. L. Picque, Phys. Rev. A., 19, 1622 (1979).

A Proposal for Optically Cooling Atoms to Temperature of 10^{-6} K

S. Chu, A. Ashkin, J.E. Bjorkholm, J.P. Gordon, and L. Hollberg

AT & T Bell Laboratories, Holmdel, NJ 07733, USA

We propose an optical method for the three-dimensional cooling of collections of atoms to extremely low temperatures. With this method it should be possible to cool sodium atoms to temperatures of 10^{-6}K or less. These low temperatures correspond to velocities less than the recoil velocity of an atom emitting a single resonant photon. Although the method is straightforward in principle, it anticipates the demonstration of optical traps capable of localizing (trapping) atoms within small regions of space (dimensions on the order of 1-100μm) and of cooling them to temperatures on the order of the "quantum limit" $kT=h\gamma/2$, where γ is the natural linewidth of the resonant transition. The recent laser cooling of sodium atoms to the "quantum limit" of 2.4 x10^{-4}K [1] suggest that this prerequisite will be met in the near future. Several schemes have been proposed previously for the cooling of magnetically trapped atoms to temperatures below $h\gamma/2$.[2], [3]. The magnetic trapping of sodium atoms [4] opens up the possibility for trying those schemes.

Our scheme is simply described as follows. Consider an ensemble of atoms tightly confined within a small region of space by a small optical trap and laser-cooled to a temperature of $kT=h\gamma/2$. At time t=0 the small, tight optical trap used to localize the atoms is turned off and is replaced by a loose trap that will allow the atoms to expand into a larger volume. The atoms are initially centered in the potential minimum of the new trap; as time proceeds they execute harmonic oscillatory motion in that trap. Assuming that the trap is an ideal harmonic potential well and that all the atoms start from the exact center of the trap, all the atoms will reach their turning points (velocity=0) at the same time, equal to 1/4 of the harmonic oscillator period, irrespective of their initial velocities. At this time the trapping light is turned off and all the atoms come to rest simultaneously at zero temperature.

There are several factors that place limits on the ultimate temperature that can be achieved by this scheme. (i) Any atom in the trap must not scatter a photon in a time corresponding to 1/4 of the harmonic oscillator period. This constraint requires the use of dipole force trap [5] where the laser is tuned far from resonance. (ii) Given an initial spread of atoms to dimensions Δx_i and an initial velocity spread of Δv_i, Liouville's Theorem states that the final velocity spread can be reduced to $\Delta v_f=(\Delta x_i/\Delta x_f)\Delta v_i$. (iii) Practical traps always have anharmonic components to the potential. so that the atoms will not come to rest simultaneously. Thus, the motion of the atoms must be confined to the central (harmonic) portion of the trap. Given a fixed amount of tunable laser power, the above conditions determine a lower temperature limit.

A three-dimensional optical trap using the dipole force of radiation pressure can be formed by three intersecting and mutually perpendicular tightly colliminated laser beams. We choose to use TEM*$_{01}$-mode laser beams having an intensity distribution given by $I(r)=I_0(r/w_0)^2\exp(-2r^2/w_0^2)$, where r is the transverse coordinate. The use of three such beams forms a nearly radially symmetric trap for displacements small compared to w_0. The use of TEM*$_{01}$ laser beams is preferred over TEM$_{00}$ beams, since the atoms are confined to regions of low light intensity, thus minimizing the problem of heating by spontaneous scattering. As a concrete example, consider a laser power of 1W and an initial atomic spread of 1.3μm. This localization can be obtained using a confining trap 4×10^2 hy deep with a radius of 37μm. We find that for a expansion trap of $w_0=130\mu$m and a laser detuning from resonance of 24.5 GHz, the final temperatures predicted by anharmonicities and by phase space considerations are both equal to 5×10^{-7} and that each atom scatters one photon on average. Similarly, for a laser power of 0.2W, the values $w_0=80\mu$m and a detuning of 17.4 GHz yield final temperatures of $\tilde{} 1\mu$K.

There are other ways to use the properties of harmonic traps to cool atoms. (1) Consider a pulsed version of the above scheme. In this techique the atoms are allowed to expand into free space when the tight small trap is turned off. At some later time the large loose trap is pulsed on for a short period time. For a harmonic trap the impulse delivered to an atom at position r is proportional to r, as is the atom's velocity. Thus by correctly choosing the pulse energy all atoms may be brought nearly to rest. Provided the pulse energy and mode structure of the pulsed laser can be controlled, cooling to temperatures below 10^{-7}K can be achieved. (2) The traps can be made more harmonic by combining trapping and antitrapping potentials analogous to optical lens correction by using both converging and diverging lenses. (3) After localizing the atoms in a small deep trap, a shallow compact trap can be used to let the high speed tail of the Boltzman distribution escape. The rms velocity of the remaining atoms is decreased. If the atoms are then allowed to cool by the harmonic expansion technique, temperatures in the nanokelvin range may be obtainable.

REFERENCES

1. S. Chu, L. Hollberg, J. E. Bjorkholm, A. Cable, and A. Ashkin, Phys. Rev. Lett. 55, 48 (1985).

2. D. E. Pritchard, Phys. Rev. Lett. 51, 1336 (1983).

3. H. F. Hess, Bull. Am. Phys. Soc. 30, 854 (1985).

4. A. L. Migdall, J. V. Prodan, W. D. Phillips, T. H. Bergeman, and H. J. Metcalf, Phys. Rev. Lett., 54, 2596 (1985).

5. A. Ashkin, Phys. Rev. Lett. 40, 729 (1978).

Stopping Atoms with Diode Lasers*

R.N. Watts and C.E. Wieman[†]

Joint Institute for Laboratory Astrophysics, University of Colorado
and National Bureau of Standards, Boulder, CO 80309, USA

We have succeeded in stopping a beam of cesium atoms using frequency-chirped diode lasers. We scan over the Doppler profile of a 100°C thermal cesium beam to bring more than 10^{10} atoms/s to a temperature of 1°K, a limit imposed by the 30 MHz linewidth of our free-running lasers. These results are preliminary,and we expect that further work will provide substantial improvements in our final temperature and density. This is an extremely simple and inexpensive way to produce cold atoms.

Two techniques have been devised for using laser light to stop atoms. The NBS group in Gaithersburg [1] used a single frequency c.w. dye laser combined with a large tapered solenoid to achieve the necessary condition that the atomic transition and the laser frequency stay in resonance as the atoms slow. Hall and co-workers [2] used an alternative method,in which the frequency of the dye laser was swept (chirped) using state-of-the-art electro-optic modulators. While both of these approaches have been shown to work well, they involve large investments of money and equipment. We have found that the frequency chirp approach can be implemented using inexpensive diode lasers and simple electronics. The frequency of a diode laser can be smoothly and rapidly varied over many GHz simply by varying the injection current. Thus, by using an appropriate current ramp, we have stopped a beam of cesium atoms.

A schematic of the apparatus is shown in Fig. 1. Cesium atoms in a 100°C oven effuse from a 0.5 mm hole and are collimated to 8 mrad. At this oven temperature, the beam has a mean velocity of 2.7×10^4 cm/s, an intensity of 3×10^{11}/s and a Doppler FWHM of 400 MHz. The atoms are stopped by bombarding them with resonant counterpropagating $6s-6p_{3/2}$ photons at 8521 Å. Each atom-photon collision slows the atom by 0.35 cm/s. Thus, it takes approximately 76,000 photons to bring an atom to a halt. For an excited state lifetime of 30 ns, this process takes 5 ms and requires about 70 cm. Because cesium has two hyperfine ground states separated by 9192 MHz, two lasers are used. The primary cooling is done by one laser tuned to the $6s(F=4)-6p_{3/2}(F=5)$ transition. For circularly polarized light, this transition is a good approximation to a leakage-free two-level system. The other laser, tuned to the $6s(F=3)-6p_{3/2}$ transition, insures that the F=3 ground state is depleted. The F=4 laser has about 5 mW of power while the F=3 laser has 0.3 mW. Both lasers have a free-running linewidth of 30 MHz and are focused to match the atomic beam. Each injection current ramp sweeps the frequencies of the two lasers 1 GHz in 15 ms. The last 10 ms are spent sweeping the full Doppler profile of

---

*Work supported by the National Science Foundation.
[†]Sloan Fellow; also with Department of Physics, University of Colorado.

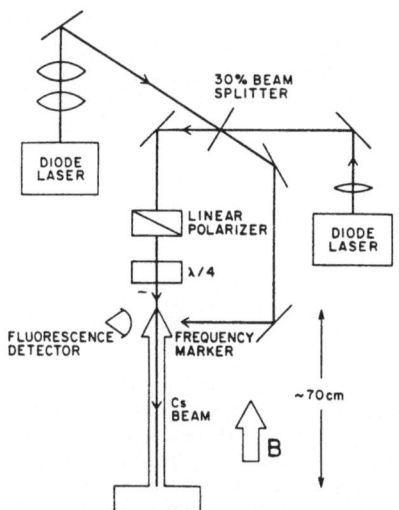

Fig. 1.   Schematic of apparatus.

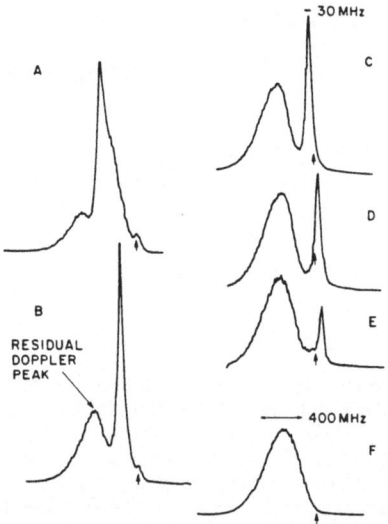

Fig. 2.   Slowed atom scans.   Curves
A–E show various amounts of cooling.
Curve F is the Doppler profile with
no cooling.   The arrows mark zero
velocity.

the cesium beam. The end of the chirp may be adjusted to bring the atoms
to any desired speed.

The same lasers are used to monitor the resulting velocity distribution. After each chirp, a portion of a much slower linear ramp (corresponding to a frequency change of 1.5 GHz in 6 s) is switched into the
laser injection currents. At the same time, a detector is gated on to
probe the slow atom fluorescence. This gate lasts for 250 µs, after which
a new chirp is started. To give a zero velocity marker, a small fraction
of the laser is sent in perpendicular to the atomic beam.

The results are shown in Fig. 2. The small bumps on A and B are the
frequency marker peaks. As the figure shows, by adjusting the end of the
frequency sweep to different points on the Doppler profile, we are able to
slow, stop, or even reverse a portion of the atomic beam. The residual
Doppler peak, which grows as more cooling is done, is due to uncooled
atoms at the probe region. The linewidth of our lasers currently limits
us to a 1°K temperature in the resulting stopped distribution.

References

1. J. V. Prodan et. al., Phys. Rev. Lett. 54 (1985) 992.
2. W. Ertmer, R. Blatt, J.L. Hall and M. Zhu, Phys. Rev. Lett. 54 (1985)
    996.

Dynamics of the Optical Piston

H.G.C. Werij, J.E.M. Haverkort, W.A. Hamel, R.W.M. Hoogeveen, and J.P. Woerdman

Molecular Physics Department, Huygens Laboratory, University of Leiden, P.O. Box 9504, NL-2300 RA Leiden, The Netherlands

1 Introduction

Recently we have demonstrated a semipermeable optical piston [1] using the principle of light-induced drift (LID) [2]. In the present work,new experimental results on the dynamics of the optical piston are reported and compared with theory. The basic idea of LID and the optical piston is as follows. Na atoms are present as trace impurity in a buffer gas. A laser beam, travelling along the z-axis and tuned near the Na D resonance, produces velocity-selective saturation of the Na atoms due to the Doppler effect; a hole will occur in the 3S ground-state velocity distribution, whereas a peak will appear in the 3P excited-state distribution. This corresponds to opposing fluxes of excited and unexcited Na atoms. In the presence of a buffer gas (in our case Ar) both velocity distributions tend to thermalize,due to velocity-changing collisions, and, since the elastic collision cross-section of an excited Na atom is larger than that of a ground-state Na atom, these fluxes do not compensate each other. As a result,the Na atoms drift in the same direction as the laser beam when the latter is tuned into the red Doppler wing. Note that in LID the laser photons act as "Maxwell demons", labelling a specific velocity class and transforming the random atomic motion (partly) into ordered motion, i.e. drift. Net momentum transfer, as in radiation pressure, does not occur: equal but opposite momenta are imparted to the Na vapor and to the buffer gas. In an optically dense system the laser intensity is strongly position-dependent; this leads to the drift velocity being spatially nonuniform and hence, in case that the laser beam completely fills the inner diameter of a tube, to the formation of a steep Na front, an "optical piston" [1,3].

2 Theory of the Optical Piston

Theoretically [4] the optical piston is represented by a soliton-type solution of a nonlinear differential equation for the laser intensity $I(z)$,

$$\frac{\partial I}{\partial z} = \frac{-n_{max} \sigma_a}{\sqrt{1+I(z)/I_{sat}}} [1-I(z)/I(0)] \; I(z) \quad , \tag{1}$$

the absorber density $n(z)$ being determined by:

$$n(z) = n_{max} [1-I(z)/I(0)] \quad , \tag{2}$$

and the maximum density in the front by:

$$n_{max} = \frac{D_g - D_e}{D_g D_e} \frac{v_L I(0)}{\hbar \omega_L (A+K_g)} \quad . \tag{3}$$

σ_a is the Doppler absorption cross-section, $I(0)$ is the laser intensity at the entrance of the cell ($z = 0$), I_{sat} is the saturation intensity, D_g and D_e are the diffusion coefficients of ground state and excited state absorbers, v_L is the axial velocity of the resonant absorber packet, A is the spontaneous decay rate and K_g is the ground state velocity thermalization rate. Equations (1-3) have been derived from the generalized Bloch equations for a one-dimensional system of two-level absorbers, in the Doppler limit [4]. In the derivation it has been assumed that $v_p << v_d(0)$, where v_p is the piston velocity and $v_d(0)$ the drift velocity at z = 0; this is a valid assumption in our experiment [1].

In order to compare our experimental results with theory, we describe the Na atoms as fictitious two-level absorbers. The 3^2P_1 and $3^2P_{3/2}$ fine structure levels are lumped together. Also the F = 1 and F = 2 hyperfine levels of the 3^2S_1 ground state are lumped together, and effects of optical hyperfine pumping are ad hoc accounted for by introducing a fictitious absorption cross-section σ_a.

3 Diagnostics of the Optical Piston

In our experiment, the beam of a single-mode cw dye laser fills the cross-section of a capillary cell, made of sodium-resistant glass (Fig. 1). The capillary (150 mm length, 2 mm inner diameter) is the connection between two reservoirs. The cell is initially

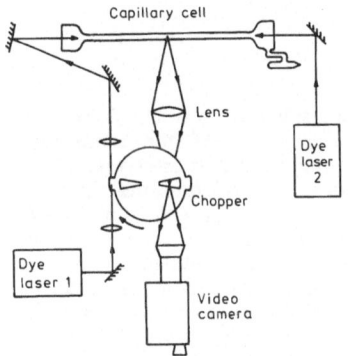

Fig. 1 Experimental set-up. Periodically (25 s^{-1}) the pump beam (dye laser 1) is intercepted for $\backsim 1$ ms; during that interception the fluorescence induced by the probe beam (dye laser 2) is monitored using a video camera

uniformly filled with Na vapor ($\backsim 1 \times 10^{12}$ cm^{-3}, i.e. optically thick) and Ar buffer gas ($\backsim 3.2 \times 10^{17}$ cm^{-3}). At this Ar density $J_{sat} = 0.27$ W/cm^2 for the Na D_2 transition; the laser intensity is typically 3 W/cm^2. When the laser frequency is tuned about 1 GHz into the red Doppler wing of the Na D_2 $^2S_1$ (F = 2) → $^2P_{3/2}$ transition an optical piston starts propagating along the capillary. We detect the Na density profile during the piston run by observing the fluorescence induced by a second dye laser beam, tuned far ($\backsim 15$ GHz) into the collision-broadened Na D_2 wing. This fluorescence is a measure of the local Na density, n(z), in the capillary. An example is shown in Fig. 2; the vertical scale has been established in a separate experiment by determining n_{max} from the transmission of a weak probe beam ($\backsim 1$ μW, $\backsim 1$ mm diameter) perpendicular to the capillary axis. We verified from the shape of the absorption spectrum that the power of this probe beam was low enough to avoid appreciable optical hyperfine pumping [5]. Substitution of typical experimental parameters into (1,2) yields a theoretical curve which is in good agreement with experiment (Fig. 2).

There is no agreement between experiment and theory regarding the value of v_p. Kinetically, the piston velocity is related to the exponential decay length, L_d, of the absorber density at the dark side of the piston: $v_p \backsim D_g/L_d$. Since in a typical experiment $L_d \backsim 2$ cm and $D_g \backsim 25$ cm^2/s theory predicts $v_p \backsim 0.1$ m/s; experimentally we find $v_p \backsim 0.001$ m/s. We suspected that the reason for this large disagreement would be adsorption at the glass wall; therefore we performed a direct

23

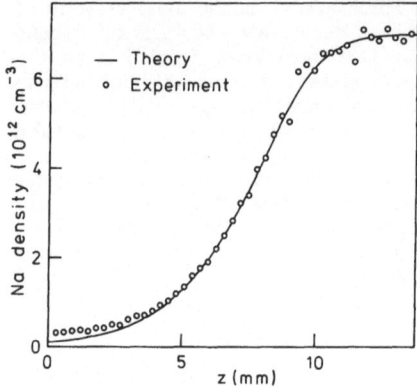

Fig. 2 The experimental Na density profile of a stationary optical piston; the argon buffer gas pressure is 10 torr, the laser is tuned in the red Doppler wing of the Na D_2 (F = 2) hyperfine component (detuning 1 GHz), exciting the velocity class v_L = -600 m/s, and the laser intensity is 1.6 W/cm^2. The theoretical curve represents the solution of (1,2) for n_{max} σ_a = 14 cm^{-1}, $I(0)$ = 1.6 W/cm^2 and I_{sat} = 0.27 W/cm^2

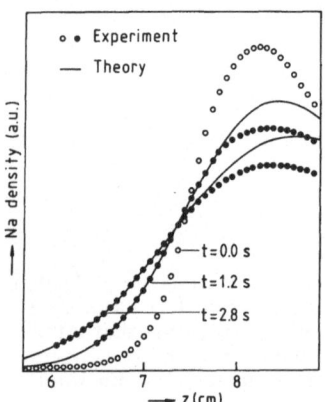

Fig. 3 Smearing of the Na front after sudden interception of the pump beam; the experimental result is compared with calculations based on the diffusion equation using $D_{g,eff}$ = 0.14 cm^2/s. The experimental data for t = 0.0 s has been used as initial condition for the calculations, leading to a prediction for the shape of the front at t = 1.2 s and t = 2.8 s

measurement of the effective diffusion coefficient of Na in our capillary cell. We suddenly intercepted the pump beam during a piston run and observed the subsequent smearing of the Na front. Analysis of the smearing yields an effective diffusion coefficient $D_{g,eff}$ ∿ 0.14 cm^2/s (Fig. 3); this value is two orders of magnitude smaller than the gas kinetic value, D_g ∿ 25 cm^2/s, indicating that indeed adsorption effects are important and explaining the anomalously low value of the piston velocity observed in the experiment.

4 Analysis of Surface Effects

The effects of Na surface adsorption can be easily incorporated into the previous analysis of the shape of the Na front [6]. Assuming that the Na in the vapor phase is in equilibrium with the Na at the surface, that the Na sticking probability is unity and that the surface monolayer coverage $\theta \ll 1$, we find that D_g and V_p are replaced by:

$$D_{g,eff} = D_g/(1 + K) \quad , \tag{4}$$

$$v_{p,eff} = v_p/(1 + K) \quad , \tag{5}$$

where $K = \bar{v}\tau_s/2R$, $\bar{v}$ being the mean thermal velocity, τ_s the sticking time and R the radius of the capillary. In fact K is the ratio of the amount of Na adsorbed at the

24

surface and the amount of Na in the vapor phase. Apart from the substitutions (4,5) the analysis given above remains valid; in particular, the shape of the Na density profile and the value of n_{max} do not change due to adsorption, it is only the velocity v_p that is reduced by a factor $(1 + K)$. In our experiments apparently $K \backsim 100\text{-}200$. This value corresponds to a sticking time $\tau_s \backsim 1$ ms, which is a reasonable value assuming an adsorption energy $Q \backsim 1$ eV.

Thus a simple physical picture emerges: during propagation of the optical piston along the capillary,the vapor-phase Na is everywhere and always in equilibrium with the adsorbed Na. We estimate that in the "piston region" the monolayer coverage $\theta \backsim 0.1$. Therefore a "ring" of adsorbed Na travels along the capillary, in pace with the optical piston. The majority of the Na travels in the ring along the surface; only a small fraction ($\backsim 1\%$) travels in the volume. Since the driving force is present in the volume only, this leads effectively to a retardation of the dynamics of the piston.

5 Conclusions

In conclusion, we have presented here new experimental results on the shape and the dynamical behavior of the optical piston. Good agreement with recent theory [4] has been obtained as far as the shape of the Na density profile is concerned. The piston velocity is anomalously low; this has been explained as a consequence of the adsorption of Na atoms at the glass wall. The theory has been extended and now incorporates such surface effects.

Systematic studies are underway in which we will investigate the dependence of LID on the nature and the pressure of the buffer gas and on the nature of the surface material. Furthermore, work is in progress on a more realistic multilevel description of LID.

Our work is part of the research program of the Foundation for Fundamental Research on Matter (FOM) and was made possible by financial support from the Netherlands Organization for the Advancement of Pure Research (ZWO).

References

1   H.G.C. Werij, J.P. Woerdman, J.J.M. Beenakker and I. Kuščer, Phys.Rev.Lett. 52, 2237 (1984).
2   F.Kh. Gel'mukhanov and A.M. Shalagin, JETP Lett. 29, 711 (1979).
3   F.Kh. Gel'mukhanov and A.M. Shalagin, Sov.Phys.JETP 51, 839 (1980).
4   G. Nienhuis, Phys.Rev. A31, 1636 (1985).
5   R. Walkup, A. Spielfiedel, W.D. Phillips and D.E. Pritchard, Phys.Rev.A23, 1869 (1981).
6   I. Kuščer, private communication.

Part II

Laser Spectroscopic Applications
to Basic Physics

Fast Beam Laser Spectroscopy and Tests of Special Relativity

M. Kaivola*, S.A. Lee†, O. Poulsen, and E. Riis

Institute of Physics, University of Aarhus, DK-8000 Aarhus C, Denmark

Tests of special relativity have been a favoured playground for physicists since the theory was introduced in 1905 [1]. This is to be contrasted with the almost unanimous view that the theory is correct. Being a physical model, based on postulates, special relativity must not be taken for granted, but should be rigorously tested in experiments. Traditionally, these tests have been performed at low velocities, but with the refinement of experiments in the field of elementary particle physics, high velocity tests of special relativity have become available.

One well-known result of special relativity is the relativistic Doppler shift, that a moving clock has its frequency modified by a factor $1/\gamma = \sqrt{1-v^2/c^2}$, where v is the velocity of the clock relative to the observer, and c is the speed of light. To study this effect, fast moving clocks are needed. One type of experiments is based on optical spectroscopy with fast atoms or molecules. The 'time' information is obtained from the frequency (or wavelength) of a particular transition in the atoms. Indeed, the first verification of the time-dilation effect of special relativity was an optical experiment performed by IVES and STILWELL [2] in 1938. They confirmed the existence of this red shift by measuring the difference in wavelength of the fluorescence emitted by a 30 keV beam of H_2^+ ions, when observed in the forward and backward directions relative to the ion beam. They reported an accuracy of 3%, although the uncertainties in the experiment were nearly 10%. A later version [3] of the experiment achieved an accuracy around 5%. As the limiting factor in these experiments was the presence of the very large first order Dopper shift in the measurement, improvements in accuracy appeared to be difficult at the time.

The advent of tunable and highly stable dye lasers has resulted in drastic improvements in the resolution achievable with optical spectroscopy. SNYDER and HALL [4] proposed using non-linear laser spectroscopy to eliminate the first order Doppler effect. They obtained an accuracy of 0.5% in the red shift by using saturation spectroscopy on a fast neon beam. A recent laser version [5] of the Ives-Stilwell experiment has an accuracy of 0.6%. But these experiments require measuring the velocity of the atoms, which would introduce major uncertainties.

A second type of experiments [6,7] uses the lifetime of decaying elementary particles to determine the time-dilation factor γ. In the most recent

---

*Permanent address: Helsinki University of Technology
Department of Technical Physics
Espoo, Finland
†Permanent address: Department of Physics
Colorado State University
Fort Collins, Colorado, USA

and most accurate determination of γ, BAILEY et al. [7] measured the decay
of muons in a storage ring. By separate measurements of the rotation
frequency of the muons and their mean radius of circulation, they obtained
an accuracy of 1×10^{-3}. Since their experiment was reported, the lifetime
of the negative muon has been revised, causing the above experimental accuracy to degrade by a factor of 3. These high velocity tests have recently
attracted some interest, because of speculations of a possible breakdown of
the Lorentz invariance at high velocities in the weak interaction [8,9].
Unfortunately, 'selection rules' suppress the Lorentz non-invariance in the
decay of the long lived $\mu^{\pm}$ particles, leaving only the decay of the shorter
lived $\pi^{\pm}$ and $K^{\pm}$ to be of interest.

The heretofore most accurate verification of time-dilation in special
relativity is of a quite different nature. While all the previous experiments have been local, VESSOT et al. [10] used a hydrogen maser, launched
vertically upward to 10,000 km, to measure the combined effects of the time
dilation and the gravitational redshift. If one assumes the correctness of
the usual expression for the gravitational redshift, the experiment obtained
an accuracy of 1.4×10^{-4} in the relativistic Doppler shift.

This status of time-dilation experiments has challenged us to enter the
race to develop a measuring scheme, capable of counting the frequency difference between two clocks, one at rest and the other moving with velocity v.
The aim is to carry out high accuracy determination of γ(v). To reduce
experimental uncertainties, the velocity v of the moving clock should be
related only to atomic parameters measurable against primary standards [11].
Furthermore, the clocks should be Doppler-free to first order to eliminate
the huge classical Doppler shifts. Simultaneous two-photon absorption (TPA)
of neon atoms in a cell [12] and in a fast beam [13,14] meets these requirements. We measured the beat frequency between two cw dye lasers, one stabilized to a fast beam TPA resonance and the other to the same TPA transition in a thermal sample in a cell. This procedure allowed us to count the
frequency difference, representing the relativistic time-dilation, with an
accuracy of 4×10^{-5}. This experiment constitutes the most accurate measurement of the time-dilation effect to date [15].

The virtue of this scheme is easily uncovered. Fig. 1(a) shows the cell
case, where the laser frequency $\nu_{L,c}$ is given by half the frequency difference $(\nu_1+\nu_2)/2$ between the 4d' final state and 3s initial state in neon. In
the fast beam case, shown in Fig. 1(b) in the atom's rest frame, the large
first order Doppler shifts allows one particular intermediate level to be
tuned into exact resonance. The $3p'[3/2]_2$ state is thus brought into resonance at an atomic velocity $v_0/c=(\nu_2-\nu_1)/(\nu_2+\nu_1)$, corresponding to a beam
energy of ∿120 keV. Experimentally, this velocity is easily located due to
a resonant enhancement of the TPA [16]. The beam energy is located to within

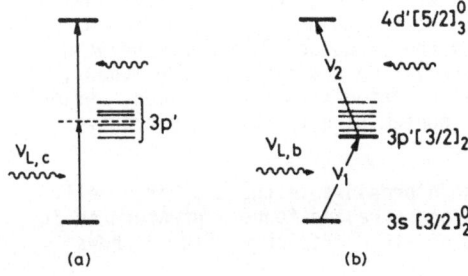

(a)   (b)

Fig. 1. Two different TPA schemes
are used in this experiment:
(a) the thermal-atom case in the
cell, where the two-photon absorption is an off-resonant process;
(b) the fast-beam case, where the
Doppler effect allows one particular
intermediate level to be tuned into
exact resonance. The atom is viewed
in its rest frame.

29

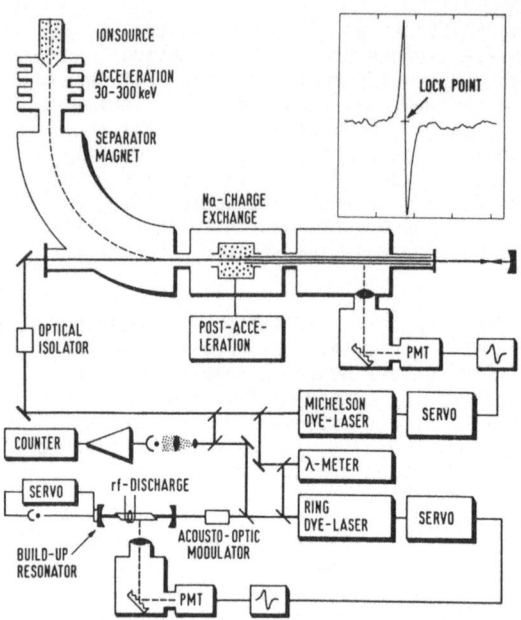

Fig. 2. Experimental apparatus used in the measurement of the relativistic time-dilation consists of two actively stabilized lasers, an accelerator, an rf-discharge cell, and frequency counting equipment. Ne$^+$ ions were produced in an rf ion source and accelerated to ~120 keV. After magnetic separation, metastable neon atoms were produced by charge exchange of Ne$^+$ in Na vapour. A post-acceleration voltage provided precise velocity control of the fast atom beam. Mu-metal shield surrounding the atom beam and Helmholtz coils surrounding the discharge cell reduced external magnetic fields to below 50 mGauss.

2 eV. On the other hand, this resonant situation is realized at the laser frequency $\nu_{L,b}=\sqrt{\nu_1\nu_2}$. Thus the frequency difference $\delta=\nu_{L,b}-\nu_{L,c}$ can be calculated by knowing only the atomic transition frequencies ν_1 and ν_2, and it can be measured independently without any specific knowledge of the beam velocity v_0.

The experiment, shown in Fig. 2, consists of three parts. The cell TPA is induced in a 80 MHz rf flow discharge of neon by a stabilized ring dye laser. To enhance the TPA signal, the neon cell was placed in a servo-controlled power-enhancement resonator. Typically 1 kW/cm^2 of circulating laser power made possible the observation of Doppler free TPA signals as narrow as 3 MHz FWHM at neon pressures as low as 0.1 torr. The insert of Fig. 2 shows the first derivative of the TPA induced fluorescence, which allows the frequency modulated ring laser to be long-term stabilized by locking it to the center of this signal. This cell TPA constituted our non-relativistic clock and was long-term stable to better than 40 kHz. The relativistic clock was realized by resonantly enhanced TPA in a well-collimated fast atom beam, the energy of which was long-term stable to 7×10^{-6}. An actively stabilized laser, with a bandwidth of 40 kHz, induced the fast beam TPA. The detected fluorescence was, as in the cell case, used to lock this dye laser to the center of the TPA signal,which had a FWHM of 3 MHz. This second clock was also long-term stable to better than 40 kHz. The third part of the experiment involved measuring the frequency difference between the two clocks, each locked to its respective TPA resonance. Both lasers, carefully mode-matched and overlapped, illuminated a fast photodiode. After amplification, the beat frequency δ was counted directly with a microwave-frequency counter.

The beat frequency δ depends on the neon pressure in the discharge cell. Thus we measured δ as a function of neon pressure,and found a pressure shift of -6.44±0.60 MHz/torr, and a broadening of -16±4 MHz/torr. Fig. 3 shows

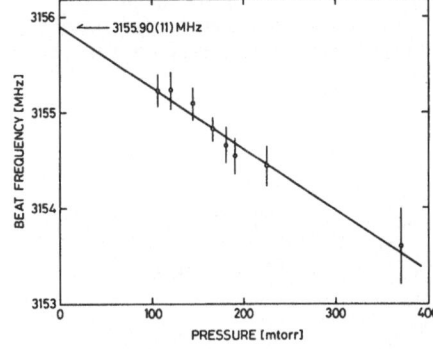

Fig. 3. Dependence of the beat frequency $\delta = \nu_{L,c} - \nu_{L,b}$ on the pressure of neon in the discharge cell. Only a limited pressure range was usable. At pressures below 0.1 torr the discharge died out, at pressures higher than 0.5 torr the pressure broadening prohibited a tight lock of the ring dye laser.

the result, which allowed us to determine a beat frequency of 3155.90±0.11 MHz at zero neon pressure.

Our final result is given in Table I, with the statistical and all systematic errors added in quadrature. A second order Doppler shift of 3235.94±0.14 MHz is obtained, in close agreement with a theoretical value of 3235.89±0.05 MHz, calculated using the known values of ν_1, ν_2 [15,17]. The accuracy of our experiment is 4×10^{-5}, which represents, by a factor of 3, the most accurate direct verification of the time-dilation due to special relativity.

The present experiment represents a new generation of precision tests of the time-dilation. The pressure shift of the cell TPA resonance introduces the most serious systematic error in our determination of δ. By substituting the cell with a thermal atomic beam, a 10x improvement is within immediate reach. This will allow us to isolate the fourth-order Doppler shift. Unfortunately, other systematic errors given in Table I will prohibit further improvements of the present experiment. The next generation of our experiment will be based on stored, laser-cooled ions. Several groups have laser-cooled ions stored in electromagnetic traps successfully [18]. These methods can be directly transferred to ion-storage rings [19]. Doppler widths around 100 kHz are predicted, allowing an unprecedented optical resolution. A small storage ring is at present being constructed at the University of Aarhus to

TABLE I: The second-order Doppler shift in neon

Measured beat frequency	
Cell-beam at zero pressure	3155.90±0.11 MHz
Corrections:	
Acoustooptic modulator	80.00±0.00 MHz
AC-Stark shift in cell	0.04±0.02
Laser locks (each laser)	±0.04
Beam crossing uncertainty	±0.02
Residual magnetic field	±0.02
High-voltage setting for TPA	±0.05
Experimental Doppler shift	3235.94±0.14 MHz
Calculated Doppler shift:	3235.89±0.05 MHz

pursue these ideas. This storage ring concept is optimized for high precision diurnal measurements of both the first-order and higher-order Doppler shifts. Thus we anticipate that relativistic time dilation will be tested to even higher accuracies in the near future.

References:

1. A. Einstein, Ann. d. Phys. 17, 891 (1905).
2. H.E. Ives and G.R. Stilwell, J. Opt. Soc. A. 28, 215 (1938).
3. H.I. Mandelberg and L. Witten, J. Opt. Soc. Am. 52, 529 (1962).
4. J.J. Snyder and J.L. Hall, in Laser Spectroscopy II, edited by T.W. Hänsch et al. (Springer Verlag, Berlin 1975).
5. P. Juncar et al., Phys. Rev. Lett. 54, 11 (1985).
6. D.S. Ayres et al., Phys. Rev. D3, 1051 (1971).
7. J. Bailey et al., Nature 268, 301 (1977).
8. L.B. Redei, Phys. Rev. 162, 1299 (1967).
9. H.B. Nielsen and I. Picek, Nucl. Phys. B211, 269 (1983).
10. R.F.C. Vessot et al., Phys. Rev. Lett. 45, 2081 (1980).
11. O. Poulsen, Nucl. Instrum. Meth. 202, 503 (1982).
12. E. Biraben, E. Giacobino and G. Grynberg, Phys. Rev. A12, 2444 (1975).
13. O. Poulsen and N.I. Winstrup, Phys. Rev. Lett. 47, 1522 (1981).
14. M. Kaivola, N. Bjerre, O. Poulsen and J. Javanainen, Opt. Comm. 49, 418 (1984).
15. Matti Kaivola, Ove Poulsen, Erling Riis and Siu Au Lee, Phys. Rev. Lett. 54, 255 (1985).
16. R. Salomaa and S. Stenholm, J. Phys. B8, 1795 (1975).
17. P. Juncar and J. Pinard, Rev. Sci. Instrum. 53, 939 (1982).
18. J.J. Bollinger, J.D. Prestage, W.M. Itano and D.J. Wineland, Phys. Rev. Lett. 54, 1000 (1985).
19. J. Javanainen, M. Kaivola, U. Nielsen, O. Poulsen and E. Riis, J. Opt. Soc. Am. B, accepted for publication (1985).

CW Two-Photon Spectroscopy of Hydrogen 1s-2s and New Precision Measurements of Fundamental Constants

C.J. Foot, B. Couillaud, R.G. Beausoleil, E.A. Hildum, D.H. McIntyre, and T.W. Hänsch

Department of Physics, Stanford University, Stanford, CA 94305, USA

1. Introduction

The 1s-2s transition of atomic hydrogen has an extremely narrow natural width which can give a resolution of 5 parts in 10^{16}. In this paper we report on the first experiment using continuous wave ultraviolet radiation at 243 nm to excite this transition by Doppler-free two-photon spectroscopy. The resolution of 5 parts in 10^9 is an order of magnitude better than achieved in any previous measurements, and the continuous wave (cw) excitation opens the way for large improvements in the future: In all experiments until now, the resolution has been limited by the bandwidth of the pulsed lasers which were used to generate the intense tunable ultraviolet (uv) radiation.

Accurate experimental measurements of transition frequencies in hydrogen and its isotopes are useful, because very precise theoretical calculations have been made for these two-body systems, so that transition frequencies can be related directly to the fundamental constants and quantum electrodynamic (QED) terms.

2. Apparatus

Although it has long been recognized that cw excitation of the 1s-2s transition would be very fruitful, the experimental realization had to await the development of a high power source of uv radiation at 243 nm. This wavelength cannot be produced directly by any tunable lasers, therefore nonlinear mixing must be used. Unfortunately, no known nonlinear crystals permit efficient 90° phase-matched second harmonic generation (SHG) at this wavelength. A group at Oxford has developed a usable source of 243 nm radiation using an angle-tuned urea crystal,[1] but at Stanford sum-frequency mixing has been used to produce higher uv powers. Up to 4 mW in a good mode have been seen, stable for many hours without damaging the crystal. Initially ADP was used as the nonlinear crystal.[2] In the present experiment [3] this was exchanged for a KDP crystal (45° z-cut, 8mmx8mmx25mm), and two slightly more widely separated fundamental wavelengths were used so that the temperature for 90° phase-matching was 62°C. With the crystal above room temperature, a windowless housing purged with nitrogen was sufficient to prevent moisture from degrading the surface polish. Also, the polish on the KDP crystal was better, so that the red fundamental beam at 790 nm could be enhanced by a factor of 30-50 using a resonant optical cavity. Typically, the input to the cavity was 500 mW from an LD700 ring dye laser (Coherent model 699-21) pumped with 8 W of red krypton ion light. The other fundamental beam, at 351 nm, had a similar power; this beam was produced by an argon ion laser (Coherent model I-20) running on a single mode and locked to an external temperature-stabilized cavity. Both fundamental lasers had a bandwidth of 1 MHz. The 351 nm beam was focused to have a waist of 30 μm radius in the crystal, so that both fundamental beams had similar confocal parameters.

The 243 nm beam was mode-matched into an enhancement cavity to give a standing-wave with a circulating intensity of 30 mW and a waist radius of 100 μm in the hydrogen cell. A dc discharge dissociated the hydrogen, and the atoms flowed down a 24" long teflon tube into the cell, so that the pressure in the interaction region was between 0.05 and 1 Torr at room temperature. The resonance was detected by counting the Lα photons emitted in the radiative decay of atoms transferred from the 2s to 2p level by collisions. These photons were observed through a MgF side window, using a solar-blind photomultiplier (EMR 541J) and a Lα interference filter.

Figure 1 shows a two-photon spectrum of hydrogen 1s-2s with the two hyperfine components. The peak signal on the large F=1 to F=1 component was 1000 counts/sec for a pressure of about 0.2 Torr and with 10 mW of 243 nm radiation inside the cavity. At lower pressures, slightly narrower resonaces have been seen with widths of 6 MHz (FWHM at 243 nm). The dominant source of line broadening is the finite transit time, but collisions and laser bandwidth also contribute.

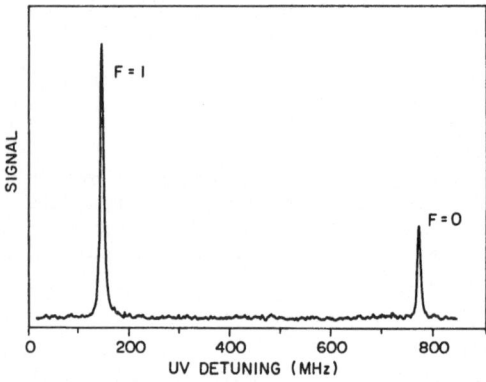

Figure 1:
Doppler-free two-photon spectrum of hydrogen 1s-2s transition.

3. The Aim of Experiments in the Near Future
Although large future improvements in resolution appear feasible, we are working on two quantitative measurements which are interesting at the present stage. Both involve a comparison of the 243 nm 1s-2s transition with a reference line at twice this wavelength. In one experiment, we are using the 486 nm Balmer-β line as a reference, in order to measure the 1s Lamb shift. In the other we are comparing 1s-2s with a transitition in the $^{130}Te_2$, whose absolute frequency has recently been measured to within 4 parts in 10^{10}.[4] This second comparision promises an improved value of the Rydberg constant.[5] In both experiments, we observe the reference line by Doppler-free saturation spectroscopy using another ring dye laser which operates with Coumarin 102 near 486 nm. An external urea frequency doubler produces a few μW of second harmonic radiation only a few GHz away from the 243 nm sum-frequency, so that heterodyne methods can be used quite readily for a precise frequency comparison.

4. Future Experiments
To make significant further improvements, the external perturbations must be reduced by observing 1s-2s in atomic beam rather than a gas-filled cell. To reduce transit broadening, such a beam can be cooled to about 4 K by mounting the escape nozzle on a liquid helium cryostat.[6] By letting the

34

light waves propagate collinearly with the atomic beam, as illustrated in Fig. 2, long interaction times can be achieved. For example, an interaction length of 1 cm would give a transit broadening of only 7 kHz, and at this temperature second order Doppler shifts, caused by time dilation, remain below 1 kHz. The excitation probability per atom would be about 10^{-7} with a uv power of 10 mW focused to a radius of 100 μm, if the laser linewidth is reduced below 1 kHz, as can be accomplished with internal or external frequency stabilizers with fast servo response.[7]

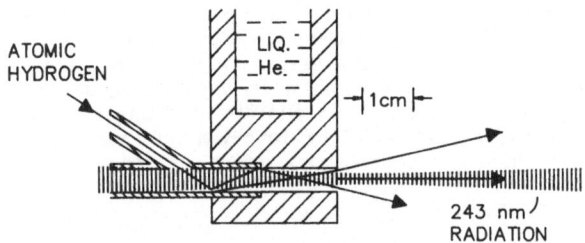

Figure 2:
Simplified scheme of a cold atomic beam.

5. Comparison with Other Transitions

In this section we will consider what can be learned from the higher resolution measurements that have been described above. Although QED provides a very good theory for the hydrogen atom, the predicted transition frequencies are limited by uncertainties in the fundamental constants and in the size of the nucleus. For the 1s-2s transition (2.5×10^{15} Hz) the largest contribution to the uncertainty is 2.5 MHz from the Rydberg constant.[5] There is also a contribution of 70 kHz from the electron/proton mass ratio [8] and 130 kHz from the charge radius of the proton. These are all much larger than the uncertainty of 4 kHz from the fine structure constant. The present QED calculations [9] have a computational uncertainty of 60 kHz but this could be reduced and should not be a limiting factor. With these uncertainties, measurement of the 1s-2s frequency can improve the value of the Rydberg constant by an order of magnitude. Nothing more would be learned from further increases in precision, without additional information, because there would then be several quantities with similar contributions to the uncertainty. Suitable additional information could be provided by a measurement of another transition frequency in hydrogen with similiar precision. Clearly, the Balmer-β transition is not ideal because it has a natural width of 13 MHz, but transitions from the 2s state to high-lying Rydberg levels can be much narrower. At this Conference, Biraben and Julien are reporting on the observation of the 2s to 8s,d transitions in a metastable beam with a resolution of 4 parts in 10^9 and improvement is expected by reducing the electric fields in the interaction region.[10] If this frequency and the 1s-2s frequency are measured accurately, then the charge radius of the proton can be determined to better than the present uncertainty of 10%.

The QED corrections are not important for the isotope shifts. Comparison of the isotope shifts in the two transitions will give an accurate value for the difference in the charge radii of the proton and the deuteron. If the isotope shifts are combined to give a composite frequency [3] so that terms proportional to $1/n^3$, e.g. nuclear size effects, are eliminated, then the electron/proton mass ratio can be improved from the present uncertainty of 5×10^{-8} to 10^{-9}, limited only by the uncertainty of α. Once the nuclear radii and m_e/m_p have been well measured, then they in turn can be used to give a more precise theoretical value for the 1s-2s transition frequency. If in the comparisons, experiment and theory reach the level at which the current uncertainty in the value of α became the limiting factor, then the Rydberg constant would be determined to 2 parts in 10^{12}.

6. Conclusion

Now that a continuous wave source of 243 nm radiation has been developed which is sufficiently powerful for two-photon spectroscopy of the 1s-2s transition, the source bandwidth can be made so narrow that it is the finite interaction time of the atoms which will determine the resolution in future experiments. An atomic beam cooled to liquid helium temperature should allow a resolution of parts in 10^{12}. In the more distant future atoms could be cooled to milliKelvin temperatures by radiation pressure cooling or by inhomogeneous magnetic fields. If such cold atoms are directed upwards they will be slowed further by gravity, and natural linewidth-limited resolution may be achievable by optical Ramsey spectroscopy of a "hydrogen fountain."[11]

This work has been supported by the National Science Foundation under Grant NSF PHY83-08721 and by the U.S. Office of Naval Research under Contract ONR N00014-C-78-0304.

References
 1 C.J.Foot, P.E.G.Baird, M.G.Boshier, D.N.Stacey, and G.K. Woodgate,
    Opt. Comm. 50, 199 (1984), and private communication.
 2 B.Couillaud, T.W.Hänsch, and S.G.MacLean, Opt. Comm. 50, 127 (1984).
 3 C.J.Foot, B.Couillaud, R.G.Beausoleil, and T.W.Hänsch,
    Phys. Rev. Lett. 54, 1913 (1985).
 4 J.R.M.Barr, J.M.Girkin, A.I.Ferguson, G.P.Barwood, P.Gill, and
    R.C.Thompson, Opt. Comm. 54, 217 (1985).
 5 S.R.Amin, C.D.Caldwell, and W.Lichten,
    Phys. Rev. Lett. 47, 1234 (1981).
 6 J.T.M.Walraven and I.F.Silvera,Rev. Sci. Instru. 53, 1167 (1982).
 7 J.L.Hall and T.W.Hänsch,Optics Letters 9, 502 (1984).
    and H.Walther, eds. (Springer-Verlag, Heidelberg, 1979), p. 130.
 8 R.S.VanDyck, Jr., F.L.Moore, and P.B.Schwinberg,
    Bull. Am. Phys. Soc. 28, 791 (1983).
 9 G.W.Erickson, J. Phys. Chem. Ref. Data 6, 831 (1977).
10 F.Biraben and L.Julien, Opt. Comm. 53, 319 (1985).
11 R.G.Beausoleil and T.W.Hänsch, submitted for publication (1985).

Atomic Parity Violation
Using the Crossed Beam Interference Technique [1]

C.E. Wieman [2], *S. Gilbert, R. Watts, and M.C. Noecker*

Joint Institute for Laboratory Astrophysics, University of Colorado
and National Bureau of Standards, Boulder, CO 80309, USA

We present a new measurement of parity violation in atomic cesium. This
is more precise than previous measurements of atomic parity violation and
is approaching the precision of the best high-energy tests of the electroweak theory. This is a preliminary result and we expect considerable further improvement.

1 Introduction

The electroweak unification theory of Weinberg, Salaam, and Glashow predicted the existence of a parity-violating (PV) neutral current interaction between electrons and nucleons. That prediction stimulated a large
number of experiments,which have now succeeded in clearly establishing the
basic structure of electroweak unification. However, the precision and
parameter space spanned by the present results are quite limited. There
is considerable motivation to work toward tests of higher precision,and to
explore different regimes where the theory may fail. As well as generally
exploring this new area of physics, there are two specific aspects which
are particularly interesting at present. These are the radiative corrections to the electroweak theory, and the possible existence of additional
neutral bosons. By measuring the radiative corrections one has a generic
check of many untested aspects of the theory simultaneously [1]. The existence of additional neutral bosons is of interest,because it has been
shown that they are required both for a supersymmetric theory of the electroweak interaction [2] and for the simplest class of grand unification
theories not presently ruled out by experimental data [3].

Atomic parity violation measurements complement experiments done with
high-energy accelerators in studying these phenomena. Because these two
types of experiments are sensitive to different combinations of electron-
quark couplings and different mass scales for the exchange of virtual
particles, the comparison of the results provides more information than
can be obtained from either individually [4].

2 Basic Experimental Concepts

This experiment shares certain basic concepts with all previous atomic
parity violation experiments, most of which were first put forth by the
BOUCHIATs [5]. A parity-violating neutral current interaction will mix
atomic eigenstates of opposite parity. Because this interaction involves

---

[1]Work supported by the National Science Foundation.
[2]Sloan Fellow; also with Department of Physics, University of Colorado.

the exchange of a massive Z_0 boson,it is localized at the nucleus and hence only mixes S and P states. The goal of the experiment is to measure the amount of this mixing. This is done by determining the electric dipole (El) transition amplitude for an S → S transition in the atom. Because El transitions are strictly forbidden between states of the same parity,this amplitude is a direct measure of the PV mixing. The obvious way to measure such an amplitude would be to simply shine a laser of the appropriate frequency for an S → S transition on an atom and measure the El transition rate. However, this is not a practical approach in this case because the transition rate is proportional to the square of the PV mixing, which is very small. This transition rate will be on the order of 10^{-35} Z^6 times a normal allowed El rate, where Z is the atomic number. Thus in any conceivable experiment, this signal is far smaller than the noise. To overcome this problem,a heterodyne approach is used. In a heterodyne measurement, a much larger transition amplitude, A_0, is provided which can interfere with the PV amplitude. The transition rate is then $R = |A_0 \pm A_{PV}|^2 = A_0^2 \pm 2A_0A_{PV} + A_{PV}^2$ and one measures the interference term $2A_0A_{PV}$. This is isolated from the much larger A_0^2 term by using the fact it is parity-violating,and hence reverses sign with any parity (i.e. mirror) reversal of the experiment. To reduce the problem of systematic errors,the experiment ideally contains several such reversals,and one looks at the component of the transition rate that modulates with all of them. These features are common to all atomic PV measurements.

In our experiment, we use a "Stark induced" amplitude for A_0, as was used in the experiments of BOUCHIAT et al. [6] and BUCKSBAUM et al. [7]. By "Stark induced" we mean the El transition amplitude that is produced by applying a static electric field to the atom which mixes S and P states. With the proper choice of field geometry and phases of the laser fields, this amplitude will interfere with A_{PV}. One of the problems in using a Stark induced amplitude is that the interference term changes sign under a reversal of the magnetic quantum number, m. This means that if the m levels are degenerate,the S → S transition rate will involve a sum over all m levels and will have no PV contribution. In this case, a parity violation is only observable as a polarization of the excited state.

In our experiment, we break the degeneracy of the m levels, which allows us to observe the parity violation directly in the transition rate. The experimental approach is discussed in detail in Ref. [8], but we will briefly review the key features here. The 6S → 7S transition is excited in a beam of atomic cesium by a narrowband dye laser that intersects it at right angles. This provides inherently narrow transition linewidths. A weak magnetic field is then applied to separate the m levels,so that transitions between particular m levels are resolved. The transition rate on a particular line of the Zeeman multiplet is then monitored as we reverse the static electric and magnetic fields and the handedness of the laser polarization. These correspond to parity reversals. As an additional test,we compare the rates for the +m and −m lines of the multiplet. The use of an atomic beam also greatly reduces background noise sources,and allows efficient detection of the excitation.

3 Experimental Design and Operation

A highly simplified schematic of the experiment is shown in Fig. 1. About 500 mW of laser light is produced at the 6S → 7S transition wavelength (540 nm) by a Spectra Physics ring dye laser. The laser linewidth is reduced to about 100 kHz by servolocking the frequency to an external refer-

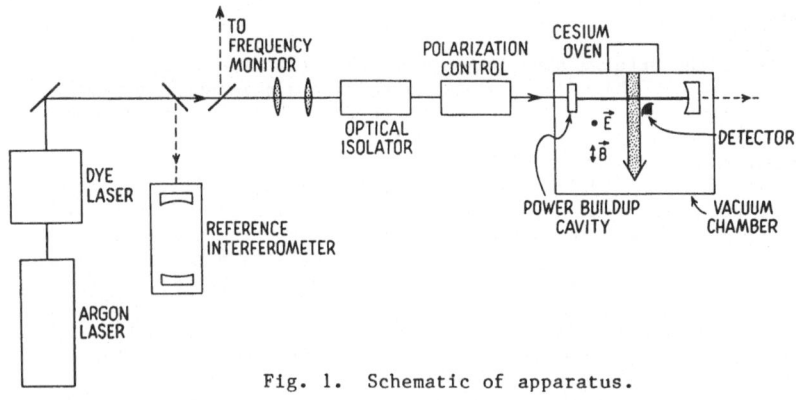

Fig. 1. Schematic of apparatus.

ence cavity. Long-term drifts are reduced to about 1 kHz by locking the
reference cavity to the atomic transition. The laser beam passes through
various optics which control the polarization and intensity and couple it
into a power buildup cavity inside the vacuum chamber. The buildup cavity
is a spherical mirror Fabry-Perot interferometer, the resonant frequency
of which is servolocked to the laser frequency. This provides a standing
wave with a field that is twenty times larger than in the incident beam.
An intense beam of atomic cesium effuses from a multislit collimator, and
intersects the standing wave field in a line 3 cm long. An electric field
(2-3 kV/cm) is produced in the intersection region by applying voltage to
an optically transparent electrically conductive plate above the region
and grounding a similar plate which is below. A 70 G magnetic field parallel to the cesium beam is provided by a Helmholtz pair. The 6S $\rightarrow$ 7S
transition rate is monitored by observing the 850 and 890 nm light that is
emitted in the $6P_{3/2,1/2} \rightarrow$ 6S branch of the 7S $\rightarrow$ 6P $\rightarrow$ 6S cascade decay.
This light is imaged onto a rectangular silicon photodiode by a cylindri-
cal gold mirror. Colored glass filters in front of the detector block the
scattered green laser light. The current from this photodiode is measured
as the experiment undergoes the various parity reversals, and the values
are stored on a computer disc.

About one-half the time the experiment is operating is spent measuring
the $6S_{F=4} \rightarrow 7S_{F=3}$ hyperfine line, and the other half is spent on the $6S_{F=3}$
$\rightarrow 7S_{F=4}$ line. The two lines are essentially identical for experimental
purposes. The data are analyzed by normalizing to the signal size and
calculating the fraction of the transition rate that modulates with all
the reversals. It takes 10 to 20 min of integration time to obtain a
±100% measurement of the parity violation.

A great deal of effort has been devoted to the study and elimination
of systematic errors but, due to space limitations, we can only briefly
summarize this here. We first considered all possible combinations of
static and oscillating fields that could contribute to the 6S $\rightarrow$ 7S transition rate. Using measurements of misalignments and imperfect reversals,
we then found which of these terms could contribute more than 1% of the
parity-violating rate after the reversals. There are three terms of this
size. Before and after each ~8 hour data run, we make a set of auxiliary
measurements which use the atoms themselves to determine all the fields
that give rise to these terms. We then subtract out their effect. The
corrections vary, but are normally less than 10% of the parity violation
and, the uncertainties in them are much less than that.

4 Results

The results of the data we have taken up to this time are Im A_{PV}/β = -1.83 ± 0.19 mV/cm for the $6S_{F=3} \rightarrow 7S_{F=4}$ transition and Im A_{PV}/β = -1.53 ± 0.18 mV/cm for the $6S_{F=4} \rightarrow 7S_{F=3}$ line, where βE is the Stark-induced amplitude. The average of the two numbers is -1.68 ± 0.13 mV/cm. This is in good agreement with the average of the two measurements by BOUCHIAT et al. [6] which is $-1.56\pm0.17\pm0.12$ mV/cm. The fractional uncertainty we have obtained is about the same size as the predicted radiative corrections to the electroweak theory. These can be tested by comparing the value of $\sin^2\theta_W$, the weak mixing angle, derived from our results with the value derived from the masses of the W and Z bosons. However, the former derivation requires a knowledge of the atomic wave functions. The calculation of these wave functions is more straightforward for cesium than for any other heavy atom because of its strongly one-electron character, but at present it appears they could still be in error by as much as 10%. When that uncertainty is added in, our results are in agreement with the predicted corrections,but do not provide a very stringent test. Significant efforts are being made to reduce the uncertainty in the calculated wave functions,and it is hoped that, particularly with our measurements to spur them on, these will soon prove successful. From the difference (or lack thereof) in the PV we obtained on the two hyperfine lines, we can find the proton spin-dependent neutral current coupling constant, C_{2p}. Our value of $\left|C_{2p}\right|$ = 2.1 ± 1.8 is a substantial improvement on the best previous limit of $\left|C_{2p}\right|$ < 100 reported in Ref. [6] and is consistent with the predicted value of 0.1.

The results presented here are quite preliminary,and we see no serious barrier to improving the accuracy by as much as a factor of 10 with improved optics and more running time. The use of an intense spin-polarized cesium beam which we have developed will allow still further improvements.

References

1. W.J. Marciano and A. Sirlin, Phys. Rev. D 27 (1983) 552.
2. P. Fayet, Phys. Lett. 96B (1980) 83.
3. R. Robinett and J. Rosner, Phys. Rev. D 25 (1982) 3036.
4. C. Bouchiat and C. Piketty, Phys. Lett. 128B (1983) 73.
5. M.A. Bouchiat and C. Bouchiat, J. de Phys. 36 (1975) 493.
6. M.A. Bouchiat et al., Phys. Lett. 134B (1984) 463.
7. P.H. Bucksbaum, E.D. Commins and L.R. Hunter, Phys. Rev. D 24 (1981) 1134.
8. S.L. Gilbert, R.N. Watts and C.E. Wieman, Phys. Rev. A 29 (1984) 137.

Quantum Interference Effect
for Two Atoms Radiating a Single Photon

A. Aspect and P. Grangier

Institut d'Optique, Université de Paris-Sud, BP 43,
F-91406 Orsay Cédex, France

J. Vigué

Laboratoire de Spectroscopie Hertzienne de l'Ecole Normale Supérieure,
24, Rue Lhomond, F-75231 Paris Cédex 05, France

In previous work [1,2] we have studied the photodissociation of a Ca_2 molecule, yielding two Calcium atoms recoiling in opposite directions, one in the first resonant $^1P_1$ state (Ca*) and one in the ground state (Ca). Either atom can actually be excited and subsequently reemit the resonance photon $\hbar\omega_0$ (λ_0 = 422,7 nm), so one must consider two paths for the whole process

$$Ca_2 + \hbar\omega_L \begin{array}{c} \nearrow Ca + Ca* \searrow \\ \searrow Ca* + Ca \nearrow \end{array} 2\ Ca + \hbar\omega_0 \qquad (1)$$

(ω_L refers to the dissociating light, here the 406.7 nm violet line of a Krypton laser).

The existence of two paths between the initial and final states suggests the possibility of a quantum interference effect between the amplitudes associated with these two paths. The observation of such an interference effect would be an evidence for the existence of the pure state.

$$|\psi\rangle = (1/\sqrt{2})\ \left[|g,e\rangle + |e,g\rangle\right] \qquad (2)$$

where $|g\rangle$ and $|e\rangle$ refer to Ca and Ca*, the first and second atoms differing by their opposite directions of flight. The existence of such a state in the intermediate stage of the process (1) is a pure quantum effect, and it is interesting to show that the system is still described by such a state (and not a mixture) even when the distance between the two atoms is very large compared to the typical size of an atom.

Two observe this interference effect, we can look at the fluorescence light $\hbar\omega_0$. We can then remark that for a fixed observer the frequency of this light will be Doppler shifted towards blue or red, corresponding to an emission by atom (1) or (2) (Figure 1).

If, according to (1), the emissions by both atoms must be taken into account simultaneously, one can hope to observe the beat note :

$$\Omega = \Omega_M \cos\theta \qquad \text{with} \qquad \Omega_M = 2\omega_0 \frac{v}{c} \qquad (3)$$

We will see later that this suggestive picture, introduced by Diebold[3], must be taken with a grain of salt, and we will rather describe the

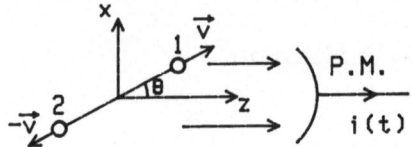

Figure 1 - Basic situation. Atoms (1) and (2) recoil with opposite velocities at an angle θ with the direction of observation of the fluorescent light.

situation of figure 1 as a Young's fringe pattern produced by two moving coherent sources at a distance $d = vt$. We thus expect an interference term proportional to:

$$1 + \cos(2\pi \cos\theta \, d/\lambda_0) = 1 + \cos \Omega t \qquad (4)$$

This reasoning leads again to the prediction of a modulation Ω in the intensity received on the detector, but equation (4) conveys more information, since we know that $t = 0$ is the time of photodissociation of the molecule (distance $d = 0$). The expression (4) can thus be interpreted as proportional to the probability $p(t)$ of detection of a photon with a delay t after the instant of the photodissociation. One can therefore perform the experiment consisting in effecting the photodissociation with a pulsed laser and looking at the time of detection of the photon $\hbar\omega_0$. By repetition of the experiment, one gets $p(t)$.

A complete quantum mechanical calculation confirms this intuitive reasoning [4,5]. Except at the very beginning of the process ($d < 100$ Å) when the resonant dipole-dipole coupling plays an important role, one finds $p(t)$ proportional to

$$e^{-\Gamma t} (1 + \cos \Omega t)$$

where Γ is the inverse of the life time of the excited atom (4.7 ns).

In a real experiment, one must take care of averaging over all the possible orientations of the molecular axis. It turns out [4,5] that this averaging does not wash out the modulation, since the excitation process-as well as the detection - make use of polarized light, which amounts to favor some directions of molecular axis.

The experimental setup (Figure 2) involves a Calcium molecular beam excited by the pulses from a mode-locked Krypton laser (406.7 nm). The fluorescent light at 422.7 nm is detected through a polarizer by a fast Multi-Channel Plate Photomultiplier Tube, used in photon counting configuration. A Time to Digital Converter allows to measure the delay t

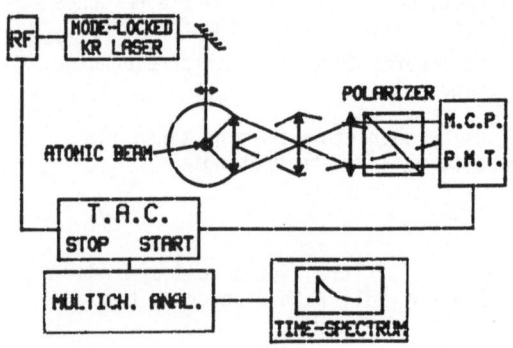

Figure 2 - Experimental set-up. The atomic beam and the laser polarization are orthogonal to the figure.

between a laser pulse and corresponding detected fluorescent photon.
With use of a Multi-Channel Analyzer, we eventually obtain a time-spectrum
proportional to p(t).

Measurements have been made in various configurations. An example of
result is given on Figure 3. The molecular density at the interaction
region was estimated to be about 5×10^7 mol/cm^3, and the average counting rate was 5×10^3 counts/s, the overall detection efficiency being
about 10^{-3}. Taking into account the measured time resolution of the
system (180 ps FWHM, including the laser pulses width, the PMT resolution, and the electronics jitter), the result can be fitted by the
theoretical prediction if we take a recoil velocity v = 550 ± 35 m/s.
This experiment can thus be considered a direct measurement of this
recoil velocity, which yields informations on the molecular states of
Ca$_2$ involved in the photodissociation process.

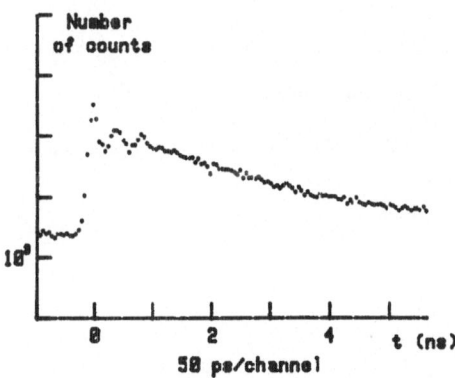

Figure 3 - Number of detected
photons as a function of the
delay after photodissociation.
The light was detected with a
polarization parallel to the
electric field of the laser.
The counting time was 300s.

Having demonstrated the effect, we want to emphasize that it is really a single photon effect, as shown by equations (1) or (2),which involve only one quantum of excitation. First, our experimental facts cannot
be explained by a process in which two excited atoms would be produced
by the photodissociation of a single molecule : this alternative expla-
nation is ruled out by our previous observation [1] that the photodissociation rate increases linearly with the power of the laser. Second,
our observations cannot be explained as an interference between the
fields emitted by two excited atoms produced by two simultaneous dis-
sociations of two different molecules : the probability of such an event
is small (less than 5 %),and the expected visibility would be
extremely small (less than 10^{-5}; see discussion in ref [3]).

It is thus hard to avoid the conclusion that we have an interference
effect in which a single photon is emitted by two different atoms, separated by a distance that may be as large as several optical wavelengths.

At this stage of the discussion, one could argue than the concept of
single photon is useless for this situation, and that a picture involving two recoiling classical dipoles emitting damped wave-packets leads
to the same results. Actually, this classical picture leads to predictions that can be distinguished from the quantum predictions involving

a single photon. In a similar experiment performed with a C.W. Laser, the classical picture predicts that the Doppler beat note evoked in the introduction could be found in the noise-power spectrum of the photocurrent produced by the fluorescence light. On the contrary, the correct quantum description predicts no peak in this spectrum, because the intensity autocorrelation function is zero for a single photon state.

In conclusion, the demonstrated effects, which are related to the most subtle features of the quantum theory, can also be used as a tool yielding informations about the molecular states involved in the photo-dissociation process. For instance, if the state (2) had an opposite parity, the phase of the modulation would be found opposite, which would be easily observable.

Reference :

1   J. Vigué, P. Grangier, A. Aspect : J. Phys. (Paris), Lett. 42, L 531 (1981)

2   J. Vigué, J.A. Besswick and M. Broyer : J. Phys. 44, 1225 (1983)

3   G. Diebold : Phys. Rev. Lett. 51, 1344 (1983)

4   P. Grangier, A. Aspect and J. Vigué : Phys. Rev. Lett. 54, 418, (1985).

5   P. Grangier et al., to be published.

Feasibility of ^{81}Br$(\nu,e^-)^{81}$Kr Solar Neutrino Experiment

G.S. Hurst*, *S.L. Allman, C.H. Chen, and S.D. Kramer*
Chemical Physics Section, Oak Ridge National Laboratory,
Oak Ridge, TN 37831, USA

J.O. Thomson
Department of Physics, University of Tennessee, Knoxville,TN 37922, USA

B. Cleveland
Chemistry Division, Brookhaven National Laboratory,
Upton, L.I., NY 11973, USA

The standard stellar model of the sun, that of BAHCALL [1], predicts the neutrino flux and energy spectrum emitted by the hot interior core where energy production occurs. It is believed that neutrinos provide the only means of observing the thermonuclear region of the sun. An experiment by DAVIS et al. [2] to measure the flux due to one of the high-energy neutrino sources (due to ^{8}B) uses the reaction ^{37}Cl$(\nu,e^-)^{37}$Ar in a radiochemical method. The standard model predicts that the flux of ^{8}B neutrinos on the earth is about 6×10^6 cm^{-2} s^{-1}. The more energetic neutrinos can interact via ^{37}Cl$(\nu,e^-)^{37}$Ar, producing a favorable atom for decay counting with $T_{1/2} = 35$ days. Another important feature of ^{37}Ar is that it is a noble gas, thus a few atoms can be recovered from tons of a chlorine-rich target for radiochemical experiments. A large tank was installed in the Homestake Gold Mine and was filled with 615 tons of C_2Cl_4, in anticipation of about 1.5 events per day, based on standard model predictions; however, the observed capture rate was only 0.4 events per day, in serious disagreement.

Several ingenious solutions have been offered for the solar neutrino problem – a defect in the solar model, the appearance of a new type of neutrino physics, the sun is no longer burning, etc. The range of these proffered solutions stresses the need for a new experiment to study the sun. The modern pulsed laser now makes possible a new solar neutrino test which examines an independent neutrino source in the sun. A recently proposed [3] experiment would use the reaction ^{81}Br$(\nu,e^-)^{81}$Kr to measure the flux of ^{7}Be neutrinos from the sun. When ^{7}Be decays by electron capture to make ^{7}Li, a neutrino is emitted at 0.862 MeV and the flux of these on the earth is about 4×10^9 cm^{-2} s^{-1}, according to the standard model. Therefore, an experiment based on ^{81}Br$(\nu,e^-)^{81}$Kr which is sensitive to these lower energy neutrinos would be of fundamental importance. To first order, the chlorine experiment detects the ^{8}B neutrinos while bromine detects the much more abundant ^{7}Be neutrino source. In practice, the proposed bromine experiment would be very similar to the chlorine radiochemical experiment, except that ^{81}Kr with a half-life of 2×10^5 years cannot be counted by decay methods. With an experiment of about the same volume as the chlorine experiment (380 m^3) filled with CH_2Br_2, the model predicts about 2 atoms of ^{81}Kr per day. The bromine experiment depends entirely on the RIS method, implemented with pulsed lasers, for its success.

---

*
Department of Physics, University of Tennessee, Knoxville, and Oak Ridge National Laboratory.

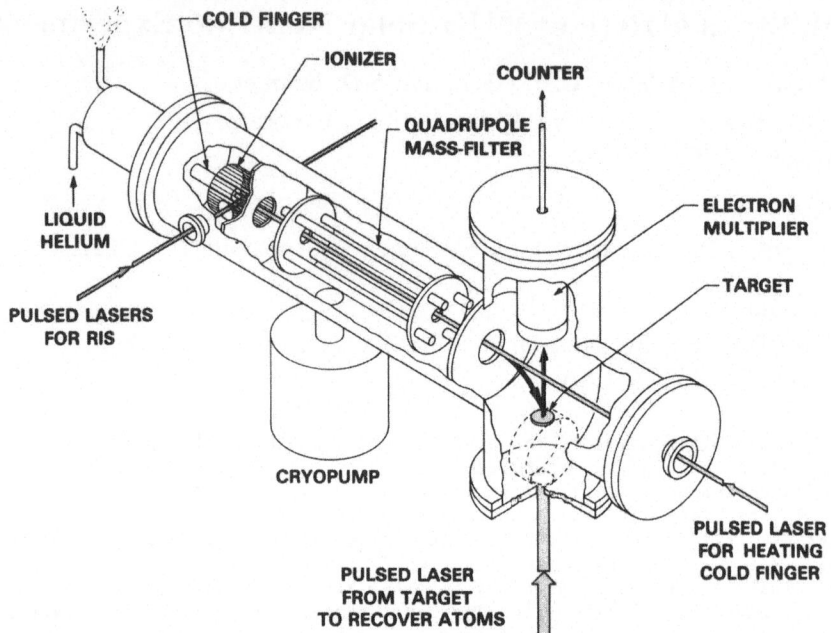

Fig. 1. Experimental schematic of "Maxwell's demon"

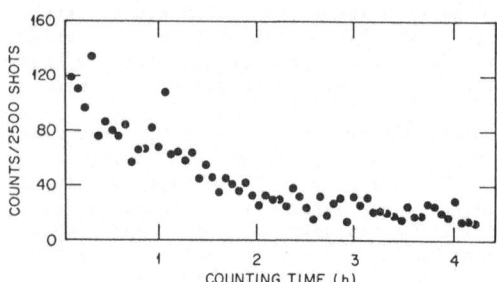

Fig. 2. Data obtained on a sample of 1000 atoms of ^{81}Kr

The importance attached to sensitive detection of noble gas atoms prompted the development of suitable laser schemes for the resonance ionization of xenon [4] and krypton [5] and has suggested a method for argon [6]. The RIS schemes developed to date for the noble gases have not been isotopically selective, hence our plan was to have Z-selection with RIS and to add A-selection with a mass spectrometer. In the Maxwell demon concept [7] (see Fig. 1), atoms meeting both Z- and A-selection are accelerated to an energy which implants ions (e.g., $^{81}Kr^+$) into a solid target such as silicon. Each atom can be counted by detecting secondary electrons – and only once because the implanted atoms stay in the target. The sorting and counting process continues until all of the atoms are counted. Figure 2 shows results for the case of about 1000 atoms of ^{81}Kr, and we take such results as proof that the present Maxwell demon is sensitive enough for the bromine experiment.

Samples of ^{81}Kr removed by helium purge of the large tank could have contributions due to cosmic rays, i.e., muon-induced proton production followed by (p,n) reactions. For this reason the bromine experiment must be located underground. Another background which could play a role in RIS counting but not in decay counting is caused by adjacent isotopes. Thus, suppose the ratio ^{82}Kr/^{81}Kr greatly exceeded the rejection ratio of the mass spectrometer tuned to ^{81}Kr. Some false counts would be recorded in the mass-81 channel. In terms of the bromine solar neutrino experiment, this means that atmospheric krypton due to leaks or outgassing must be controlled. However, some of these problems can be dealt with in the the demon itself. Suppose the rejection ratio for ^{82}Kr when tuned to ^{81}Kr is 10^4; then a 10% error would be encountered for 10^6 atoms of ^{82}Kr when the level of ^{81}Kr is 10^3. By laser annealing of the target (using a single pulse of a suitable excimer laser), nearly all of the krypton atoms will be returned to the gas phase, and a repeat of the RIS counting would give another factor of 10^4 isotopic enrichment. Thus, ^{82}Kr at a level of about 10^{10} atoms would be acceptable. Space charge limits the number of atoms which can be ionized in one laser pulse without affecting ion optics to about 10^5; therefore, for larger samples of ^{82}Kr it is advisable to do an external isotopic enrichment [8] before introducing the sample to the demon. Presently [9], it is found that a purge of the neutrino tank at Homestake (filled with C_2Cl_4) yields about 5×10^{12} atoms of ^{82}Kr. In Fig. 3 we show a mass scan of a krypton sample removed from a well in Switzerland [10], after having been enriched with a quadrupole mass spectrometer. Clearly, the ^{81}Kr peak can be resolved even without the final stage of enrichment laser annealing of the silicon target. Initially, the sample had about 1.5×10^3 atoms of ^{81}Kr and more than 10^{14} atoms of ^{82}Kr, a more demanding isotope ratio than expected in the solar neutrino experiment.

RIS counting of ^{81}Kr atoms makes possible a solar neutrino experiment using ^{81}Br as a target to measure the abundant ^{7}Be source in the sun. The same detector (^{81}Kr) is also useful for groundwater hydrology studies, for ice-cap dating, and for the study of meteorites.

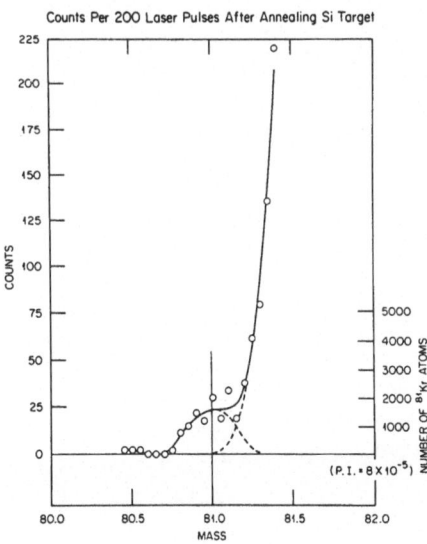

Fig. 3. Mass scan of the krypton sample removed from a well in Switzerland and isotopically enriched with a quadrupole mass spectrometer

This research was sponsored by the Office of Nuclear Physics, U.S. Department of Energy under contract DE–AC05–840R21400 with Martin Marietta Energy Systems, Inc.

References

1. J.N. Bahcall: Rev. Mod. Phys. 54, 767 (1982)
2. R. Davis, Jr., B.T. Cleveland, and J.K. Rowley: in Science Underground, eds. Michael Martin Nieto, W.C. Haxton, C.M. Hoffman, E.W. Kolb, V.D. Sandberg, and J.W. Toevs, AIP Conference Proceedings No. 96 (American Institute of Physics, New York 1983).
3. G.S. Hurst, C.H. Chen, S.D. Kramer, Raymond Davis, Jr., Bruce Cleveland, Fletcher Gabbard, and F.J. Schima: Phys. Rev. Lett. 53, 1116 (1984)
4. C.H. Chen, G.S. Hurst, and M.G. Payne: Chem. Phys. Lett. 75, 473 (1980)
5. S.D. Kramer, C.H. Chen, M.G. Payne, G.S. Hurst, and B.E. Lehmann: Appl. Optics 22, 3271 (1983)
6. S.D. Kramer: in Resonance Ionization Spectroscopy 84, eds. G.S. Hurst and M.G. Payne, Conference Series No. 71 (The Institute of Physics, Bristol 1984)
7. C.H. Chen, S.D. Kramer, S.L. Allman, and G.S. Hurst: Appl. Phys. Lett. 44, 640 (1984); for a complete account, see also G.S. Hurst, M.G. Payne, S.D. Kramer, C.H. Chen, R.C. Phillips, S.L. Allman, G.D. Alton, J.W.T. Dabbs, R.D. Willis, and B.E. Lehmann: Reports Prog. Phys. (in press)
8. C.H.Chen, R.D. Willis, and G.S. Hurst: Vacuum 34(5), 581 (1984)
9. Raymond Davis, Jr. (University of Pennsylvania), private communication
10. B.E.Lehmann, H. Oeschger, H.H. Loosli, G.S. Hurst, S.L. Allman, C.H. Chen, S.D. Kramer, M.G. Payne, R.C. Phillips, R.D. Willis, and N. Thonnard: J. Geophys. Res. (in press)

Laser-Induced Nuclear Orientation of 1 μs ^{85m}Rb

G. Shimkaveg, D.M. Smith, M.S. Otteson, W.W. Quivers, Jr., R.R. Dasari, C.H. Holbrow, J.E. Thomas, D.E. Murnick, and M.S. Feld

George R. Harrison Spectroscopy Laboratory and
Department of Physics, Massachusetts Institute of Technology,
Cambridge, MA 02139, USA

1. Introduction

Laser-Induced Nuclear Orientation (LINO) [1] experiments employ intense laser optical pumping to orient unstable nuclei, and detect the subsequent angular anisotropy of the nuclear gamma ray emission. The profiles of anisotropy versus laser tuning yield atomic resonance frequencies from which nuclear moments and other structure parameters are derived. Prior to the present study, a number of experiments were performed using lamps as light sources, and were limited to nuclides with half-lives >1 s. LINO type experiments were also performed on-line at accelerator facilities [2,3] on nuclides whose lifetimes were in the range of milliseconds.

We report here the results for ^{85m}Rb (1 μs), successfully carried out in a table top experiment to yield values for the isomer shift of -52 ± 9 MHz and the nuclear magnetic moment of (6.046 ± 0.010) μ_N [4]. The lifetime of this nucleus is a thousand times shorter than that of any nucleus previously studied by optical methods, and the average density of ^{85m}Rb is less than $1/cm^3$ demonstrating the potential of the technique. Knowledge of the structure of this nucleus adds to the systematic information of the Rb chain obtained by other investigators. Furthermore, this experiment provides an important test system for many on- and off-line studies proposed or in progress based on laser interactions with short-lived radioactive species.

2. Experiments

Most of the details on the cell configuration, detector locations, nuclear electronics and lasers employed are described in our recent publication [4]. Briefly, the cell is a sealed pyrex cylinder filled with 4 Torr of Kr (one third of which is ^{85}Kr) and a small sample of natural rubidium. ^{85m}Rb is produced from ^{85}Kr through a 0.4% β^- decay branch. Two dye lasers (circularly or linearly polarized) with single-mode powers in the range 50-150 mW at 795 nm are employed to pump F=4 and F=5 D_1 ^{85m}Rb hf ground states.

3. Theory

The magnitude of the anisotropy signal depends on the interplay between a number of processes. The parent Kr gas provides velocity-changing collisions, allowing the single mode laser radiation to interact with the entire Doppler distribution of ^{85m}Rb atoms. Natural Rb is required for charge exchange neutralization, but causes absorption of the laser power and spin-exchange-induced loss of polarization. Increasing the ^{85}Kr density increases the count rate, but broadens the natural Rb absorption, thereby decreasing the available laser power. For a given pump laser power, the optimum configuration can be estimated using a simple model of the optical pumping,which is briefly described in this section.

Generally, the angular anisotropy in the gamma ray count rate depends upon the population of each ground electronic state magnetic sublevel. In ^{85m}Rb, there are 20 individual M_F sublevels in the ground state and gamma decay involves $\Delta I=2$. It is easy to show from the gamma ray angular distribution functions that the sublevels of maximum $|M_F|$ produce the most anisotropic radiation patterns. All other M states contribute very little to the anisotropy,and hence the exact population distribution of those other states is relatively unimportant, and can be taken to be isotropic. Rate equations relating population transfer between sublevel groups with like responses to optical pumping can be written and solved analytically for the various steady-state group populations. Simplifying the resulting expression for the specific case of equal pumping of both ground state hyperfine levels,the anisotropy A can be written as:

$$A = P_M \left(\frac{c}{1+c} \right) \left\langle \frac{I}{I+I_{op}} \right\rangle_{CELL} \left\langle \Phi \right\rangle_{DET.}$$

50

The factor P_M is the limiting maximum sample polarization obtainable, which occurs as a result of the finite number of optical pumping cycles possible in a short nuclear lifetime. The factor C/1+C results from the steady-state balance of charge exchange neutralization and nuclear decay. The optical pumping factor $I/I+I_{op}$ must be averaged (brackets) over the variations in laser intensity occurring within the cell volume. Finally, Φ represents the sample gamma-ray angular radiation distribution, which must be averaged over detector solid angle. The signal-to-noise function (S/N) is given by $A\sqrt{N}$, where N is the number of counts recorded during a measurement. All the factors comprising (S/N) are dependent on the experimental operating parameters, such as rubidium density, Kr pressure, cell radius and length, beam diameter, and detector solid angle; therefore, one can maximize (S/N) with respect to these parameters to yield optimum conditions.

The excellent agreement between theory and experimental results (Fig. 1) shows that for the first time we understand various mechanisms involved in laser optical pumping of short-lived isomeric atoms. Fig. 2 shows observed gamma anisotropy as a function of laser frequency for circular and linear polarizations. The observed resonances are used in the determination of nuclear magnetic moment and isomer shift for

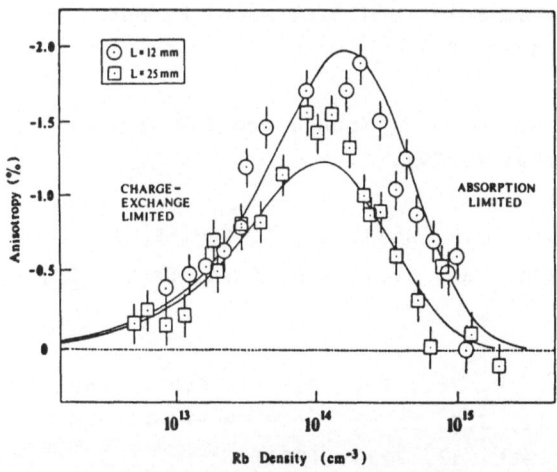

Figure 1. Gamma anisotropy versus rubidium density. Data points are for cells of two different lengths and solid curves are rate equation model predictions with no free parameters.

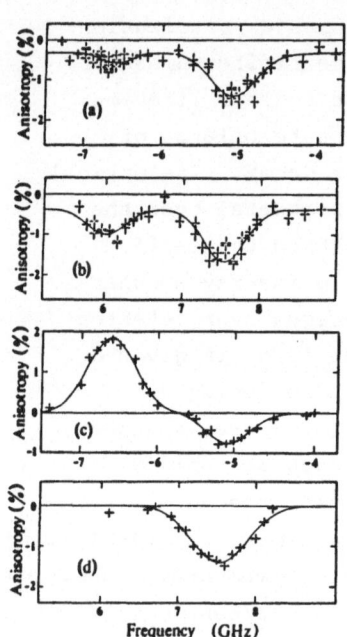

Figure 2: ^{85m}Rb D$_1$ line anisotropy spectra; frequencies shown are relative to ^{85}Rb line center.

^{85m}Rb. This work was performed at the MIT Laser Research Center, an NSF supported regional instrumentation facility.

References

1. M. Burns, P. Pappas, M.S. Feld, and D.E. Murnick, Nucl. Instrum. Methods <u>141</u>, 429 (1977).

2. C.E. Bemis, Jr., J.R. Beene, J.P. Young, and S.D. Kramer, Phys. Rev. Lett. <u>43</u>, 1854 (1979).

3. D.E. Murnick, H.M. Gibbs, O.R. Wood, II, L. Zamick, P. Pappas, M. Burns, T. Kuhl, and M.S. Feld, Phys. Lett. <u>88B</u>, 242 (1979).

4. G. Shimkaveg, W.W. Quivers, Jr., R.R. Dasari, C.H. Holbrow, P.G. Pappas, M.A. Attili, J.E. Thomas, D.E. Murnick, and M.S. Feld, Phys. Rev. Lett. <u>53</u>, 2230 (1984).

Optical Searches for Fractional Charges and Superheavy Atoms

W.M. Fairbank, Jr., W.F. Perger, and E. Riis

Physics Department, Colorado State University, Fort Collins, CO 80523, USA

G.S. Hurst

Oak Ridge National Laboratory, P.O. Box X, Oak Ridge, TN 37831, USA

J.E. Parks

Atom Sciences, Inc., 114 Ridgeway Center, Oak Ridge, TN 37830, USA

We are using a new ultrasensitive analysis technique called SIRIS (Sputter-Initiated Resonance Ionization Spectroscopy) to search for new elementary particles in stable matter surviving as relics of supernovae or the Big Bang. This allows us to do high energy physics in the large mass range of current interest,while bypassing the enormous cost of next-generation accelerators. In our method we look for exotic atoms formed when these new particles bind by strong or electric forces to nuclei. In this paper we report on our progress to date on these experiments.

In the SIRIS method (Fig. 1) a small cloud of vaporized atoms is created by sputtering with a microsecond pulse of argon ions from a focussed ion source ($40\mu A$). Shortly thereafter,pulsed laser beams from a Nd:YAG/dye laser system resonantly ionize the unusual atoms of interest in a selective manner, which leaves the other atoms and molecules in the cloud almost completely untouched. These ions are accelerated,passed through an electrostatic analyzer to remove unselected ions from the sputtering sprocess (SIMS ions), and mass analyzed in a time-of-flight mass spectrometer. The latter is used so that data can be collected on all masses at the same time.

In experiments conducted to date we have searched for hydrogen-like quark atoms ($Z=2/3$) in niobium and tungsten samples,as well as superheavy isotopes of lithium in Li compounds. For the quark search we tune the excitation laser to the $n=1$ to $n=3$ transition at 2306Å of a $Z=2/3$ hydrogen-like atom. The 1.06 micron Nd:YAG funadmental is used for the ionizing step. In our superheavy atoms searches we tune our exciting laser to the isotopically shifted

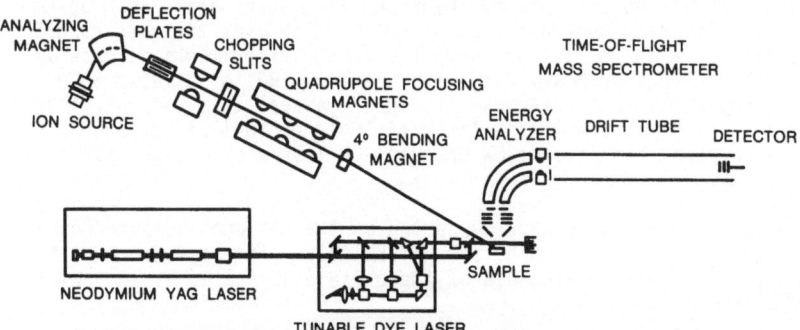

Fig. 1. The SIRIS exotic atom detector featuring sputtering, resonance ionization, and time-of-flight analysis

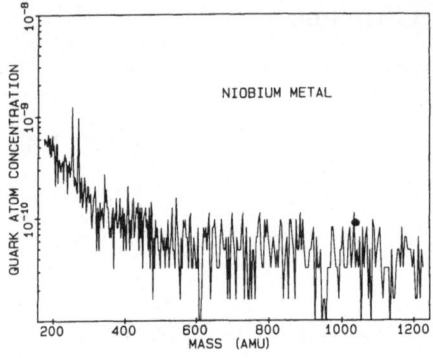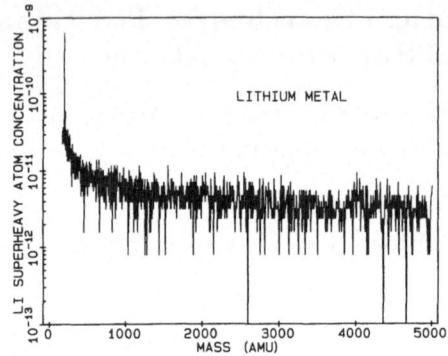

Fig. 2. Sample data from searches for fractionally-charged atoms in niobium metal (left) and heavy lithium isotopes in lithium metal (right)

wavelength (∿100 GHz shift) for a heavy Li isotope. We have been using 2475Å (2s-6p) excitation with 1.06 micron ionization. The typical results shown in Fig. 2 illustrate our accomplishments to date,as well as some of the back-grounds which we are working to reduce. One can see that over a wide range of masses we find no interesting exotic atom candidates at a concentration limit of 10^{-10} to 10^{-11}. Similar spectra have been taken with other samples.

The three interesting sharp peaks in these spectra have been checked by UV wavelength scans and other tests for the expected spectral characteristics of exotic atoms. In all cases the UV spectra were broad, indicating molecular origins. For example, the peaks at 254 and 270 amu in the quark spectrum are thought to be UO and UO_2 from prior experiments with uranium samples. The 197 amu peak in the lithium spectrum is probably TaO sputtered from a tantalum electrode near the sample. These peaks can be eliminated in the future by greater care in the experimental procedure. The broad back-grounds in these spectra come from SIMS ions which are inadequately rejected at the sample and in the energy analyzer. We are working on apparatus changes such as additional energy analysis, new electrode designs, electrode pulsing, and aperturing to further reduce these backgrounds. The order of magnitude improvement in the lithium spectrum compared to the quark spectrum was due to a recent apparatus change.

In summary, we find no evidence for hydrogen-like quark atoms in niobium and tungsten samples at a concentration of about 10^{-10} over the mass range 200 to 1200 amu and also in separate magnetic sector experiments at mass 1/3 amu. We have found no superheavy lithium atoms from 180 to at least 3000 amu (some uncertainty on detector response at higher masses) at a concentration of about 10^{-11}. With improvements,a limit of 10^{-13} or 10^{-14} is expected in the present apparatus. The use of CW or quasi-CW lasers offers promise for even lower limits.

This material is based upon work supported by the National Science Foun-dation under Grants Nos. PHY-8106763 and PHY-8210835. The assistance of the Atom Sciences, Inc. staff and the use of their SIRIS apparatus is appreciated.

Part III

Rydberg States

Rydberg Atoms and the Test
of Simple Quantum Electrodynamical Effects

P. Dobiasch, G. Rempe, and H. Walther

Max-Planck-Institut für Quantenoptik and Sektion Physik
der Universität München, D-8046 Garching, Fed. Rep. of Germany

1. Introduction

Rydberg Atoms represent an ideal testing ground for some of the most fundamental models and predictions of low-energy quantum electrodynamics (QED).
The following are reasons and examples:

(a) The matrix elements for electric dipole transitions between neighbouring Rydberg states scale as n^2, where n is the principal quantum number.
For high enough n, stimulated effects overcome spontaneous emission
already for very small photon numbers. As a consequence, Rydberg atoms
are very sensitive e.g. to blackbody radiation (see Ref. [1] and [2] for
recent reviews).

(b) The transitions to neighbouring levels are in the region of millimeter
waves, therefore it is possible to physically modify the nature of the
environment into which they decay, using for example conducting walls.
Introducing conductors imposes boundary conditions on the electromagnetic field, and leads back to a descrete spectrum in the case of a finite volume enclosed in a cavity. In principle, there are essentially
two distinct cases to be discussed. First, the situation of an atom
in close proximity to a conducting plate [3-8]. The induced image
charges give rise to extra contributions of a van der Waals-type force
to the inner atomic forces, thus leading to position-dependent level
shifts. Second, there are effects from a discrete mode structure of
the electromagnetic field inside a cavity. due to its geometry. Of
course, it is not possible to consider one of these phenomena without
the other, but in most cases only one of the two produces the major
influence. Consequences of the discrete mode structure of a cavity for Rydberg atoms are: the rate of the spontaneous emission is enhanced or dimi-
nished, depending upon the cavity being tuned on or off resonance with
a transition frequency [9-13], as well as modifying the Lamb shift of
Rydberg levels [14].

(c) For cavities with high quality factors, the photon emitted by an atom
in a Rydberg state remains stored inside the resonator long enough to
be reabsorbed by the same atom with a finite probability. In this way,
it is possible to realize a single-atom maser [15]. A single Rydberg atom
inside a low-loss, single-mode resonator is an experimental realization of
the Jaynes-Cummings model [16], describing the interaction between a single
two-level atom and a single mode of the electromagnetic field. This model
has been the object of considerable attention in the past, and a number of
purely quantum mechanical predictions on the dynamics of this system have
been made. These include the collapses and revivals in the dynamics of the
atomic population. Rydberg atoms will for the first time offer the possibility to test these predictions [16-18].

In the following, the results on a single-atom maser will be described
and the extension of this experiment to test the collapses and revivals in
Rabi nutations will be discussed. In the second part of the paper, theoreti-

cal results on the modification of the Lamb shift of hydrogen Rydberg atoms in a waveguide will be reported.

2. The Single-Atom Maser

For the Rydberg maser experiment [15],an atomic beam of highly excited Rydberg atoms was used. The beam oven is carefully heat-shielded from the cavity by copper plates cooled by water, liquid nitrogen and liquid helium; the atoms pass through small apertures into the liquid helium-cooled part of the apparatus. There they are excited to the upper maser level and enter the cavity. Behind the cavity the atoms in Rydberg states were monitored by field ionization.

The Rydberg states were populated using the frequency-doubled radiation of a continuous wave ring dye laser. The constant stream of Rydberg atoms is ionized in an inhomogeneous dc electric field of a plate capacitor. The atoms attain the point of maximum field strength in front of a hole in the anode through which the ejected electrons pass and reach a channeltron multiplier. If the field strength is adjusted properly mainly the atoms in the upper maser level are monitored. Transition from the initially prepared state to the lower maser level are thus detected by a reduction in the electron count rate. The cylindrical cavity (diameter 24.8 mm, length 24 mm) was manufactured from pure niobium rods. It is enclosed in a cryoperm shield to reduce the influence of ambient magnetic fields on the Q value due to frozen-in flux. The temperature of the cavity could be varied from 4.3 to 2.0 K, corresponding to Q factors of 1.7×10^7 and 8×10^8, respectively. The atomic beam passes through the cylindrical cavity along its axis, where only the TE_{1np} and TM_{1np} modes possess a nonvanishing transversal electric field. For our experiment the TE_{121} mode was used. This mode has a plane field distribution and is doubly degenerate in an ideal cylindrical cavity. The degeneracy is removed by a slight deformation of the circular cross-section into an oval shape, which then determines the direction of polarization of the field mode. The deformation is achieved by squeezing the cylinder with a screw and a piezoelectric transducer for fine tuning (0.5 MHz/1500V). The upper maser level was the $63p_{3/2}$ level of ^{85}Rb. The fine structure splitting between $63p_{3/2}$ and $63p_{1/2}$ amounts to 396 MHz (see Fig. 1). It is, therefore no problem to excite a single fine structure level with the narrow-band ultraviolet radiation ($\Delta\upsilon \cong 2$ MHz).

To demonstrate maser operation, the cavity was tuned over the $63p_{3/2}$ - $61d_{3/2}$ transition by changing the voltage of the piezoelectric transducer; the field ionization was recorded simultaneously. Transitions from the initially prepared $63p_{3/2}$ state to the $61d_{3/2}$ level (21.50658 GHz) are detected by reduction of the electron count rate.

In a recent experiment also,the transition $63p_{3/2}$-$61d_{5/2}$ at 21.456 GHz (Fig. 1) was observed in a single atom maser. The transition matrix element for this transition is about 7 times larger than that for $63p_{3/2}$ - $61d_{3/2}$. The results on these measurements will be published elsewhere [19].

In the case of measurements at a cavity temperature of 2K [15], a reduction of the $63p_{3/2}$ signal could be clearly seen for atomic fluxes as small as 800 atoms/s. An increase in flux caused power-broadening and finally an asymmetry and a small shift. This shift is attributed to the ac Stark effect, caused predominantly by virtual transitions to the $61d_{5/2}$ level, which is only 50MHz away from the maser transition (Fig. 1). The fact that the field ionization signal at resonance is independent of the particle flux (between 800 and 22×10^3 atoms/s) indicates that the transition is saturated. This, and the observed power broadening, show that there is a multiple exchange of photons between Rydberg atoms and the cavity field.

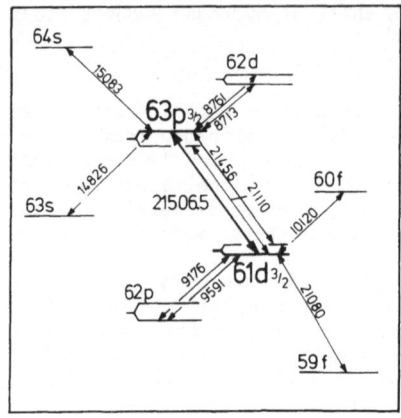

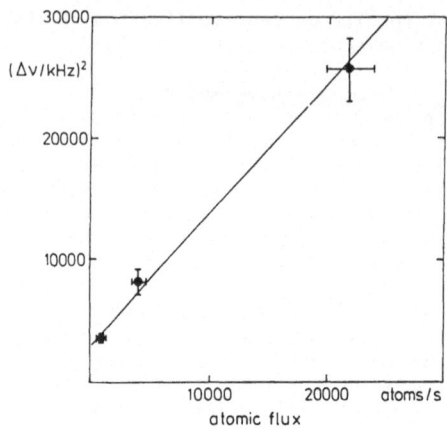

Fig. 1:   (left side) Rubidium level scheme with the maser transition.
Fig. 2:   (right side) Squares of the halfwidths of the maser resonances
          versus atomic flux. From the intersection of the straight line
          with the $(\Delta\nu)^2$ axis, the number of blackbody photons in the cavi-
          ty can be evaluated [15].

        With an average transit time of the Rydberg atoms through the cavi-
ty of 80 μs one calculates for a flux of 800 atoms/s a probability of
0.06 of finding a Rydberg atom in the cavity. According to Poisson statistics, this implies that more than 99% of the events are single atom.
This clearly demonstrates that single atoms are able to maintain continuous oscillation of the cavity.
        Since the transition is saturated, half of the atoms initially ex-
cited in the $63p_{3/2}$ state leave the cavity in the lower $61d_{3/2}$ maser level. The decay to other levels can be neglected for the average transit
time of 80 μs. The energy radiated by those atoms is stored in the cavity for its decay time, increasing the average field strength. The average number of photons left in the cavity by the Rydberg atoms is given
by $n = \tau_d N/2$ where τ_d is the characteristic decay time of the cavity and
N the number of Rydberg atoms entering the cavity in the upper maser level per unit time. For the highest particle flux used in our experiment
$N = 22 \times 10^3$ atoms/s, one finds $n \cong 55$ photons at 2 K ($\tau_d \cong 5$ ms) and
$n \cong 1.4$ photons at 4.3 K ($\tau_d \cong .13$ ms). This last value is smaller than the
average number of blackbody photons $n_{b\ell} \cong 4$ at 4.3 K. For $N \cong 800$ atoms/s
one obtains $n \cong 2$ at 2 K, which means that the radiation generated by the
Rydberg atoms in the cavity has about the same energy as the blackbody radiation ($n_{b\ell} \cong 1.5$).
        When the squares of the halfwidth $\Delta\nu$ of the signal curves are plot-
ted versus the Rydberg atom flux, a straight line is obtained as expected (Fig. 2). This line intersects the $(\Delta\nu)^2$ axis at a finite value, from
which the number of blackbody photons originally in the cavity can be
evaluated. The result (3 ± 1) is in reasonable agreement with the value
given above. It follows that as the atomic flux decreases, the thermal
radiation becomes the dominant part of the field.

3.   Collapses and Revivals in the Rabi Nutation
Experiments with single atoms in high Q superconducting cavities
are suitable to test the Jaynes-Cummings model [16-18, 20-22] describing
the interaction between a single two-level atom and a single mode of the

electromagnetic field. This model is amenable to an exact analytical solution, and has been the object of extensive theoretical analysis. A
number of purely quantum-mechanical effects, whose experimental verification would be of considerable interest, have been predicted. These include
collapses in the evolution of the population inversion [16,17,20], as well
as its partial revival for longer interaction times [18].

    An experimental setup such as described in the preceding section can
be used to test the predictions of the Jaynes-Cummings model. An important
requirement is, however, that the atoms of the beam have a homogeneous velocity, so that it is possible to observe the Rabi nutation in the cavity
directly. In a modified setup shown in Fig. 3 we have now inserted a Fizeau
type velocity selector between atomic beam oven and cavity, so that a fixed
atom-field interaction time is obtained. Changing the selected velocity
leads to a different interaction time, and leaves the atom in another phase
of the Rabi cycle when it reaches the detector. First measurements are presently carried out with this setup. The results will be published elsewhere
[19]. Because in the one-atom maser there is no coherent field injected
into the cavity from the beginning, but it rather builds-up from noise due
to the masing process, it is not at all clear whether collapses and revivals can be observed in this case. To answer this question, Filipowicz et
al. [23] recently derived a simple theory of the one-atom maser. The result is that for small enough atomic velocity spreads ($\Delta v/v \cong 10^{-2}$) collapses and revivals should be observable (see also ref. [24]).

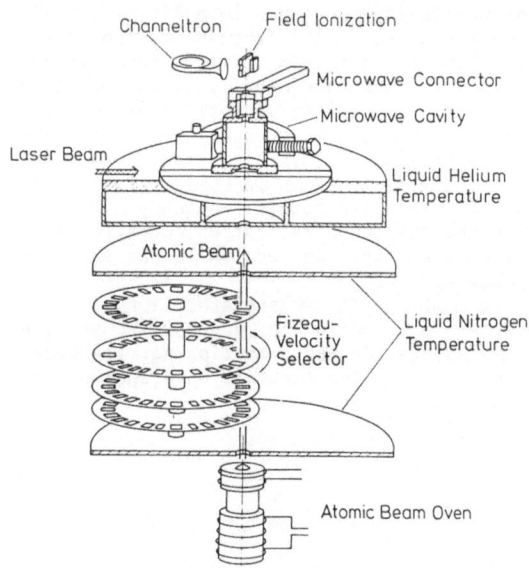

Fig. 3:   Vacuum chamber with the atomic beam arrangement and the
           microwave cavity

4. Quantum Electrodynamic Effects in Finite Space

Now we are going to discuss the question to what extent radiation corrections like the Lamb shift can be altered if they are calculated under the
restriction of a certain mode structure due to a cavity [14]. For the calculations, we assume that the cavity only affects the propagation of real
and virtual photons, leaving the atomic system unchanged. This can be justi-

fied by comparing the extension of the atomic wave function ($\cong$ 1µm even for highly excited Rydberg states) to the size of the cavity. Under this assumption, the QED corrections have to be calculated as usual, but the conventional propagation function of the photon has to be replaced by that of the photon in the cavity. It is obvious that this procedure causes tiny "apparatus-dependent" deviations from the high-precision predictions of QED, as recently pointed out by Fischbach and Nakagawa [25] for the anomalous magnetic moment of the electron.

These deviations are expected to depend on both the shape and size of the surrounding cavity. It has been shown for rectangular cavities [26] that the photon propagator in ordinary space has the same form as for an unaltered vacuum; however, changes appear in the momentum space representation, because of the discrete structure of the modes. Essentially, the integral over the wave vector is replaced by a sum over the mode wave vectors. For a certain high-frequency cutoff Λ, the cavity is assumed to become transparent, so that the mode structure only enters for $\omega < \Lambda$, whereas beyond Λ the usual integration is performed [25]. In the case of the Lamb shift, this argument excludes cavity effects on the vacuum polarization and on the high-frequency part of the self-energy diagram for one virtual photon emission.

In the calculation of the low-frequency part of the Lamb shift, only those intermediate states will contribute that accompany the emission of a virtual photon with momentum appropriate to the cavity. Since there are broadening mechanisms both for the atomic levels of the atomic system due to the natural line width, and for the cavity modes due to a finite quality factor Q, the coincidences between atomic transition frequencies and eigenfrequencies of the cavity are appreciably larger than it might be expected. A special case is the waveguide for which the mode density approaches that of the undisturbed vacuum just above its fundamental cutoff. Therefore the calculation of the modification of the Lamb shift is especially simple for this case, and our calculation was restricted to this situation. That means that in the Lamb shift calculation, those terms were omitted corresponding to energies $\hbar\omega$ of the virtual photons below the cutoff frequency ω_c of the waveguide. It should be noted that the mass renormalization counterterm is also changed owing to the cavity, however, this correction is the same for all levels with the same main quantum number.

The results for the change of the Lamb shift of s-states $\delta E_{n,0}$ due to the model waveguide outlined above are plotted in Fig. 4 versus n and the cutoff wavelength $\lambda_c = 2\pi c/\omega_c$. Some points are worth mentioning. The overall effect is extremely small, even for very short cutoff wavelengths. For

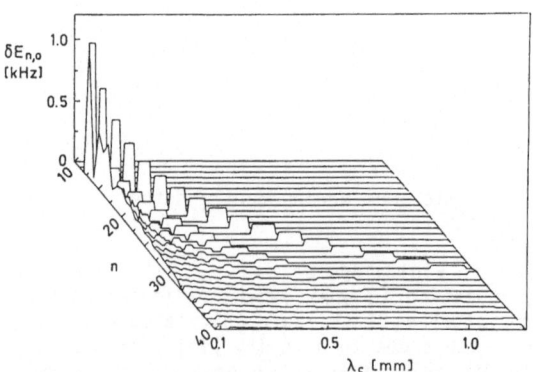

Fig. 4:   Absolute change of the Lamb shift of hydrogen s-states as a function of principal quantum number n and cut-off wavelength λ_c [14]

a fixed λ_c, the largest $\delta E_{n,\ell}$ occurs for the state with the smallest quantum number n^* that is just above the threshold for a suppressed transition to the $(n^*+1)p$ state. As λ_c is decreased, also n^* gets smaller. For values $n > n^*$, two facts tend to decrease $\delta E_{n,\ell}$: First, the general n^{-3} behaviour of the Lamb shift itself; second, there may be virtual cancellation of contributions to $\delta E_{n,\ell}$ resulting from transition to neighbouring states as e.g. $n^*s \to (n^*+1)p$ and $n^*s \to (n^*-1)p$ since they enter with opposite signs.

Since the change of the Lamb shift of an hydrogen atom in a waveguide shows a stong variation with the main quantum number n, it is not necessary to perform a measurement in the optical region in order to demonstrate the phenomenon experimentally. It has been shown in ref. [14] that the change of the Lamb shift can be measured by comparing transitions of the kind $n^*s \to (n^*+1)p$ with $(n^*+2)s \to (n^*+1)p$. To achieve the necessary precision, the Ramsey method, with two spatially separated microwave fields [27] has to be used. In the space between the microwave fields the atoms are moving between the conducting plates of a waveguide (for details see ref. [14]).

1 S. Haroche, J.M. Raimond: Advances Atomic and Molecular Physics, Vol. 20, Eds. D. Bates and B. Bederson, pp. 350-411 (Academic Press, New York 1985).
2 J.A. Gallas, G. Leuchs, H. Walther, H. Figger: Advances in Atomic and Molecular Physics, Vol. 20, Eds. D. Bates and B. Bederson, pp. 413-466 (Academic Press, New York 1985).
3 V.B. Berestetskii, E.M. Lifshitz, and L.P. Pitaevskii: "Quantum Electrodynamics" (Pergamon Press, Oxford 1982)
4 G. Barton: Proc. Roy. Soc. London A320, 251 (1970)
5 P. Stehle: Phys. Rev. A2, 102 (1970)
6 K.H. Drexhage: Progress in Optics, Vol. 12, Ed. by E. Wolf (North Holland, Amsterdam 1974)
7 P.W. Milonni, and P.L. Knight: Opt. Commun. 9, 119 (1973)
8 E.A. Power, and T. Thirunamachandran: Phys. Rev. A25, 2473 (1982)
9 E.M. Purcell: Phys. Rev. 69, 681 (1946)
10 D. Kleppner: Phys. Rev. Lett. 47, 233 (1981)
11 A.G. Vaidyanthan, W.P. Spencer, and D. Kleppner: Phys. Rev. Lett. 47, 1592 (1981)
12 G. Gabrielse, H. Dehmelt (to be published); G. Gabrielse, R. van Dyck, Jr., J. Schwinberg, H. Dehmelt: Bull. Ann. Phys. Soc. 29, 926 (1984)
13 P. Goy, J.D. Raimond, M. Gross, S. Haroche: Phys. Rev. Lett. 50, 1903 (1983)
14 P. Dobiasch, and H. Walther: Annales No6 - Alfred Kastler Symposium (Editions de Physique 1985) in print
15 D. Meschede, H. Walther, G. Müller: Phys. Rev. Lett. 54, 551 (1985)
16 E.T. Jaynes, and F.W. Cummings: Proc. IEEE 51, 89 (1963)
17 P. Meystre: PhD Thesis, Ecole Polytechnique Fédéderale Lausanne (1974); P. Meystre, E. Geneux, A. Quattropani, A. Faist, Nuovo Cimento 25B, 521 (1975)
18 J.H. Eberly, N.B. Narozhny, J.J. Sanchez-Mondragon: Phys. Rev. Lett. 44, 1323 (1980), and references therein
19 G. Rempe, G. Babst, H. Walther, N. Klein: publication in preparation
20 T. von Foerster: J. Phys. A8, 95 (1975)
21 S. Stenholm: Phys. Rep. 6, 1 (1975)
22 P.L. Knight, and P.W. Milonni: Phys. Rev. C66, 21 (1980)
23 P. Filipowicz, J. Javanainen, P. Meystre, to be published
24 P. Filipowicz, P. Meystre, G. Rempe, H. Walther: Optica Acta (1985), in print
25 E. Fischbach, and N. Nakagawa: Phys. Lett. 149B, 504 (1984)
26 E. Ledinegg, Acta Phys. Austr. 51, 85 (1979)
27 N.F. Ramsey: Phys. Rev. 76, 996 (1949)

Atomic and Field Fluctuations in Rydberg Masers: A Potential Source of Squeezed Radiation

S. Haroche, C. Fabre, P. Goy, M. Gross, J.M. Raimond, A. Heidmann, and S. Reynaud

Laboratoire de Physique de l'Ecole Normale Supérieure, 24, rue Lhomond, F-75231 Paris Cedex 05, France

1. ATOMIC and FIELD FLUCTUATIONS in a RYDBERG MASER

Rydberg atom masers, made of a small number of very excited atoms radiating at millimeter wavelengths in high Q cavities, have proven to be interesting devices to test various simple quantum electrodynamics effects [1.2.3]. The emission properties of these masers in pulsed operation have been studied in detail [4] as well as their use as high-gain amplifiers of small external field [5]. The limiting regime of these masers when the number of atoms and photons goes down to unity has also been studied. Changes in the spontaneous emission of a single atom induced by the cavity have recently been observed [6], as well as continuous maser action when the inverted medium is reduced to a single atom at a time in the cavity [7].

The fluctuations of atomic and field observables in a pulsed Rydberg maser are particularly interesting to analyze [2]. These fluctuations originate from the quantum and thermal noise acting on the inverted atomic medium at initial time, and correspond in fact to a highly non-linear amplification of this noise. Simple experiments, in which the atomic energy distribution in the Rydberg atom medium was recorded as a function of time have recently allowed us to measure the atomic energy fluctuations of these systems, that we have found to be in good agreement with theory [2 - 8]. The field fluctuation characteristics of these Rydberg masers seem to be of even greater potential interest. It has recently been shown that the statistical dispersion on one component of the radiation field must - some times- be reduced below the value it takes in a coherent field [9]. The detection of this field "squeezing" does not seem to be out of reach of state of the art millimeter wave technology. Rydberg atom masers thus appear to be promising candidates to develop useful sources of squeezed radiation.

A very attractive feature of these systems, as opposed to other optical non-linear devices discussed in the literature, is their great conceptual simplicity and the possibility of actually realizing in the laboratory a system very close to the theoretical model exhibiting the squeezing effect. Rydberg masers realize indeed the quasi-ideal case of an ensemble of N two-level atoms coupled to a single radiation mode [1]. The atomic and field fluctuations of these systems can be computed to any desired accuracy from first Q.E.D. principles. Furthermore a simple semi-classical model (Bloch pendulum) can be used to describe the system evolution as soon as N ≫ 1 and the main results of the quantum theory, including the squeezing, can be understood in a quite illuminating way from this model. In particular, simple relations appear to exist between the yet to be observed field fluctuation and the already detected atomic fluctuations of these systems, thus enhancing the that the field fluctuation characteristics should be observable as well as the atomic ones...

2. QUANTUM MECHANICAL AND SEMI-CLASSICAL THEORETICAL MODELS

The Rydberg atom maser operating in the pulsed regime is a very basic quantum mechanical system corresponding to the coupling of an angular momentum with an harmonic oscillator. The oscillator (eigenstates $|n>$ with energy $n\hbar\omega$ corresponding to n photon state) describes the cavity field mode. On the other hand, the angular momentum represents the atomic medium : all levels of the atom but the two connected by the transition resonant with the cavity can be neglected.and we can describe the medium as an N two-level atom system evolving -under the coupling with the field- within a subspace of states symmetrical with respect to atom exchange. This subspace is known [1] to represent an angular momentum J (J = N/2), the various degrees of excitation of the collective atomic system being the $|M>$ eigenstates (energy $M\omega_0$) of the J_z component of J. The atomic (J_+, J_-) and field (a^+, a) operators respectively raise and lower the atom and field excitation by one unit. The coupling between these two systems is described by the interaction $V = \hbar\Omega$ [$aJ_+ + a^+J_-$] where Ω is the atom-field coupling constant depending upon the atomic transition moment and the cavity mode geometry. Various initial states can be considered for this system at time t = 0, corresponding to different kinds of closely related experiments : (i) $|\psi(0)> = |M = N/2$; n = 0 > represents the case of a pulsed Rydberg maser starting on pure spontaneous emission; (ii) $|\psi(0)> = \sum_M C_M |M$; n=0 > (M $\sim$ N/2) describes a "superradiant" maser in which a small macroscopic atomic polarization has been prepared at time t=0; (iii) $|\psi(0)> = \sum_n C_n |M=N/2$; n > (n $\ll$ N) represents the case of a "triggered" maser, in which the atomic system is initially fully inverted with a small coherent field impinging in the cavity at time t=0. Initiation of the maser by thermal radiation can also be accounted for by describing the system initial state (and subsequent evolution) in terms of a density operator instead of a wave function [1]. After time t=0, this system evolves under the action of the coupling V and damping mechanism (dissipation of the field in the cavity with a quality factor Q). Many basic features of the system evolution (namely all phenomena appearing on atomic scale shorter than Q/ω) can be qualitatively understood within a simplified dissipation-free model. Solving the Schrödinger equation for this model can be performed directly with the help of a computer for N not too large (N $\lesssim$ 150). For large atomic samples, approximate calculations involving cumulants provide 1/N expansions of the solution [10]. The cumulant method is necessary, even for small N's if field damping is to be taken into account [11]. With these methods, various physical quantities can be computed as a function of time: $< J_z(t) >$ and $< a^+a(t) >$ which represent respectively the atom and field mean energies, $< J_x >$ and $< a_1 >$, $< a_2 >$ ($a_1 = a+a^+$; $a_2 = i(a-a^+)$) which describe respectively the two $\pi/2$ out-of-phase components of the atomic polarization and e.m field, ΔJ_z, $\Delta(a^+a)$, ΔJ_x and Δa_1 which are the fluctuations of these quantities (second order moments)... One can also compute the distribution P(M,t) and P(n,t) of the atomic and field system along their respective energy ladders, which are quantities containing in fact much more information than the second-order statistical moments...

Before analyzing the results of these calculations and their comparison with preliminary experiments, let us introduce the semi-classical Bloch vector model of the Rydberg maser, valid for large N's, which will allow us to give an interesting mechanical analogy explaining most of the system features. In this picture [1-4] , the Rydberg maser is a vector evolving in an abstract space and pointing at time t in a direction $\theta(t)$, $\phi(t)$ ($\theta=\pi$ and $\theta=0$ correspond to the fully inverted and fully deexcited system, ϕ is the phase of the atomic polarization). The atomic energy J_z corresponds to

-N/2 cosθ, the atomic polarization components J_x and J_y to N/2 sinθ cosφ
and N/2 sinθ sinφ and the e.m field component a_1 to $\sqrt{N}\beta cos\phi$ where β=-θ̇ is
the Bloch vector angular velocity. This vector obeys to an equation analo-
guous to the one of a gravitational pendulum, whose potential and kinetic
energies become the atom and field excitations respectively. To analyze the
fluctuations in this point of view [1-2] , one has to describe the initial
noise in term of angular dispersion (atomic polarization fluctuations) and
velocity dispersion (field vacuum or thermal fluctuations) of the pendulum
at t=0. Solving the equation for an ensemble of initial conditions corresponding to these dispersions, one gets a statistical ensemble of pendulum
position and velocities at any later time from which one can reconstruct
the atomic and field fluctuations of the Rydberg maser.

3. EXPERIMENTAL OBSERVATION of ATOMIC ENERGY MEAN VALUE and FLUCTUATIONS

The experiments we have performed so far on Rydberg masers have allowed us
to measure as a function of time the atomic energy [12] and the field envelope of the radiated pulse [13] . Only the atomic measurements, based
on the delayed field ionization method of the Rydberg levels [1-2] have
reached a precision enabling us to measure in any details the system fluctuations. The experiment consists in counting, for each maser shot, the
number of atoms in the upper and the lower level of the Rydberg transition
at a fixed delay t after the system preparation (pulsed laser pumping), and
to reconstruct from a large ensemble of maser pulses with identical initial
condition the average atomic energy $\langle J_z(t)\rangle$ and the atomic distribution function P(M,t) [1-2-3] .

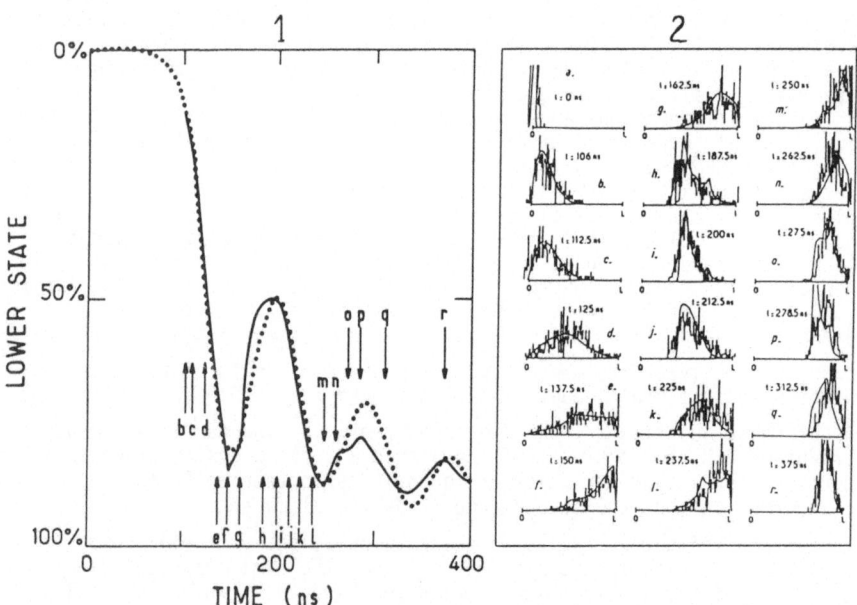

Figure 1 : Experimental (full line) and theoretical (dotted line) evolution
of the average atomic energy in a transient Rydberg maser operating on the
33S→32P transition in Na (N=2.6 10^5; Ω $\sqrt{N}$=7.10^7 s^{-1}).
Figure 2 : Probability distribution P(1-2M/N,t) at various times indicated
by arrows on figure 1. Experimental histograms compared to theory.

Figure 1 shows in full line the measured atomic energy mean value $\langle J_z(t)\rangle$ as a function of time for a <u>spontaneous</u> maser operating on the 33S→ 32P transition in Na (In dotted line on the same figure, the theoretical prediction for $\langle J_z(t)\rangle$). The atomic energy undergoes oscillations characteristic of the pendulum motion. The agreement between experiment and theory is good. Figure 2 shows for the same maser the time development of the atomic distribution function P(1-2M/N,t). Each histogram corresponds to a time marked by an arrow on figure 1. It clearly appears that the atomic energy variance (width of each histogram) also undergoes oscillations. The J_z fluctuation is maximum around the time when the atomic energy mean value reaches its first minimum (fig. 2e-f), then decreases and goes through a relative minimum when $\langle J_z\rangle$ has bounced back to its second maximum (fig.2i) and so on... These results are again in agreement with theory (full line curves superposed on the experimental histograms [14]). They can also be understood from the pendulum model : a pendulum undergoing large oscillations has a period depending upon its initial amplitude and velocity (non-isochromism). A small initial $\Delta\theta_0$ or $\Delta\beta_0$ fluctuation (due in this case to thermal radiation noise) will result in a large dispersion of the arrival time of the pendulum at its potential energy minimum ($\theta\sim0$). This time fluctuation in turn entails large energy dispersions at any given time around this position. When the pendulum bounces back, its velocity decreases and the J_z dispersion is reduced,since the slowest pendulums in the statistical ensemble start to catch the fastest ones,which reduce their velocity first.

4. <u>FIELD FLUCTUATION PREDICTIONS : SQUEEZING</u>

We now turn to the study of the <u>field amplitude</u> fluctuation. Obviously, the variances Δa_1 and Δa_2 of the two-field components of a "spontaneous" maser are always equal, since such a system cannot statistically discriminate between the phases. Furthermore, Δa_1 and Δa_2, which are the fluctuations of two canonically conjugate quantum mechanical quantities obey the Heisenberg uncertainty relation $\Delta a_1.\Delta a_2 \geqslant 1$. It thus follows that $\Delta a_1 \approx \Delta a_2 \geqslant 1$ and no squeezing is expected from this system. The symmetry between the two phases is obviously broken when the maser operates in a superradiant or triggered regime (case of preparation ii or iii above). The quantum mechanical calculation then shows that the field component in phase with the injected field or the initial polarization has a variance Δa, which becomes smaller than 1 at some point during the system evolution. Figure 3a shows Δa_1^2 as well as the atomic energy as a function of time for a maser with N=100 starting with an initial polarization corresponding to an angle $\theta_0 = 3\pi/4$. The field variance first increases above one, then decreases and reaches a minimum smaller than one at the

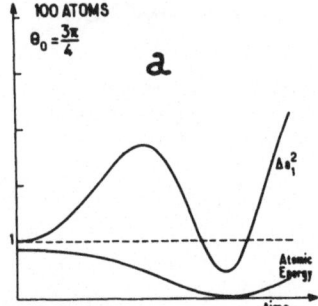
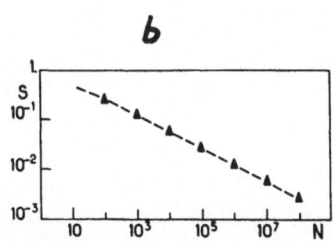

Figure 3a) : Evolution of field variance Δa_1^2 for a superradiant Rydberg maser (N=100; $\theta_0 = 3\pi/4$); b) maximum attainable squeezing $S = \Delta a_1^2$ as a function of N.

65

time corresponding roughly to the minimum of atomic energy (maximum of field emission). The amount of squeezing thus obtained (minimum Δa_1 value) depends upon the initial state of the system. For a given atom number N, the squeezing around $\theta \sim 0$ is optimum for a small but finite value of the initial atomic polarization (value of θ_0 slightly different from π) or for a small but finite value of the injected field. This optimum squeezing furthermore increases with N sensibly as $N^{-1/3}$, as shown on fig.3b representing the optimum squeezing as a function of the atom number. A minimum value of Δa_1 very close to zero.is possible as soon as N is larger than a few thousand atoms. All these results, obtained from the quantum theory,are remarkably explained in the Bloch pendulum point of view. The necessity of triggering the maser corresponds to the requirement of fixing a phase ϕ for the pendulum, so that the projection of its velocity along a given direction will not widely fluctuate. Once this phase is chosen, the pendulum motion can be simply described in the θ, β phase space. If one neglects the system damping, θ and β are classically conjugate quantities whose dispersion product $\Delta\theta\Delta\beta$ must be conserved (Liouville theorem). We have recalled above that a characteristic feature of the pendulum motion is a large increase of $\Delta\theta$ when the potential energy reaches its minimum. According to Liouville theorem, $\Delta\beta$ must be accordingly reduced, which corresponds to a decrease of the field amplitude noise... Taking into account the dissipation of the field is possible [11] and shows that the squeezing remains significant as long as the Bloch pendulum is effectively allowed to swing ($\omega/Q \ll \Omega \sqrt{N}$).

The possibility of actually detecting such squeezed states of microwave radiation generated by triggered Rydberg masers is presently considered in our laboratory.

1. S. Haroche in "New Trends in Atomic Physics" (Les Houches Session 38), R. Stora and G. Grynberg editors, North Holland, Amsterdam
2. S. Haroche and J.M. Raimond in Advances in Atomic and Molecular Physics Vol. 20, 1985, Academic Press Inc.
3. S. Haroche, C. Fabre, J.M. Raimond, P. Goy, M. Gross and L. Moi, Journal de Physique, 43, C2-265 and C2-275 (1982).
4. L. Moi et al. Phys. Rev. A27, 2043
5. P. Goy et al. Phys. Rev. A27, 2065
6. P. Goy, J.M. Raimond, M. Gross and S. Haroche, Phys.Rev. Lett. 50,1903 (1983)
7. D. Meschede, H. Walther and G. Muller, Phys. Rev. Lett. 54, 551 (1985)
8. Y. Kaluzny, Thèse de 3ème cycle, Paris VI (1984), unpublished
9. A. Heidmann, J.M. Raimond and S. Reynaud, Phys. Rev. Lett. 56, 326 (1985)
10. A. Heidmann, J.M. Raimond, S. Reynaud and N. Zagury, Opt. Comm. (1985) to be published
11. A. Heidmann, J.M. Raimond, S. Reynaud and N. Zagury, Opt. Comm. 54, 54 (1985)
12. Y. Kaluzny, P. Goy, M. Gross, J.M. Raimond and S. Haroche, Phys. Rev. Lett. 51, 1175 (1983)
13. L. Moi et al. Opt. Comm. 33, 47 (1980)
14. In this experiment, the Rydberg Maser emits on two orthogonally polarized transitions and P(1-2M/N,t) is the convolution product of two probability distributions associated to two identical Bloch pendulums (see references 2 and 8)

Doppler-Free Two-Photon Spectroscopy of Hydrogen Rydberg States

F. Biraben and L. Julien

Laboratoire de Spectroscopie Hertzienne de l'ENS, Tour 12 EO1,
4 place Jussieu, F-75230 Paris Cedex 05, France

1. Introduction

We have observed the 2S - nS and 2S - nD (n = 8, 9, 10) transitions in atomic hydrogen by continuous wave Doppler-free two-photon spectroscopy [1,2].
The first purpose of this experiment is to improve the precision of the
Rydberg constant measurement. The recent values of this constant have been
derived from the study of the Balmer-α line, either in a gas discharge
[3,4] or in an atomic beam [5]. The natural width of the 3P level is 30 MHz
and the relative width of the Balmer-α line is 6.5×10^{-8}. With this method
the precision of the Rydberg constant measurement is limited to 10^{-9} [5].

Another way to measure the Rydberg constant is to study the Doppler-free
two-photon 2S - nS or 2S - nD transitions. In this case, the natural width
of the nS and nD level can be very small. For example, the 10S natural
width is 80 kHz and the relative linewidth 10^{-10}. Thus, this method can lead
to an improvement of the Rydberg constant precision by a huge factor. At
first sight a better way is the 1S-2S Doppler-free two-photon experiment
[6]. Lately Hänsch and his collaborators have performed this experiment
with a continuous-wave laser [7]. Unfortunately, the precision of the Rydberg
constant measurement is limited by the 1S Lamb-shift uncertainty. From this
point of view, our experiment offers two advantages. Firstly, the Lamb-shift
decreases as n^3 and the Lamb-shift contribution to the 2S - nS or nD interval is relatively smaller than the one to the 1S - 2S interval. Secondly,
the 2S Lamb-shift has been measured with a high precision [8,9]. Consequently the 1S Lamb-shift uncertainty (including the QED corrections and
the nuclear-size effect) limits the precision of the hydrogen Rydberg constant measurement to 6×10^{-11} for the 1S - 2S experiment. In contrast, in
the 2S - nS or nD case, the same effects limit the precision to only
1.2×10^{-11}.

2. Description of the Experiment

The principle of the method is to induce the 2S - nS or nD Doppler-free
two-photon transitions using a metastable atomic beam collinear with two
counterpropagating laser beams. Thus, the line-broadening due to the finite
transit time of atoms in the laser beams is very small. The metastable
$2S_{1/2}$ atomic beam is obtained by electronic excitation of a ground-state
atomic beam. The inelastic collisions with the electrons deviate the atoms
and the metastable atomic beam makes an angle of 20° with the incident
atomic beam. The metastable beam is then collinear with the two laser beams.

A schematic overview of the apparatus is given in Figure 1. Molecular
hydrogen is dissociated by a 50 Watt RF discharge at 29 MHz. Atomic hydro-

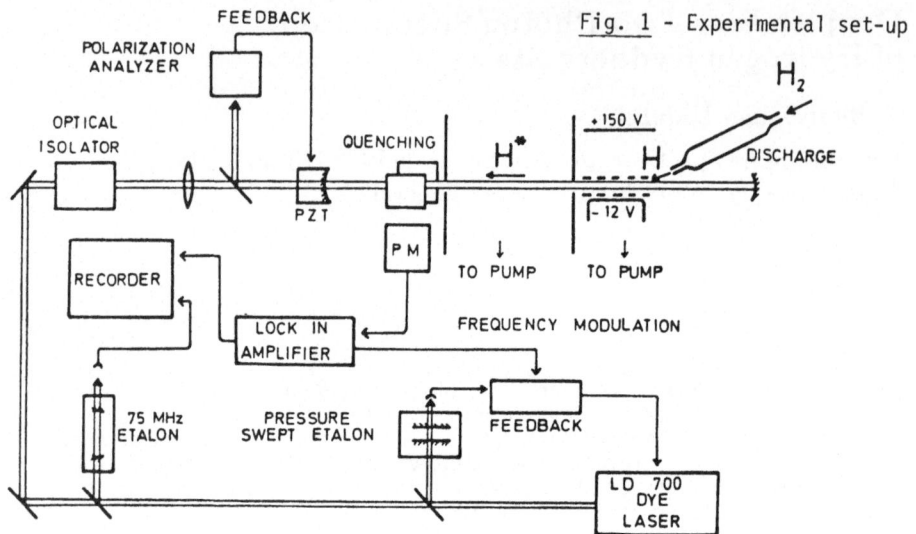

Fig. 1 - Experimental set-up

gen flows into a first vacuum chamber through a 4 cm long tube, 3 mm in diameter. The electron gun consists of three electrodes :
(i) a double-grid at the ground potential delimits an equipotential volume along the metastable beam axis
(ii) a heated tungsten wire operated at - 12 eV emits an electronic current of 5 mA
(iii) an anode at + 150 V collects the electrons.

The second vacuum chamber is an equipotential volume where stray electric fields are reduced at best. In this chamber, the metastable atomic beam is delimited by two holes which are 21 cm apart and whose diameters are 5 and 7 mm.

The metastable atoms are detected in the third chamber. Two electrodes quench the $2S_{1/2}$ state and a photomultiplier (Hamamatsu R 1459) detects the Lyman-α fluorescence. Taking into account the photomultiplier quantum efficiency and the detection solid angle, we detect in that way 1.7 per cent of the metastable atoms. Measuring the photomultiplier current, we estimate a metastable beam intensity to be about 10^7 atom.s^{-1}.

To perform the 2S - nS and 2S - nD two-photon excitations with n > 8, we need wavelengths in the range 730 - 780 nm which are easily obtained using LD 700 dye. We use a home-made cw ring dye laser [10] pumped by the red lines of a krypton ion laser. It delivers a power of about 700 mW at 777.8 nm (wavelength of the 2S - 8S and 2S - 8D transitions). Two feedback loops lock the laser on the side of the transmission peak of a pressure-swept Fabry-Perot etalon. We then obtain a laser linewidth of about 0.15 MHz.

To enhance the two-photon excitation efficiency, the metastable atomic beam is placed inside a Fabry-Perot cavity. Inside the cavity the light power is about 44 Watt and the beam waist 375 μm (i.e. the power density is 100 W/mm^2).

3. Linewidths

After a two-photon excitation from the 2S metastable, the nS and nD states undergo radiative cascade to the 1S ground state in a proportion of about 90 %. The two-photon transitions can thus be detected by observing the modification of the 2S beam intensity. In Fig. 2 a typical recording shows the 2S-8D two-photon transition. The largest signal amplitude $(2S_{1/2}(F=1) \to 8D_{5/2}$ transition) corresponds to a 10 per cent decrease of the metastable beam intensity. The experimental linewidth (in the laser frequency) is about 700 kHz. This corresponds to a relative linewidth of 2×10^{-9}. The natural width of the 8D level is 550 kHz. There are several reasons for the broadening and the shift of the line.

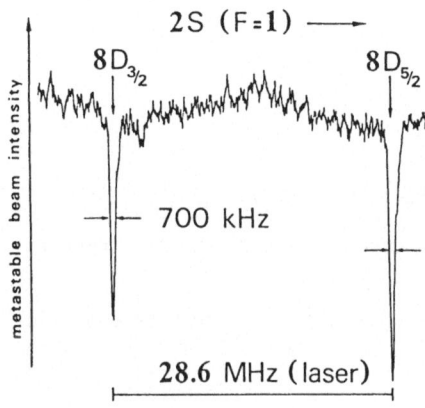

Fig. 2 - Recording of the $\overline{2S_{1/2} \to 8D_J}$ two-photon transition in hydrogen.

(i) Finite Transit Time. The finite transit time in the laser beams should give a line broadening of 120 kHz for a mean angle of the atomic trajectory with respect to the laser beam.
(ii) Second-Order Doppler Effect [6]. For an atomic beam of 3.2 km/s mean velocity, the second-order Doppler effect decreases the line frequency by 40 kHz and broadens it by about 60 kHz.
(iii) Light-Shift. In the case of a 2S nS or nD transition, the light shift of the 2S level is predominant,since the 2S-3P oscillator strength is much larger than the 3P-nS and 3P-nD ones. In our experimental conditions, for an atom at rest in the center of the laser beams, the light-shift is about +600 kHz. After an average over the light intensities seen by the atoms, one obtains a mean light-shift of + 400 kHz and a broadening of 300 kHz.

Taking into account the laser linewidth, these various broadening effects result in a total width of about 800 kHz. Thus other stray effects (as Stark effect or pressure broadening) play a role. For example,an electric field of 30 mV/cm produces a Stark splitting of 1 MHz in the $8D_{5/2}$ level and a shift of 40 kHz. We are presently working to reduce these effects.

4. Spectroscopic Results

To eliminate the metastable beam intensity fluctuations, the laser frequency can be modulated and the lineshape becomes a derivative trace. In Fig. 3 a typical recording shows the 2S-8S and 2S-8D transitions for the deuterium. The laser frequency being monitored with a 75 MHz free spectral range planar focal Fabry-Perot, we have measured the hydrogen fine struc-

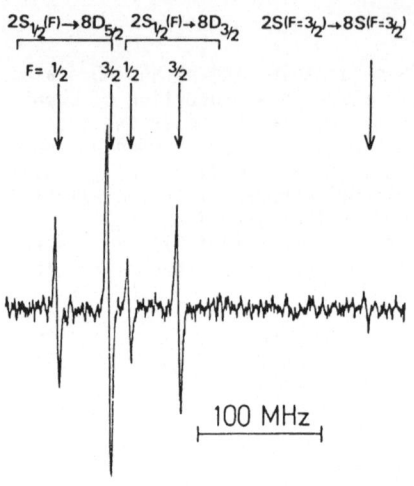

Fig. 3 - Recording of the $2S_{1/2} \to 8D_J$ and $2S_{1/2}$ two-photon transition in deuterium.

ture interval : $8D_{5/2}-8D_{3/2}$= 57.2±1.9 MHz ; $8D_{3/2}-8S_{1/2}$= 153.9±0.9 MHz and the $2S_{1/2}-8D_{5/2}$ isotopic shift Δ of hydrogen and deuterium : Δ = 209692.2±2.2 MHz. Our results are in good agreement with the theoretical predictions.

In conclusion, we have performed an experiment in atomic hydrogen, allowing the measurement of optical transition wavelength with a high precision. We have obtained a relative linewidth of 2×10^{-9}. This relative linewidth is the smallest that has been observed to date on an optical transition in atomic hydrogen. We are working to reduce the linewidth, and we hope reduce the systematic shift to less than 100 kHz. This precision should give the Rydberg constant with an uncertainty of about 10^{-10}.

1 F. Biraben and L. Julien : C.R. Acad. Sci. 300, 161 (1985)
2 F. Biraben and L. Julien : Opt. Comm. 53, 319 (1985)
3 J.E.M. Goldsmith, E.W. Weber and T.W. Hänsch : Phys. Rev. Lett. 41, 1525 (1978)
4 B.W. Petley, K. Morris and R.E. Shanvyer : J. Phys. B13, 3099 (1980)
5 S.R. Amin, C.D. Caldwell and W. Lichten : Phys. Rev. Lett. 47, 1234 (1981)
6 B. Cagnac, G. Grynberg and F. Biraben : J. Physique 34, 845 (1973)
7 C.J. Foot, B. Couillaud, R.G. Beausoleil and T.W. Hänsch : Phys. Rev. Lett. 54, 1913 (1985
8 S.R. Lundeen and F.M. Pipkin : Phys. Rev. Lett. 46, 232 (1981)
9 Yu.L. Sokolov and V.P. Yakovlev : Sov. Phys. JETP 56, 7 (1982)
10 F. Biraben and P. Labastie : Opt. Comm. 41, 49 (1982)

Highly Excited Hydrogen in Strong Electric Fields

*M.H. Nayfeh, K. Ng, and D. Yao**

Department of Physics, University of Illinois at Urbana-Champaign,
1110 W. Green Street, Urbana, IL 61801, USA

We have studied theoretically and experimentally the photoionization of
hydrogen in strong electric fields F in the energy region between the classical point E_c = $-2\sqrt{F}$ and E=0 and for positive energies. We used two-photon
excitation from the ground state to selectively excite distinct charge distributions of n=2 states: a charge distribution that is extended upfield,
(m_ℓ=0 blue state) or a charge distribution that is extended downfield (m_ℓ=0
red state). An additional one-photon excitation resulted in the excitation
of highly excited ionizing states having giant dipole charge distributions.
Only in hydrogen pure low-lying excited parabolic states that have these
distinct charge distributions can be prepared using laboratory fields. Our
results show that the percentage of the charge distribution that can be
molded into the giant dipole from the otherwise free electronic charge of
the final state can be enhanced dramatically by using multistep excitation
(multistep molding) via resonant intermediate states whose charge is
partially extended instead of a single step excitation from the ground
state. This enhancement in the molding is manifested in the enhancement of
the depth of the Stark-induced resonances in the region above E=0 over the
depth observed in the case of excited complex atoms.

We showed that it is possible to create, at lower energies, moderately
giant dipoles of distinct charge distributions with much longer life times
(on the order of ns), thus perhaps opening the way for new types of applications. We also found that the oscillator strength from the m_ℓ=0 parabolic
states of n=2 to the rapidly ionizing m_ℓ=0 states of the giant dipoles concentrate near the two thresholds. Such concentration has not been observed
in similar experiments in complex atoms. Moreover, we found the systematics
of the ionization life times: lines of fixed principle quantum number n get
sharper as their parabolic energy quantum number n_1 increases, and lines of
fixed n_1 get wider as their quantum number n increases. The collisional
interaction of these giant dipoles is discussed.

1. Theory of Photoionization from Parabolic States

Several quantum mechanical calculations have been performed on the effects
of a strong electric field on the highly excited states of atomic hydrogen
excited from the ground state,or from excited spherical states! The initial
states used were good states of angular momentum (for example the 3p state).
Since detailed comparison with the experiment to be described will require
the determination of the cross-sections for photoionization from the n=2
parabolic excited states, the need for a new calculation was seen.

Recently,we theoretically studied excitations from the parabolic (Stark)
states in a given low-lying n state in hydrogen to the highly excited states

*On leave from the Graduate School of the University of Science and
Technology of China, Beijing, People's Republic of China.

near E=0 and between E_c and E=0 in the presence of a strong electric field.[2] We presented calculations of the photoionization cross-section as a function of the quantum number n of the initial state and as a function of its parabolic quantum numbers n_1, n_2, and m_ℓ and gave the dependence of the depth of the Stark-induced resonances on these quantum numbers.

The percentage of the depth of the broad modulations at E > 0 for excitation from some states are presented in Table I. Moreover, the results for excitation from some spherical states are shown in the same table.

Table I - Effect of initial state.

n	parabolic states $(n_1 \quad n_2 \quad m_\ell)$			%	spherical states	%
1	1	0	0	20.4	1s	20.4
2	1	0	0	57	2s	20.4
	0	1	0	2.3	2p	38
3	2	0	0	89.7	3s	18.8
	0	2	0	.3	3p	39.6
	1	1	0	11.0		
4	3	0	0	123.9		

Comparison between excitation from spherical and parabolic states is interesting,since the spherical case is applicable to excitation in complex atoms. Our calculations in Table I show that for excitation from the ground states both cases give essentially the same result. However, for excitation from excited states we have drastic differences between the two cases. We find an enhancement of the depth of the Stark-induced modulation in the region E > 0 when the initial excited state is a pure m_ℓ=0 blue state, and disappear almost completely when the initial state is a pure m_ℓ=0 red state.

2. Experimental

Simultaneous absorption of two photons from a single tunable pulsed laser beam at 243 nm results in excitation from 1s to n=2, and some photoionization of the resulting n=2 population.[3] A second pulsed beam excites states near the continuum from the n=2 state. An atomic beam is formed by effusion from a Wood discharge tube through a multicollimator assembly composed of 25 small glass capillaries. The beam is loosely collimated, but produces a density of about 10^{11} H°/cm^3; the background gas density is on the order of $10^{12}/cm^3$. One of the Stark plates has a 3 mm x 10 mm slot cut into it and covered by a fine mesh to allow the passage of ions. This gives a limit of 1 μs on the detection time. Ions travel through a 100 cm long, field-free drift tube,which provides mass analysis and are detected using an 18 stage venetian blind electron multiplier capable of single ion detection. A fraction of the second harmonic of a YAG laser at 532 nm is used to pump one dye laser producing a beam at 630 nm, which is frequency doubled to 315 nm by a KDP crystal and then summed with the residual YAG fundamental by a KDP crystal.resulting in a beam at 243 nm of pulse length of about 10 ns, a bandwidth of about 1.5 cm^{-1} and pulse energies on the order of 10 microjoules. A second dye laser produces a beam at about 555 nm which is summed with part of the YAG fundamental to produce a beam with pulse length near 10 ns, bandwidth of .8 cm^{-1}, pulse energies of a few tenths of a millijoule and a wavelength near 365 nm. The data are collected and analyzed using an LSI-11 computer system.

3. Experimental Results Near E=0

We tested the performance of the overall system by exciting high Rydberg states of hydrogen in a field-free, atomic beam environment. For this test both field plates were grounded until one microsecond after the passage of the laser pulses, thereby field-ionizing the highly excited atoms and collecting the ions. The spectrum is shown in Fig. 1. Figure 2 was taken at higher fields, where the Stark splitting is large enough to allow selective excitation of the parabolic states. In Fig. 2a, the 243 nm beam was selectively exciting the $m_\ell=0$ blue state, and the ionization beam is tuned across the E=0 region and is also of π polarization, whereas in Fig. 7b the $m_\ell=0$ red state was selectively excited. The results are in very good agreement with Table I. These observations can also be understood from the following arguments that are based on the charge distribution of the broad resonances, and that of the parabolic states of n=2 state. From the Stark theory, it is known that the electronic charge distribution of the broad resonances is highly polarized along the axis of the electric field and predominantly up-field with respect to the nucleus. On the other hand, each Stark state of n=2 state has a different charge distribution along the field, being mostly up-field if the shift is positive (blue states) and down-field if the shift is negative (red states) as shown in Fig. 3. Therefore we expect to have a large overlap between the wave function of the $m_\ell=0$ blue state of n=2 and that of a broad resonance, because both distributions are elongated up-field. On the other hand, we expect very little overlap between the wavefunctions of the $m_\ell=0$ red state and a broad resonance.

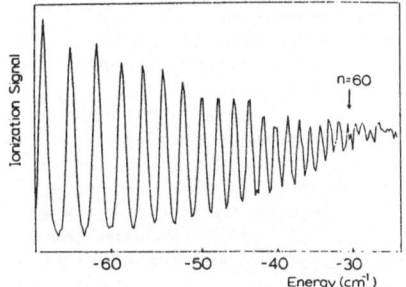

Fig. 1 Field-free Rydberg spectrum

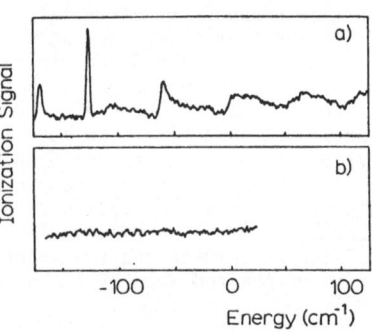

Fig. 2 Spectrum in the presence of 16.8 kV/cm from $m_\ell = 0$ (a) blue state and (b) red state of n = 2

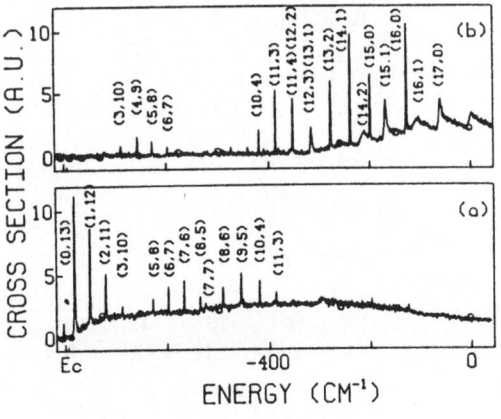

Fig. 3 Same conditions as fig. 2

73

The best system in which to test the dependence of the spacing of resonances near E=0 on the field is hydrogen, since the enhancement of the resonance by excitation from pure parabolic states leads to large resonance and very good signal to background ratios. The resulting best-fit gives a power law of .751 ± .02, in excellent agreement with the semiclassical result, which gives 3/4 power law. Thus this analysis has verified the semiclassical result for the field-dependence of the spacing to the highest precision reported to date, and hence rules out the competing theoretical work that predicts a 2/3 power law.[4]

4. Experimental Results Between E=0 and E_c

Figure 3a-b give the photoionization yield from the n=2,m_ℓ=0 blue state and from n=2,m_ℓ=0 red state as a function of energy in the energy range between E_c and E=0. The electric field imposed is 16.8 kV/cm, and the polarization of the ionizing laser beam is π polarization. The same spectrum was also studied by Welga et al. at lower fields.[5] The figures show that concentration of the ionization activity occurs at and near the two thresholds. This is so for both the sharp states and the smooth underlying background. We identified the peaks and calculated their ionization cross-sections numerically. Their parabolic quantum numbers are given in Fig. 3. Figure 4 gives the ratio of the cross-sections from the red state and blue state of n=2 as a function of Z_1, the fraction of the charge that drives the bound motion. Because our instrumental resolution Γ_i is 1.5-2 cm^{-1}, then our measurement gives the integrated cross-sections over energy for states whose widths Γ are much less than Γ_i. However for states of $\Gamma \gg \Gamma_i$, the line shape is essentially unaffected. Convolution analysis of the calculated spectrum shows good agreement with the experiment. We also studied the width of the Stark ionizing states as a function of their quantum numbers. Figure 5 gives plots of the half-width versus the parabolic quantum number n_1 (for fixed n) of three Stark n manifolds in the photoionization spectra of the m_ℓ=0 state of n=2 using π polarization and at 16.7 kV/cm. Figure 6 on the other hand gives plots of the half-width versus the principle quantum number n (for fixed n_1) of the Stark n_1 manifolds in the photoionization spectrum of the $\bar{m}_\ell$=0 state of n=2 using π polarization and at 16.7 kV/cm. The open circles and the solid circles are those of the experiment and the exact numerical calculations. We used the half-width of the red wing of the wide asymmetric profiles, and the straight

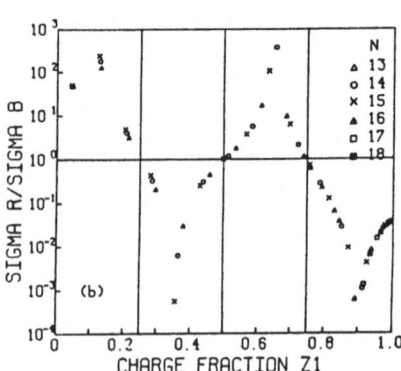

Fig. 4 Ratio of cross-sections from m_ℓ = 0 red and m_ℓ = 0 blue state of n = 2 as a function of Z_1

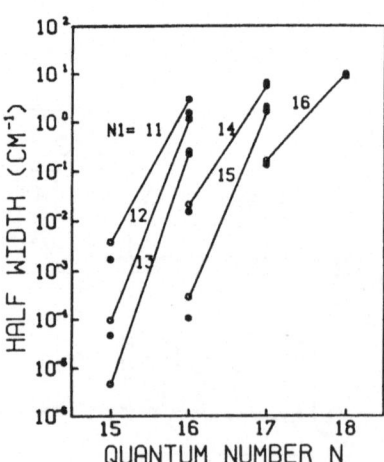

Fig. 5 Ionization width as a function of N_1 for fixed N

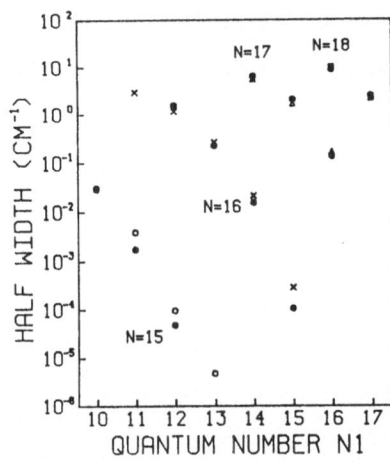

Fig. 6 Ionization width as a function of N for fixed N_1

lines drawn between the calculated points are just used to indicate the grouping of the points. The figures show that the states get narrower as n_1 increases for a fixed n, whereas they get wider as n increases for a fixed n_1.

5. Simultaneous Presence of Collisions and DC Electric Fields

In the presence of strong external fields, giant dipoles can be produced with ps life times. Moderately giant dipoles on the other hand can be produced with life times on the order of ns. How do collisions affect a cigar-like giant dipole? What is the role of the geometrical cross-section of the atom? We have recently made some investigations of the simultaneous interactions of highly excited atoms with external dc electric fields and depolarizing collisional interactions with electrons, ions, or neutral atoms.[2] We find for the first time that the electric field enhances by many orders of magnitude the depolarization cross-section. The interaction is so long-range that the excitation duration (pulse width) governs the time over which the interaction takes place. Although this long-range dipole lives less than 10^{-11} seconds and the excitation pulses are less than 10 ns, the interaction is strong enough to cause appreciable depolarization even at number densities as low as 10^8/cc. In other words, the electric field renders the geometrical cross-section irrelevant as an indicator of the collisional activity. An alternative indicator of the collisional activity is the distance from the nucleus to the classical tuning point.

The work was supported by the National Science Foundation Grant No. NSF-PHY-81-09305.

References

1. E. Luc-Koenig and A. Bachelier, J. Phys. B 13, 743 (1980); V. D. Knodratovich and V. N. Ostrovsky, Zh. Eksp. Teor. Fiz. 4, 1256 (1982) [Sov. Phys. JETP 56, 719 (1982)]; R. J. Damburg and V. V. Kolosov, J. Phys. B 9, 3149 (1976); D. A. Harmin, Phys. Rev. A 24, 2491 (1981); U. Fano, Phys. Rev. A 24, 619 (1981).

2. M. H. Nayfeh, K. Ng and D. Yao, in Atomic Excitation and Recombination in External Fields, M. Nayfeh and C. Clark eds., Harwood Academic Publishers, New York (1985); W. L. Glab and M. H. Nayfeh, Phys. Rev. A 31, 530 (1985); W. L. Glab, K. Ng, D. Yao and M. H. Nayfeh, Phys. Rev. A, June (1985).

3. W. L. Glab and M. H. Nayfeh, Opt. Lett. $\underline{8}$, 30 (1983).

4. I. I. Fabrikant, Zh. Eksp. Teor. Fiz. $\underline{79}$, 2070 (1980); [Sov. Phys. JETP $\underline{52}$, 1045 (1980).]

5. H. Rottke, A. Hole and K. H. Welge, in Atomic Excitation and Recombination in External Fields, M. Nayfeh and C. Clark eds., Harwood

Laser Investigations of Electron Correlations: Doubly Excited States of Ba

R.R. Freeman, L.A. Bloomfield, and J. Bokor

AT & T Bell Laboratories, Murray Hill, NJ 07974, USA

W.E. Cooke

University of Southern California, Los Angeles, CA 90007, USA

We have been conducting a series of laser spectroscopic investigations into the consequences of electron correlations in doubly excited electron systems. It is the purpose of this presentation to summarize our results to date and to suggest directions for future work.

We have adopted the notation of RAU [1], and distinguish two types of doubly excited m l n' l' states: those for which $m << n$, yet both m and n large, we call "excited core Rydberg states"; those for which $m \sim n$ we call "Wannier" states. For the excited core Rydberg states, m and n are approximately good labels, while for Wannier states the electrons are so strongly correlated that this independent electron labeling scheme breaks down completely. Recently, BUCKMAN et. al [2] reported evidence for Wannier states in electron impact studies on He⁻. Here we show some surprising results in designated, laser-resolved states in Ba for both cases.

Figure 1 shows the method of "Isolated Core Excitation" used to prepare an msns doubly excited state in Ba. Two lasers are used to excite the Rydberg electron, then two photons of a third laser drive the 6s-->ms "core" transition. For lower values of m, the autoionization of the msns state is primarily to the 6s Ba^+ ion, and the resulting fast electrons are detected. For larger values of m, autoionization is primarily to excited nl Ba^+ states, and Ba^{++} ions, arising from the photoionization of the Ba^+ nl state, are detected.

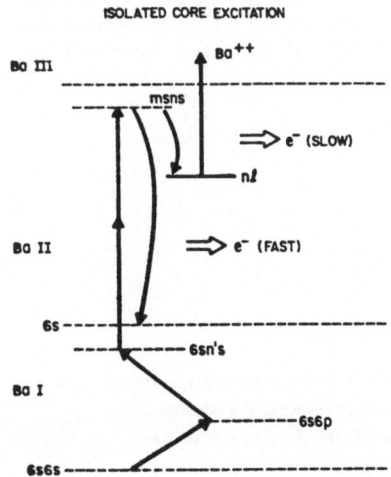

ISOLATED CORE EXCITATION

Fig. 1 Principle of "Isolated Core Excitation" to prepare an msna doubly excited state of Ba

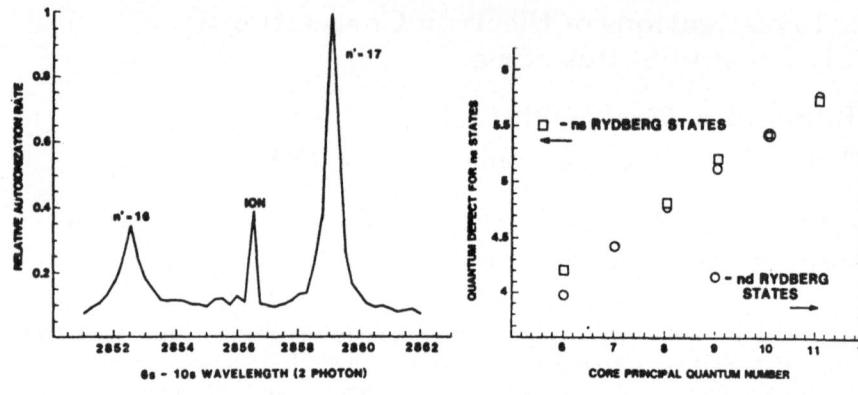

Fig. 2 Spectrum obtained when a 6s16s-->10sn's "core" transition is driven

Fig. 3 Quantum defects of ns and nd states of Ba as a function core size

Figure 2 shows a typical spectrum when such an isolated core excitation transition is performed. Because the excited ms core is larger than the original 6s core, the Rydberg electron is "shaken-up" into nearby Rydberg states of the new ms core. JOPSON *et. al* [3] have discussed the dynamics of this process and its dependence upon detuning from intermediate states. If the quantum defe change from the 6s core to the ms core were exactly integral, then there would be only one line in the spectrum centered on the ionic transition. By measuring the displacement of these peaks from the ionic line, the quantum defects of both nd and ns states as a function of core size may be extracted. This data is shown in Fig. 3. Note that the slope of increase in the quantum defect with m is linear.

We have also investigated the dependence of the autoionization rate (Γ) of msns and msnd states as a function of m, again in the range of $m << n$. It is well known that for a given core size, the product (Γn^3) (called the "scaled autoionization") is a constant, a consequence of the normalization of the Rydberg wavefunction at the origin. Figure 4 summarizes our data on the dependence of the scaled autoionization rate on m. The surprising result is that while nd scaled autoionization rates increase

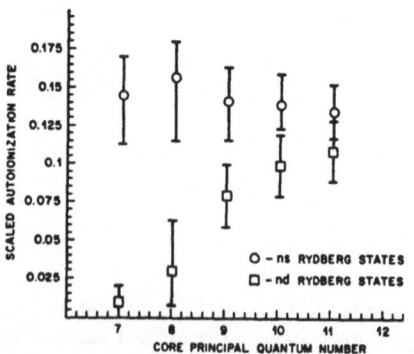

Fig. 4 Dependence of scaled autoionization rate of ns and nd states as a function of core principal quantum number

over and order of magnitude between m=6 and 11, ns state remain virtually unchanged over the same interval. Calculations using frozen-core Hartree-Fock wavefunctions yield results qualitatively in agreement with our data for the nd states. Although Bloomfield *et.al* [4] describe a classical analog that suggests why the scaled autoionization rates for ns states should be independent of m, we cannot reproduce the results for ns states with quantum mechanical calculations based on single particle formalisms (e.g., Hartree-Fock).

We have also investigated the spectrum of msns states when m~n. In general we found that when m*~n*, the signal becomes weaker, and we have not been able to detect any state in which m*=n*. Here we show results of 6s15s-->11sn's, 6s14s--
>11sn's and 6s13s-->11sn's. Because of the changes in the quantum defects with increasing core, the 6s13s-->11sn's is a transition in which the final n* is approximately m*+1. Figure 5 shows the results of these transitions where only the lowest energy peak in the final state shake-up is shown. As m* approaches n* the isolated, relatively symmetric line becomes a complicated line shape with prominent side structure. Figure 6 shows that these states have several nearly degenerate series with which to mix (e.g.,6gng, 6hnh, 6jnj). These spacings are quite suggestive of the equally spaced levels of the "vibrator-rotator" spectrum that a Wannier atom is predicted to evidence [1].

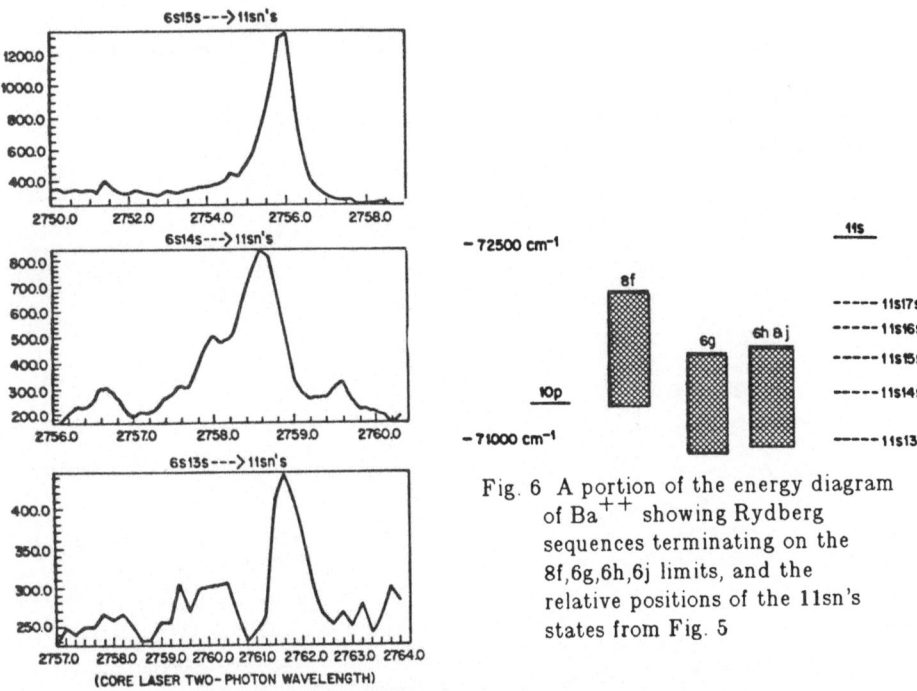

Fig. 6 A portion of the energy diagram of Ba^{++} showing Rydberg sequences terminating on the 8f,6g,6h,6j limits, and the relative positions of the 11sn's states from Fig. 5

Fig. 5 Dependence of the line shape of the transition 6sns-->11sn's as a function of n

Although there is not enough space here to present any significant portion of our data, we summarize our findings: when a state has $m^{*\sim}n^*$, it will readily mix with other nearby states of equal total angular momentum and parity. Moreover, our data show that the preferential mixing is to states in which the angular momenta of the individual electrons is maximized (e.g. a 6jnj rather than a 8fnf, although both states satisfy all other requirements for mixing with the 11s13s state). We can state this result by formulating the rule that the transition to true Wannier states will occur when $m^{*\sim}n^*$, and there exists a very large manifold of high angular momentum states with which the "target" state can mix.

We have found that although there exists substantial theory for Wannier states in He-like systems (where all the angular momenta are degenerate) [1], there is precious little theory in the transition region we are discussing here. To push the experiments further into the regime of many available states requires higher excitations; this, in turn, opens up many more modes of decay, with a resulting loss in detection efficiency. Optical detection of excited ions after autoionization of the Wannier state may help overcome this problem.

REFERENCES

1. A.R.P. Rau *IX ICAP*, Seattle, July 1984 (to be published)

2. S.J. Buckman, P. Hammond, F.H. Read and G.C. King, J. Phys. B *16*,4039 (1983)

3. R.M. Jopson, R.R. Freeman, W.E. Cooke, and J. Bokor, Phys. Rev. Lett. *51*, 1640 (1983)

4. L.A. Bloomfield, R.R. Freeman, W.E. Cooke, and J. Bokor, Phys. Rev. Lett. *53*, 2234 (1984)

Laser Spectroscopy of Highly Excited Hydrogen Atoms in Strong Magnetic Fields

A. Holle and K.H. Welge

Fakultät f. Physik, Universität Bielefeld, D-4800 Bielefeld 1, Fed. Rep. of Germany

1. INTRODUCTION

In the physics of highly excited atoms around the ionization limit in external electric and magnetic fields [1], that is under strong field mixing [2] conditions, the H atom naturally plays a fundamental role. However, while the atom has served as basis for extensive theoretical studies, little experimental work has been performed with it. Electric field experiments have been carried out recently [3,4,5] showing agreement with well developed theory [1,2,6], as must be expected on account of the separability of the nonrelativistic problem. No experiments are known with H in magnetic fields, except most recent ones [4], which also include the first observation of quasi-Landau resonances above the ionization threshold. Different from the electric field case, the H atom with diamagnetic interaction described in the simplest form by the Hamiltonian

$$H_{diam}^{coul} = -1/r + 1/8 \, \alpha^2 B^2 r^2 \sin^2 \Theta,$$ still presents basic, open problems, theoretically and experimentally.

Here we report some of the new results obtained in our magnetic field experiments with H atoms using, like in our previous electric field studies [3,4], the resonant two-step excitation, $H(n = 1) + VUV \rightarrow H(n = 2) + UV \rightarrow H^*$, by pulsed tunable VUV and UV laser radiation. Since only a fraction of the work can be covered here, we focus on some results in the spectral region around the ionization limit, i.e., the quasi-Landau region, which is of most interest.

2. EXPERIMENTAL

A crossed laser-atom beam set-up (Figure 1) is used, with the axis of the three beams perpendicular to each other, and the atomic beam axis parallel to the magnetic field. The electric field arrangement with the two parallel electrodes serves for the detection of H^* atoms, ionized spontaneously or by the field. Ionization is monitored through the electrons by means of a surface barrier detector. Experiments have been carried out with a weak constant field of the order of 1 V/cm, continuously present. This field mainly served to extract the electrons from the excitation region. It was strong enough also to field-ionize

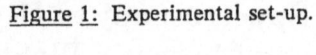
Figure 1: Experimental set-up.

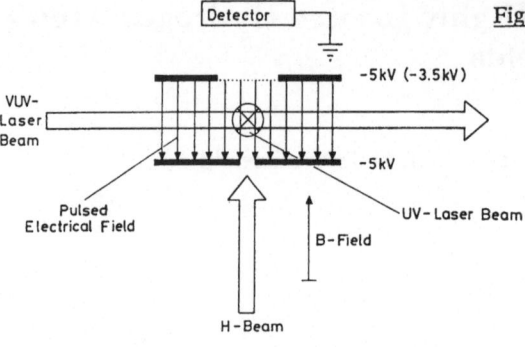

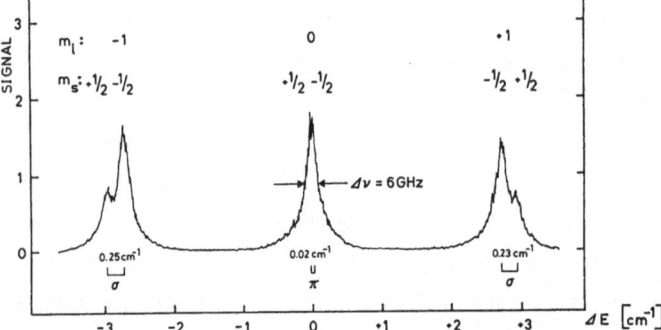

Figure 2: Paschen-Back splitting of $(n=1) \rightarrow (n=2)$ transition of the H atom. Excitation with linearly polarized VUV laser radiation at ~45° angle to the B-field direction. Field strength B=6T.

H* atoms at energies in the region $E \gtrsim$ -20cm^{-1}. On the other hand the electric interaction was still small enough in comparison with the diamagnetic one at the magnetic fields applied (B≤6T).

3. RESULTS

At fields applied (3T≤B≤6T) the first excitation step, H(n=1) → H(n=2), is governed fully by the Paschen-Back splitting, yielding levels $m_\ell = 0, \pm 1$ in n=2. Figure 2 shows the splitting at B=6T, obtained with linearly polarized VUV at ~45° to the B-field. The resolution of ~6 GHz is given mostly by the VUV laser bandwidth and to a small part by the residual Doppler width.

In the actual experiments the VUV was either parallel (π) or perpendicular (σ) polarized, leading, respectively, to $m_\ell = 0$ or $m_\ell = \pm 1$ levels. Applying the UV also π or σ polarized, spectra with final states of even parity and $m_\ell^f = 0, \pm 1, \pm 2$ are excited, with no essential difference between states of positive and negative sign.

82

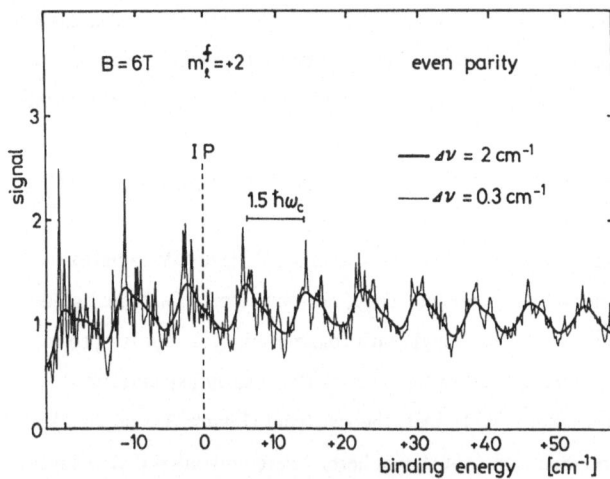

Figure 3: Excitation-ionization spectrum (light line) of the H atom around the ionization threshold (E=0) in a magnetic field B=6T. Excitation from Paschen-Back | n = 2, m_ℓ = + 1> state with tunable UV radiation, polarized perpendicular (σ) to the B=field. Heavy line: light line spectrum averaged by Gaussian of 2cm⁻¹ FWHM.

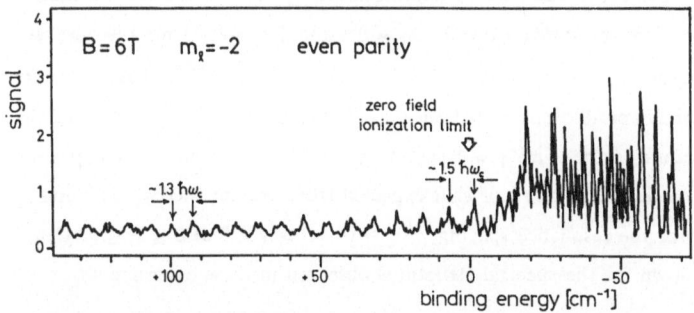

Figure 4: Excitation-ionization spectrum of the H atom previously obtained [4]. Resolution ~2cm⁻¹, field strength B=6T, excitation from | n = 2, m_ℓ = − 1> state with σ polarized UV radiation.

We have observed all three types of spectra. Figure 3 shows, as an example, one with m_ℓ^f = +2, excited at B=6T from the | n = 2, m_ℓ = +1> state with σ polarized UV. This may be compared with the spectrum (Figure 4) previously obtained [4] with substantially lower resolution (~2cm⁻¹ compared to 0.3cm⁻¹) and poorer signal stability and quality. The essential difference is the spectral fine structure (Figure 3) extending through the threshold into the quasi-Landau regime, i.e., E≥0. In fact, the width of some line structures at E>0 is apparently equal to the experimental bandwidth, indicating that the actual line width may well be less than 0.3cm⁻¹ and the ionization lifetime of states at E>0 accordingly ≥10⁻¹¹ sec.

The spectrum in Figure 3 was recorded with incremental wavelength steps of an effective channel width of ~0.1cm^{-1}. When the signal amplitude of each channel is represented by Gaussian function of 2cm^{-1} FWHM the heavy-line smooth curve is obtained. It exhibits the structureless quasi-Landau oscillation with the energy spacing 1.5 × $\hbar\omega_c$ at E=0, as expected from, and in agreement with WKB [2] and time-dependent wavepacket [7] treatments.

The excitation from $| n = 2, \ | m_\ell | = 1>$ states with σ polarized UV radiation where the light beam crosses the B-field perpendicularly, not only $| m_\ell^f | = 2$ final states are reached. It also leads to $m_\ell^f = 0$ states, though with roughly only 0.1 relative probability. The presence of $m_\ell^f = 0$ states is presumably causing the slightly asymmetric shape of the smoothed quasi-Landau oscillations. In fact, the spectrum (Figure 3) contains line structures that are present in the spectrum (not shown here) where only $m_\ell^f = 0$ states were excited.

The spectra with final states $m_\ell^f = 0$ and $| m_\ell^f | = 1$, not presented here, also show fine structure in the quasi-Landau regime again with lines of widths given by the experimental resolution limit, i.e., ~0.3cm^{-1}. These spectra were obtained by excitation with UV radiation of π polarization from the states $| n = 2, m_\ell = 0>$ and $| n = 2, \ | m_\ell | = 1>$, respectively. Averaging the $m_\ell^f = 0$ spectrum with a Gaussian of 2cm^{-1} FWHM again results in smooth quasi – Landau oscillations with $1.5\hbar\omega_c$ spacing at E=0. However, applying the same smoothing procedure to the $| m_\ell^f | = 1$ spectrum resulted, surprisingly, in a spacing ~$0.65\hbar\omega_c$, i.e., somewhat less than one half that expected from present theory. Existing theory accounts only for the resonance spacing $\gamma\hbar\omega_c$ with $1.5 \geq \gamma \geq 0$ at $0 \leq E \leq \infty$. Otherwise no theory is known for the spectral structures observed in these experiments.

REFERENCES

[1] The field is reviewed extensively. For recent reviews see for example:
  (a) J. Physique, Colloque C-2, 43 (1983) on "Atomic and Molecular Physics Close to the ionization Threshold in High Fields".
  (b) C.W. Clark, K.T. Lu, and A. F. Slarace: in "Progress in Atomic Spectroscopy", Part C, ed. H.J. Beyer and H. Kleinpoppen; Plenum (1984).
  (c) J.C. Gay: in "Progress in Atomic Spectroscopy", Part C, ed. H.J. Beyer and H. Kleinpoppen; Plenum (1984).
  (d) D. Kleppner, M.G. Littman, and M.L. Zimmerman: in "Rydberg State of Atoms and Molecules", ed. R.F. Stebbings and F.B. Gunning; Cambridge Univ. Press, (1983).

[2] A.R.P. Rau: Phys. Rev. A16, 613 (1977); A.R.P. Rau and K.T. Lu: Phys. Rev. A21, 1057 (1980); A.R.P. Rau: J. Phys. B12, L193 (1979).

[3] K.H. Welge and H. Rottke: in "Laser Techniques in the Extreme Ultraviolet", ed. S.E. Harris and T.B. Tucatorto, AIP Conference Proceedings, No. 119 (1984), pp. 213-219; H. Rottke and K.H. Welge: Phys. Rev. A, in press (1985).

[4]   H. Rottke, H. Holle, and K.H. Welge: in "Atomic Excitation and Recombination in External Fields", ed. M.H. Nayfeh and C.W. Clarke, Harwood Publ. (1985); H. Holle, H. Rottke, and K.H. Welge: in "Fundamentals of Laser Interaction", ed. F. Ehlotzky, "Lecture Notes in Physics" Vol. 229, Springer Verlag (1985).

[5]   W.L. Glab and M.H. Nayfeh: Phys. Rev. $\underline{A31}$, 530 (1985); W.L. Glab, K. Ng, D. Yao, and M.H. Nayfeh: Phys. Rev. $\underline{A31}$, 3677 (1985); M.H. Nayfeh, K. Ng, and D. Yao: in "Atomic Excitation and Recombination in External Fields", ed. M.H. Nayfeh and C.W. Clark, Harwood Publ. (1985).

[6.] H.J. Silverstone and P.M. Koch: J. Phys. $\underline{B12}$, L537 (1979); R.J. Damburg and V.V. Kolosov: J. Phys. $\underline{B9}$, 3149 (1976: ibid. $\underline{B14}$, 829 (1981); E. Luc-Koenig and A. Bachelier: J. Phys. $\underline{B13}$, 1743 (1980; ibid. 1769 (1983); V.D. Kondratovich and V.N. Ostrovsky: Sov. Phys. JETP $\underline{52}$, 198 (1980; J. Phys. $\underline{B17}$, 1981 (1984); ibid. 2011 (1984); D.H. Harmin: Phys. Rev. $\underline{A24}$, 2491 (1981); ibid. $\underline{A26}$, 2656 (1982).

[7]   W.P. Reinhardt: J. Phys. $\underline{B16}$, L635 (1983).

High Resolution Laser Spectroscopy of Rydberg States with Principal Quantum Numbers n>100

R. Beigang

IBM Thomas J. Watson Research Center, Yorktown Heights, NY 10598, USA

A. Timmermann, P.J. West, and E. Matthias

Freie Universität Berlin, D-1000 Berlin, Germany

The development of new techniques to excite and detect high Rydberg states of atoms and molecules has stimulated an increasing number of new experiments (see e. g. /1/ and references therein). High resolution spectroscopic techniques allow for very sensitive detection of various types of interactions between different or within the same Rydberg series and the results of these experiments lead to an improved understanding of interacting Rydberg series. In particular, the investigation of Rydberg states with extremely high principal quantum numbers n offers the unique possibility to study interactions which are too weak and therefore negligible at low n.

Doppler free two-photon spectroscopy in combination with thermionic detection was applied to investigate high Rydberg states of strontium. The measurements were made possible by use of an improved space charge limited thermionic diode which allows for field-free excitation and detection of high Rydberg states in a cell geometry up to principal quantum numbers n < 250 /2/. The diode basically consists of a highly symmetric arrangement of the detecting filaments with respect to the exciting laser beam and thus compensating for the perturbing electric fields produced by the space charge itself.

As a typical example , the magnetic hyperfine interaction of Sr Rydberg states was studied systematically over a wide range of principal quantum numbers (5 < n < 250). The change of the relative strength of the magnetic hyperfine interaction with n compared to Coulomb and spin orbit interaction, results in various types of perturbations. In an energy range where the hyperfine interaction is comparable to or even exceeds the Coulomb interaction(n >100) a hyperfine-induced n-mixing was observed for the first time /3/. This effect is comparable to the anticrossing of Stark levels or the avoided crossings induced by external magnetic fields. It should be pointed out, however, that in the case of hyperfine- induced n-mixing the internal magnetic field of the nucleus alone produces the interaction leading to the anticrossing. It is thus an intrinsic feature of the atom and can be expected to occur in high Rydberg states of all atoms with two valence electrons and nonzero nuclear spin. Typical two-photon excitation spectra for three selected pairs of principal quantum numbers, demonstrating the hyperfine-induced n-mixing for 5snd states, are shown in the figure. Above n > 115 the relative position of singlet and triplet hyperfine components is inverted as indicated in the lower spectrum (different isotopes are labelled by their mass number). The experimental results are in good agreement with calculations using the hyperfine interaction as the dominant perturbation in this energy range.

The investigation of Rydberg states with extreme n also facilitates the measurements of effects which scale with high powers of n. Taking advantage of the n dependence of the diamagnetic interaction, the quadratic Zeeman effect and the associated l-mixing of high n Rydberg states were measured in magnetic fields of a few hundred Gauss /4/.

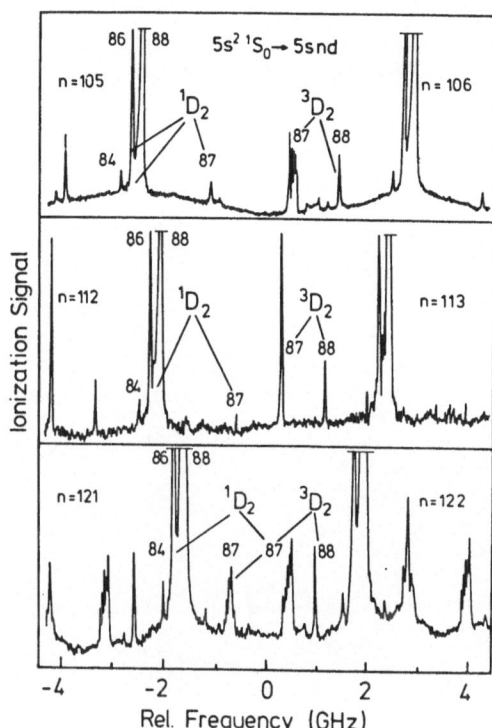

Figure 1: Doppler-free two - photon excitation spectra of 5snd Rydberg states of Sr for three selected pairs of principal quantum numbers n.

This work was supported by the Deutsche Forschungsgemeinschaft, Sonderforschungsbereich 161.

References:

1. F. B. Dunning and R. F. Stebbings (eds.): Rydberg States of Atoms and Molecules (Cambridge, London 1983)

2. R. Beigang, W. Makat , and A. Timmermann: Opt. Commun. 49, 253 (1984)

3. R. Beigang, W. Makat, A. Timmermann, and P. J. West: Phys. Rev. Lett. 51,771 (1983)

4. R. Beigang, W. Makat, E. Matthias, A. Timmermann, and P. J. West: J. Phys. B: At. Mol. Phys. 17, L475 (1984)

Part IV

Atomic Spectroscopy

Ion Formation in Sodium Vapor Containing Laser Selected Rydberg Atoms

M. Allegrini, S. Gozzini, and L. Moi

Istituto di Fisica Atomica e Molecolare del C.N.R.,
I-56100 Pisa, Italy

C.E. Burkhardt, M. Ciocca, R.L. Corey, W.P. Garver, and J.J. Leventhal

Department of Physics, University of Missouri, St. Louis, MO 63121, USA

Ion formation in interactions involving laser-produced sodium Rydberg atoms, Na^{**}, has been experimentally studied. Three heavy body collisional mechanisms leading to ionic products were observed:

$$Na^{**} + Na \rightarrow Na^+ + Na + e \qquad \text{collisional ionization} \qquad (1)$$

$$Na^{**} + Na \rightarrow Na_2^+ + e \qquad \text{associative ionization} \qquad (2)$$

$$Na^{**} + Na \rightarrow Na^+ + Na^- \qquad \text{ion pair formation} \qquad (3)$$

It was found however that the dominant source of ions was photoionization of Na^{**} by blackbody radiation, the rates of which were measured for various principal quantum numbers, n. These measured rates were found to be in excellent agreement with theoretical predictions.

Study of the photoionization yield as a function of n leads to the important conclusion that self-1-mixing, that is, mixing of angular momentum state of a given n in Na^{**}/Na collisions, rapidly destroys the state selectivity provided by laser-production of the Rydberg atoms. The cross-section for this process is deduced to be of order of 10^5Å^2.

Cross-section for collisional ionization and associative ionization were determined to be $\sim$600 Å^2 and $\sim$150 Å^2 respectively for n=18. Since these cross-sections are orders of magnitude lower than that for 1-mixing, they must be understood to be associated with a statistical distribution of 1-states within a given n-state. The observed strong decrease of the associative ionization cross-section with increasing n is interpreted to mean that high angular momentum states are less reactive than lower ones.

Current experiments are directed toward determination of the cross-section for ion pair formation by examination of the Na^- signal as a function of both atom density and laser power density.

The experiments were conducted in apparatus which permits mass analysis of ions produced in sodium vapor at $\sim$$10^{11} cm^{-3}$. Rydberg atoms were prepared by two-step pulsed laser excitation, $3s \rightarrow 3p \rightarrow nl$, where 1 is either 0 or 2.

Atomic ions were formed by two processes: collisional ionization reaction (1), which depends on the square of the atom density N, and photoionization by blackbody radiation which depends only on the first power of N. By examining the N dependence of the Na^+ signal it was deduced that at $4 \times 10^{11} cm^{-3}$ only about 17% of the Na^+ are formed by collisional ionization. Therefore, at $1 \times 10^{11} cm^{-3}$ the contribution of reaction (1) to the observed signal may be neglected.

The data show that at 10^{11} cm^{-3} the Na^+ yield is very nearly independent of n, the principal quantum number, over the range $18 \leqslant n \leqslant 30$. This signal is proportional to the photoionization rate, the initial concentration of the Rydberg atoms and the lifetime of the Rydberg atom population. The initial concentration of Rydberg atoms is proportional to n^{-3} and, over this range of n's, the photoionization rate is proportional to $n^{-1.4}$. Therefore, to be consistent with the observed constancy of the Na^+ signal, the lifetime must vary approximately as $n^{4.4}$, which is very close to the $n^{4.5}$ dependence of the lifetime of a given n-state (for hydrogen atom) averaged over all angular momentum state [1]. This contrast to the n^3 dependence of the lifetime of a specific nl state. Thus we conclude that self-l-mixing of the initially produced ns or nd states partially destroys the state selectivity provided by laser excitation. Furthermore, comparison with the rate of radiative decay suggests that self-l-mixing cross-section is 10^{11} $\overset{\circ}{A}^2$.

Having determined the effective lifetime of the Rydberg population it is now possible to measure collisional and associative cross-sections. For n = 18 it is found that they are ~600 $\overset{\circ}{A}^2$ and ~150 $\overset{\circ}{A}^2$ respectively. While the large Na^+ signal from blackbody photoionization makes it impossible to determine the n-dependence of collisional ionization, it was found that the associative ionization cross-section decreases strongly, roughly as $n^{-7.5}$, with increasing principal quantum number. This strong dependence suggests that high angular momentum states are less reactive than the lower ones, a reasonable conclusion since interaction between the Na^+ - Na^- collision complex and the Rydberg electon is necessary to eject the electron with sufficient kinetic energy to stabilize the product Na_2^+. High n-states require that the Rydberg electron remains remote from the collision complex, minimizing this interaction.

In addition to the above we have directly determined blackbody photoionization rates and found excellent agreement with theory. We have also observed Na^- formation, and are in the process of analyzing the data in detail. Most probably these ions are formed by charge exchange between Na^{**} and $Na(3S)$, i.e. ion pair formation, reaction (3).

[1] . H.A.Bethe and S.A.Salpeter, The Quantum Mechanics of One and Two Electron Atoms (Academic, New York, 1957)

Resonance Ionization of Cesium Atoms in Presence of Buffer Gas Collisions

E. Arimondo and F. Giammanco

Dipartimento di Fisica, Universita di Pisa, I-56100 Pisa, Italy

A. Sasso and M.I. Schisano

Dipartimento di Fisica Nucleare, Struttura della Materia e Fisica Applicata, Universita di Napoli, I-80015 Napoli, Italy

In resonant laser ionization spectroscopy an atomic or molecular system is ionized by absorption of several laser photons and the resulting electrons and ions collected by an externally applied electric field. However, if heavy body collisions are significant, the number of collected electrons and ions may be modified. We have investigated, in a cell containing cesium atoms and buffer gases at different pressures, the influence of heavy body collisions on the ionization signal. Two-photon ionization of the cesium atoms occurred by absorption of laser photons at the wavelength of the second resonance transition. Broadband pulsed laser excitation was used and all hyperfine levels of the ground and excited states were uniformly populated. It was found that the laser ionization rate depends on collisions because: i) the population in the intermediate resonant state is determined by the rate of m_J mixing; ii) the electron-ion recombination rate is affected by collisions.

We have measured the cesium ionization rate via the intermediate $7^2P_{1/2}$ and $7^2P_{3/2}$ states as a function of the laser intensity and buffer gas pressure. In these measurements an electric field was continuously applied to the electrode plates for charge collection. We have observed that under π-polarized laser excitation to the $7^2P_{3/2}$ state collisional m_J-mixing enhances the ionization rate. π-Polarized laser excitation produces an alignment component in the population distribution of the $7^2P_{3/2}$ state, but m_J-mixing destroys this alignment. This process has been described in terms of an electronic disalignment cross-section [1] because the ionization rate depends on the population distribution in the m_J-levels. We have compared cesium two-photon ionization via the intermediate $7^2P_{3/2}$ state, where the disalignment mixing occurs, with two-photon ionization via the $7^2P_{1/2}$ state for which mixing cannot occur. Because the ionization signals in the presence of buffer gas are affected by recombination and attachment, the results have been analyzed by introducing the ratio R between the atoms photoionized via the intermediate $7^2P_{3/2}$ and $7^2P_{1/2}$ states:

$$R = n_i^{3/2} / n_i^{1/2}$$

Here n_i^J ($J = 3/2, 1/2$) denotes the number of ions produced by a laser pulse tuned to one of the intermediate states. Figure 1 shows the measured ratio as a function of the buffer gas pressure in a cell containing cesium atoms at 10^{10} atoms/cm^3 and for a 57.5 MW/cm^2 laser intensity in a 12 nsec pulse. The collision m_J-mixing disalignment cross-section in the $7^2P_{3/2}$ state has been determined from these measurements making use of a rate-equation model [1].

In order to investigate electron-ion recombination, a time delay between the laser pulse and the collecting electric field was introduced and the ion yield measured as a function of this time delay. Figure 2 shows a typical result at different argon pressures with cesium density 10^{12} atoms/cm^3. A 50 MW/cm^2

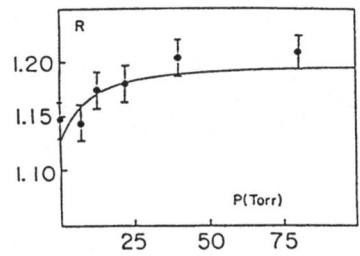

Figure 1. Experimental points and theoretical curve for the ratio R vs. the argon buffer gas pressure.

laser intensity ionized the cesium atoms in the 1.2 mm^3 focus volume. The dependence of the ion signal on the time delay provides information on collisional processes involving cesium atoms prior to collection. Thus, the recombination processes may be directly monitored. The exponential decay observed in the data of Fig. 2 shows that atomic recombination is not the dominant loss mechanism for atomic ions. In effect the charge collection depends on the conversion of atomic ions to molecular ions through collisions involving cesium or argon atoms as a third body. These molecular ions then recombine with electrons in a dissociative process. This conversion of atomic ions into molecular ions follows an exponential decay much slower than the subsequent molecular ion recombination. Thus, the exponential decay provides a direct measurement of the rate of destruction of atomic ions [2].

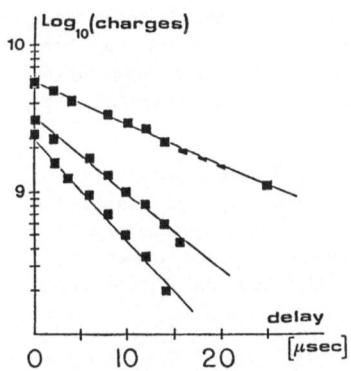

Figure 2. Experimental points and theoretical curves for the dependence of the ion signal on the time delay for different argon gas pressures, 0, 40, and 300 mTorr from top to bottom.

Figure 2 shows that introduction of the buffer gas modifies the ion signal even in an experiment with zero time delay. The conversion mechanism discussed above does not however explain this effect so that further investigation is required.

References

1. A. Sasso, M. I. Schisano, B. Tescione and E. Arimondo: Optics Comm. 53, 324 (1985)
2. E. Arimondo, F. Giammanco, A. Sasso and M. I. Schisano: to be published

Multiphoton Ionization Spectroscopy of Atomic Fluorine

W.K. Bischel and L.E. Jusinski

Chemical Physics Laboratory, SRI International,
Menlo Park, CA 94025, USA

The study of the electronic spectrum of atomic fluorine is extremely difficult since all the single-photon transitions from the ground state require wavelengths that are below the LiF window cutoff at 106 nm. This situation presents an ideal application for the techniques of multiphoton excitation, followed by either fluorescence or ionization detection. We report here the first observation of resonantly enhanced multi-photon ionization (REMPI) of atomic fluorine in a 3+2 photon process.

The excited states that are three-photon resonant in this experiment are the $^2P_{3/2,1/2}$ states at 104731.0 cm^{-1} and 105056.3 cm^{-1}. Since the ground state fine-structure splitting between the $^2P^o_{3/2,1/2}$ states is 404.1 cm^{-1} (3/2 lower), we expect that the REMPI spectrum would have four transitions if both these states were populated. Thus the observation of these four transitions would form unambigious evidence that atomic fluorine has been observed. The four transitions correspond to vacuum dye laser wavelengths of: 287.56 nm (1/2→3/2), 286.67 nm (1/2→3/2), 286.45 nm (3/2→3/2), and 285.56 nm (3/2→1/2).

Since the ionization limit occurs at 140524.5 cm^{-1}, the possibility of 4+1 REMPI process also exists. Starting from the $^2P^o_{3/2}$ ground state, the equivalent energy of four uv laser photons accesses Rydberg states that are 883 cm^{-1} below the ionization limit. These states will be observed if there are near coincidences with the uv wavelengths for the three-photon resonant states.

The experiment consists of focusing a doubled YAG pumped dye laser at 286 nm into a flowing cell containing a mixture of 10% F_2 and the balance He with a total pressure of 1 Torr. In the focal volume, the uv radiation dissociates the F_2 to form 2F and simultaneously ionizes the atomic fluorine. The ions are collected with a pair of biased electrodes, the charge is amplified, and the signal is then averaged with a boxcar averager. We estimate that it is necessary to produce approximately 10^4 ions in the focal volume to give a 1:1 signal/noise ratio given our background signals.

A low resolution MPI spectrum for a scan of the vu laser from 291-278 nm (equivalent three-photon energy of 103,000-108,000 cm^{-1}) is given by the solid line in Figure 1a for tight focussing conditions (f=15 cm). The laser energy is given by the dashed line. There are four resonances in this spectrum that correspond to the above three-photon resonant transitions in F, thus confirming that F atoms are being observed. These are labeled

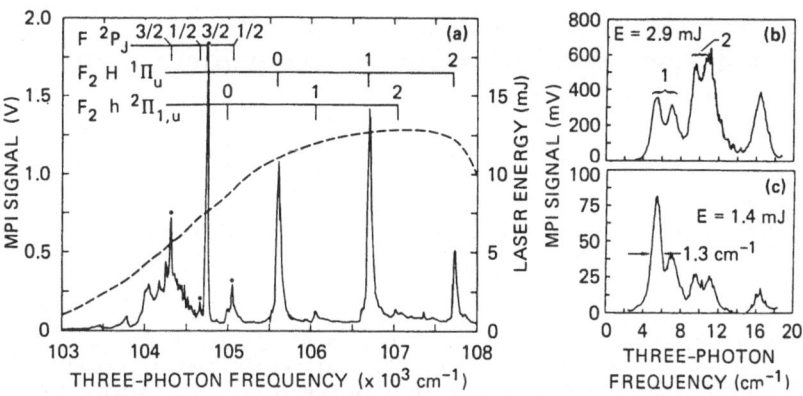

Fig. 1 REMPI spectrum of F and F_2 with 1b and 1c showing high resolution
scans near 104,731 cm^{-1}.

in the figure by the J of the excited ^2P state. The other features
observed in the spectrum correspond to transitions to the H $^1\Pi_u$, and the
h $^3\Pi_{1,u}$ states in F_2 that have previously been observed in single photon
absorption [1]. All the observed transitions are severely broadened due to
the AC Stark effect. The laser intensity had to be reduced in order to
increase the resolution.

Fig. 1b,c is the REMPI spectrum at reduced intensity for a 20 cm^{-1} scan
near the strong $^2P^o_{3/2}\rightarrow^2P_{3/2}$ transition at 104,731 cm^{-1} for two different
laser energies. Here 1 mJ of energy corresponds to ~800 MW/cm^2. Note from
Fig. 1c that the resolution is approximately 1.3 cm^{-1} limited by the linewidth of the visible dye laser (~0.25 cm^{-1}). We observe five different
peaks (instead of one). We interpert this observation as a combination of
the 3+2 and 4+1 ionization processes discussed above. In addition, the ion
signals labeled by 1 and 2 in Fig. 1b have different dependences on the
laser intensity due to the fact that the two REMPI processes saturate differently. This complicates any determination of effective cross sections.

We can, however, make estimates of the F-atom detection sensitivity in
the present experiment. The slope of the ion yield curve for feature 2 in
Figure 2a is 6.0 mV/(mJ)^5. At 1 mJ of laser energy, we estimate that we
have a ground state density at the time of peak laser intensity of approximately 10^{15} F atoms/cm^3. For this density, we can obtain a
signal/background ratio of greater than 1000:1 giving a F-atom detection
sensitivity of approximately 10^{12} cm^{-3}. This is unoptimized with respect
to a number of parameters and thus could be significantly improved.

In conclusion, we have unambiguously demonstrated for the first time the
detection of atomic fluorine using REMPI. We anticipate that these results
can be utilized in a number of research areas requiring the sensitivity
detection of spatially and temporally resolved F-atom distributions.

This work was supported by AFOSR under Contract No. F4620-85-K-0005.

1. E. A. Colbourn, M. Dagenais, A. E. Douglas, and J. W. Raymonda: Can. J.
    Phys. 54, 1343 (1976).

The Dynamic Stark Effect in a $J=0\rightarrow1\rightarrow0$ Three-Level System: A Systematic Experimental and Theoretical Investigation

P.T.H. Fisk, H.-A. Bachor, and R.J. Sandeman

Department of Physics and Theoretical Physics,
Australian National University, GPO Box 4, Canberra, ACT 2601, Australia

We report the results of an experiment on the Dynamic Stark Effect in a $J=0\rightarrow1\rightarrow0$ three-level system in ^{138}Ba (Fig. 1). This experiment involved tuning an intense laser field to resonance with the transition between the middle (6s6p $^1P_1$) and upper (6p^2 $^3P_0$) levels of the three-level system. A probe laser of sufficiently low intensity so as not to introduce any further nonlinear effects was scanned through resonance with a transition connecting the ground state (6s^2 $^1S_0$) with the middle level. The populations of the middle and upper levels were simultaneously but separately monitored as a function of probe laser frequency by recording fluorescence resulting from transitions out of these levels. The barium atoms were prepared in an atomic beam, and the linewidths of the counterpropagating laser fields were less than 1MHz RMS.

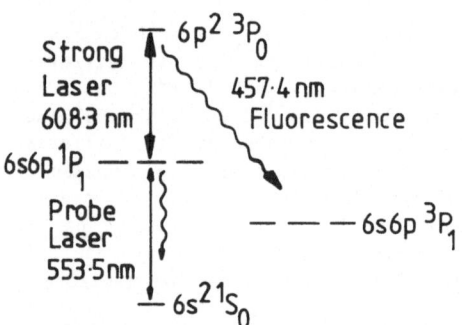

Fig.1 Three-level system in Ba^{138}. The lasers were both linearly polarized in the same direction, so that only the $\Delta m=0$ transitions were driven

This configuration of strong and probe lasers is the reverse of previous experiments[1,2,3], and it allowed us to observe the effect of the strong laser on the populations of both the middle and upper levels of the system for a range of probe laser and strong laser detunings.

Natural barium, which was used in this experiment, consists of 72% ^{138}Ba. The signal due to ^{138}Ba could be easily identified against the background due to the less abundant isotopes.

The Bloch equations for this system were solved analytically in the weak probe regime using the method of Ben-Reuven[4]. The relative sizes and Gaussian intensity profiles of the laser beams, and the four other isotopes of barium contributing to the observed profiles were included in the theoretical treatment. For two of these isotopes, due to their non zero nuclear spin, a five-level treatment was required to include the three hyperfine components of the middle level. The agreement between the theoretical and experimental profiles was excellent for all strong laser detunings and intensities investigated. An example is shown in Fig. 2.

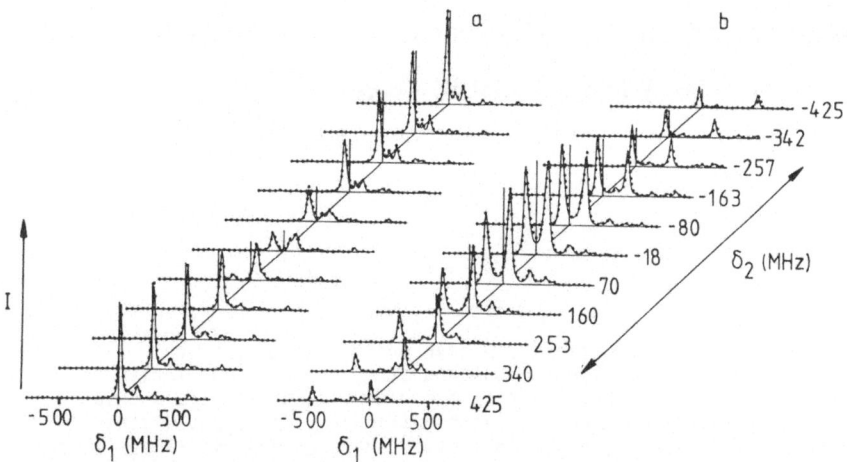

Fig. 2 Comparison of experimental and theoretical profiles over a range of strong laser detunings, with a strong laser Rabi frequency of 200 MHz. Probe laser detuning $= \delta_1$, strong laser detuning $= \delta_2$, fluorescence intensity $= 1$. a) Middle level; b) Upper level. Solid lines = theory, dotted lines = experiment. The smaller peaks are due to isotope structure, and hyperfine structure within the odd isotopes

Using the measured diameter and power of the strong laser beam, and the observed Rabi frequency of the interaction, the absorption oscillator strength of the $6s6p$ $^1P_1 \rightarrow 6p^2$ $^3P_0$ transition was calculated to be 0.007 ± 0.001.

References

1. J.E. Bjorkholm and P.F. Liao: *J. Phys. B 9*, 132 (1977)
2. H.R. Gray and C.R. Stroud: *Opt. Commun. 25*, 359 (1978)
3. J.L. Picque and J. Pinard: *J. Phys. B 9*, L77 (1976)
4. A. Ben-Reuven and L. Klein: *Phys. Rev. A 4*, 753 (1971)

The Effect of Electric Fields on Autoionizing Resonances

D.E. Kelleher, E.B. Saloman, and J.W. Cooper

National Bureau of Standards, Gaithersburg, MD 20899, USA

We have observed electric fields to have pronounced effects on autoionization resonances in the continuum region above the first ionization threshold. These effects are observed between field-coupled levels of opposite parity which have significantly different widths in zero-field. For example, [1] a relatively sharp even parity level which is nearly degenerate with a much broader odd level causes an interference dip in the broad level which deepens as the field is increased, until at higher fields the broad level is split into two components. The two components become "plus and minus" mixtures of the zero-field states, and repel one another as the field increases further. See Figures below.

Sharp levels in the wings of broad opposite parity levels lead to field-induced Fano-Beutler type interference patterns in the wings, whose amplitude increases as the field is increased. The narrow levels

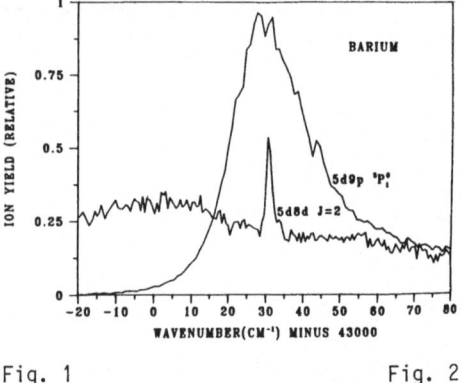

Fig. 1

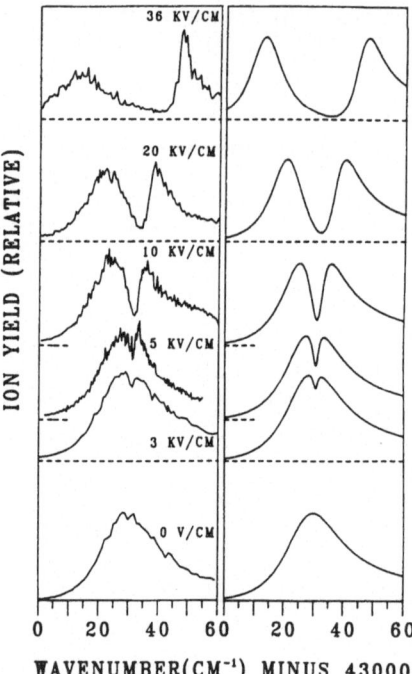

Fig. 2

The above figure shows the zero-field spectra of the even and odd spectra obtained by two and three laser excitation schemes, respectively. The center of the sharp even parity line lies 3.2 cm^{-1} above the peak of the broad odd parity level. The figure at right shows the changes in the "odd" parity spectrum as the field is increased. The profiles on the right are computed using a two-level model.

themselves broaden roughly quadratically with the field, up to a point where the "forbidden" component associated with the broad opposite parity line becomes significant. The broadening of the narrow level saturates at these higher fields and in the two-level case "plus and minus" will form here also. The theoretical treatment we have developed to describe is these phenomena is formally equivalent to that of Mies [2]; it is described briefly in Ref. (1).

We have also observed the Stark effect for high angular momentum autoionizing levels, e.g. the $5d_{3/2}$ 8ℓ, in the zero quantum defect region for n=8 w.r.t. the $5d_{3/2}$ limit. The spin coupling with the $d_{3/2}$ core gives rise in this case to eight-fold more Stark components (not all resolvable) than normally observed in a hydrogenic Stark spectrum, [3,4]. This superabundance of comparably broadened levels gives rise to a relatively smooth distribution of oscillator strength at high fields over a broad region of the spectrum.

In collaboration with C. Clark of NBS, we have used measured [5] fine structure splittings to measure ion quadrupole moments. The electrostatic f.s. splitting of certain high angular momentum states (no external field) is proportional to the quadrupole moment of the ion core to the extent that the orbital does not penetrate the core. As an example, we estimate the quadrupole moment of the Ba^+ $5d_{5/2}$ ion from measurement of the energies of the $5d_{5/2}$ng (J=1,2) autoionizing states.

References

1. Scheduled for publication in July 8, Phys. Rev. Lett. (1985)
2. F. Mies, Phys. Rev. 175, 164 (1968).
3. R. R. Freeman and G. C. Bjorklund, Phys. Rev. Lett. 40, 118 (1978).
4. M. L. Zimmerman et al., J. Phys. B. Atom. Molec. Phys. 11, L 11 (1978).
5. P. Camus et al., Physica Scripta 27, 125 (1983).

Raman Heterodyne Detection
of Radio-Frequency Resonances in Sm Vapor:
Effects of Velocity-Changing Collisions

J. Mlynek, Chr. Tamm, E. Buhr, and W. Lange

Institut für Quantenoptik, Universität Hannover, Welfengarten 1,
D-3000 Hannover 1, Fed. Rep. of Germany

We present results on Raman heterodyne detection of rf-induced sublevel coherence in a Doppler-broadened optical transition under conditions of velocity changing collisions (VCC). As a specific example, we report studies on Zeeman resonances in a $J=1 \rightarrow J'=0$ transition in atomic samarium vapor in the presence of rare-gas perturbers [1].

The basic rf-induced coherent Raman process for a Zeeman split $J=1 \rightarrow J'=0$ transition is shown in Fig. 1 together with the schematic of our experimental arrangement. The laser field $\bar{E}_0$ only drives the optical π-transition and thereby optically pumps the ground-state Zeeman sublevels. A longitudinal rf-field of frequency ω_H can now resonantly excite $|\Delta m|=1$ coherences and, via a two-photon process, the simultaneous presence of the light field $\bar{E}_0$ gives rise to coherent Raman sidebands $\bar{E}\pm$ with frequencies $\omega_E\pm\omega_H$. Behind the sample the total light field consists of orthogonally polarized carrier ($\bar{E}_0$) and Raman sideband ($\bar{E}\pm$) components with modulation frequency ω_H. With m_{AM} and m_{FM} denoting the modulation depth of the amplitude and frequency modulation of the resulting field $\bar{E}_D(t)$ due to the Raman sidebands, the heterodyne beat at the detector is of the form

$$|E_D|^2_{beat} = 2E_0\sin\phi\,\cos\phi\,[E_0\cos\Theta\{Re(m_{AM})\cos\omega_H t + Im(m_{AM})\sin\omega_H t\}$$
$$-E_0\sin\Theta\{Re(m_{FM})\cos\omega_H t + Im(m_{FM})\sin\omega_H t\}]$$

Here Θ denotes a constant phase-shift introduced by the optical retardation plate behind the sample cell. For $\Theta=0$ (no retardation plate) the FM-Raman heterodyne signal(RHS) cancels out as expected; for $\Theta=\pi/2(\lambda/4$-plate), the FM-RHS can be monitored and the AM-RHS vanishes. Physically, the AM- and FM-Raman signals are due to velocity-selective optical absorption and dispersion, respectively [2].

Experiments were performed on the SmI line $\lambda=570.6$ nm($4f^6 6s^2\ ^7F_1 \rightarrow 4f^6 6s6p\ ^7F_0$). Our measurements clearly reveal that in the presence of rare-gas perturbers, VCC determine the characteristics of the coherent rf-laser double resonance signals by strongly affecting their line shapes and linewidths. Two examples are given in Fig.2 which displays the measured rf-linewidths of the AM- and FM-RHS for a) xenon and b) helium in the pressure range up to 4 mbar. The full curves correspond to fits that are based on a theory including VCC [2]. From the type of data displayed in Fig.2 we can derive rate constants γ_{VCC} for VCC and γ_{col} for collisions destroying the Zeeman coherence: Theory predicts that

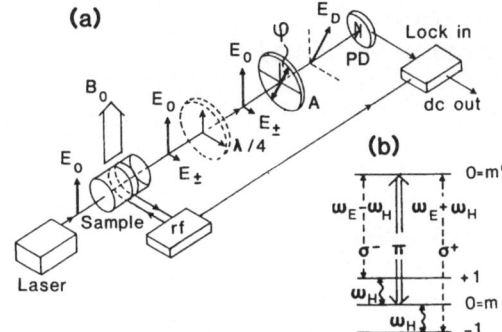

Fig.1 a) Experimental scheme: B_O, static magnetic field; A, polarization analyzer. The $\lambda/4$ plate is used for detection of the FM-Raman heterodyne signal. b) Energy level diagram for a Zeeman split $J=1\rightarrow J'=0$ transition, showing the coherent Raman process.

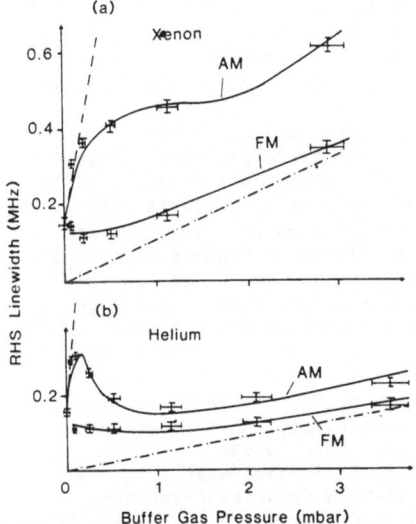

Fig.2:Variation of the rf linewidth of the AM and FM- -Raman heterodyne signals with buffer gas pressure for a) xenon and b) helium.

the asymptotic slope of the FM-RHS yields γ_{col} (dash-dotted line in Fig. 2) whereas the inital slope of the AM-RHS yields γ_{VCC} (dashed line in Fig. 2).

Most surprisingly, recent experiments seem to indicate that in the absence of VCC, the FM-RHS shows pronounced contributions originating from velocity-changes due to photon momentum trans- fer during optical pumping.

This work is supported by the Deutsche Forschungsgemeinschaft.

1. J. Mlynek, Chr. Tamm, E. Buhr, N.C. Wong, Phys.Rev.Lett. 53, 1814 (1984)
2. Chr. Tamm, E. Buhr, W. Lange, J. Mlynek, to be published

Absolute Wavelength Measurement and Fine Structure Determination in ^{7}Li II [*]

E. Riis[1], H.G. Berry[2], and O. Poulsen

Institute of Physics, University of Aarhus C, DK-8000 Aarhus, Denmark

S.A. Lee and S.Y. Tang

Department of Physics, Colorado State University,
Fort Collins, CO 80523, USA

The energy levels of two-electron atoms continue to provide rigorous tests of relativistic quantum theory, and of correlation effects within a multi-particle system. These interactions are determined perturbatively, with several approximations, and theoretical results often differ. It is critical to provide precise measurements of absolute wavelengths connecting these atomic energy levels to obtain a resolution of the precision of the different parts of such complex calculations.

Recently, high-precision measurements have been carried out in high nuclear charge (Z) ions using various fast beam spectroscopy techniques, principally to study the higher order Z-correction terms in the relativistic Breit and QED interactions. The measurements have been made principally of the 1s2s ^{3}S - 1s2p ^{3}P transitions for ions of Z=10 up to 26. The best of these measurements test the QED corrections at a level of 2 parts in 10^4, with absolute wavelength accuracies of about 5 parts in 10^6. For lower Z systems, the relativistic corrections are smaller, since they scale as Z^4 and higher, and non-relativistic correlations become larger, since they scale with Z^{-1} and lower. Relativistic correlations can scale with all powers of Z, and hence some should be measurable for many different ranges of Z. The application of laser techniques can lead to much higher precision in atomic wavelength measurements. Clearly, these are most easily applied for low Z ions where the 1s2s ^{3}S - 1s2p ^{3}P transitions are above 2000Å.

In this work, we report a high precision optical measurements in the 1s2s ^{3}S - 1s2p ^{3}P multiplet of Li II using fast-beam laser spectroscopy. A collinear interaction using both parallel and antiparallel laser and ion beams allows both for precise elimination of large Doppler shifts, and for a strong kinematic narrowing of the observed resonances, as compared with thermal beam experiments. The wavelengths of the observed resonance fluorescence radiation are determined by comparing them with simultaneously recorded saturated absorption profiles of molecular iodine hyperfine components. In turn, the absolute wavelengths of the iodine lines are obtained from precisely calibrated Fabry-Perot etalon fringes in a separate experiment. The final precision of the Li II wavelengths is 5 parts in 10^9, which is at a level of precision of 80 ppm of the QED corrections in the transition.

A schematic of the fast beam experimental arrangement is shown in Fig. 1. Data were taken in several series of measurements of σ_+ and σ_-, the fluorescence wavenumbers in the parallel and antiparallel geometries. The alternating sets of data were corrected for a slow temperature-dependent drift of the accelerator voltage, which was of the order of 1-2 volts per hour in 50 kilovolts. The

---

[*] Work supported by the NATO Research Exchange Grant No. 238/83.
[1] Present address: Department of Physics, Colorado State University, Fort Collins, CO 80523. Supported by a NATO Science Fellowship Program.
[2] On leave from the Physics Division, Argonne National Laboratory, Argonne, IL 60439. Supported by the U.S. Dept. of Energy (Office of Basic Energy Sciences) under Contract W-31-109-38.

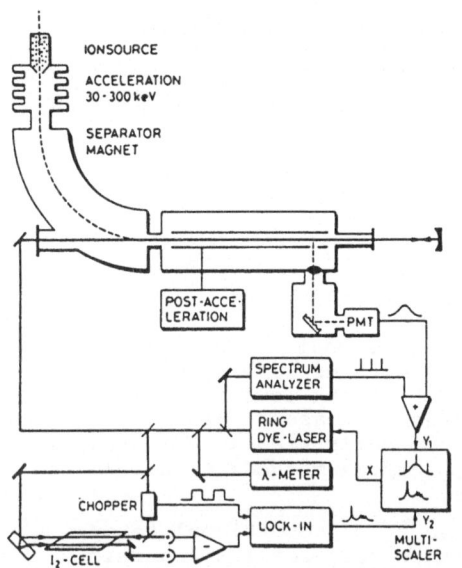

IONSOURCE

ACCELERATION
30 - 300 keV

SEPARATOR
MAGNET

POST-ACCE-
LERATION

PMT

SPECTRUM
ANALYZER

RING
DYE - LASER

λ - METER

CHOPPER

LOCK - IN

I₂ - CELL

Y₁

X

Y₂

MULTI-
SCALER

Fig. 1. The experimental arrangement for the laser excitation of the fast (50 keV) Li^+ beam. The ring dye laser is tuned successively to the resonance fluorescence for parallel and anti-parallel excitation, all other parameters remaining fixed, with outputs from the spectrum analyzer, the I_2 absorption cell and the λ-meter being measured simultaneously.

resonance wavenumber is then given by $\sigma_0 = \sqrt{\sigma_+ \cdot \sigma_-}$. This is independent of the ion beam velocity, provided that it is stable to the required precision during the time of the measurements. The absolute values of the I_2 lines used in the experiment (one for each σ_+ and σ_+ measurement) were measured using a 0.5m and a 1.0m temperature-stabilized (within 10^{-3}K) interferometer.

Since the hyperfine structure is calculated to a precision better than our experimental accuracy (±3 MHz), we can use theoretical values to derive three separate parameters from our data. These three are the two fine structure intervals, Δ_{02}, and Δ_{21}, of the 1s2p ^3P state, and the absolute wavelength of the transition from the center of gravity of the 1s2s ^3S state, and the center of the 1s2p $^3P_0$ state. Our preliminary results are shown below, and compared with other experimental and theoretical values.

Table I. All measurements in cm^{-1}, with accuracies given in parentheses. The first 5 figures are omitted in the last 3 columns for σ_0.

Interval	This work	Ref. 1	Ref. 2	Ref. 3 (Theory)
σ_0	18231.30200(10)	-.3028(8)	-.3030(12)	-.313
Δ_{02}	3.10265(10)	3.1028(8)	3.1051(12)	3.118
Δ_{21}	2.08730(10)	0.0906(8)	2.0897(12)	2.086

We conclude that our measurements are consistent with previous measurements. The improved accuracy is due mostly to narrower line profiles in the fast beam geometry and the absolute calibration against I_2. Theory is also consistent, but at a much lower level of accuracy, being limited both by the non-relativistic energy calculations (± 0.001 cm^{-1}), and also by a relativistic correlation term (± 0.01 cm^{-1}). We expect that similar techniques are applicable for neighboring Z elements.

1. R. Bayer et. al. Z. f. Phys. A292, 329 (1979).
2. R. A. Holt et. al., Phys. Rev. A22, 1513.
3. We thank Gordon Drake for communication of these results.

Laser Spectroscopy in Highly-Precisely Determined Electric Fields: Application to Sodium D_1

L. Windholz and C. Neureiter

Institut für Experimentalphysik, Technische Universität Graz,
Petersgasse 16, A-8010 Graz, Austria

In recent years dopplerfree laserspectroscopic methods made it possible to investigate hyperfine structures with a resolution limited only by the natural linewidth of the levels involved. These techniques are very well suited to measure relative frequency differences , e.g. by means of calibrated marker etalons. However, it was obvious to use dopplerfree methods to determine the Stark shift of resonance lines.

The polarizibility of atomic levels is proportional to app. n^6 or n^7. Therefore, resonance lines show very small Stark shifts compared with higher members of spectral series. In opposition to classical investigations of emission lines (see e.g. [1]) generally high resolution methods like interferometry (e.g. [2]) or laser spectroscopy have to be used in this case. Fields of remarkable strength (> app. 25 kV/cm) are necessary to produce shifts large enough for a precise measurement.

To determine the polarizibility of a spectral line not only the shift had to be known. A defined homogeneous electric field is also a necessary condition. The problem to evaluate the absolute value of $E = U/d$ by a measurement of the field voltage U and the spacing d with high precision is an old one. To produce high fields, d should be as small as possible (app. 0,7 mm). To get a value for d , the field assembly was formed as a Fabry-Perot interferometer [3]. The spacing was determined by a measurement of the free spectral range by use of a cw dye laser and a lambdameter, tuning the laser to two transmission maxima app. 150 Å apart. To avoid thermal drifts of the spacing during the measurements - and therefore drifts of the transmission wavelengths - the interferometer was locked to a helium neon laser controlling the temperature of the ground body of the field assembly. Performing the methods described in [4] the spacing could be determined with a relative accuracy of 2.10^{-4}. The field voltage was measured using a calibrated voltage divider and a digital multimeter to 5 parts in 10^{-5}. With this assembly field strengths of up to 130 kV/cm were generated.

By use of laser-atomic-beam technique we investigated the D_1-line in a field of app. 90 kV/cm. The shift of the hyperfine transitions $^2P_{1/2}$, F"=2, 1 — $^2S_{1/2}$, F'=2 were measured relative to the field-free component F"=2 — F'=2 switching off the field voltage during registration.

The result for the scalar polarizibility is

$$\alpha_o = 48,986(112) \text{ kHz.(kV/cm)}^{-2}.$$

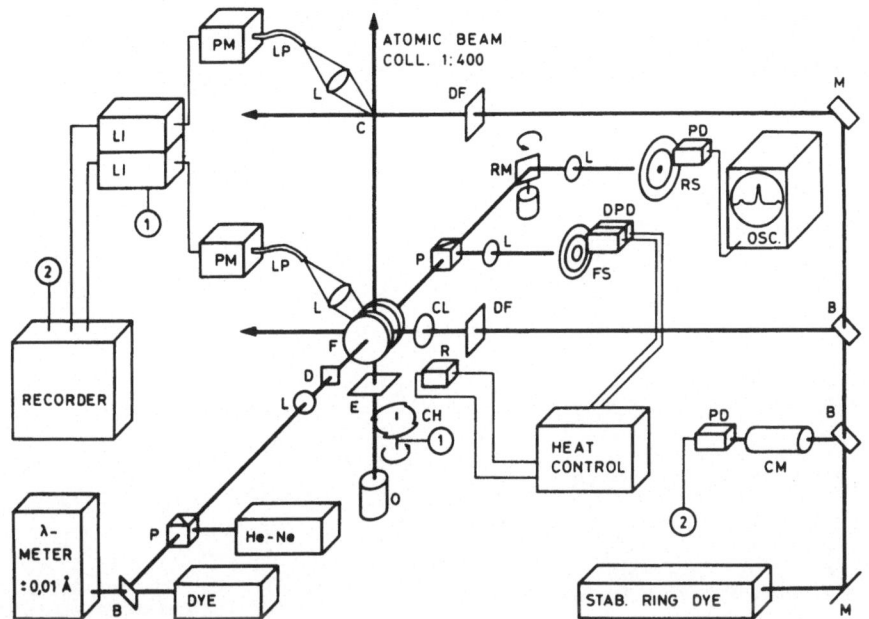

Experimental arrangement

O oven; CH chopper; E entrance aperture of the atomic beam;
F field plates, forming a Fabry-Perot interferometer, adjustable
by micrometer screws, spacing app. 0,7 mm; C field-free reference crossing; M mirrors; B beam splitters; CM confocal marker
etalon, FSR = 197,50(8) MHz, oven stabilized; PD photodiodes;
DF density filters; CL cylindrical lens; L lenses; LP light
pipes; PM photomultipliers; LI lock-in amplifiers; P polariz-
ing beam splitter cubes; D diffusing disc; FS fringe system
due to the He-Ne laser. A difference photodiode DPD mounted off-
axis detects changes in the diameter of an interference ring. The
signal of DPD was used to control the temperature of the ground
body of the field assembly heated by the resistor R; RS fringe
system due to the dye laser. These fringes are moved over the
photodiode PD by the rotating mirror RM. The connected oscilloscope shows an intensity profile of the fringe system in order to
find reproduceably the wavelengths of transmission maxima.

The error is mainly determined by the evaluation of the frequency
shift and could be decreased by use of side-band spectroscopy. In
fields up to 300 kV/cm no splitting or broadening of the hyperfine components can be observed. The frequency difference $F''=2$
— $F''=1$ (app. 189 MHz) does not change its value, too. The quadratic interdependence of shift and field strength was proved in
a former work [5].

1 L.Windholz; Physica Scripta 21, 67 (1980)
2 H.Kopfermann, W.Paul; Zs.f.Phys. 120, 545 (1943)
3 L.Windholz, C.Neureiter; J.Phys.E 17, 186 (1984)
4 L.Windholz, C.Neureiter; Phys.Lett.A 109, 155 (1985)
5 W.Danielczyk; Diplomarbeit, Inst.f.Exp.Phys., TU Graz (1984)

Molecular and Ion Spectroscopy

Internal Dynamics of Simple Molecules
Revealed by the Superfine and Hyperfine Structures
of Their Infrared Spectra

Ch.J. Bordé, J. Bordé, Ch. Bréant, Ch. Chardonnet, A. van Lerberghe, and Ch. Salomon

Laboratoire de Physique des Lasers, LA 282, Université Paris-Nord, Avenue J.-B. Clément, F-93430 Villetaneuse, France

This paper reports some of the progress achieved with our 10 µm saturation spectrometer since our 1979 FICOLS presentation [1] with a special emphasis on recent results in molecular physics. Let us first summarize the improvements of the spectrometer acquired during that period :

A. The absolute frequency calibration is now based : /1/ on a grid of OsO_4 frequency markers which is connected to the Cesium primary frequency standard with accuracies ranging from 50 Hz to a couple of kHz [2,3] and which has been extended to include many identified lines of the ν_3 band of various isotopic species : $P(30)A_1^1(g)$ and $A_1^1(u)$ of $^{188}OsO_4$, $P(63)A_1^1(u)$, $P(56)A_1^2(g)$, $P(49)A_1^3(u)$ and $A_1^3(g)$, $R(26)A_1^0(g)$ and $A_1^0(u)$, $R(45)A_1^0(u)$ and $R(64)A_1^2(u)$ of $^{189}OsO_4$, $R(40)A_1^0(u)$, $A_1^0(g)$ and $A_1^1(g)$ of $^{190}OsO_4$, $P(46)A_1^2(u)$, $P(39)A_1^3(u)$ and $A_1^2(g)$, $R(23)A_1^1(u)$ and $A_1^0(g)$, $R(36)A_1^0(g)$ and $A_1^0(u)$, $R(55)A_1^1(g)$ of $^{192}OsO_4$ (the previous point group label A_2 is replaced by $A_1(u)$ where u is the overall parity [4]); Fig.1 illustrates two such $^{192}OsO_4$ markers and their positions in the Fourier transform spectrum (a) of the ν_3 band [5] /2/ on the saturation peaks of CO_2 itself, observed directly in the absorption cell with a half-width as low as 2.1 kHz.

These two grids have been connected for the P(12) CO_2 line, as shown in Fig.1, and for the P(14) CO_2 line, and excellent agreement (within our 1kHz uncertainty) has been found with the values determined by the saturated fluorescence technique [6] .

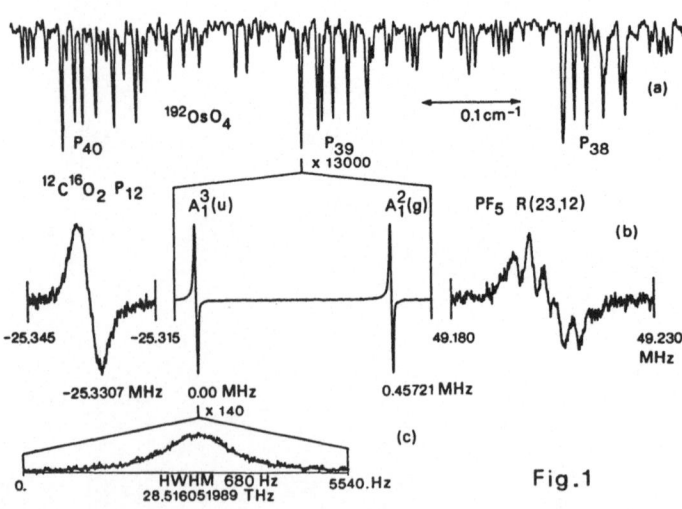

Fig.1

B. The resolving power times tuning range product as well as the quality of frequency calibration have benefited from phase-locking waveguide CO_2 lasers to a conventional reference laser, resulting in a 10 Hz spectral purity over 600 MHz tuning ranges [2,7] and also from high quality corner-cube optics (which is a by-product of our Ramsey fringes studies [8]) ; the linewidth is transit-limited and corresponds to the 4 cm beam waist radius. Fig.1 illustrates this resolving power for a single OsO_4 line as well as in the case of a typical hyperfine structure of PF_5.

C. The detectivity (S/N) has also benefited : /1/ from long-term computerized frequency control (e.g. the CO_2 saturation peak is obtained by integration over 20 minutes periods in turn with 200 seconds periods for OsO_4 and this process is repeated 6 to 7 times to determine accurately frequency drifts of the reference laser (< 10 Hz/minute)) /2/ in some cases from high frequency modulation applied to the laser piezoelectric transducer (97 kHz in the case of Fig.1c).

1. SUPERFINE AND HYPERFINE STRUCTURES IN THE ν_3 BAND OF OsO_4

The doublet in Fig. 1 is a very simple example of superfine structure : the clustering of these lines results from a spontaneous symmetry-breaking of the point group T_d into a lower symmetry (C_3) and the splitting corresponds to a tunnelling rate between equivalent C_{3v} distorted configurations [9] . Other dynamical aspects are visible through hyperfine effects, for which a remarkable test system is the isotopic species with a central ^{189}Os nucleus, since a spin 3/2 probes both the electric field gradient and the magnetic field at the centre of the molecule. Our previous saturation spectroscopic studies [1] have revealed quadruplet structures which could be attributed to an electric quadrupole interaction. The existence of such structures results from the breakdown of spherical symmetry of the molecule by vibration-rotation distortions which induce an electric field gradient at the Osmium nucleus site. In a first step, these structures have been interpreted as a purely vibrational effect and two corresponding constants (respectively for a scalar and a tensor interaction) were derived in [10] . On the other hand, it was suggested that pure rotation could induce similar effects and two corresponding constants χ_s^R and χ_t^R were introduced [11] . This effect was confirmed by further theoretical studies [12,13] , but since the observed spectra usually correspond to a difference between excited and ground state structures ($\Delta F = \Delta J$ transitions), this rotational effect cancels almost completely and requires high precision spectroscopy to be observed experimentally. We have therefore reinvestigated the structures corresponding to the 8 vibration-rotation transitions listed above. The magnetic dipole hyperfine structure of these lines results from purely scalar spin-rotation and spin-vibration interactions and can be easily separated from the symmetric purely quadrupolar quadruplet as shown in [1] . The high accuracy (< 100 Hz) of the measured lines(HWHM∿1kHz) is such that they cannot be fitted with a single eQq in the excited state but·require an important value for this quantity, also in the ground state. The set of values for eQq in the ground vibrational state is fully consistent with : $\chi_s^R \approx -760$ Hz and $\chi_t^R \approx -35$ Hz. This conclusion has been fully confirmed by a study of crossover resonances for the two R(26) and the two P(49) lines and is illustrated in Fig.2 for $R(26)A_1^O(g)$. These crossover peaks are less than 1/1000 times weaker than the main peaks and measured with an accuracy of 200 Hz (HWHM ∿6kHz). They give a direct and independent access to hyperfine energies in both vibrational levels, and yield an accurate value of the spin-rotation interaction constant in the $v_3 = 0$ state : $c_a = -21.69$kHz. The overall analysis of the magnetic dipole interactions also provides values for a change in this constant with vibration : $\delta c_a \approx 18$ Hz and for a scalar spin-vibration interaction constant : $A \approx -2.76$ kHz. Finally let us note that the ratio of A and c_a is quite close to $\zeta_3 = 0.127$ as could be anticipated.

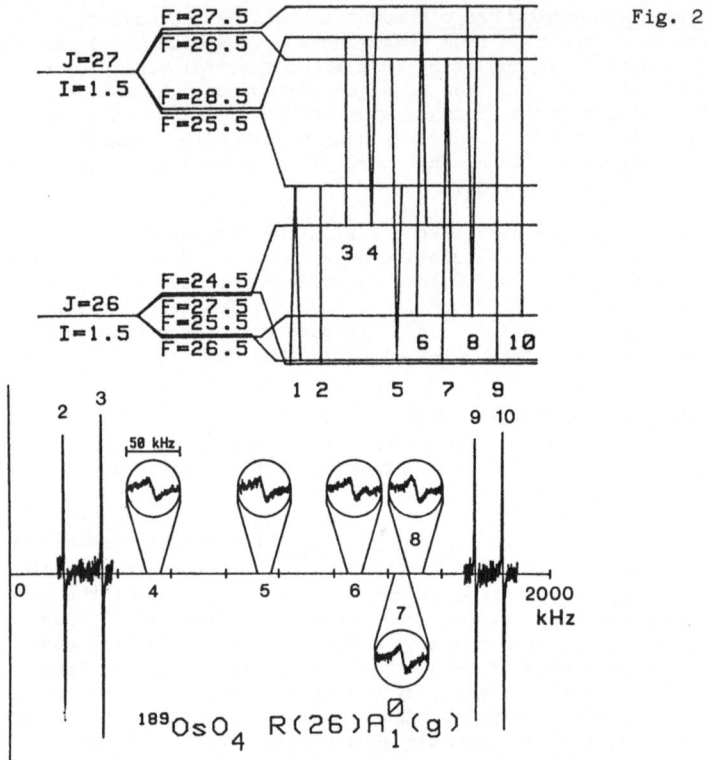

Fig. 2

$^{189}OsO_4$ R(26)A_1^0(g)

2. SUPERFINE, HYPERFINE AND SUPERHYPERFINE STRUCTURES IN THE ν_3 BAND OF $^{32}SF_6$

A much richer variety of spectra is offered by SF_6 for which all possibilities of superfine patterns have been observed [1,14] and fully demonstrate the spontaneous symmetry breaking of O_h into C_3 or C_4 subgroups, as well as the tunnelling effects described by W. HARTER and coll. [9] . A full account of the fine and superfine structures with a 30 kHz standard deviation has been given recently [16] for 136 transitions ranging from P(84) to R(94). Hyperfine structures themselves display superclusters $O_h \uparrow S_6$ and have been observed for all symmetry species of the O_h group. As illustrated in Fig. 3,4,5, the observed spectra agree beautifully with those calculated from the theory presented in [14,15] with the following set of constants (in kHz) :

$c_a = -5.27$, $c_d = -4.60$, A= 4.57, A_t = 6.75 (*), d_1 = 9.82, d_2 = 3.47

respectively for the scalar spin-rotation, tensor spin-rotation, scalar spin-vibration, tensor spin-vibration and spin-spin interactions. In addition, we have determined three coupling constants associated with three formal operators given in [12] and which correspond to vibrational corrections to the scalar and tensor spin-rotation interactions. As a result, 85 % of the structures (i.e. for 108 vibration-rotation transitions) have been perfectly reproduced, whereas 15 % have still one or two hyperfine components off by quantities of the order of one kilohertz. As studied in detail in previous papers

(*) with this value the second member of equation Eq. (2) of reference [15] should be multiplied by $\sqrt{2}$.

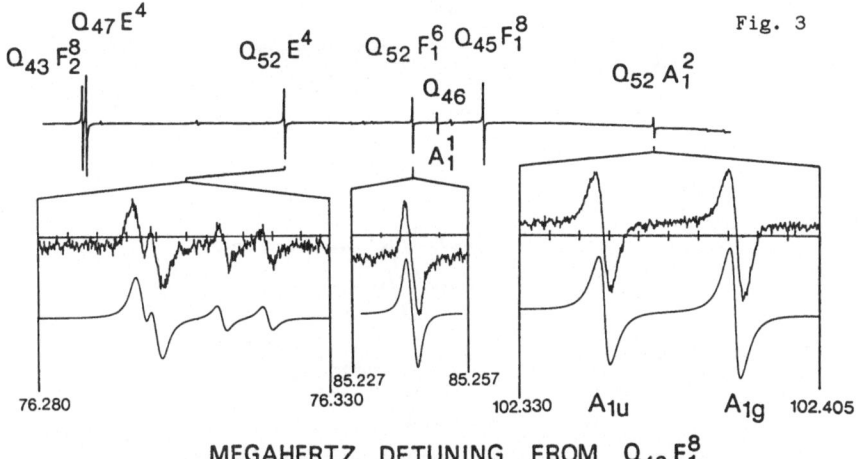

Fig. 3

$Q_{43} F_2^8$ $Q_{47} E^4$ $Q_{52} E^4$ $Q_{52} F_1^6$ $Q_{45} F_1^8$ Q_{46} $Q_{52} A_1^2$

A_1^1

76.280   85.227   85.257   102.330 A_{1u} A_{1g} 102.405
76.330

MEGAHERTZ DETUNING FROM $Q_{43} F_1^8$

[14] hyperfine interactions break the point group symmetry. A spectacular manifestation of this mixing can be seen as a splitting of the inversion doublet $A_{1u} - A_{1g}$ in Fig. 3 (only the tensor spin-rotation interaction in $v_3 = 0$ contributes noticeably to this splitting which yields directly a value for $c_d = -4.63$ kHz). As vibration-rotation states get closer and closer within rovibrational clusters this hyperfine mixing gets stronger and the individual hyperfine structures collapse into a single superhyperfine structure [9] . Fig.4 illustrates this evolution in the case of $F_1 - F_2$ lines, for which the final stage of this mixing is the complete breakdown of the S_6 symmetry into $S_4 \times S_2$. States are then labelled by broken Young tableaux [9].A remarkable and unpredicted feature is the quadruplet organization of super-hyperfine tetragonal clusters,which can be understood from the large difference in axial and equatorial spin-rotation interactions. It can be shown that in the lower state,the two axial and equatorial interaction coupling constants are precisely the $c_{//} \simeq -2.2$ and $c_{\perp} \simeq -6.8$ kHz introduced in [14] . Fig. 5 displays an example of the 4 possibilities of axial spin Young tableaux associated with each choice of equatorial spin Young tableau (the S_2 and S_4 parts of the broken tableaux are respectively given above and below the experimental and theoretical spectra).

3. HYPERFINE STUDIES OF SYMMETRIC TOPS

We have undertaken similar hyperfine studies for a number of other molecules belonging to other symmetry groups and whose fundamental bands fall within the CO_2 (or N_2O) laser tuning ranges e.g. the v_2 band of ammonia or the v_3 band of PF_5 (Fig.1). In the case of the v_2 band of ammonia,both electric quadrupole effects [17] and magnetic dipole interactions have been shown to depend strongly on internal degrees of freedom : inversion, rotation [18,19] and vibration. A recent study of the asQ(8,7) line (HWHM $\simeq$7kHz) has not only confirmed the important change in eqQ (Δ eqQ $\simeq$ -350 kHz) in good agreement with a theoretical prediction of V. SPIRKO [20] but also revealed important variations in the nitrogen and hydrogen spin-rotation constants R and S with vibration and inversion (ΔR $\simeq$ 0.25 kHz and ΔS $\simeq$ -0.59 kHz). As another example, Fig.6 displays a comparison between theory and experiment for the asR(2,0) line of the same band of $^{15}NH_3$. The calculated asR(2,0) spectrum offers a nice confirmation of the low-field theory of intensities in saturation spectroscopy presented in [22] .

111

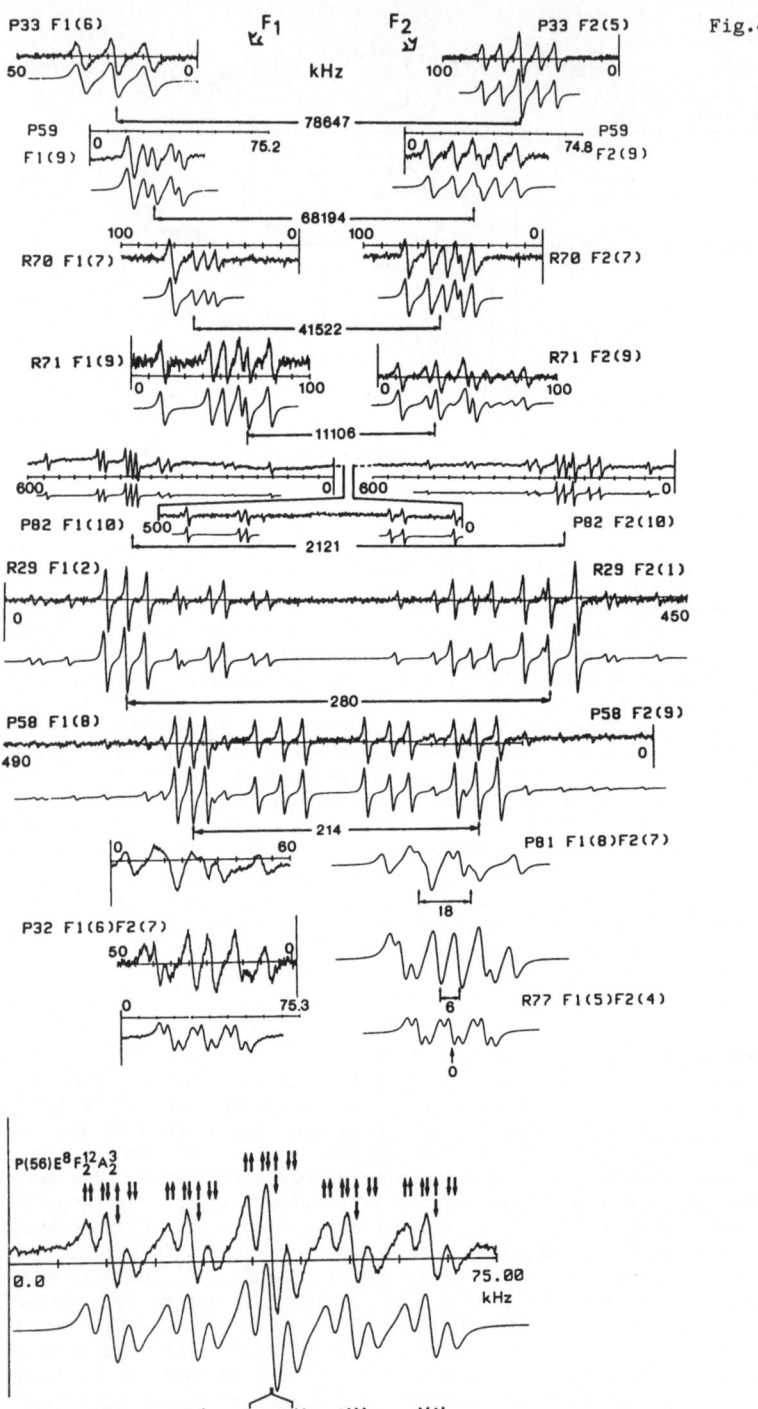

Fig.4

Fig. 5

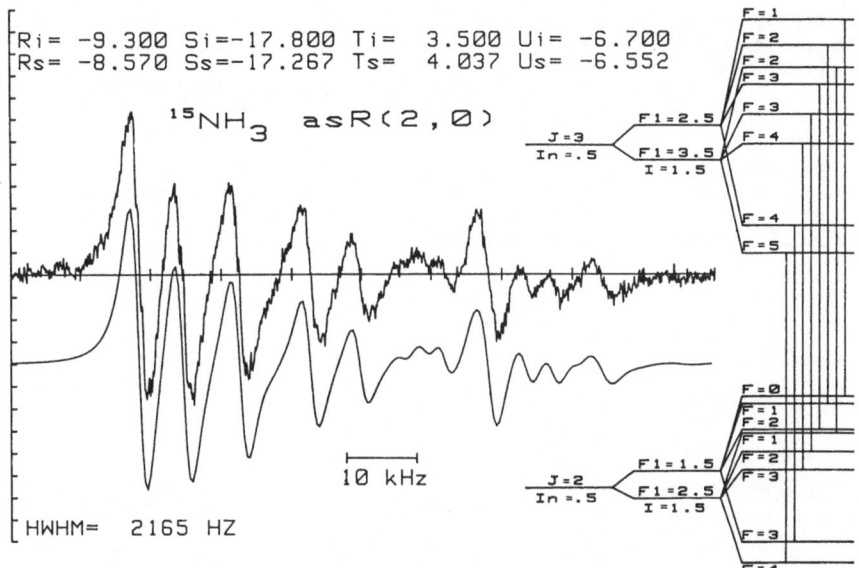

Fig. 6

In this case, the high accuracy of the data requires to introduce also a change in the nitrogen-hydrogen and hydrogen-hydrogen spin-spin interaction constants T and U which are directly related to a change in the mean geometry of the molecule. This is another illustration of the role that hyperfine studies can play as a probe of large amplitude motions within molecules.

REFERENCES

1. Ch.J. BORDÉ, M. OUHAYOUN, A. VAN LERBERGHE, Ch. SALOMON, S. AVRILLIER, C.D. CANTRELL and J. BORDÉ, in Laser Spectroscopy IV, edited by H. WALTHER and K.W. ROTHE, Springer Verlag, p. 142 (1979).
2. Ch.J. BORDÉ, Revue du Cethedec-Ondes et Signal NS 83-1, 1 (1983) and references therein.
3. A. CLAIRON, A. VAN LERBERGHE, Ch. BREANT, Ch. SALOMON, G. CAMY and Ch.J. BORDÉ, J. de Phys. Coll. 42, C8-127 (1981).
4. H. BERGER, J. Physique, 38, 1371 (1977).
5. L. HENRY and A. VALENTIN (Private communication).
6. F.R. PETERSEN, E.C. BEATY and C.R. POLLOCK, J. Mol. Spectr. 102, 112 (1983) and references therein.
7. Ch. SALOMON, Ch. BREANT, A. VAN LERBERGHE, G. CAMY and Ch.J. BORDÉ, Appl. Phys. B, 29, 153 (1982).
8. Ch.J. BORDÉ, Ch. SALOMON, S. AVRILLIER, A. VAN LERBERGHE, Ch. BREANT, D. BASSI and G. SCOLES, Phys. Rev. A 30, 1836 (1984) and references therein.
9. W.G. HARTER Phys. Rev. A 24, 192 (1981) and references therein.
10. J.T. HOUGEN and T. OKA, J. Chem. Phys. 74, 1830 (1981). F. SCAPPINI, W.A. KREINER, J.M. FRYE and T. OKA (to be published).
11. M.L. PALMA and J. BORDÉ, J. Physique 42, 1239 (1981).
12. F. MICHELOT, J. Mol. Spectr. 106, 77 (1984) and references therein.
13. M.R. ALIEV and J.T. HOUGEN, J. Mol. Spectr. 106, 110 (1984).
14. J. BORDÉ and Ch.J. BORDÉ, Chem. Phys. 71, 417 (1982).
15. J. BORDÉ and Ch.J. BORDÉ, Chem. Phys. 84, 159 (1984).

16. B. BOBIN, Ch. BREANT, J. BORDÉ and Ch.J. BORDÉ, paper FB1, Fortieth Symposium on Molecular Spectroscopy, Columbus, (June 1985) and to be published.
17. M. OUHAYOUN, Ch.J. BORDÉ and J. BORDÉ, Mol. Phys. 33, 597 (1977).
18. S.G. KUKOLICH, Phys. Rev. 172, 59 (1968).
19. J.T. HOUGEN, J. Chem. Phys. 57, 4207 (1972).
20. V. SPIRKO, Molecular Physics, 38, 1761 (1979).
21. Ch. SALOMON, Ch. CHARDONNET, A. VAN LERBERGHE, Ch. BREANT and Ch.J. BORDÉ J. de Phys. Lettres, 45, L1125 (1984).
22. J. BORDÉ and Ch.J. BORDÉ, J. Mol. Spectr. 78, 353 (1979).

Infrared Laser Kinetic Spectroscopy

H. Kanamori, J.E. Butler, T. Minowa, K. Kawaguchi, C. Yamada, and E. Hirota

Institute for Molecular Science, Okazaki 444, Japan

1 Introduction

High-resolution infrared spectroscopy has been low in sensitivity and thus it has been almost impossible to observe vibration-rotation transitions of transient molecules. Most infrared studies of such species have thus been conducted in low-temperature inert matrices, whereby structural information has been lost to a large extent. The sensitivity has, however, recently been much improved thanks to introduction of lasers. Infrared diode laser spectroscopy, for example, now allows us to detect molecules as small in number as 10^{10}, and has in fact been successfully applied to high-resolution spectroscopic studies of transient molecules, free radicals, and molecular ions, which were generated mainly by electrical discharges [1]. The recent development of high-power lasers such as excimer laser and carbon dioxide laser makes it possible to employ such a laser for photochemical generation of transient molecules from an appropriate precursor; an excimer laser pulse typically contains 10^{17} photons. In the present study infrared cw diode laser spectroscopy is used to probe the nascent distribution of photofragments, mostly transient molecules, generated by the excimer laser photolysis.

There are three motivations for this method. First, we expect to generate transient molecules or excited states of molecules that are difficult to obtain by other methods. The second reason is that we anticipate some detailed information on the initial process of photochemical reactions to be extracted from the nascent distribution of photofragments over vibrational and rotational states. Finally we might be able to follow relaxation processes following the photolysis or, in some cases, secondary reactions.

2 Spectroscopic System

The infrared diode laser spectrometer employed is identical to that used for cw spectroscopic experiments, except for the absorption cell and the signal detection circuitory [2]. A window is added to a White-type multiple reflection cell to introduce the excimer laser (UV) beam into the cell. The IR(diode) and UV beams are aligned so as to overlap each other, otherwise it would be difficult to observe spectra of short-lived species and to shorten the response time of the system. The response time, which is about 0.5 μs, is mainly determined by the IR detector. The detected signal, after being amplified, is fed to a transient recorder with 1024 8-bit memory points; the maximum sampling speed of the recorder is 50 ns/word. The recorder is triggered by a pulse which is synchronized with another pulse to fire the excimer laser. A microcomputer, which transfers the output of the recorder to a minicomputer, allows us to choose this data acquisition system either as a dual gated integrator or as a signal averager. The former function is useful in recording the spectra of photofragments and the latter in examining the transient behaviors of infrared signals, either absorption or emission.

3 Results

The method described above has been applied to a few systems summarized in Table 1, which lists the precursors, the excimer laser lines employed for photolysis, and the photofragments for which infrared spectra have been observed.

The CCO radical is an example of molecules which are difficult to generate by other methods; we have attempted to produce it by electrical discharge in carbon suboxide, but the infrared signal was barely seen. The response time of our system is short enough to examine the nascent distribution of photofragments, provided that the sample pressure is sufficiently low. In the case of SO generated from SO_2, about 70 % were prepared in v=2, 20 % in v=1 with the rotational distributions very much shifted toward high N levels. For CS from CS_2 the rotational levels of J as large as 100 were found to be populated. An interesting non-thermal distribution was observed among spin triplets of SO[2]. A careful recording of the line shape allows us to determine the velocities of photofragments, i.e. the

Table 1. Molecular Systems Investigated by IR Laser Kinetic Spectroscopy

Precursor	Excimer Laser Line [nm]	Photofragment
$(CH_3)_2CO$	193	$CH_3(\tilde{X}^2A''_2)\ \nu_2$
CH_3I	249	
SO_2	193	$SO(X^3\Sigma^-)$
C_3O_2	193,249	$CCO(\tilde{X}^3\Sigma^-)\ \nu_1$
B_2H_6, BH_3CO	193	$BH_3\ \nu_2(?)$
NH_3	193	$NH_2(\tilde{X}^2B_1)\ \nu_2$
CS_2	193	$CS(X^1\Sigma^+)$
$SOCl_2$	193	$SO(X^3\Sigma^-)$, $Cl(^2P)$
Cl_2	351	$Cl(^2P)$
$HNCO$, C_2H_5NCO	193	$NCO(\tilde{X}^2\Pi)\ \nu_1$
SCl_2	193	$Cl(^2P)$

translational energy distributions, which depend on the vibrational and rotational states. It is a little more difficult to extract information on relaxation processes following photolysis, because only information averaged along the infrared beam is obtainable, but still such data are of considerable significance in examining the dynamical behaviors of the system.

[1] E. Hirota: High-Resolution Spectroscopy of Transient Molecules (Springer, Heidelberg 1985).
[2] H. Kanamori, J. E. Butler, K. Kawaguchi, C. Yamada, E. Hirota: J. Chem. Phys. in press.

High-Resolution Laser Threshold Photoelectron Spectroscopy of Nitric Oxide and Benzene

K. Müller-Dethlefs, M. Sander, and L.A. Chewter

Institut für Physikalische und Theoretische Chemie
der Technischen Universität München, Lichtenbergstrasse 4,
D-8046 Garching, Fed. Rep. of Germany

I. Introduction

As is well known, photoelectron spectroscopy (PES) by conventional techniques (i. e. using a geometrical analyser or the time-of-flight method for the photoelectron energy analysis /1, 2/) suffers from the drawback of an inadequately low resolution compared to the line-width available by commercial pulsed dye lasers. The use of conventional PES with its resolution of somewhat better than 10 meV (80 cm^{-1}) is hence restricted to the study of vibronic ion states, and rotational ion structure cannot be studied except for cases of very large rotational level spacing e. g. for H_2^+ and D_2^+ /3, 4/ or levels of very high J^+ in NO^+ /5/. In order to make PES a suitable method for the study of ion fine structure and to utilize better the small line-widths of laser sources, a substantial improvement on photoelectron energy resolution is demanded. For this goal, we have developed the technique of threshold photoelectron spectroscopy (TPES), which we have termed photoionization resonance spectroscopy (PIRS). This method is based on the detection of photoelectrons of <u>zero kinetic energy</u> which are found only, if the ionizing photon(s) match exactly the energy difference between an ionic and a molecular state (i. e. if ionic and molecular state are in resonance).

II. TPE Spectrum of NO

In our first two-colour experiments using a steradiancy analyser, with static electric fields, we achieved a resolution of 1.5 meV (12 cm^{-1}) which allowed for the observation of rotational states of NO^+ at the ionization threshold using the ionizing four-photon transition (NO^+) $X^1\Sigma^+(v^+=0,J^+)$ ←ω_2 $(NO)C^2\Pi_r(v=0,\ T_y(3/2))$←$3\omega_1$$(NO)X^2\Pi_{1/2}(v=0,J=5/2)$. The transitions into rotational states of NO^+ were not yet fully separated but observed as a convolution /6/.

The full separation of rotational states of NO^+ at the ionization threshold ($B_o \simeq 2$ cm^{-1} for $(NO^+)X^1\Sigma^+$) was only made possible by a further improvement of threshold photoelectron energy resolution by more than one order of magnitude. The details of the skimmed supersonic beam apparatus, and the pulsed electric field - time discrimination method employed are given in ref. 7. The result is shown in Fig. 1. The $(NO)C^2\Pi_r(v=0,\ T_y(1/2),\ N=1)$ level was chosen as the intermediate resonant molecular level. The $T_y(1/2)$ (average of lambda doublet: 52372.6 cm^{-1} /11/) was populated via the $P_y(3/2)$ C←X three-photon transition and the ionizing photon wavenumber scanned around the ionization threshold. In order to exclude any coherent ionization processes from combinations of frequencies w_1 and w_2 the pulse from laser 2 was delayed by 10 ns. Also the number of ions produced per laser shot was kept below five to exclude any Coulomb effects or Rydberg-Rydberg collisions /8/. An inspection of Fig. 1 shows that the TPE resolution is 1.2 cm^{-1} and hence the rotational levels of NO^+ at the ionization threshold are fully separated. The observed spacings between the states of $J^+=N^+=0,1,2$ and 3 namely $F^+(1)-F^+(0)=4$ cm^{-1}, $F^+(2)-F^+(1)=8$cm^{-1} and $F^+(3)-F^+(2)=12$ cm^{-1} make the assignment of the NO^+ rota-

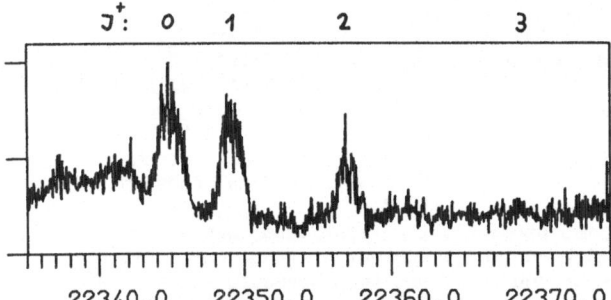

Fig. 1: Threshold photoelectron signal from
(NO^+) X $^1\Sigma^+$ $(v^+=0, J^+) \xleftarrow{-1}$ (NO) C $^2\Pi_r$ $(v=0, T_y(1/2), N=1)$
transition. Scale in cm^{-1}

tional structure unambiguous. The measured intensity ratios for the transitions into $J^+=N^+=0,1,2$ and 3 are 1,1,0.5 and 0.08 respectively. This
kind of intensity information which was not available before can be used
as a verification of the multichannel quantum defect theory (MQDT) /9/,
a theory which takes the same parameters e. g. eigenquantum defects for
the treatment of discrete or continuous channels. A very recent MQDT calculation for the transition probability from the (NO) C $^2\Pi_r$ $(v=0, T_y(1/2)$
N=1) state into the (NO^+) X $^1\Sigma^+$ $(v^+=0, J^+)$ state gives an intensity distribution of 1:0.82:0.74:0.1:0.04 for $J^+=N^+=0,1,2,3,4$ respectively /10/,
which is in good agreement with our results. Our observed intensity ratios
can also be compared to the transition probability one would expect for a
transition from a pure Hund's coupling case (b) to a pure Hund's case (d)
thereby approximating the system ion + free electron as a Hund's case (d).
If a Rydberg electron of $\ell=L=1$ is assumed for the intermediate C state,
the transition probability is proportional to $(2N+1)(2N^++1)\begin{pmatrix} N & N^+ & 1 \\ -1 & 0 & 1 \end{pmatrix}^2$
which leads to an intensity relation for the transitions from N=1 to $J^+=$
$N^+=0,1,2,3$.of 2/3:1:1/3:0. This is also in reasonable agreement with the
experimental results, but the MQDT treatment is certainly the much more
adequate because it is able to explain the $J^+=N^+=3$ peak, which is very
small in the experimental spectrum. It is important to note that these
results show the importance of total angular momentum <u>selection rules</u>
for ionizing transitions; a priori the free electron could take any
angular momentum and hence selection rules cannot intuitively be anticipated. However, the observed rather strong selection rules $\Delta J^+=\pm1/2$,
$\pm 3/2$ are mainly an indication of the fact that the ionizing transition
takes place from a Rydberg state (with a nearly well defined orbital angular momentum of the Rydberg electron) into the continuum. For valence
state - continuum transitions also higher ΔJ^+ contributions have to be
expected /12/.

III. TPE Spectra of Benzene

To obtain TPE spectra of benzene we chose a two-photon, two-colour process to reach the ionization threshold with the $S_1, \nu_6(v=1, J'=1,2;K)$
state as intermediate molecular resonance. The small rotational spacing
in Benzene $(B\approx 0.18$ cm$^{-1})$ made it necessary to use intracavity etalons
and pressure tuning for both frequency doubled dye lasers (Quanta-Ray
PDL, line-width 0.1 cm^{-1}, pumped by Nd:YAG). Laser wavelength calibration was carried out by simultaneously displaying the laser line and
well known emission lines from a uranium hollow cathode lamp on an OMA
screen mounted in the exit plane of a 1 m double pass monochromator.

119

By this we could determine the laser wavenumber to an absolute accuracy of 0.1 cm^{-1}. First the $2w_1$ total ion signal from the $S_1 \leftarrow S_0$, 6_0^1 one-photon transition was measured by scanning the first laser only; the result is shown in Fig. 2. From our laser wavelength calibration we find a 6_0^1, ν_{00} value of 38606.2 cm^{-1} in agreement with the value of 38606.16 cm^{-1} from a 200 MHz resolution experiment /13/ but with a discrepancy to /14/. The rotational structure of the $S_1 \leftarrow S_0$ transition could be fitted quantitatively with the known rotational constants /13, 14, 15/ and a rotational temperature of 3K.

For the two-colour TPES measurements, the first laser was set to either the R(0) or R(1) transition and the frequency drift of the laser was carefully monitored. The second laser was then pressure tuned around the ionization potential. The results are shown in Fig. 3 for the J'=1 and in Fig. 4 for the J'=2 intermediate state.

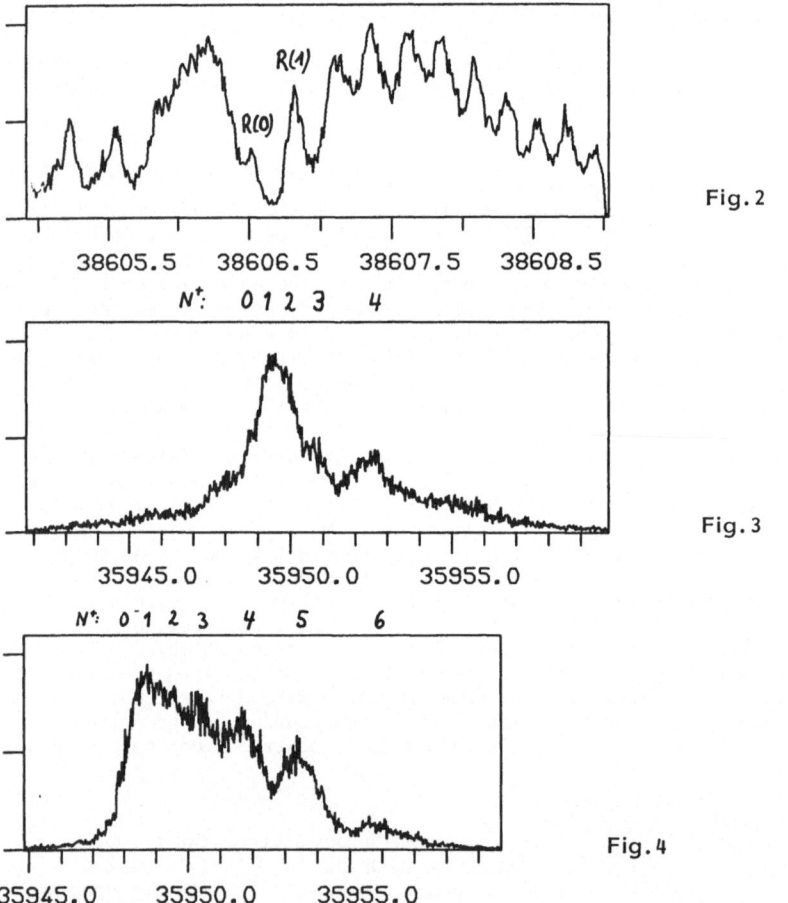

Fig. 2

Fig. 3

Fig. 4

Fig. 2: Two-photon ion signal from one-photon resonant $S_1 \leftarrow S_0$, 6_0^1 transition in benzene, T = 3K. Scale in cm^{-1}

Fig. 3: Threshold photoelectron signal: transitions from S_1 (v=1), J'=1 of benzene into ion ground state

Fig. 4: Threshold photoelectron signal: transitions from S_1 (v=1), J'=2 of benzene into ion ground state

As expected, the TPE spectrum for J'=2 (Fig. 4) is shifted by 0.75 cm^{-1} to the red compared to J'=1 (Fig. 3). With our termvalues for J'=1 and J'=2 of 38606.52 and 38607.24 cm^{-1} we obtain an ionization potential of 74555.1$_5$ cm^{-1} for benzene. Some care has to be taken as regards the intensity distribution because of the unresolved K structure. In the intermediate level mainly J'=1, K'=0 or J'=2, K'=0 are populated /13/.

An inspection of Figs. 3 and 4, where the rotational levels of $C_6H_6^+$ have been marked for N$^+$, K=0, shows clearly that ionization into other K sub-levels is important. However, the different N$^+$ levels are separated for N$^+$=3 (leading to an estimated value of B$^+$=0.20 cm^{-1}) and ionizing transitions up to N$^+$=6 (corresponding to Δ N$^+$=N$^+$-N'=4) are observed. Still very strong are the ΔN$^+$=3 transitions which could be explained by a strong contribution from the l=3 electron partial wave.

In conclusion, we have shown that our method of high resolution TPES is applicable to the study of rotational structure even of large molecular ions and that the angular momentum transfer differs markedly from Rydberg state-continuum to valence state-continuum transitions /16/.

Acknowledgement: We thank the Deutsche Forschungsgemeinschaft (DFG) for sponsoring this research

/1/ K. Siegbahn, Some Current Problems in Electron Spectroscopy,
   Uppsala University, Institute of Physics
   Report UUIP-1074, Uppsala (1982)
/2/ P. Morin, I. Nenner, P. M. Guyon, O. Duduit and K. Ito
   J. Chim. Phys. 77 (1980) 605
/3/ J. E. Pollard, D. J. Trevor, J. E. Reutt, Y. T. Lee and D. A. Shirley
   J. Chem. Phys. 77 (1982) 34
/4/ W. Peatman, F.-P. Wolf and R. Unwin, Chem. Phys. Lett., 95 (1983)453
/5/ W. G. Wilson, K. S. Viswanathan, E. Sekreta and J. P. Reilly
   J. Phys. Chem. 88 (1984) 672
/6/ K. Müller-Dethlefs and R. Frey in: Laser spectroscopy VI,
   Springer Series in Optical Sciences, eds. H. P. Weber and W. Lüthy
   (Springer, Berlin 1983) p. 367
/7/ K. Müller-Dethlefs, M. Sander and E. W. Schlag
   Z. Naturf. 39a (1984) 1089
/8/ K. Müller-Dethlefs, M. Sander and E. W. Schlag
   Chem. Phys. Lett. 112 (1984) 291
/9/ A. Giusti-Suzor and Chr. Jungen
   J. Chem. Phys. 80 (1984) 986
/10/ Chr. Jungen and F. Masnon, priv. comm.
/11/ C. Amiot and J. Verges, Physica Scripta 25 (1982) 302
/12/ K. Müller-Dethlefs, to be published
/13/ E. Riedle, priv. comm.
/14/ J. H. Callomon, T. M. Dunn and I. M. Mills, Phil. Trans. Roy. Soc.
   (London), 259 (1966) 499
/15/ A. Kiermeier, priv. comm.
/16/ K. Müller-Dethlefs, M. Sander, L. A. Chewter and E. W. Schlag
   to be submitted

Vibrational Predissociation Spectroscopy of Hydrogen Cluster Ions

M. Okumura, L.I. Yeh, and Y.T. Lee
Materials and Molecular Research Division,
Lawrence Berkeley Laboratory, and Department of Chemistry,
University of California, Berkeley, CA 94720, USA

Although molecular ion spectroscopy has progressed rapidly[1], there has been very little spectroscopy done of weakly bound ionic clusters[2,3]. These clusters play an important role in atmospheric chemistry and have been studied extensively in thermo-chemical and kinetic experiments over the past decade[4]. Much of our understanding of the dynamics and structure of cluster ions to date has relied on the results of ab initio quantum calculations. Stimulated by the theoretical work, we have begun experiments on the infrared spectroscopy of cluster ions, and have recently made the first spectroscopic observation of the protonated hydrogen clusters H_n^+ (n=5,7,9,11,13 and 15).

According to CI calculations[5], the hydrogen cluster ions are H_3^+ ions with hydrogen molecules complexed around it, with the first three molecules coordinated to the corners of the H_3^+. Several experiments[6-8] have measured equilibrium constants for $H_5^+ \rightarrow H_3^+ + H_2$ and $H_7^+ \rightarrow H_5^+ + H_2$. The most recent results[6,7] yield dissociation energies of approximately 6 kcal/mol (2100 cm^{-1}) and 3.8 kcal/mol (1100 cm^{-1}) respectively, in agreement with the CI calculations of Yamaguchi, Gaw and Schaefer[5]. Yamaguchi et al have also calculated the vibrational frequencies for H_5^+, H_7^+ and H_9^+. The highest frequency modes are red-shifted from the free H_2 frequency(4161 cm^{-1}) by 100 to 270 cm^{-1}, and correspond to stretching of the H_2 moieties. Next is a mode corresponding to the 3220 cm^{-1} symmetric stretch of H_3^+. Although transitions to these modes are IR forbidden in the free molecules, they are predicted to be strong absorbers in the complex[9]. Upon excitation of these modes, the clusters will vibrationally predissociate, and the absorption can be detected by monitoring the products.

The experimental apparatus is shown in Fig. 1, and will be described in more detail elsewhere[10]. A molecular beam containing clusters of hydrogen molecules was formed by expanding hydrogen (stagnation conditions 130°K, 20 to 30 atm) through a 10 μ nozzle. Electron impact ionization of the neutral clusters produced cluster ions which corresponds to $H_3^+(H_2)_n$. The ion beam passed through a magnetic mass filter to select the parent cluster to be studied. The ions were then decelerated and stored in a radio frequency octupole ion trap for 0.5 ms, after which a tunable infrared laser was fired into the trap. After 1 ms, the ions were released from the other side of the trap into a quadrupole mass spectrometer to select and count only a specific product ion. The sequence was then repeated without the laser, and the signal was obtained by subtracting this background. Spectra were taken for each dissociation channel by counting product ions as a function of laser frequency.

A $LiNbO_3$ optical parametric oscillator, pumped by the far field of a Quanta Ray Nd:YAG laser, was tuned between 3800 and 4200 cm^{-1} to find

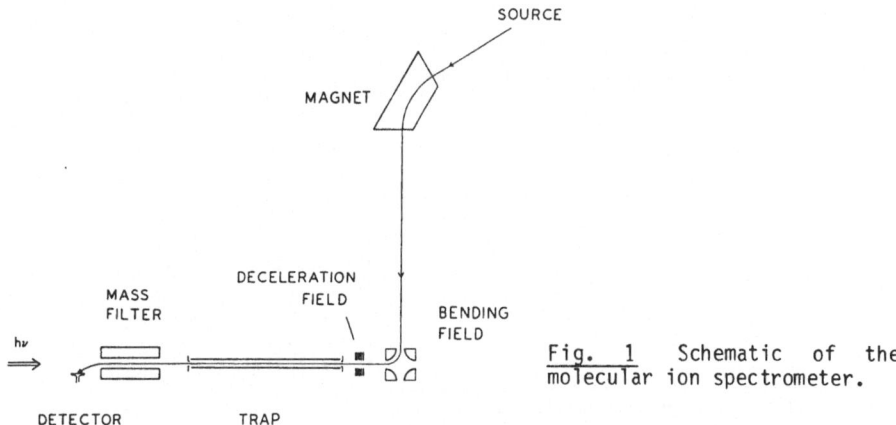

Fig. 1 Schematic of the molecular ion spectrometer.

the H_2 stretching modes. The pulses were typically 4 to 6 mJ in energy and 8 ns in duration at a 10Hz repetition rate. For these preliminary experiments, there were no line-narrowing elements in the OPO cavity, so the linewidth was approximately 10 cm^{-1}. The Doppler width of the trapped ions was estimated to be <0.2 cm^{-1}. In additional experiments, a Burleigh F-center laser was used, with and without an intracavity etalon, at resolutions of $\Delta\nu \sim 3 \times 10^{-5}$ and 0.5 cm^{-1} respectively. In these experiments, the ion beam was not trapped, but passed through the interaction region at energies of 1 to 100 eV to vary the Doppler width.

The peaks of the bands for H_5^+, H_7^+ and H_9^+ are listed in Table I. The frequencies calculated for these transitions, after a semi-empirical correction[5], agree with our results to within 70 cm^{-1}. One must interpret the peak positions with caution, given the uncertainty in locating the true origin.

Table I   Infrared absorption frequencies in cm^{-1} for the cluster ions H_5^+, H_7^+ and H_9^+.

	Experimental[a]	Calculated origin[b]
H_5^+	3532	3464[c]
	3910	3906[c]
H_7^+	3980	4025[d]
H_9^+	4020	4020[e]

a. This work. Peak of the observed band, to within 10 cm^{-1}.
b. ab initio calculations, from Ref. 5 and 9. The frequencies reported in Ref. 5 are harmonic frequencies obtained by analytic gradient SCF and CI calculations. The frequencies in this column have been corrected for anharmonicity and systematic deviations by subtracting the difference between the harmonic frequency calculated for H_2 and the experimental H_2 (1←0) origin.
c. (6s2p/4s2p) basis set, CISD. See Ref. 9.
d. DZ+P basis set, CISD. See Ref. 5.
e. DZ+P basis set, SCF. See Ref. 5.

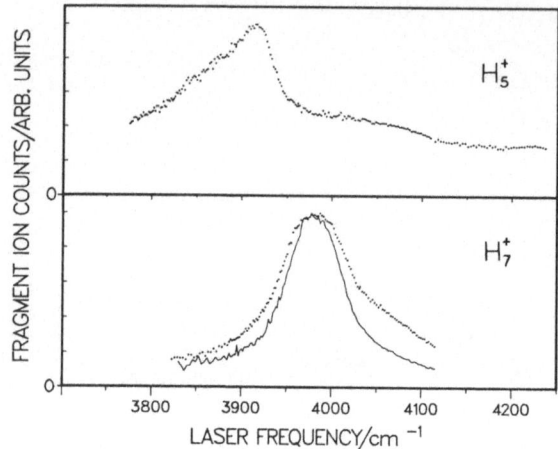

Fig. 2 The infrared absorption spectra of hydrogen cluster ions of the H_2 stretching bands at a resolution of ~10 cm^{-1}. The spectra were recorded by detecting predissociation product ions as a function of laser frequency. Upper plot: the spectrum of H_5^+ detecting H_3^+ fragments. Lower plot: the spectrum of H_7^+ detecting H_5^+ fragments (solid line) and H_3^+ fragments(points).

The spectrum of H_5^+ is shown in Fig. 2. No rotational structure has been resolved, although the rotational spacing for the $\Delta J = \pm 1$ transitions is estimated from the CI geometry to be 6 to 7 cm^{-1}. The lack of structure may be due to i) the low resolution of the laser, ii) spectral congestion caused by the high internal temperature of the molecule, and/or iii) homogeneous broadening. Experiments using the F-center laser with and without the intra-cavity etalon yield essentially the same spectra, ruling out (i). The predissociation lifetimes are unknown, but we do expect the spectra to be congested. Upon ionization of the neutral cluster, the reaction $H_2^+ + H_2 \rightarrow H_3^+ + H$ occurs rapidly within the cluster, accompanied by release of 40 kcal/mol into the internal modes. The excess energy is dissipated as the cluster fragments, but dissociation leaves the cluster ion in a wide distribution of excited rovibrational states. In the near future, we hope to eliminate the spectral congestion by using high pressure discharge sources to create internally cold cluster ions.

Another feature of the spectrum is the long tail to the blue. This may arise from combination bands involving the low frequency stretching of the H_3^+ and H_2 moieties. The CI calculations[5] indicate that this motion involves intramolecular proton exchange, $H_3^+ \cdot H_2 \rightarrow H_2 \cdot H_3^+$ with little or no barrier. Because the center of charge oscillates across the center of mass, the transition moment will be very large[9].

The spectra of H_7^+ shown in Fig. 2 are considerably narrower than that of H_5^+, with a very small shoulder to the blue. We have observed both product channels, H_3^+ and H_5^+, and have obtained spectra for each. The spectra are similar, although the dissociation to H_3^+ is broader, with more intensity in the wings.

Spectra for larger hydrogen clusters (up to H_{15}^+) have been observed as well. The predissociation spectra of H_9^+ and larger are similar in appearance to those of H_7^+; the spectrum of H_5^+ is qualitatively different. This result may reflect the different structure and larger binding energy of H_5^+, as predicted by theory[5]. In particular, the intramolecular proton exchange motion should occur exclusively in H_5^+.

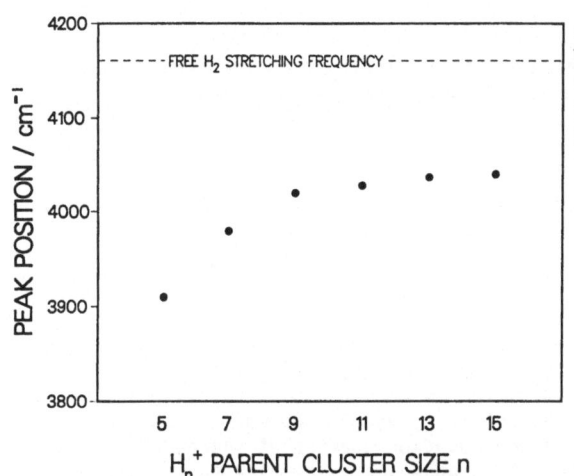

Fig. 3 Frequencies of hydrogen stretching band absorption maxima as a function of cluster size. Uncertainty in peak position is $\Delta\nu \sim \pm 5$ cm^{-1}.

As shown in Fig. 3, the absorption peaks shift very little after H_9^+. This result suggests a model in which the three H_2 molecules complexed at the corner act as chromophores in the 4000 cm^{-1} region. Additional H_2 molecules are bound weakly to the H_9^+ core, causing only small shifts in the absorption. Because they are only slightly perturbed, "outer shell" H_2's are poor infrared absorbers. This model is consistent with theoretical[11] and experimental[8] results which show that the energy required to remove one H_2 molecule from H_{11}^+ or higher clusters is about half that for H_9^+.

This work was supported by the Director, Office of Energy Research, Office of Basic Energy Sciences, Chemical Sciences Div. of the U.S. Dept. of Energy under Contract No. DE-AC03-76SF00098. L.I.Y. acknowledges support of a NSF Predoctoral Fellowship. The F-center laser was on loan from the San Francisco Laser Center, a NSF Regional Instrumentation Facility, NSF Grant No. CHE79-16250 awarded to the University of California at Berkeley in collaboration with Stanford University.

References
1. R.J. Saykally and C.S. Gudeman, Ann. Rev. Phys. Chem. 35, 387 (1984).
2. H.A. Schwarz, J. Chem. Phys. 67, 5525 (1977); 72, 284 (1980).
3. M. Heaven, T.A. Miller and V.E. Bondeybey, J. Chem. Phys. 76, 3831 (1983); L.F. DiMauro, M. Heaven, and T.A. Miller, Chem. Phys. Lett. 104, 526 (1984).
4. P. Kebarle, Ann. Rev. Phys. Chem. 28, 445 (1977); E.E. Ferguson, F.C. Fehsenfeld, and D.L. Albritton, in Gas Phase Ion Chemistry (M.T. Bowers, ed.) 1, 45 (Academic Press, New York, 1979); T.D. Mark and A.W. Castleman, Jr., Adv. At. Mol. Phys. 20, 66 (1985).
5. Y. Yamaguchi, J.F. Gaw and H.F. Schaefer III, J. Chem. Phys. 78, 4074 (1983), and references therein.
6. R.J. Beuhler, S. Ehrenson and L. Friedman, J. Chem. Phys. 79, 5982 (1983).
7. M.T. Elford, J. Chem. Phys. 79, 5951 (1983).
8. K. Hiraoka and P. Kebarle, J. Chem. Phys. 2267 (1975).
9. Y. Yamaguchi, J.F. Gaw, and H.F. Schaefer III, unpublished results.
10. M. Okumura, L.I. Yeh and Y.T. Lee (in preparation).
11. S. Raynor and D.R. Herschbach, J. Phys. Chem. 87, 289 (1983); K. Hirao and S. Yamabe, Chem. Phys. 80, 237 (1983); H. Huber, Chem. Phys. Lett. 70, 353 (1980).

Intracavity Far Infrared Laser Spectroscopy of Supersonic Jets: Direct Measurement of the Vibrational Motions in van der Waals Bonds

D. Ray, R. Robinson, D.-H. Gwo, and R.J. Saykally

Department of Chemistry and Materials and Molecular Research Division
Lawrence Berkeley Laboratory, University of California,
Berkeley, CA 94720, USA

Molecules held together by van der Waals forces generally posess bond strengths which are ~0.1-1.0% of normal chemical bonds, i.e. 0.1-1.0 Kcal/mole (.35-350 cm^{-1}). Such extremely weak bonds exhibit vibrational frequencies which lie in the far-infrared, eq. 10-100 cm^{-1} (1000-100 μm). In order to measure vibration-rotation spectra of van der Waals bonds, one must devise a method which possesses very high sensitivity in this region of the spectrum, and which preferably also has very high resolution capabilities, such that the hyperfine interactions, which can provide important information on the potential surface of these clusters, can be resolved. In this paper we report the development of such a new spectroscopic technique for directly measuring vibrational absorption spectra of van der Waals bonds. In this method, van der Waals molecules are produced in a supersonic free jet expansion located inside the optical cavity of an optically pumped far infrared laser. Electric field tuning of dipole-allowed vibration-rotation transitions into coincidence with the laser frequency produces an extremely sensitive detection method which also possesses very high (~1 MHz) resolution. We shall describe this new technique and its initial application for the measurement of the bending (v_2) vibration in ArHCl near 34 cm^{-1}.

A schematic diagram of the spectrometer, is given in Figure 1. An optically pumped far infrared laser is the radiation source; by pumping ~60 different molecules with the lines from a CO_2 laser, approximately 2000 discrete frequencies can be obtained throughout the range 10-200 cm^{-1}. The high finesse (f ≈ 100) optical cavity is divided into a gain cell and a sample region by a rotatable polypropylene beam splitter placed at the Brewster angle; the gain cell contains two parallel mirrors which are used to multipass the CO_2 pump laser in a transverse configuration, and an output coupler, which directs a portion of the laser radiation into a cryogenic Ga:Ge or InSb detector. The sample region consists of a large vacuum chamber pumped by a 10" diffusion pump and contains a pair of 56 cm diameter polished aluminum electrodes and mounted to be parallel with a set of 5.0 cm Macor spacers. One of the electrodes is connected to a programmable power supply, which is swept from 0 to 100 kV, while the other is connected to an audio oscillator, amplifier, and transformer combination which can provide modulation voltages up to 1 kV at frequencies up to 100 kHz. The DC voltage is controlled by a microcomputer system, and is measured with a voltage divider and digital voltmeter. With this configuration, electric fields as high as 20 kV/cm are produced with high homogeneity and measured with an absolute accuracy of 0.02%. Both the far infrared laser and the CO_2 pump laser are locked to the cavity of the FIR laser for suitable frequency stability.

A 1/4" Pyrex tube with the tip drawn to a 10-100 μm orifice serves as a supersonic nozzle, which is inserted into the center of the electrodes

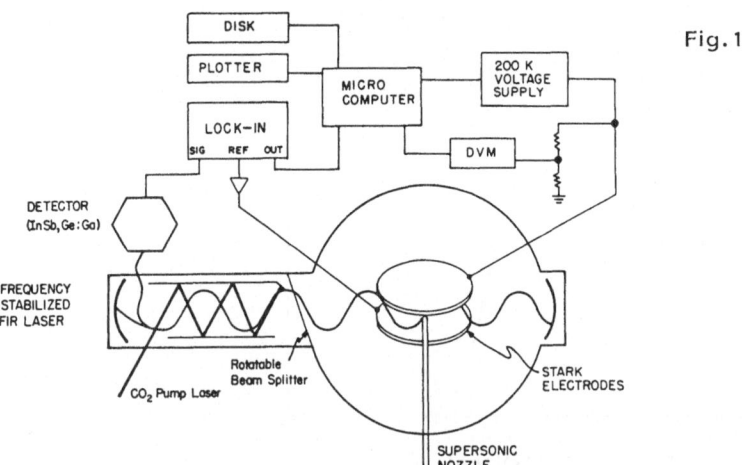

Fig. 1

INTRACAVITY FAR-INFRARED LASER STARK SPECTROMETER

through an XZ translation stage. Typical backing pressures employed range from 1 to 10 atmospheres, which result in pressures up to 1 mTorr in the vacuum chamber, above which dielectric breakdown becomes incapacitating. With this supersonic free jet configuration, Mach numbers as high as 33 can be achieved, yielding average total number densities near 10^{16} cm^{-3} and translational temperatures near 10°K in the region of the intracavity FIR laser beam. Absorption of the intracavity laser radiation by vibration-rotation transitions in van der Waals molecules produced in the supersonic expansion are demodulated by a lock-in amplifier, and displayed as a function of the electric field strength.

We have observed extensive far infrared spectra of ArHCl in several of the regions where transitions are predicted[1,2] to occur. Typical examples are presented in Figures 2 and 3, which are assigned to Stark-hyperfine components of Q(1) of the bending fundamental. Optimization of the experimental conditions (3% HCl in Ar, 6 atm backing pressure) produced signal-to-noise ratios as high as 10^4 for the 35 cm^{-1} transitions. Moreover, because the signals were measured in the intracavity configuration, saturation dips could readily be observed. Approximately 120 transitions have been measured in this region and assigned to Q(1), Q(2), and R(0) components. A preliminary least squares analysis of these transitions yields values for the ArH^{35}Cl rotational constant, ℓ-type doubling constant, dipole moment, and axial and perpendicular quadrupole coupling constants for the first excited bending state, and for the band origin of the bending vibration, which are given in Table 1. While a large number of transitions have been observed in the region where the parallel bending (24 cm^{-1}) and the stretching vibrations (33 cm^{-1}) are predicted to occur, detailed assignments and analyses of these have not yet been achieved. Similarly, a detailed analysis of spectra belonging to ArH^{37}Cl isotope has not been carried out at this time.

Interpretation of the spectroscopic information presented in Table 1 indicates that the average bending angle (of the HCl bond relative to the A inertial axis) increases from 41.2° in the ground state[3] to nearly 76° in the first excited bending state. Because the dipole moment of the ArHCl molecule is essentially the projection of the HCl dipole onto the

127

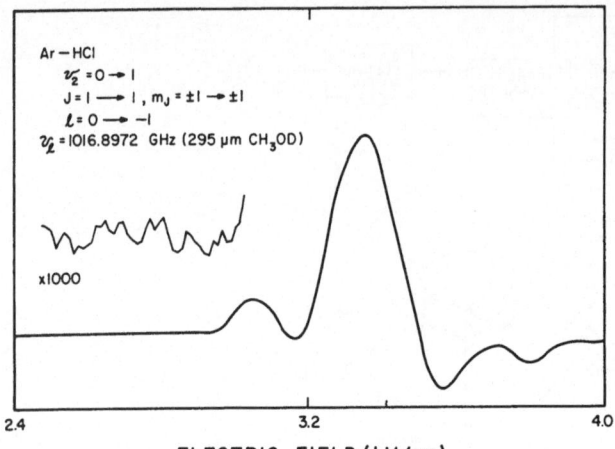

Fig. 2. High Sensitivity Spectrum Showing Signal-to-noise ratio of >10^4

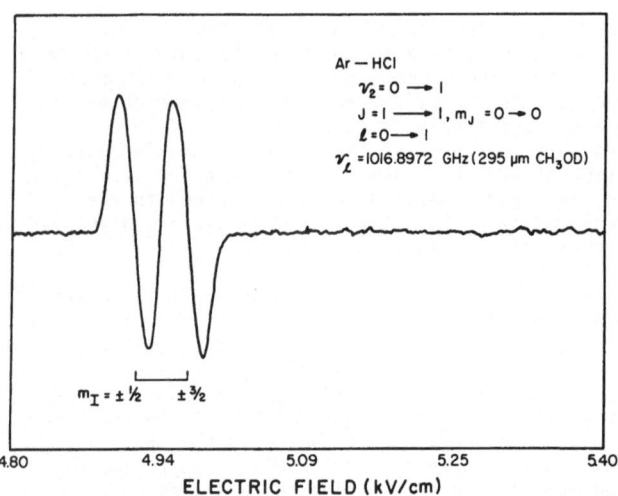

Fig. 3. High-Resolution Spectrum Exhibiting Resolved Hyperfine Splitting

Table 1. Effective Parameters for the ν_2 = 1 State of ArHCl

B = 1707.0(2) MHz	eqQ_a = 10(2) MHz
ν = 33.98127(1) cm^{-1}	$eqQ_b - eqQ_c$ = -75(2) MHz
q = -50(3) MHz	μ = 0.267(1) D

symmetry axis, excitation of the bend clearly produces a large dipole derivative, partially accounting for the very large signals observed for
these transitions. Similarly large bending transitions moments are likely
to occur quite generally for weakly bound molecules, as a result of the

large amplitude motions they undergo. Moreover, although the stretching vibration should, by analogy, produce a relatively small dipole derivative in these molecules, a very large Coriolis coupling between the van der Waals stretching and bending vibrations is indicated by the anomalously large ℓ-type doubling constant; hence ,intensity borrowing from the bend should make van der Waals stretches generally accessible by direct absorption spectroscopy as well. One of the most interesting short-term prospects for these experiments is the probable assignment of many of the transitions observed near 25 cm^{-1} to the parallel bending vibration ($\Delta\ell = 0$). This transition correlates to the bending overtone ($\Delta v_2 = 2$) in the high anisotropy limit, but will actually occur at lower frequencies than the perpendicular bend ($\Delta\ell \pm 1$), which correlates to the fundamental ($\Delta v_2 = 1$), when the potential surface exhibits two attractive wells[1,2]. Assignment of the 25 cm^{-1} transitions to the parallel bend would constitute the first experimental confirmation of the secondary minimum at ArClH configuration if this simple desorption is applicable to the potential surface.

Finally, we wish to point out that the simultaneous development of far-infrared laser-molecular beam electric resonance techniques by the Klemperer group at Harvard University constitutes a powerful complimentary technique for the study of vibrational spectra of van der Waals molecules. The combination of these two new approaches to the study of van der Waals bonding should add considerably to our knowledge of this important phenomenon over the next several years.

Acknowledgements

This work was supported by the Director, Office of Basic Energy Sciences, United States Department of Energy

References

1. J.M. Hutson: J. Chem. Phys. 81, 2357 (1984)
2. J.M. Hutson, and B.J. Howard: Mol. Phys. 45, 769 (1982)
3. S.E. Novick, P. Davies, S.J. Harris, and W. Klemperer: J. Chem. Phys. 59, 2273 (1973)

Photodetachment Spectroscopy of $^-CH_2CN$

K.R. Lykke, D.M. Neumark, T. Andersen*, V.J. Trapa, and W.C. Lineberger

Joint Institute for Laboratory Astrophysics, University of Colorado
and National Bureau of Standards and Department of Chemistry,
University of Colorado, Boulder, CO 80309, USA

1. Introduction

Spectroscopic studies of the structure of negative ions exhibit two major
limitations. Negative ions can be obtained in only very small densities
compared to positive ions and neutrals ($\lesssim 10^6$ cm^{-3} for negative ions) [1].
In addition, the extra electron is usually bound by less than a few eV so
that, in general, bound electronically excited states which would otherwise permit an electronic spectroscopic study of the ground state do not
exist. The first problem can be avoided by using single particle detection of a transition, the second by several different techniques: a)
threshold photodetachment to the neutral [2]; b) infrared vibration-
rotation spectroscopy [3]; c) excitation to an electronically excited
negative ion above the detachment continuum and subsequent autodetachment
[4-6]. The fact that any dipolar neutral molecule ($\gtrsim 2$ Debye) can support
an electronic state [7] has greatly facilitated the study of the ground
and excited states of negative ions [5,6].

The first rotationally resolved study of a polyatomic molecule was reported in 1934 [8] and yielded the structure of formaldehyde. Much work
has been done since then on neutrals and positive ions [9], but no polyatomic negative-ion spectrum was rotationally resolved until 1984 [5].
If the rotational constants are obtained and the molecule is sufficiently
symmetric, the structure of the molecule is determined. The following is
a preliminary report on the electronic spectroscopy of $^-CH_2CN$.

2. Experimental

The coaxial laser-ion beam spectrometer used for the present work has been
previously described in detail [4,5]. An ~500 pA beam of $^-CH_2CN$ (cyano-
methyl anion) is formed by extraction from a hot cathode discharge source
containing CH_3CN and NH_3, mass selection with a 90° sector magnet, and acceleration to 2650 eV. The ions are then bent 90° by a transverse electric
field and merged with the output of a home-built tunable ring dye laser
pumped by all lines of an Ar II laser. The neutrals that are formed by
detachment are separated from the remaining ions by another transverse
electric field and strike a KDP plate. This produces secondary electrons
which are then detected by an electron multiplier and counted to yield the
total cross-section. The electrons that are detached in the interaction
region are collected by a weak solenoidal field (~5 Gauss) and counted
with another electron multiplier.

---

*1984-85 JILA Visiting Fellow. Permanent address: Institute of Physics,
University of Aarhus, Denmark.

The data were taken by scanning the laser in frequency (measured with a λ-meter) while monitoring the neutrals and electrons formed as a function of photon energy and normalized to the ion current and laser power. The laser using styryl 9 dye operates from about 790 to 870 nm with ~400 mW broadband and ~150 mW single-mode output when pumped with ~5 W all lines from the Ar II laser. The dye laser is easily configured in either a standing wave, broadband mode (birefringent tuner only, $\Delta\nu \sim 1$ cm^{-1}) or in single mode ($\Delta\nu < 1$ MHz). The resolution of this spectrometer is <20 MHz, limited by the Doppler spread in the kinematically compressed ion beam [4].

3. Results

The broadband scan of the total photodetachment cross-section of $^-$CH$_2$CN is shown in Fig. 1. There are ~10^4 points in this scan, 1 sec integration time per point with an effective resolution of ~1 cm^{-1} (standing wave laser with stepper-motor driven birefringent tuner). There are two basic processes at work here; direct photodetachment of the anion [process a in Eq. (1)] and excitation to an excited state of the anion followed by autodetachment [process b in Eq. (1)]

$$^-CH_2CN + h\nu \longrightarrow CH_2CN + e^- \hspace{3cm} \underline{a}$$

$$^-CH_2CN + h\nu \longrightarrow (^-CH_2CN)^* \longrightarrow CH_2CN + e^- \hspace{1cm} \underline{b} \hspace{1cm} (1)$$

The gentle rise in the cross-section is dominated by process a whereas process b gives the sharp structure. This sharp structure is clustered around two different frequency regions centered at ~12,000 cm^{-1} and ~12,400 cm^{-1}, corresponding to transitions to an electronically excited state of the ion from a vibrationally excited ground state and the vibrationless ground state, respectively.

The low-resolution scans give only approximate transition frequencies, and most of the region between 11,900 and 12,600 cm^{-1} was scanned in single mode with the effective resolution of ~20 MHz. Figure 2 shows a high-resolution scan of a K = 4←3, $\Delta J = 0$, $^rQ_3$ branch. The rotationally-resolved structure of this near-prolate rotor [$\kappa \equiv (2B-A-C)/(A-C) \cong -0.998$] consists of $\Delta J = 0, \pm1$ and $\Delta K = \pm1$ (⊥ band, transition moment in the c-axis) [9], where J is the total angular momentum and K is the component of the angular momentum around the a-axis. Combination differences [9] of the P, Q, and R sub-branches ($\Delta J = -1, 0, +1$, respectively) of each P and R-form branch ($\Delta K = -1, +1$, respectively) have yielded preliminary rotational constants for both the upper (A' = 9.46 cm^{-1}, $\frac{1}{2}$ (B'+C') $\cong$ 0.3346 cm^{-1}) and lower (A" = 9.24 cm^{-1}, $\frac{1}{2}$ (B"+C") $\cong$ 0.3325 cm^{-1}) elec-

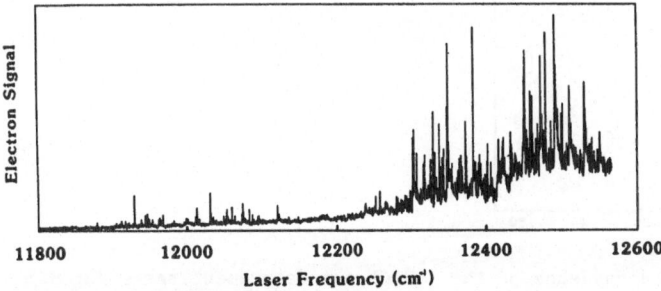

Fig. 1.   Low resolution (~1 cm^{-1}) scan of the photodetachment spectrum of $^-$CH$_2$CN.

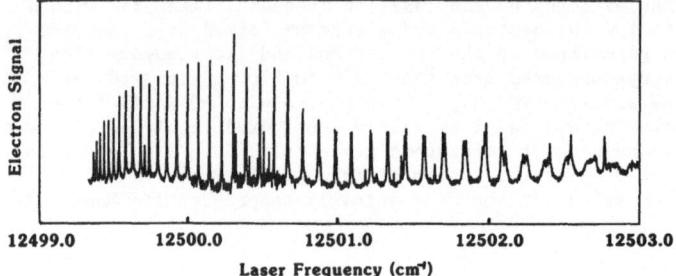

Fig. 2. High resolution (~20 MHz) scan of the $^rQ_3$ branch of the K = 4←3, ΔJ = 0 transition.

tronic states. A computer fit to an asymmetric rotor Hamiltonian is in progress, and should yield more precise constants.

Linewidths of autodetachment resonances are related to the rate of auto-detachment (inverse lifetime) by the Heisenberg uncertainty principle. Shown in Fig. 3 is the autodetachment rate versus the total rotational quantum number J for different K stacks. The most surprising result is the close similarity between the various K-stacks and the fast rise in rate at nearly the same J, even though K = 8 is located about 600 cm^{-1} above K = 0.

The dipole moment in CH_2CN is directed along the a axis. Since the dipole is the binding force for the excited electron, any motion that will tend to change the dipole's direction in space will also decouple the elec-tron. Therefore, we assume that the decoupling of the electron is due to the end-over-end rotation of the molecule (J-dependent) rather than to the rotation about the dipole axis (K-dependent).

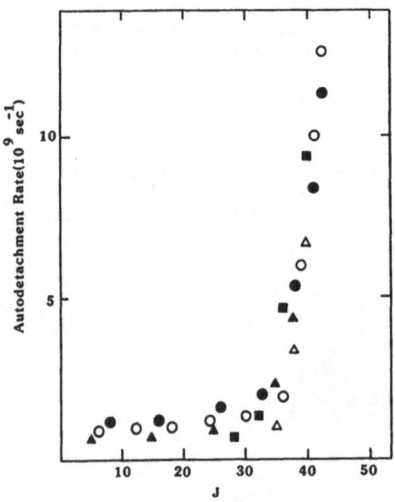

Fig. 3. Rotational dependence of the $^-CH_2CN$ autodetachment rates for dif-ferent K values of the excited state, K = 0: ■, K = 2: Δ, K = 4: ▲, K = 6: ○, K = 8: ●.

Support for this work was provided by National Science Foundation grants PHY82-00805 and CHE83-16628 to the University of Colorado. TA acknowledges a grant from the Danish Natural Science Foundation.

References

1. H. Massey, Negative Ions (Cambridge University Press, 1976).
2. P. A. Schulz, R. D. Mead, P. L. Jones and W. C. Lineberger, J. Chem. Phys. 77, 1153 (1982).
3. D. M. Neumark, K. R. Lykke, T. Andersen and W. C. Lineberger, J. Chem. Phys., submitted.
4. U. Hefter, R. D. Mead, P. A. Schulz and W. C. Lineberger, Phys. Rev. A 28, 1429 (1983).
5. K. R. Lykke, R. D. Mead and W. C. Lineberger, Phys. Rev. Lett. 52, 2221 (1984); R. D. Mead, K. R. Lykke, W. C. Lineberger, J. Marks and J. I. Brauman, J. Chem. Phys. 81, 4883 (1984).
6. T. Andersen, K. R. Lykke, D. M. Neumark and W. C. Lineberger, J. Chem. Phys., to be submitted.
7. W. R. Garrett, J. Chem. Phys. 73, 5721 (1980); ibid., 77, 3666 (1982); R. L. Jackson, P. C. Hiberty and J. I. Brauman, J. Chem. Phys. 74, 3705 (1981).
8. G. H. Dieke and G. B. Kistiakowsky, Phys. Rev. 45, 4 (1934).
9. G. Herzberg, Molecular Spectra and Molecular Structure (Van Nostrand, New York, 1966), Vol. 3.

Laser Double Resonance Studies of
Rotational Relaxation in a Supersonic Molecular Beam

T. Shimizu and F. Matsushima

Department of Physics, Faculty of Science, University of Tokyo,
Hongo, Tokyo 113, Japan

The technique of the supersonic molecular beam and that of the infrared laser Stark spectroscopy are combined in the present work and applied to the study of the elementary processes of collision-induced transitions among vibration-rotation energy levels of molecules. The molecules are prepared in a spectroscopically and also kinematically "pure" state. The velocity distribution in the molecular beam is quite monochromatic, and the molecules are populated mostly in several low-lying energy levels. Various controls to the molecular system can be introduced by using the external fields and the laser radiations. The detection sensitivity is quite high because of high intensity of molecular flux. These allow us to obtain detailed and precise knowledge on molecular dynamics. Three examples of the recent investigations are shown below.

1. Measurement of state-to-state cross-section of collision-induced transition using a collision chamber

The setup for the experiment is shown in Fig.1(a). The NH_3 molecules in the beam produced by a nozzle (diameter 300 µm) and a skimmer (aperture 1 mm) are subjected to collisions,with the target molecules confined in the collision chamber. The relevant energy levels of NH_3 are shown in Fig.2. The ν_2 saQ(J=1,K=1,M=1) transition is pumped by the N_2O laser R(37) line. The excess and depleted populations of the upper and lower levels of the pumped transition are transferred to other levels by the collisions with the target molecules. The changes in the populations of the $(V_2=0,J=2,K=1)$ and $(V_2=1,J=2,K=1)$ levels are probed by measuring the absorption intensity of the ν_2 asQ(J=2,K=1) transition. The absorption intensities of the probe laser beam with and without pumping radiation are denoted by I_p and I, respectively. The change in absorption intensity $(\Delta I=I_p-I)$ normalized to I is plotted against the density of the target gas in Fig.3.

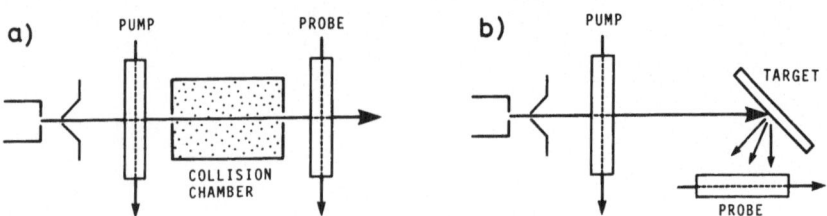

Fig. 1   The experimental setup

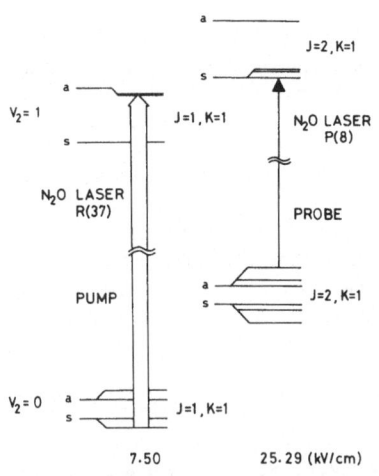

Fig. 2 The relevant energy levels of NH$_3$

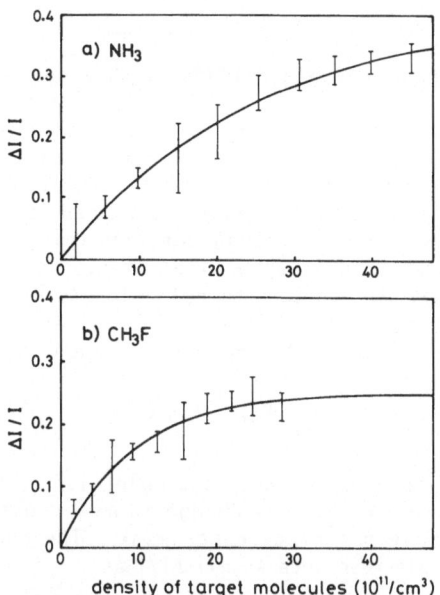

Fig.3 The observed and the calculated values of double resonance signal

The pumped levels and the probed levels are directly connected by the collison-induced transitions;

(a) $(V_2=0, J=1, K=1, a) \longrightarrow (V_2=0, J=2, K=1, s)$

(b) $(V_2=1, J=1, K=1, s) \longrightarrow (V_2=1, J=2, K=1, a)$.

The rate equations for the populations of the relevant levels are solved. The equations include the cross-sections σ_g of the process (a) and σ_e of the process (b). The cross-sections are determined by fitting the calculated values of $\Delta I/I$ (solid curve in Fig.3) to those obtained experimentally. The cross-sections $\sigma_g + \sigma_e$ for various target molecules are listed in Table 1. With the aid of a non-empirical calculation of the cross-sections based on the Anderson theory, the observed cross-sections $(\sigma_g + \sigma_e)_{obs}$ were separated to σ_g and σ_e. Variations in the cross-sections among the collisions with the target molecules are well understood from the consideration of the intermolecular interactions and the rotational energy level structures of the molecules.

Table 1   Observed and calculated cross sections [A^2]

target	calc.			obs.		
	σ_g	σ_e	$\sigma_g + \sigma_e$	$\sigma_g + \sigma_e$	σ_g	σ_e
NH$_3$	18.6	119.8	138.4	87±14	12	75
CH$_3$F	85.1	107.9	193.0	167±28	74	93
CH$_3$Cl	44.0	144.5	188.5	148±28	35	113
N$_2$O	9.9	19.1	29.0	29±16	10	19

2. Collision with solid surfaces

When a molecule collides with a solid surface, the internal state of the molecule is perturbed strongly. The change in the rotational levels before and after the collision is investigated by an experimental setup shown in Fig.1(b). A bare glass surface and an Al-coated surface are employed as the targets. The population distributions in the incident beam and for the scattered molecules are obtained from the absorption intensities of IR transitions. Though the temperature, which is characteristic of the population distribution, is raised by the collision with surface, the temperature after the collision is still below that of the surface (260K). The population in the (J,K) level after the collision, n_{JK}', is related to the population before the collision, n_{JK}, as

$$n_{JK}' = \sum_{J'K'} P_{JKJ'K'} n_{J'K'} \quad . \quad (1)$$

In the present work, the molecules in the $(J=1, K=1)$ level are pumped by an amount of n_p. The change in n_{JK}' caused by the pumping is studied by the double resonance experiment. The coefficient P_{JK11} is related to the double resonance signal $\Delta I/I$ as

$$P_{JK11} = (n_{JK}'/n_p)/(\Delta I/I) \quad . \quad (2)$$

The observed values of P_{JK11} are listed in Table 2 in the cases of the glass and aluminum targets. In the collisions with gas target molecules it is well known that the collision-induced transitions which change the quantum number K seldom occur [1]. In the collision with a solid-surface, the appreciable probabilities of the K-changing collisions are observed, as well as relatively large probabilities of the $\Delta K=0$ transitions. However, it is interesting to note that the preference in the $\Delta K=0$ transition is still preserved in the NH_3-glass collision.

Table 2   Observed values of the transfer probability

target	$(J,K)=(1,1)$	$(2,1)$	$(2,2)$	$(3,2)$
glass	0.25	0.42	0.08	0.05
Al	0.20	0.14	0.10	0.09

3. Rotational cooling in the supersonic molecular beam

The rotational population distribution of the molecules is cooled down in the supersonic beam [2-8]. The excitation temperatures among the low-lying rotational levels of NH_3 were measured from the absorption intensities of IR transitions. In Fig.4, the rotational excitation temperatures measured in the upstream (10 cm apart from the nozzle) and in the downstream (60 cm apart from the nozzle) are shown. It is found that the rotational levels are not yet in thermal equilibrium. These temperatures are well reproduced by a computer simulation of the cooling process. The rate equations for the populations in the (J,K)levels are given by

$$\frac{d}{dt} P(J,K,{}^{s}_{a},t) = \pm \Gamma_\beta(J,K) \, P(J,K,a,t)$$
$$+ \Gamma_\alpha(J+1,K) \, P(J+1,K,{}^{a}_{s},t) - \Gamma_\alpha(J,K) \, P(J,K,{}^{s}_{a},t)$$
$$+ \Gamma_K \, [\ P(J+1,K-3,{}^{s}_{a},t) + P(J,K-3,{}^{s}_{a},t)$$
$$+ P(J+1,K+3,{}^{s}_{a},t) - 3P(J,K,{}^{s}_{a},t) \ ] \qquad (3)$$

where

$$\Gamma_\beta(J,K) = \Gamma_{\beta 0} \, K^2 / [J(J+1)]$$
$$\Gamma_\alpha(J,K) = \Gamma_{\alpha 0} \, (J^2 - K^2) / [J(2J+1)] \ .$$

The factors Γ_β, Γ_α, and Γ_K represent the rate constants for the β-transition ($\Delta J=0, \Delta K=0, s \longleftrightarrow a$), α-transition ($\Delta J=\pm 1, \Delta K=0, s \longleftrightarrow a$), and the $\Delta K=3$ transition, respectively. When we neglect Γ_K the observed features are not reproduced. From the calculated and the observed excitation temperatures, we find the order of magnitude of the small rate Γ_K to be $10^{-5}\Gamma_\beta$.

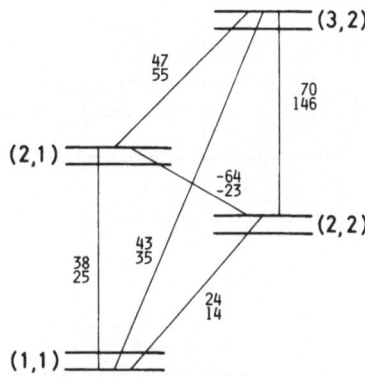

Fig. 4   The excitation temperatures among the rotational levels of NH_3. The upper and the lower figures show the temperatures (K) observed at the upstream and the downstream in the supersonic molecular beam, respectively.

References:

1. T. Oka: Adv. At. Mol. Phys. 9, 127 (1973)
2. D.N. Travis, J.C. McGurck, D. McKeown, and R.C. Denning: Chem. Phys. Lett. 45, 287 (1977)
3. D. Bassi, A. Boschetti, S. Marchetti, G. Scoles, and M. Zen: J. Chem. Phys. 74, 2221 (1980)
4. D.L. Snavely, S.D. Colson, and K.B. Wiberg: J. Chem. Phys. 74, 6975 (1981)
5. B. Antonelli, S. Marchetti, and V. Montelatici: Appl. Phys. B28, 51 (1982)
6. Y. Mizugai, H. Kuze, H. Jones, and M. Takami: Appl. Phys. B32, 43 (1983)
7. K. Veeken and J. Reuss: Appl. Phys. B34, 149 (1984)
8. D. Boscher and J.P. Martin: Chem. Phys. Lett. 113, 225 (1985)

Rabi Oscillations and Ramsey Fringes in CH_3F Using CO_2 Laser-Stark Spectroscopy of a Molecular Beam

A.G. Adam, T.E. Gough, N.R. Isenor, and G. Scoles

Centre for Molecular Beams and Laser Chemistry, University of Waterloo, Waterloo, Ontario, Canada N2L 3G1

APPARATUS

The apparatus show in Figure 1 is designed to observe infra-red spectra of molecules in supersonic beams using the Stark effect to tune molecular transitions into resonance with the lasing transitions of a frequency-stabilised carbon dioxide laser. The bolometer provides a signal proportional to the flux of molecules excited by the laser radiation. Collimation of the molecular beam is ± 150 microns, and the laser stability and electric field homogeneity are sufficient that transit time broadening determines the observed line width (WO = 3.4mm, H.W.H.M. = 100 KHZ). With this apparatus the effects of coherent excitation of IR transitions may be observed.

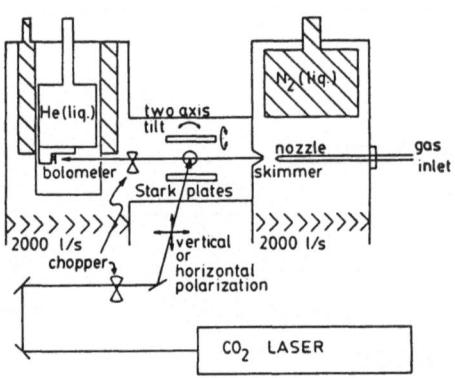

FIG 1

MOLECULAR BEAM SPECTROMETER

RABI OSCILLATIONS

Rabi oscillations are observable in a macroscopic sample when all absorbers in the sample experience identical excitation conditions. The experimental conditions must be such that the observed spectrum is homogeneously broadened (eg. transit limited). For a two-level system passing orthogonally through the waist of a Gaussian laser beam, the Rabi angle is given by

$$\Theta = \mu \frac{2}{v \hbar} \left(\frac{W}{c \, \epsilon_0}\right)^{\frac{1}{2}}$$

where μ is the transition dipole, W is the laser power in watts, and v the speed of the molecule. The result is independent of the spot size of the laser.

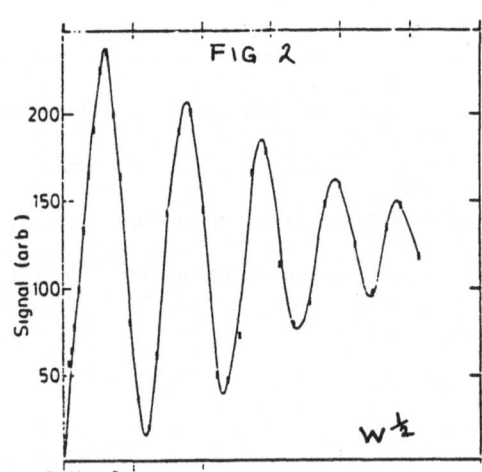

FIG 2

A plot of bolometer signal versus $W^{\frac{1}{2}}$ is shown in Figure 2 for the Q(1,1) component of the ν_3 stretch of fluoromethane. From such plots the transition dipole moment was obtained; Table 1 gives the results of such measurements together with the results of other measurements and calculations.

TABLE 1: Comparison of measured and estimated vibrational transition dipole moments.

Value	Source
0.144 D	E. Arimondo and T. Oka, Phys. Rev. A26, 1494 (1982).
0.182 D	S. Kondo and S. Saeki, J. Chem. Phys. 76, 809 (1982).
0.195 D	S. Kondo and s. Saeki, J. Chem. Phys. 76, 809 (1982).
0.17 D ($C^{13}H_3F$)	T. Sun, M.K. Kim and E.L. Hahn, Cleo'85 conference abstract.
0.22 D	B. Schlegel, Wayne State U., Private Communication 1985.
0.21 $\pm$ 0.01 D	This work.

RAMSEY FRINGES

Ramsey's method of two separated oscillatory fields for M.B.M.R. experiments has been previously adapted for laser spectroscopy by including extra fields. These extra fields remove the effects of molecular beam divergence. The sensitivity of bolometric detection allows the use of highly collimated molecular beams, suggesting that two laser crossings are sufficient to observe Ramsey fringes.

Accordingly, we have irradiated our molecular beam through pairs of 1mm slits spaced by 3, 5 and 10mm. Each slit produces adiabatic following, while the pairs of slits produce well-resolved Ramsey fringes as shown in Figure 3. The damping of the experimental fringes is well simulated by including the calculated broadening from beam divergence (HWHM = 25 KHZ). Our S/N ratio suggests fringes of HWHM < 10 KHZ should be observed by collimating the molecular beam to $\pm$ 50 microns.

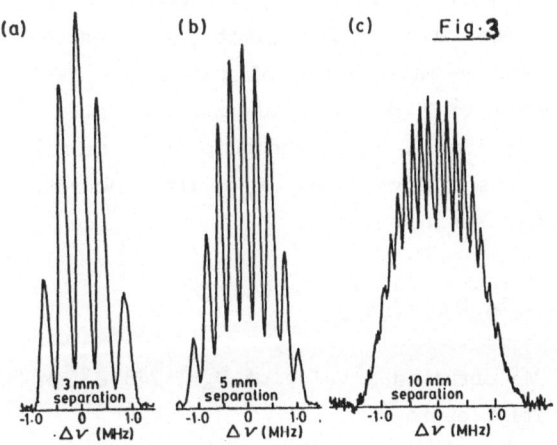

Fig.3

Two-Step Polarization Spectroscopy of Bi$_2$

G. Ehret and G. Gerber

Fakultät für Physik, Universität Freiburg,
D-7800 Freiburg, Fed. Rep. of Germany

The spectroscopy of the Bi$_2$ molecule has recently been the subject of several studies, since it forms the active medium of an optically pumped vapor laser. Interest is also focussed on the spectroscopy of small bismuth clusters, which can easily be produced in an adiabatic expansion. For the investigation of excited electronic states of Bi$_2$ we employed the sensitive method of double resonance polarization spectroscopy.

The experiments were performed in the gas phase with pulsed dye lasers. An amplified cw single-mode dye-laser served as narrowband (Δv = 250 MHz) pulsed pump laser inducing the Bi$_2$ A-X transition v''=5, J''=103 - v'=0, J'=104 while a grazing incidence-type dye-laser was used as a probe laser (Δv=5GHz). We have observed polarization signals from different double resonance schemes like lower level labeling, upper level labeling and also from two-step labeling. In our polarization spectra we observed and identified several excited electronic states of Bi$_2$. A particularly interesting spectrum obtained with two-step excitation consists of a series of doublet lines which converge to the dissociation limit of a yet unobserved electronic state. Part of the spectrum for probe wavelengths around 470 nm is shown in fig. 1. The dissociation limit is determined within a few cm^{-1} by the energy calibration of the probe laser spectrum, using frequency marks of Fabry-Perot and optogalvanic lines, both displayed on top of fig. 1. Since the pump transition is accurately known, the unknown dissociation energy D$_e$ of the electronic ground state X(0$_g^+$) is readily determined by

$$D_e = E(v'',J'') + E(pump) + E(probe) - 2 \cdot E(^2D_{3/2} - ^4S_{3/2})$$

as can be seen from fig. 2. We obtained a value of D$_e$=16778 ± 5cm^{-1} for the ground state dissociation energy.

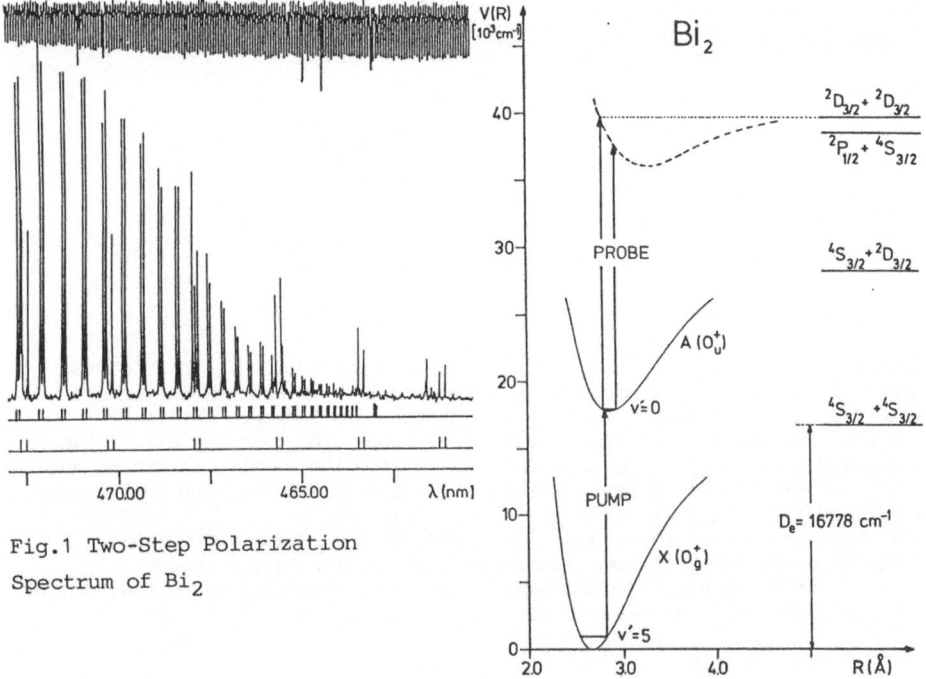

Fig.1 Two-Step Polarization
Spectrum of Bi$_2$

Fig.2 Excitation scheme and potential
curves of Bi$_2$

By scanning the probe laser to the red, we observed strongly broadened probe transitions due to predissociation of the upper electronic state into Bi$^*(^2P_{1/2})$ and Bi $(^4S_{3/2})$ atoms. However, tuning the probe laser further to the red, the probe transitions become narrow again when the total excitation energy is below the $^2P_{1/2} + {}^4S_{3/2}$ dissociation limit.

The strongly predissociated rotational states show a J-dependent broadening, indicating a heterogeneous predissociation. Since we observed only P and R probe transitions connected with the lower $A(O_u^+)$ state, we assume a O_g^+ symmetry for the upper electronic state, which dissociates into two $^2D_{3/2}$ atoms and which is predissociated into $^2P_{1/2} + {}^4S_{3/2}$ atoms. The potential curve of this upper state has been determined from the term energies obtained with two-step polarization spectroscopy.

Electronic Structure of Alkaline Earth Monohalides from Laser-Microwave Double Resonance Spectroscopy

W.E. Ernst, J.O. Schröder, and J. Kändler

Institut für Molekülphysik, Freie Universität Berlin, Arnimallee 14, D-1000 Berlin 33, Germany

1. Introduction

Among the diatomic radicals, the alkaline earth monohalides have attracted the interest of many research groups. With their single metal centered electron outside two closed shell ions these radicals represent in some respect a molecular analogon to the alkali atoms. The internuclear distances and potential energy curves are very similar for the electronic ground and first excited states. Hyperfine structure (hfs) arises from the weak inter- action of the single electron with the nuclear spin of the halogen nucleus. The heavy species of the group, i.e. strontium and barium monohalides, exhibit particularly complex spectra, and the interpretation became only possible with the help of our recently developed laser-microwave (mw) double resonance methods. The results can only be summarized in this paper.

2. Cell Experiments

For all investigations in reaction cells, optical polarization detection schemes were used. Microwave optical polarization spectroscopy (MOPS) was applied for studying the small hyperfine splittings in the rotational spec- trum of the $X^2\Sigma^+$ states of alkaline earth monohalides, most recently of SrCl [1] and SrBr. Large parts of the $B^2\Sigma^+$-$X^2\Sigma^+$ systems of SrCl [2] and SrBr [3] were investigated with Doppler-free laser polarization spectros- copy. Microwave modulated polarization spectroscopy (MMPS), a labeling technique, was helpful for the assignment of the dense spectra. In the study of SrBr several local perturbations were revealed in the spectra and attrib- uted to the interaction of the $A^2\Pi_{1/2}$ and $B^2\Sigma^+$ states. Because of the accurate identification of the perturbed lines it was possible to derive the rotational constant of the so far unknown $A^2\Pi$ state, as well as interaction matrix elements. In an investigation of microwave transitions in an excited state MOPS was applied to the $A^1\Sigma$ state of BaO as a first candidate [4].

3. Molecular Beam Experiments

If the alkaline earth monohalide radicals are produced in a molecular beam, extremely high resolution can be obtained by using the molecular beam, laser-mw, double-resonance technique. A single frequency laser is split into a pump and a probe beam, depleting and probing the population of a specific ground state rotational level in places A and B, respectively, of the molecular beam. Between A and B, microwaves induce a rotational transi- tion repopulating the depleted level and causing an increase in the la- ser-induced fluorescence at B. Microwave transitions are observed with linewidths of 10-50 kHz, mainly owing to time-of-flight broadening. In a cooperation project with the group of Prof. Zare at Stanford we determined the hfs parameters of the $X^2\Sigma^+$ state of BaI this way. This has helped to interpret the hfs in the optical spectra of BaI observed at Stanford.

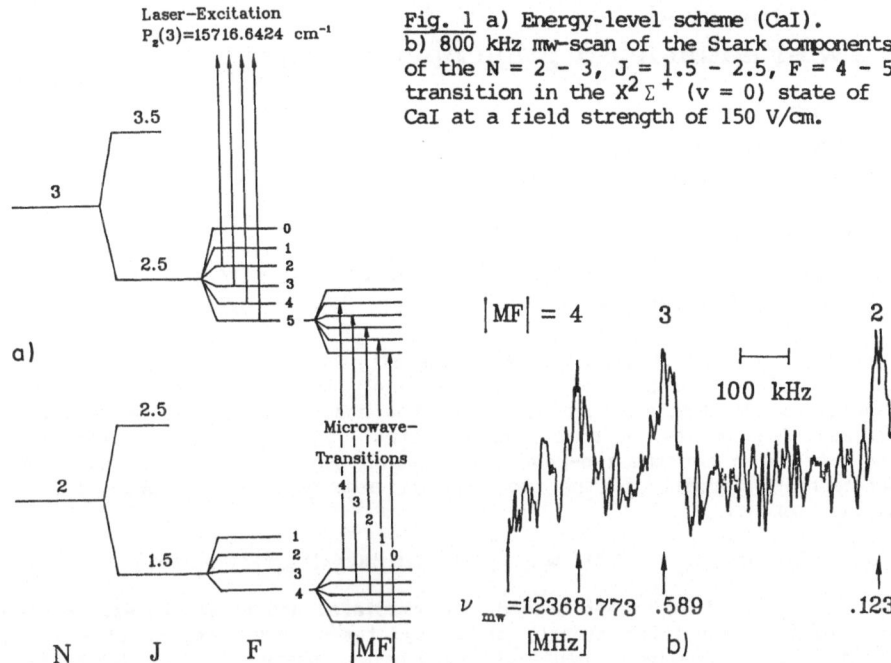

Laser-Excitation
$P_2(3) = 15716.6424$ cm^{-1}

Fig. 1 a) Energy-level scheme (CaI).
b) 800 kHz mw-scan of the Stark components of the N = 2 - 3, J = 1.5 - 2.5, F = 4 - 5 transition in the $X^2\Sigma^+$ (v = 0) state of CaI at a field strength of 150 V/cm.

100 kHz

$|MF| = 4$ 3   2

$\nu_{mw} = 12368.773$ .589   .123
[MHz]   b)

N   J   F |MF|

Microwave-Transitions

Applying an electric field in the mw interaction region, we performed the first precise Stark effect measurements on alkaline earth monohalide radicals [5]. The line splittings and shifts of hfs components were recorded at different electric field strengths. Figure 1a shows a level scheme including the transitions observed in a study of CaI. The laser frequency was fixed to the $P_2(3)$ line of the $B^2\Sigma^+$-$X^2\Sigma^+$ (0,0) band. Microwaves polarized parallel to the electric field-induced transitions with $\Delta M_F = 0$. Three different M_F-components of the F = 4 - 5 transition at a field strength of 150 V/cm are shown in a 800 kHz mw scan in Fig. 1b. The measured Stark shifts were fitted using a diagonalization treatment of the complete energy matrix, and electric dipole moments were determined for CaCl, CaBr, CaI, SrF, and BaI. All values can be well reproduced by our recently developed ionic bonding model [6].

Helpful discussions with Prof. T. Törring and financial support of the Deutsche Forschungsgemeinschaft (Sfb 161) are gratefully acknowledged.

References

1. W.E. Ernst, J.O. Schröder, T. Törring, Chem.Phys.Lett. 109, 175 (1984)
2. W.E. Ernst and J.O. Schröder, J.Chem.Phys. 81, 136 (1984)
3. J.O. Schröder and W.E. Ernst, J.Mol.Spectrosc. 1985 (in press)
4. W.E. Ernst and J.O. Schröder, Phys.Rev. A 30, 665 (1984)
5. W.E. Ernst, S. Kindt, T. Törring, Phys.Rev.Lett. 51, 979 (1983)
6. T. Törring, W.E. Ernst. S.Kindt, J.Chem.Phys. 81, 4614 (1984)

Newly Observed Selection Rule
in Two-Photon Absorption Spectroscopy

B.A. Garetz[1]

Department of Chemistry, Polytechnic Institute of New York,
Brooklyn, NY 11201, USA

C. Kittrell

Department of Chemistry and George R. Harrison Spectroscopy Laboratory,
Massachusetts Institute of Technology, Cambridge, MA 02139, USA

We report the first observation of an *identity-forbidden* transition, between
the $I^1\Sigma^-$ and $X^1\Sigma^+$ states of carbon monoxide. McCLAIN has shown that such
transitions are dipole-allowed only for the absorption of two photons with
different frequencies and polarization states [1]. In the general perturbation expression for the two-photon transition tensor, there are two terms,
corresponding formally to the two possible orderings for the absorption of
the two photons:

$$S_{ab} = \Sigma_k[\langle g|r_a|k\rangle\langle k|r_b|f\rangle/(\omega_{kg}-\omega_1) + \langle g|r_b|k\rangle\langle k|r_a|f\rangle/(\omega_{kg}-\omega_2)], \qquad (1)$$

where g, k, and f represent ground, intermediate, and final states. For an
antisymmetric transition tensor, these two terms have opposite phases, so
that if the photons have identical frequencies, there is *total destructive
interference* between the two terms, and the tensor vanishes [2]. Such is the
case for the CO I-X transition, and a two-color experiment must be performed
to observe it.

The I-X transition was excited using non-identical two-photon excitation
(NTPE). Two ultraviolet laser sources were used: the fourth harmonic of a
Nd:YAG laser at 266 nm and a frequency-doubled Nd:YAG-pumped dye laser near
294 nm. These two beams were focused collinearly from opposite directions
into a cell containing 3 Torr of carbon monoxide. Parallel and perpendicular
polarization configurations were obtained by rotating a multiorder quartz
half-wave plate in the path of the 266 nm beam. The I state is metastable,
so that no direct fluorescence was observed. Collisional energy transfer to
the $A^1\Pi$ state leads to vacuum ultraviolet fluorescence, which was detected
with a solar-blind photomultiplier tube [3]. Some of the undoubled dye-laser
light was passed through an iodine fluorescence cell for wavelength calibration. The signals were amplified and averaged using a boxcar integrator.

The NTPE bandhead region of the 7-0 band of the I-X transition is shown in
Fig. 1. The I-X lines are distinguishable by 1) their characteristic longer
lifetime, due to the collisional mechanism of their observation, and 2) their
sensitivity to the polarization of the excitation sources. Their intensities
are maximized with perpendicular polarization, and they nearly vanish with
parallel polarization. This observed polarization-dependence uniquely
identifies the transition as $\Sigma^+ - \Sigma^-$. Only P and R branches are observed, as
is expected by parity considerations. Preliminary analysis of the I-X (7-0)
band yields a band origin ν_0 of 71,612.1 cm^{-1} and a rotational constant B_v of
1.142 cm^{-1}.

---

[1]Alfred P. Sloan Fellow

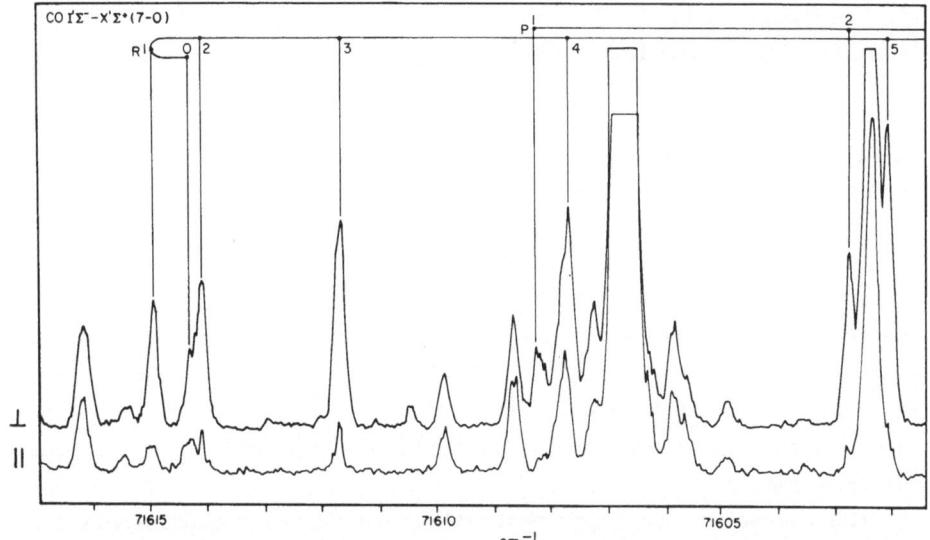

Fig. 1. NTPE spectrum of the bandhead region of the (7-0) band of the $I^1\Sigma^-$ - $X^1\Sigma^+$ transition in CO. The upper trace is with perpendicular polarization; the lower trace with parallel polarization. The two off-scale lines are, from right to left, the P(18) and R(25) lines of the A-X (5-0) band [4].

Raman scattering, also a two-photon process, displays analogous transitions with antisymmetric scattering tensors. These become active only for resonance Raman scattering and exhibit *inverse* polarization, i.e., the polarization of the scattered radiation is rotated through 90° from the incident polarization. Such transitions have been observed by SPIRO and STREKAS in solutions of square-planar porphyrin molecules [5].

In summary, an identity-forbidden transition has been observed for the first time. NTPE provides a means of studying a new class of symmetries (Σ^-, Σ^-_g, A_2, A_{2g}, T_{1g}) of molecular excited electronic states. Many of these states are metastable or dark states, which are likely to act as energy reservoirs in reactive systems, such as those found in atmospheric and interstellar chemistry. In addition, the production, using NTPE, of such states as the $I^1\Sigma^-$ state of CO has possible applications in the study of gas-phase collisional processes and gas-surface interactions, and as intermediates in the study of highly-lying excited states. This work was partially supported by National Science Foundation Grant No. ECS-8218326. Part of this work was performed while B.A.G. was a Visiting Scientist at the Massachusetts Institute of Technology Laser Research Center, which is a National Science Foundation Regional Instrumentation Facility.

References
1. W.M. McClain: J. Chem. Phys. **55**, 2789 (1971)
2. B.A. Garetz and C. Kittrell: Phys. Rev. Lett. **53**, 156 (1984)
3. C. Kittrell, S. Cameron, L. Butler, R.W. Field, and R.F. Barrow: J. Chem. Phys. **78**, 3623 (1983)
4. J.D. Simmons, A.M. Bass, and S.G. Tilford: Astrophys. J. **155**, 345 (1969)
5. T.G. Spiro and T.C. Strekas: Proc. Nat. Acad. Sci. USA **69**, 2622 (1972)

Photodissociation of Na_2
Studied by Doppler Spectroscopy of Atomic Fragments

G. Gerber and R. Möller

Fakultät für Physik, Universität Freiburg,
D-7800 Freiburg, Fed. Rep. of Germany

We report on the photodissociation of neutral diatomic molecules in a molecular beam experiment where all the relevant quantities are well defined. The velocity and angular distributions and the internal states of the atomic fragments were determined by Doppler spectroscopy.

In the course of our experiments three different photodissociation processes were observed and identified: Direct photodissociation by excitation from the X $^1\Sigma_g^+$ ground state to the dissociation continuum of the B $^1\Pi_u$ state, dissociation of quasi-bound levels of the B state and induced two-photon dissociation of Na_2 leading to $Na^*(nl)$ fragments.

The basic experimental set-up consists of a collimated sodium molecular beam crossed by two laser beams, all three being perpendicular to each other. Different Ar^+ laser lines, multi-mode and single mode, were used to dissociate the Na_2 molecules. The fragmentation was analyzed by a single frequency Rh6G dye laser, which was tuned over the Na $3^2P_{3/2}-4^2D_{3/2,5/2}$ atomic transition at 568.82 nm. The velocity component v_z perpendicular to the molecular beam was determined from the Doppler shift $\Delta v = v \cdot v_z/c$. The atomic excitation was monitored by the emission of the $4\,^2P_{1/2,3/2} - 3\,^2S_{1/2}$ line at 330 nm, part of the 4D - 4P - 3S cascade. This detection scheme is shown on the right-hand side of fig. 1 together with the schematic representation of the dissociation of a quasi-bound level by tunneling through the potential barrier of the B $^1\Pi_u$-state. The quantum numbers of the quasi-bound state v'=31, J'=42, excited by a single-mode Ar^+ laser at 457.9 nm, were determined by double-resonance spectroscopy. The linewidth of this state was found to be ≈ 100 MHz corresponding to a lifetime of 1.5 ns.

The Doppler spectrum of the atomic fragments arising from the quasi-bound level v'=31, J'=42 which was excited by a Q-transition

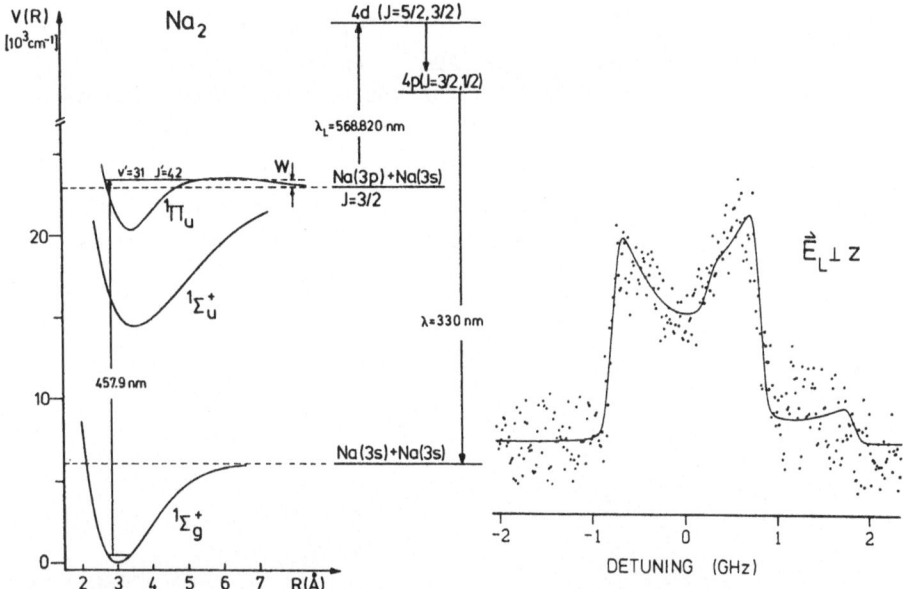

Fig.1   Photodissociation and
Detection Scheme

Fig.2   Doppler Spectrum of the
Na* 3p J = 3/2 Fragments

from v"=8, J"=42 is shown in fig. 2. It can be described by a
distribution calculated with the parameters v=465 m/s (corresponding
to W=400 cm^{-1} kinetic energy) and β = -1, if the angular distribution
is written in the form $f(\vartheta) \sim 1 + \beta \cdot P_2(\cos\vartheta)$ (solid line in
fig.2). Since the two fine-structure components 3 $^2P_{3/2}$ - 4 $^2D_{3/2,5/2}$
of the monitor transitions are only 1.02 GHz apart, their two
overlapping Doppler spectra have to be taken into account. The
obtained value W=400 cm^{-1} ± 20 cm^{-1} agrees well with the energy of
the quasi-bound level above the dissociation limit, E=397.7 cm^{-1}.
The angular distribution $f(\vartheta)$ of the Na* $^2P_{3/2}$ fragments characterized
by β = -1, which is equivalent to $f(\vartheta) \sim \sin^2\vartheta$, is identical to the
anisotropy of the excitation process. This is to be expected, since
for a perpendicular Q(ΔJ=0)-excitation the electric vector $\vec{E}_L$ of
the polarized light is parallel to $\vec{J}$ and the initial anisotropy is
not changed by the rotation about the symmetry axes.

Direct photodissociation from the $X^1\Sigma_g^+$ ground state to the
continuum of the B $^1\Pi_u$ with wavelengths between 450 nm and 490 nm
yields only Na* $^2P_{3/2}$ excited fragments, indicating that there are no
(or very weak) nonadiabatic interactions between molecular states

converging to the Na* 3 $^2P_{3/2}$ and Na* 3 $^2P_{1/2}$ states. Our measurements do not support the reported $^2P_{3/2}$ and $^2P_{1/2}$ fragment emission.

By two-photon dissociation with the 488.0 nm and the 476.5 nm Ar$^+$ laser lines we obtain Na* 3 $^2P_{1/2}$ excited fragments but with completely different angular distribution, kinetic energy and temperature-dependence. The best agreement with our experimental data is found for a $^1\Pi_g$ state dissociating into Na* 3P and Na* 3P atoms.

Intracavity Triple Resonance Spectroscopy in CH$_3$OH FIR Lasers

N. Ioli, A. Moretti, G. Moruzzi, P. Roselli, and F. Strumia

Dipartimento di Fisica dell'Università di Pisa and GNSM del CNR, Sezione di Pisa, Piazza Torricelli 2, I-56100 Pisa, Italy

A very interesting aspect of FIR laser spectroscopy is its contribution to the physical insight of the laser active molecules themselves. In fact, a molecule can display several FIR laser emissions only if its spectrum is dense and extended enough so as to have high coincidence probabilities with the available pump lines (usually CO_2). This usually means serious difficulties in the spectrum assignments, since experimental and computational inaccuracies lead to very frequent ambiguous cases. On the other hand, new line assignments are a continuous challenge to further improvements in the molecular model. The use of optical pumping for the excitation of the FIR lasers allows the direct application of external CW and oscillating fields to the active medium, thus making, for instance, Stark effect and triple resonance (IR-FIR-RF) experiments possible. The study of the stark behavior of a FIR laser emission can give valuable hints for the J and K quantum numbers involved in the laser cycle [1]. Triple resonance experiments have proved to be very important for the study of the A symmetry states of CH$_3$OH [2,3], the richest laser_ active molecule known up to now. Here RF transitions are induced between the components of K- splitting doublets involved in the laser cycle. According to the position of the doublet in the cycle, an RF resonance is detected as a power decrease (common upper level of the pump and laser transitions) or increase (lower level either of the pump or of the laser transition) in the FIR emission, see Fig. 1. Our experimental has been described elsewhere [2]. It consists essentially of a Pyrex waveguide CO_2 pump laser, two waveguide FIR lasers, a sweeper oscillator connected to the FIR lasers via an RF amplifier, a frequency meter, Golay detectors for the FIR output and a lock-in amplifier. Comparison of the resonance RF frequency to the expression

$$\delta(K;J) = \frac{(J+K)!}{(J-K)!} [S(K)+J(J+1)T(K)] \tag{1}$$

leads to prove or disprove the assignment and to a better (by orders of magnitude) determination of the S and T coefficients. Here we present the results for some CH$_3$OH states with torsional quantum number n=0 and CO-stretch quantum number v=0 and 1. Our results for the splittings are shown in Table I and the corresponding best fit S and T parameters in Table II. The S and T coefficients for the v=0, K=3 states are actually obtained from Fourier Transform data (3), since it is obviously not possible to obtain two parameters from the single TR measurement available up to now.

TABLE I

v	K	J	Splitting [MHz]
0	3	27	10527.0 (3)*
0	4	25	426.5 (5)
0	4	26	577.5 (5)
1	3	24	5532.2 (2)*
1	3	25	7001.8 (2)*
1	3	27	10894.9 (3)*
1	3	28	13410.4 (2)*
1	4	14	6.3 (3)
1	4	23	209.8 (3)
1	4	24	283.2 (2)
1	4	25	370.6 (2)

TABLE II

v	K	S [MHz]	T [MHz]
0	3	-2.610×10^{-5}	2.24×10^{-9}
0	4	2.635×10^{-9}	-2.69×10^{-13}
1	3	-2.786×10^{-5}	3.15×10^{-9}
1	4	3.725×10^{-9}	-2.20×10^{-12}

* new measurements

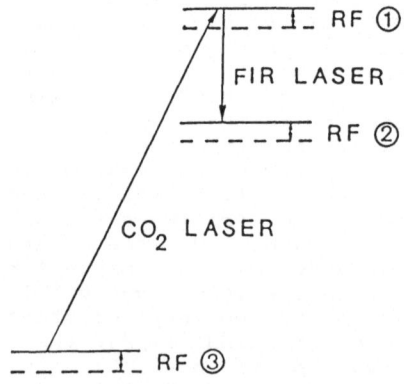

Fig.1: Laser cycle and possible positions of RF resonances

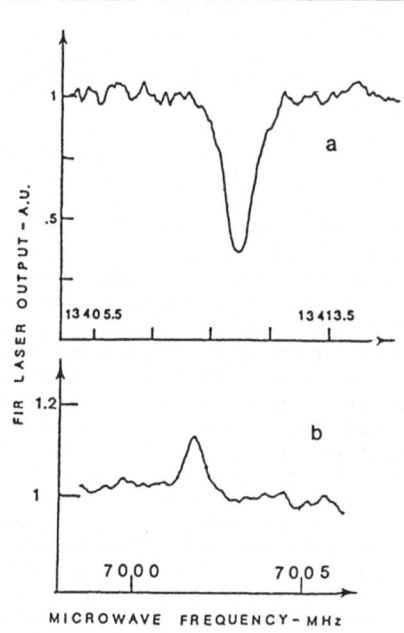

Fig.2: Triple resonance signals
a) RF resonance in level 1
b) RF resonance in level 2
These signals refer to v=1, K=3
J=28 and 25, see Table I

However, the agreement between these parameters and the measurement of row 1 of table I is well within the experimental errors.

1) M.Inguscio, A.Moretti, G.Moruzzi, F.Strumia; Int. J. Infrared and mm Waves, 2, 943 (1981)
2) N.Ioli, A.Moretti, G.Moruzzi, P.Roselli, and F.Strumia J. Molec. Spectrosc., 105, 284-298 (1984)
3) N.Ioli, A.Moretti, F.Strumia and I.Longo; Optics Lett. July 1985
4) G.Moruzzi, F.Strumia, F.Colao; Infrared Phys., 25, 251-3 (1985)

Multiphoton Ionization of NO-Rare Gas van der Waals Species

J.C. Miller

Chemical Physics Section, Health and Safety Research Division,
Oak Ridge National Laboratory*, Oak Ridge, TN 37831, USA

To date, most spectroscopic studies of electronic states of clusters involving small molecules and rare gas (RG) atoms have employed laser-induced fluorescence techniques. Consequently, such studies (e.g., Levy and co-workers on I_2-RG species [1]) have been limited to the lowest lying electronic states, which are usually valence in character. Furthermore, the identification of the cluster often must be inferred from spectral shifts, pressure dependence, etc.

We have applied the technique of multiphoton ionization (MPI) to the study of electronic spectroscopy of NO-RG van der Waals species and extended the previous experiments of SATO et al. [2]. These works show several of the advantages of the MPI technique relative to the LIF method. Specifically, mass analysis can provide identification of the new species formed, as well as allowing discrimination against other species (the parent monomer or higher clusters) while recording spectra. Furthermore, the use of MPI photoelectron spectroscopy can provide data on the states of the cluster ion as well [2].

The MPI technique has another advantage which is particularly important in the study of van der Waals molecules. This is the ability to study Rydberg states much more easily than LIF or other techniques. Because most excited states are highly polarizable, van der Waals forces are enhanced. Furthermore, for high Rydberg states, the influence of the positively charged core allows the stronger dipole forces to become more important contributors to the binding energy. Thus, the binding energy of molecule-RG complexes should increase in the order: ground state < valence state < Rydberg state < ionic ground state. Nitric oxide is the "alkali atom" of molecules, having one unpaired electron outside of filled molecular orbitals. Consequently, its spectroscopy is dominated by Rydberg transitions which have been very well characterized.

The apparatus is a modified version of that used previously for MPI mass analysis [3] and MPI photoelectron spectroscopy [4]. The apparatus consists of a pulsed supersonic nozzle, a short, low-resolution time-of-flight mass spectrometer and an excimer-pumped dye laser. Typically, the laser (1-5 mJ) is focused by a 38-mm lens into the gas pulse (5% NO in Ar at a total backing pressure of 2 atm) after a variable delay relative to the nozzle opening. The time (and hence mass) gated ion signal is averaged with a boxcar integrator, and recorded as a function of wavelength. Mass spectra showing the production of $NeNO^+$, $ArNO^+$ and $KrNO^+$ are shown in Fig. 1.

MPI spectra are presented for several NO-RG van der Waals molecules. Structured excitation spectra of the ArNO cluster associated with the $B^2\pi$

*Operated by Martin Marietta Energy Systems, Inc. under contract DE-AC05-84OR21400 with the U.S. Department of Energy.

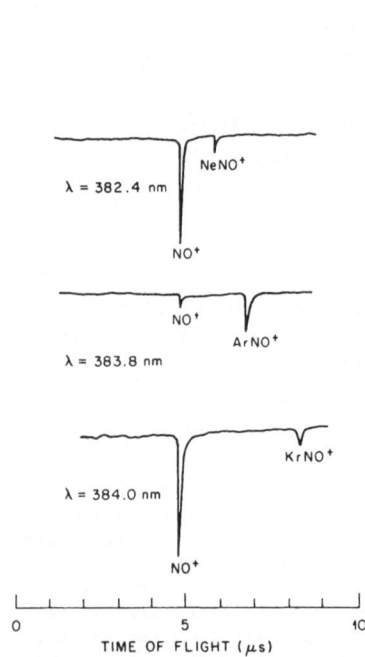

Fig. 1. Time-of-flight mass spectra showing the MPI detection of NeNO, ArNO, and KrNO van der Waals molecules. The figure is traced from photographs of the oscilloscope image.

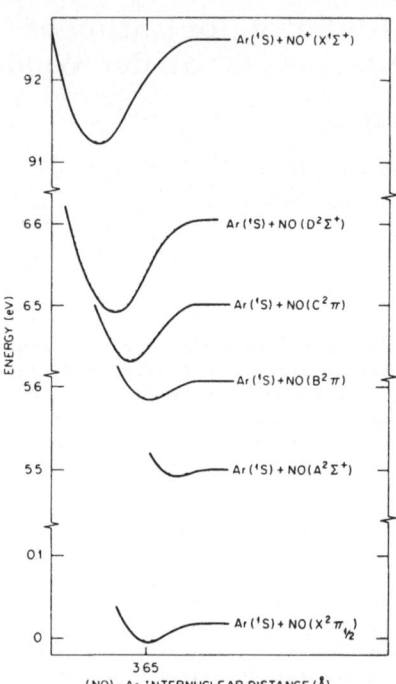

Fig. 2. Schematic representation of the potential energy curves for ArNO.

(v=0), $C^2\pi$ (v=0,1,2,3), and $D^2\Sigma^+$ (v=0) states of NO are observed as well as an unstructured, presumably dissociative, spectrum near the $A^2\Sigma^+$ (v=0) state. The C state was also observed for NeNO and KrNO. The B state represents a valence transition and the A, C, and D states are Rydberg states. Details of these studies may be found in a recent paper [5].

From the observed spectral shifts and analysis of the spectra, excited state vibrational constants and bond dissociation energies can be extracted. These parameters are summarized in Fig. 2 for ArNO.

References

1. D. H. Levy in **Photoselective Chemistry** (I. J. Jortner, R. D. Levine, and S. A. Rice, Eds.) Advances in Chemical Physics **XLVII**, 323 (1981).
2. K. Sato, Y. Achiba, and K. Kimura, J. Chem. Phys. **81**, 57 (1984).
3. C. D. Cooper, A. D. Williamson, J. C. Miller, and R. N. Compton, J. Chem. Phys. **73**, 1527 (1980).
4. J. C. Miller and R. N. Compton, J. Chem. Phys. **75**, 22 (1981); J. Chem. Phys. **75**, 2020 (1981); Chem. Phys. Lett. **93**, 453 (1982).
5. J. C. Miller and W. C. Cheng, J. Phys. Chem. **89**, 1647 (1985).

New Identifications of Highly Excited States of the H$_2$ Molecule Accessible by Multiphoton Spectroscopy

P. Senn, P. Quadrelli, and K. Dressler

Physical Chemistry Laboratory, ETH-Zentrum, CH-8092 Zürich, Switzerland

L. Wolniewicz

Institute of Physics, N. Copernicus University, PL-87-100 Toruń, Poland

The first excited singlet gerade state of H$_2$, EF $^1\Sigma_g^+$, has a double-minimum potential function arising from an avoided crossing of the E(1sσ 2sσ) and the F(2sσ)^2 configurations [1]. This state can be selectively populated through two-photon absorption processes as first demonstrated by KLIGLER and RHODES [2]. This fact has been exploited in various three-photon double-resonance excitation schemes which recently have led to the discovery of previously unobserved rotation-vibration levels in the EF state [3], and also have provided novel information about autoionizing and field-ionizing Rydberg levels of $^1\Sigma_u^+$ and $^1\Pi_u$ symmetry [4]. Emissions following the excitation of highly excited *gerade* states have been used to determine radiative and collisional electronic relaxation rates [5].

The most extensive analysis of the energy levels in H$_2$ has been reported by DIEKE [6]. His analysis was based mainly on high resolution spectra from electric discharges [7]. We have previously reinterpreted some of Dieke's findings in the light of nonadiabatic *ab initio* calculations [8] and we are presently engaged in a search for unobserved energy levels of *gerade* symmetry. This search has already yielded the spectroscopic identifications of the hitherto unknown lowest levels localized in the outer minimum, F,v=0 and v=1 of the EF state [9] and of the levels EF,v=14 to v=19 above the maximum of the EF potential curve which interact with the lowest levels of the GK$^1\Sigma_g^+$ state. Most of the newly discovered levels are shown in Table I.

Our search is based mostly on the original emission spectra listed by DIEKE [7] supplemented by recent measurements in the infrared [9,10]. This analysis also profited from the recent accurate determinations of the *ungerade* energy levels of H$_2$ by DABROWSKI [11]. In addition, the present analysis draws on virtually every available source of information, such as calculated intensity distributions [12]. We also expect as an outcome of major importance a deeper understanding of the effects responsible for the remaining discrepancies between *ab initio* calculations and experiment.

1. W. Kołos and L. Wolniewicz: J. Chem. Phys. 50, 3328 (1969)
2. D.J. Kligler and C.K. Rhodes: Phys. Rev. Lett. 40, 309 (1978)
3. E.E. Marinero, R. Vasudev, and R.N. Zare: J. Chem. Phys. 78, 692 (1983)
4. N. Bjerre, R. Kachru, and H. Helm: Phys. Rev. A 31, 1206 (1985)
5. R.L. Day, R.J. Anderson, and F.A. Sharpton: J. Chem. Phys. 71, 3683 (1979)
6. G.H. Dieke: J. Molec. Spectrosc. 2, 494 (1958)
7. H.M. Crosswhite: The Hydrogen Molecule Wavelength Tables of Gerhard Heinrich Dieke (Wiley, New York 1972)
8. K. Dressler, R. Gallusser, P. Quadrelli, and L. Wolniewicz: J. Molec. Spectrosc. 75, 205 (1979); cf. also M. Glass-Maujean, P. Quadrelli and

K. Dressler: J. Chem. Phys. <u>80</u>, 4355 (1984)
9. P. Senn, P. Quadrelli, K. Dressler, and G. Herzberg: J. Chem. Phys., in press (<u>83</u>, 1985)
10. G. Herzberg: Private communication
11. I. Dabrowski: Can. J. Phys. <u>62</u>, 1639 (1984)
12. M. Glass-Maujean, P. Quadrelli, and K. Dressler: Atom. Data Nucl. Data Tabl. <u>30</u>, 273 (1984)

Table I. Upper entry: Spectroscopic term value T. Lower entry: $T_{n.a.}-T$, where $T_{n.a.}$ is the nonadiabatic *ab initio* result. All values in units of cm^{-1}.

J	F,v=0	F,v=1	F,v=2	F,v=3	EF,v=14
0	99363.92 1.34	100558.92 1.43	101698.93 1.58	102778.28 1.69	108793.55 5.11
1	99376.04 1.47	100570.81 1.55	101710.80 1.63	102790.09 1.80	108814.84 5.07
2	99400.52 1.48	100594.82 1.53	101735.03 1.63	102813.85 1.82	108857.39 5.11
3	99437.16 1.54	100630.71 1.58	101768.53 1.68	102849.33 1.86	108921.60 5.07
4	99485.98 1.58	100678.52 1.63	101816.10 1.73	102896.49 1.86	109007.77 5.13
5	99546.83 1.69	100738.19 1.65	101875.04 1.58	102955.12 1.96	109116.60 5.21

J	EF,v=15	EF,v=16	EF,v=17	EF,v=18	EF,v=19
0	109493.90 5.73	110163.38 5.82	110794.19 5.22	111370.69 7.79	112106.09 6.12
1	109514.70 5.70	110185.12 5.84	110815.23 5.22	111387.20[a] 7.69	112126.21[b] 6.02
2	109555.93 5.68	110228.21 5.91	110857.53 5.31	111420.72 7.59	112167.83 6.35
3	109617.12 5.61	110291.91 5.95	110921.66 5.37	111472.60 7.42	112230.94 6.67
4	109697.67 5.55	110374.77 6.07	111007.66 5.55	111544.97 7.31	112315.97 7.30
5	109797.51 5.43	110475.26 6.03	111114.66 5.78	111641.89 6.99	112421.49 8.28

[a]This is the V0,J=1 energy level reported in ref. 7 (cf. ref. 8 above)
[b]This is the V1,J=1 energy level reported in ref. 7 (cf. ref. 8 above)

Predissociation in the $B^3\Pi_{0^+}(\nu'=3)$ State of ICl: Study Using a Sub-Nanosecond Laser Fluorometer

T. Suzuki and T. Kasuya

Institute of Physical and Chemical Research,
Hirosawa, Wako, Saitama 351-01 Japan

The laser-selected single-rovibronic study was performed on the $B^3\Pi_{0^+}$ – $X^1\Sigma^-$ system of ICl with interest in the predissociative interaction of the bound B with the repulsive 0^+ state[1-4]. The interaction is certainly reflected on the absorption spectra, but it is hardly analyzed from the absorption line profiles because of the substantial influence of unresolved iodine hyperfine structure. The measurement of fluorescence decay is, therefore, essentially required. We observed fluorescence from the highly predissociative $\nu'=3$, $B^3\Pi_{0^+}$ level, which has never been available due to its serious diffuseness[3], and measured single-rovibronic decay-rate with the subnanosecond time resolution[5].

The experimental apparatus is as follows[6]. A nitrogen-laser-pumped dye laser(2 ns pulse duration, 0.5 GHz bandwidth, and 30kW peak power) is used as a light source. Time-profiles of fluorescence decay are recorded with a transient digitizer(Tektronix, 7912AD) controlled by a micro-computer(DEC, V-03). The fluorescence light from sample cell is detected with a photomultiplier of 2.8 ns rise time. The waveform data of fluorescence decay are averaged over 64 laser shots. The fluorescence sample cell is then replaced by the reference cell, which contains cigarette smoke as light scatterer. The procedure is repeated for the scattered laser light to find the instrumental response. As the duration of excitation pulse and the response time of the detection system are of comparative order of magnitude to the fluorescence decay-time, the observed fluorescence profiles are fitted to the convoluted integral waveform by a least-squares analysis. A typical example of the time profile of the fluorescence decay is shown in Fig. 1. The lifetime measurement in the range of 4.2 - 0.2 ns was calibrated to ±0.05 ns against the fluorescence decay of rhoda-

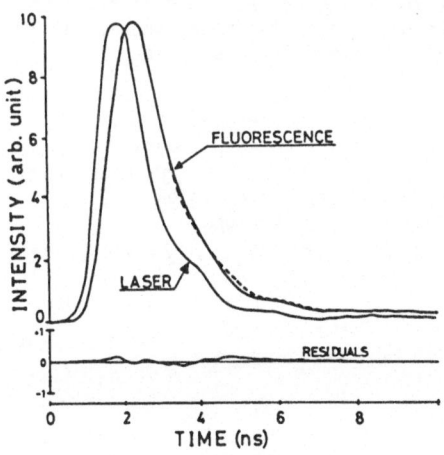

Fig. 1. Time profile of fluorescence decay of ICl. Superposed on the observed fluorescence decay(dotted curve) is the least squares fit(solid curve) to the experimental data. The bottom trace describes the residuals between the calculated and the experimental data

155

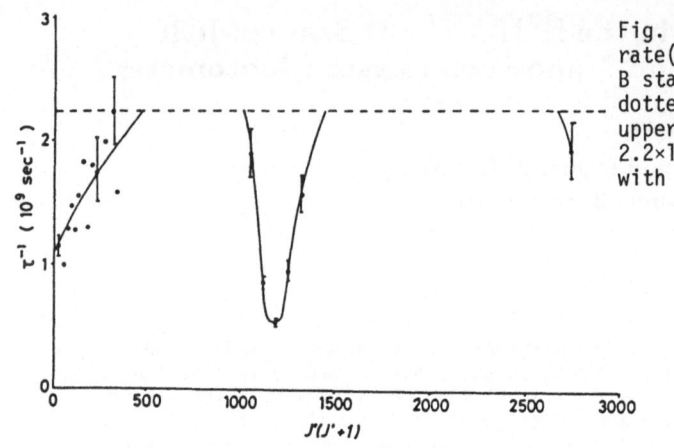

Fig. 2. Collisionless decay rate($1/\tau_0$) for v'=3 of the B state of $I^{35}Cl$. The dotted line indicates the upper limit of decay-rate, $2.2\times10^9s^{-1}$, measurable with our present system

mine 6G solution with pottasium iodide as fluorescence quencher[6]. The decay rates extrapolated to zero pressure are plotted in Fig. 2 as a function of $J'(J'+1)$. Since levels in v'=3 should decay radiatively at nearly the same rate as v'=1 and 2, the remarkable shortening of lifetime observed for v'=3 certainly reflects the predissociation of the $B^3\pi_0{}^+$ state.

Our present experiment offers new findings in the following respects. The longest observed lifetime of 2 ns for the J'=34 level in v'=3 is ten times longer than that expected from the absorption data [4]. The shorter lifetime at lower J' means that the predissociation is more pronounced at lower J' levels. Gordon et $al.$[4] considered a heterogeneous(rotation-orbit) interaction between the B and some Ω = 1 states in their qualitative interpretation of the decay-rate variation. The linewidth caused by the heterogeneous interaction is proportional to $J'(J'+1)$. The proposed interaction is inconsistent with our observed results. The observed v'=3 levels lie below the homogeneous B-Y crossing, and are little influenced by this intersystem interaction[2]. Both the heterogeneous and homogeneous interactions cited above predict a weaker perturbation at lower J' as opposed to our observed results. While, the observed short lifetimes at low J' levels and the oscillatory J' dependence of the decay rates are very suggestive of a homogeneous interaction below the energy of J'=34. Furthermore, we have observed several new fragmentary bands which are not ascribed to the hitherto proposed adiabatic potential. The curve-crossing model should, therefore, be revised for the overall behavior to be fully explained.

1. W.G. Brown and G.E. Gibson: Phys. Rev., 40, 529 (1932)
2. M.S. Child and R.B. Bernstein: J. Chem. Phys., 59 5916 (1973)
3. M.A.A. Clyne and I.S. McDermid: J. Chem. Soc. Faraday 2, 73, 1094 (1977)
4. R.D. Gordon and K.K. Innes: J. Chem. Phys., 71, 2824 (1979)
5. T. Suzuki and T. Kasuya: J. Chem. Phys., 81, 4818 (1984)
6. T. Suzuki, K. Yamada and T. Kasuya: Jpn. J. Appl. Phys., 22, 522 (1983)

Observation of Accidental Infrared Double Resonance Transitions in PH_3 Using a CO_2 Laser and Their Application to Electric Field Standards in Laser Stark Spectroscopy

K. Takagi

Department of Physics, Toyama University, Gofuku, Toyama 930, Japan

T. Tanaka and K. Tanaka

Department of Chemistry, Faculty of Science, Kyushu University 33, Hakozaki, Higashiku, Fukuoka 812, Japan

The electric field in the CO_2 laser Stark spectroscopy is usually calibrated by using resonant electric fields of CH_3F[1] as standards. However, the inaccuracy of resonant fields due to the frequency instability of the laser causes the largest error in the calibration.

During the study of the $^qP(5,3)$ line in the ν_2 fundamental band of PH_3 by laser Stark spectroscopy using the 10P(18) CO_2 line, three unexpected Doppler-free lines were observed at the Stark field of $\sim$3000V/mm[2], which are shown in Fig.1 as a, b, and c. In this observation two counter-propagating laser beams were overlapped in a Stark cell placed outside the laser cavity. Lines a, b, and c have the same sign as that of the usual absorption (the opposite sign to that of the Lamb dips, shown as 1, 2, and 3 in Fig.1). The resonant voltage of line a depends on the laser frequency, whereas those of b and c are independent of the laser frequency-like level crossing signals.

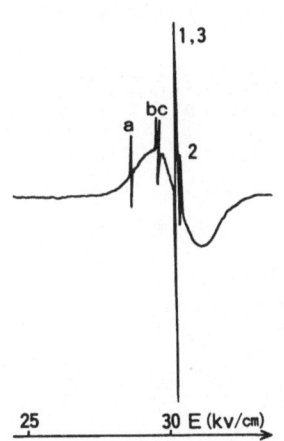

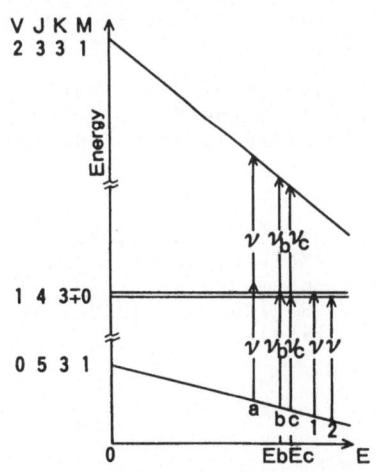

Fig.1 Laser Stark spectrum of ν_2 $^qP(5,3)$ near the Stark field of 3000V/mm using the CO_2 10P(18) line. Lines 1-3 are Lamb dips (Lines 1 and 3 are overlapped)

Fig.2 Energy levels for $(\nu_2,J,K, M) = (2,3,3,1)$, $(1,4,3,\mp0)$ and $(0,5,3,1)$ against applied electric field and assignment of the observed signals. ν is the laser frequency. Line 3 is not shown

They are explained by the accidental overlapping of the ^qP (5,3) line in the ν_2 fundamental band and the $^qP(4,3)$ line in the $2\nu_2 - \nu_2$ hot band. Figure 2 shows the assignment of the signals a, b, c, 1 and 2. Line a is a Doppler-free two photon absorption. Lines b and c are the double resonance lines when the frequencies of the transitions $\nu_2 \leftarrow 0$ (J,K,M) = (4,3,$\mp$0) $\leftarrow$ (5, 3,1) are equal to those of $2\nu_2 \leftarrow \nu_2$ (J,K,M) = (3,3,1) $\leftarrow$ (4,3,$\mp$0), where M = ±0 denotes the M = 0 components of the A_1A_2 doublet. Lines 1 and 2 are fundamental band lines. The frequency difference between the laser and transition b or c is about 20 MHz, which is smaller than the half Doppler width (30 MHz). The double resonance Stark fields depend only on the level structure of the molecule and not on the laser frequency.

The double resonance fields were measured precisely using the Stark Lamb-dip and laser microwave double resonance (LMDR) spectrometer reported already [3]. The observed linewidth was about 0.60 V/mm when the laser power was reduced to 15 mW. The resonant electric fields were calibrated using the LMDR signal of OCS[3]. The LMDR measurement was performed immediately before and after each run of the observation of the PH_3 double resonance signals. The effective spacings between the Stark plates obtained from the preceding and following LMDR measurements were averaged and used to determine the resonant electric fields of the PH_3 signals. The experiment was repeated on three different days within a week and the whole procedure was followed again a month later. The results of the measurements show a good reproducibility. The average values of the resonant electric fields are

$$b: 2965.556 \pm 0.03 \text{ V/mm}$$
$$c: 2972.408 \pm 0.03 \text{ V/mm}.$$

We conclude that the double resonance signals of PH_3 are conveniently used as precise standards for the electric field calibration in CO_2 laser Stark spectroscopy.

This work was supported in part by Grant-in-aid for Scientific Research from the Ministry of Education, Science and Culture and by Shimadzu Science Foundation.

References

1. S. M. Freund, G. Duxbury, M. Römheld, J. T. Tiedje, and T. Oka, J. Mol. Spectrosc. 52, 38 (1974).
2. K. Takagi, Chem. Phys. Letters 112, 302 (1984).
3. K. Tanaka, H. Ito, K. Harada, and T. Tanaka, J. Chem. Phys. 80, 5893 (1984).

Vibrational Energy Transfer Study of Benzene by IR-UV Double Resonance

X.M. Wang, G.Y. Wu, P.X. Ding, J.D. Chen, W.B. Gao, H.T. Ji, and S.H. Liu

Anhui Institute of Optics and Fine Mechanics, Academia Sinica, P.O. Box 25, Hefei, People's Republic of China

The collision-induced vibrational energy transfer of Benzene (C_6H_6) was studied by means of infrared-ultraviolet double resonance. The time-dependent population of vibrational mode ν_{18} was measured after excitation of vibrational mode ν_4. The rate constant for V-V energy transfer of $\nu_{14} \rightarrow \nu_{18}$ was obtained. The experiment was also carried out using the rare gases as the collision partner. The simplified model for the vibrational energy transfer was discussed.

The molecular formation of C_6H_6 belongs to point group D_{6h} with a high symmetry, and has been extensively studied. It has 20 fundamental vibrational modes and only 4 modes of them are IR active. The vibrational mode ν_{18} is IR inactive. The V-T energy transfer of C_6H_6 have been previously studied by ultrasonic absorption [1] and the measurements near room temperature gave relaxation time of about $p\tau$=50nsatm [2] . A partial energy level scheme of the vibrational levels in the electronic ground state and first electronic excited singlet state of C_6H_6 is shown in Fig. 1.

The line-tunable TEACO$_2$ laser was used and operated on either P(30) or P(28) line in 9.6 μm band to excite the ν_{14} mode of C_6H_6 ($^1A_{1g}$). The probing light was provided by a high pressure Hg lamp (at λ=2537Å) which is resonant with $18_1^2 \, 2_0^1$ absorption line in $^1A_{1g} - ^1B_{2u}$ UV band. The IR and UV beams were adjusted in the sample cell coaxially.

IRUVDR signals were observed following excitation of the ν_{14} mode of C_6H_6 by a TEACO$_2$ laser. After exciting a signal corresponding to increasing absorption was produced, caused by the populating of the monitored state. The sample benzene, with a purity of 99.9% and the rare gases, with a purity of 99.99%.

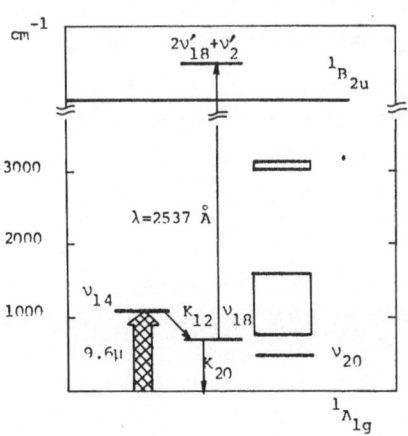

Fig. 1 The simplified schematic of vibronic level of C_6H_6

The rate constants obtained are: $K^{12}_{C_6H_6-C_6H_6}=87.4\pm 5.2ms^{-1}torr^{-1}$, $K^{20}_{C_6H_6-C_6H_6}$ $=20.7\pm 2.5ms^{-1}torr^{-1}$, $K^{20}_{C_6H_6-Ne}=3.1\pm 0.4ms^{-1}torr^{-1}$, $K^{20}_{C_6H_6-Ar}=2.4\pm 0.3ms^{-1}torr^{-1}$ were obtained.

The probability that a polyatomic molecule will change its vibrational qantum state during a collision can be calculated using modified SSH theory, including the breathing sphere approximation [3,4] . The SSH probability is given by

$$P_{ij}=P_0(a)P_0(b)Q_j|U_{ij}|^2 I(\Lambda E, T, \mu, L, \varepsilon) \tag{1}$$

The formula (1) was used to calculate the probabilities for V-V and V-T energy transfers with various values of L and colliders. The calculated probabilities for the processes are as follows:

$$C_6H_6(\nu_{14}) + C_6H_6(0) \longrightarrow C_6H_6(\nu_{18}) + C_6H_6(0) + 432cm^{-1} \tag{2}$$

and

$$C_6H_6(\nu_{14}) + C_6H_6(0) \rightleftharpoons C_6H_6(\nu_{18}) + C_6H_6(\nu_{20}) + 30cm^{-1} \tag{3}$$

The deactivation of vibrational state ν_{18} can be accomplished through the following processes:

$$C_6H_6(\nu_{18}) + M \longrightarrow C_6H_6(\nu_{20}) + M + 202cm^{-1} \tag{4}$$

$$C_6H_6(\nu_{20}) + M \longrightarrow C_6H_6(0) + M + 404cm^{-1} \tag{5}$$

$$C_6H_6(\nu_{18}) + M \longrightarrow C_6H_6(0) + M + 606cm^{-1} \tag{6}$$

where M are C_6H_6 or rare gas atoms and processes (4) and (5) constitute a cascade process.

The experimental result of rate constant $K^{V-T}_{C_6H_6-C_6H_6}=20.7\pm 2.5ms^{-1}torr^{-1}$ is in good agreement with Cheng [2] and the following conclusions have been obtained:

1. The vibrational mode ν_{18} is excited by a near-resonant energy transfer $\nu_{14}\longrightarrow\nu_{18} + \nu_{20}$.

2. The vibrational mode ν_{18} is deactivated through two channels, one of which is V-T/R process from ν_{18} and the other is a multiple process consisting of V-V ($\nu_{18}\longrightarrow\nu_{20}$) and V-T/R from ν_{20} process. In the V-T/R processes, when He, Ne as colliders, the V-T processes are more important than the V-R processes, when Ar, Kr, Xe and C_6H_6 as colliders, the V-R processes are more important than V-T processes.

References
1. T. L. Cottrell and J. C. Mccoubrey, Energy Transfer in Gases (Butter worths, London, 1961)
2. L. M. Cheng, J. Chem. Phys., 19, 693 (1951)
3. R. N. Schwartz, Z. I. Slawsky and K. F. Herzfeld, J. Chem. Phys., 20, 1591 (1952)
4. F. I. Tanczos, J. Chem. Phys., 25, 439 (1956)

Part VI

VUV and X-Ray,
Sources and Spectroscopy

Quasi-Metastable Energy Levels and Applications [1]

S.E. Harris, J.F. Young, A.J. Mendelsohn, D.E. Holmgren [2], *K.D. Pedrotti, and D.P. Dimiduk*

Edward L. Ginzton Laboratory, Stanford University,
Stanford, CA 94305, USA

The paper summarizes the properties of certain radiating levels of the column I metals. XUV emission studies and an absorption experiment on the 1091 Å line of neutral Cs are described. A technique for correlating pico-second, broadband XUV pulses is discussed.

1. Introduction

In general, core-excited energy levels of atoms and ions which lie above a lower continuum have autoionizing rates of between 10^{12} and 10^{14} sec^{-1}; and therefore, in the extreme ultraviolet (XUV) portion of the spectrum have very poor radiative yields. Recently, while using the RCN/RCN atomic physics code [1] to study the alkali atoms, we noted the presence of a number of important exceptions, that is of levels which autoionize slowly and have radiative yields of about 50%.

These radiating levels are quartet levels, which have the additional distinguishing property that the spin-orbit selection rules only allow non-zero matrix elements to doublet basis levels which are themselves prohibited, by simultaneous conservation of parity and angular momentum, from autoionizing [2]. For alkali atoms and alkali-like ions, the condition for quasi-metastability is that a level have both odd or even parity and orbital angular momentum, that $|J-L| = 3/2$, and that the level not be a pure (highest possible J) quartet [2]. We have termed levels of this type as quasi-metastable.

The insert of Fig. 1 is an energy level diagram showing the neutral $Cs(5p^55d6s)^4P_{5/2}$ level. This level has calculated autoionizing and radiating rates of 1.0 x 10^8 and 5.4 x 10^7, respectively. It is the lowest level in the quartet manifold, and radiates predominantly to the $(5p^65d)^2D_{5/2}$ valence level. Figure 1 also shows the emission spectrum [3] of neutral Cs taken in a pulsed hollow-cathode discharge. The 1091 Å emission from $(5p^55d6s)^4P_{5/2}$ dominates the neutral spectrum, and has an intensity equal to about one part in ten of the strong Xe-like ion lines.

2. Recent Experiments

During the last year we have performed two experiments to delineate and clarify the properties of the alkali quasi-metastable levels. These were an absorption experiment [4] in neutral Cs and a microwave emission study [5] of Na, K, Rb and Cs.

[1] The work described here was supported by the Army Research Office and the Air Force Office of Scientific research.

[2] Now at the Institute of Optics, University of Rochester, Rochester, New York 14627.

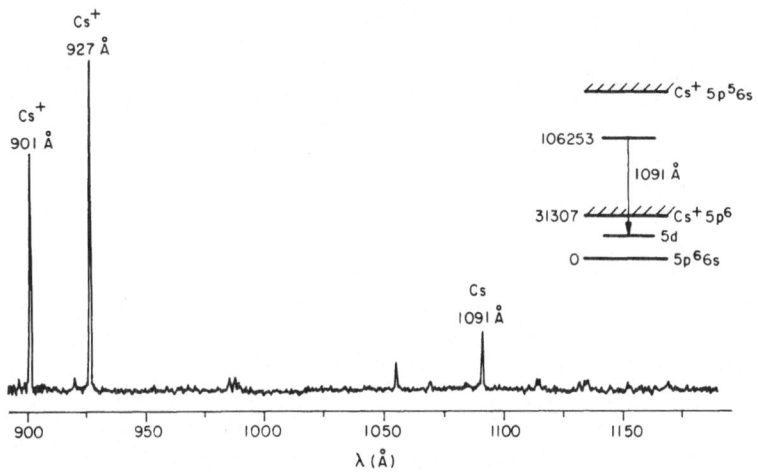

Fig. 1. Partial energy level diagram and emission scan of neutral Cs.

A. 1091 Å Absorption in Neutral Cs

To ascertain the identity of the 1091 Å emission and also to measure its oscillator strength, we have generated tunable XUV radiation and measured the absorption of neutral Cs in a pulsed hollow cathode. The tunable XUV radiation was generated in Zn vapor by the process $\omega_{XUV} = 2\omega_{UV} + \omega$, where ω_{UV} = 2712 Å and ω = 5583 Å . The hollow-cathode discharge populated the valence levels of the atom, in particular both fine structure components $(5p^{6}5d)^{2}D_{5/2}$ and $(5p^{6}5d)^{2}D_{3/2}$. The identification of 1091 Å was established by the verification of absorption on both fine structure components and by the tracking of the absorption with the measured $(5p^{6}5d)^{2}D_{5/2}$ population. The measured oscillator strength on the $^{2}D_{5/2}$ component is f = 0.003 , which agrees to within a factor of two to that of the RCN/RGN code.

B. Microwave Emission Scans

We have used a high pulsed power (500 kW) x-band magnetron to take the emission spectrum of Na, K, Rb and Cs. The results of this study are summarized in Table 1. In each case, the emission of the quasi-metastable dominates the neutral spectrum. As in the earlier work [3] of Holmgren on Cs, the observation of the fine structure components in emission gives credence to the identity of the observed lines. We note the early work of Aleksakhin, et al. in the identification of many of these lines [6].

3. Applications of Quasi-Metastable Levels

A. Spectroscopy

In each of the alkali atoms except Li, the lowest level of the quartet manifold is a quasi-metastable level, and radiates in the XUV. Using tunable visible lasers, other core-excited levels of both the doublet and quartet manifold may be positioned relative to the quasi-metastable level, thereby positioning all of these levels with respect to ground. We use two

163

Table 1. Observed Quasi-Metastable Emissions

Element	Upper Level	Lower Level	λ (nm)	Upper Level Energy (cm^{-1})
Na	$2p^53s3p\ ^4S_{3/2}$	$2p^63p\ ^2P_{3/2,1/2}$	40.52	263789 ± 120
K	$3p^54s4p\ ^4S_{3/2}$	$3p^64p\ ^2P_{3/2}$	67.39	161426 ± 60
		$3p^64p\ ^2P_{1/2}$	67.36	
	$3p^54s3d\ ^4F_{3/2}$	$3p^63d\ ^2D_{3/2,5/2}$	69.17	166092 ± 60
	$3p^54s3d\ ^4P_{5/2}$	$3p^63d\ ^2D_{3/2,5/2}$	72.10	160227 ± 60
Rb	$4p^55s5p\ ^4S_{3/2}$	$4p^65p\ ^2P_{3/2}$	82.37	134220 ± 40
		$4p^65p\ ^2P_{1/2}$	82.20	
		$4p^66p\ ^2P_{3/2,1/2}$	90.53	
	$4p^55s4d\ ^4P_{5/2}$	$4p^64d\ ^2D_{5/2,3/2}$	85.18	136756 ± 40
Cs	$5p^56s5d\ ^4P_{5/2}$	$5p^65d\ ^2D_{5/2}$	109.10	106256 ± 30
		$5p^65d\ ^2D_{3/2}$	108.98	
		$5p^66d\ ^2D_{5/2,3/2}$	119.58	
		$5p^67d\ ^2D_{5/2,3/2}$	124.72	

techniques: In the first, demonstrated by Holmgren, et al. [7] population stored in the quasi-metastable $(2p^53s3p)^4S_{3/2}$ and the $(2p^53s3p)^4D_{7/2}$ levels of Na was transferred to other quartet levels.and the resulting fluorescence was observed. This has allowed the identification of thirty new transitions.and the development of a partial Grotrian diagram of neutral Na. In the other technique, quasi-metastable XUV radiation will be monitored while tuning a visible laser to access other core-excited levels. As these levels are accessed, the XUV emission will be reduced, thereby giving the position and linewidth, and thereby the autoionizing time, of these levels.

B. Correlation of Picosecond X-Ray Pulses

There is now interest in generating ultrashort pulses of incoherent XUV radiation, either by spontaneous anti-Stokes [8] scattering or by direct generation in laser plasmas. The quasi-metastable levels of column II provide a means of measuring the pulse length of such radiation. The idea is shown in Fig. 2. Broadband incoherent XUV radiation will photoionize Ca, thereby producing $(3p^63d^2)^2P$ ions. These autoionize in about 1 ps. A picosecond tunable laser will transfer these ions to a radiating quasi-metastable level only when the broadband XUV pulse and tunable laser pulse overlap in time. For the Ca system of Fig. 2, the calculated saturation energy for the visible transfer laser is 7.5 mJ/cm^2. The energies and transfer wavelengths in this figure are code values. We expect to begin an experimental study of the column II quasi-metastables in the next several months.

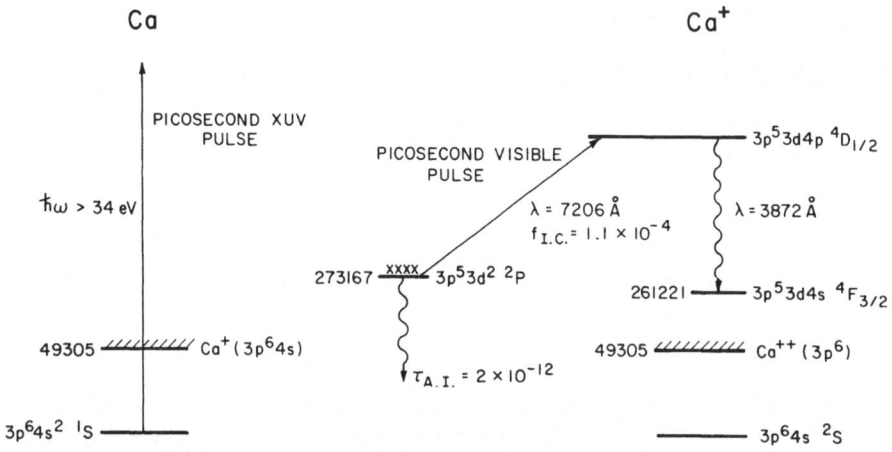

Fig. 2.   Correlation of picosecond XUV pulses

C. Lasers

The Doppler broadened gain cross-sections of the quasi-metastable transitions in the alkali atoms range from 4.8×10^{-14} cm^2 in Cs at 1091 Å to 1.5 $\times 10^{-16}$ cm^2 in Na at 405.2 Å. In each case the lower level of the possible laser transition is a valence level of the atom. We expect that the upper level will be populated by both charge transfer collisions and hot electrons, both in turn produced by laser generated soft x-rays [9,10]. We expect to reduce the lower level population by laser transfer to Rydberg levels or to the continuum.

The authors thank Richard Caro and Roger Falcone for many helpful discussions.

1. Robert D. Cowan, The Theory of Atomic Structure and Spectra (University of California, Berkeley, 1981), Secs. 8-1, 16-1, and 18-7.
2. S. E. Harris, D. J. Walker, R. G. Caro, A. J. Mendelsohn, and R. D. Cowan, Opt. Lett. 9, 168 (May 1984).
3. D. E. Holmgren, D. J. Walker, and S. E. Harris, in Laser Techniques in the Extreme Ultraviolet, S. E. Harris and T. B. Lucatorto, eds. (New York, AIP, 1984), pp. 496.
4. K. D. Pedrotti, D. P. Dimiduk, J. F. Young, and S. E. Harris, "Identification and Oscillator Strength Measurement of the 1091 Å Transition in Neutral Cs" (in preparation).
5. A. J. Mendelsohn, C. P. J. Barty, M. H. Sher, J. F. Young, and S. E. Harris, "Emission Spectra of Quasi-Metastable Levels of Alkali Atoms" (in preparation).
6. I. S. Aleksakhin, G. G. Bogachev, I. P. Zapesochnyl, and S. Yu. Ugrin, Sov. Phys. JETP 53, 1140 (1981).
7. D. E. Holmgren, D. J. Walker, D. A. King, and S. E. Harris, Phys. Rev. A 31, 677 (1985).
8. S. E. Harris, Appl. Phys. Lett. 31, 498 (1977); L. J. Zych, J. Lukasik, J. F. Young, and S. E. Harris, Phys. Rev. Lett. 40, 1493 (1978).
9. R. G. Caro, J. C. Wang, R. W. Falcone, J. F. Young, and S. E. Harris, Appl. Phys. Lett. 42, 9 (1983); R. G. Caro, J. C. Wang, J. F. Young, and S. E. Harris, Phys. Rev. A 30, 1407 (1984).
10. J. C. Wang, R. G. Caro, and S. E. Harris, Phys. Rev. Lett. 51, 767 (1983).

High Order Nonlinear Processes in the Ultraviolet

T.S. Luk, U. Johann, H. Egger, K. Boyer, and C.K. Rhodes

Department of Physics, University of Illinois at Chicago,
P.O. Box 4348, Chicago, IL 60680, USA

1. Abstract

Studies of multiphoton ionization of atoms have revealed several unexpected characteristics. The confluence of the experimental evidence involving studies of ion production and electron energy spectra leads to the hypothesis that the basic character of the atomic response involves highly organized coherent motions of entire atomic shells. This physical picture provides a basis for the expectation that stimulated emission in the x-ray range can be produced by direct highly nonlinear coupling of ultraviolet radiation to atoms.

2. Discussion

The use of picosecond and femtosecond ultraviolet laser technology allows a new regime of laser-atom interaction to be explored in which the laser electric field strength is comparable to or stronger than an atomic unit e/a_0^2. For processes that occur under these conditions, one of the areas of substantial interest involves the possibility of producing atomic excitation suitable for the generation of stimulated emission in the x-ray range. In recent experiments, a multiphoton scheme [1] of excitation has proved to be a viable and relatively efficient approach for the generation of coherent radiation in vacuum ultraviolet region. Furthermore, for irradiation with a field strength comparable to an atomic unit, it is expected that multiphoton ionization will be an important channel.[2]

In order to study the fundamental nature of the multiphoton interaction in the intensity range of 10^{14} - 10^{17} W/cm^2, several experiments have been performed to (1) analyze charge state production,[2] (2) determine the electron energy spectrum,[3] and (3) measure the radiation emitted.[2,4] The results of these studies cannot be reconciled [5] with the standard descriptions of multiphoton processes such as stepwise ionization,[6] but rather indicate that an ordered many-electron motion may play an important role [2] in the basic nature of strong field interaction.

One of the simplest experiments that gives direct evidence of the coupling strength of multiphoton processes is the measurement of ion production in the generic process

$$X + N\hbar\omega \rightarrow \sum_q p_q (X^{q+} + qe^-) \tag{1}$$

in which p_q is the probability of production of the X^{q+} from atom X. Experiments of this type were performed with a focused near-diffraction-limited picosecond ArF* (192 nm) laser [7] having $\sim$ 4 GW peak power, $\sim$ 5 ps pulse duration and $\sim$ 10 cm^{-1} bandwidth. Overall, the measurements were conducted in the range of intensities spanning 10^{14} - 10^{17} W/cm^2 and the charge states

were analyzed by the time-of-flight technique.[2,5] The material density
and the ion extraction voltage were varied to ensure that avalanches and
collisions did not play a role in the process studied. Figure (1) summarizes
the results of the total energy absorbed and the correspondingly highest
charge state produced, using various atoms at an intensity of $\cong 10^{16}$ W/cm^2.
The salient features of these results are (1) the large and unexpected scale
of the energy transfer which, in the case of U, reaches a magnitude of $\sim$633 eV,
(2) the relatively weak dependence of the coupling strength with the number
of photons, and (3) the distinct enhancement in the coupling strength for
$Z \geq 18$ and, specifically, (4) the complete removal of atomic shells, such as
that represented by the rare gases Ar, Kr and Xe. As noted above, it is found
that neither the standard perturbative treatments nor the nonperturbative
Keldysh-type descriptions reasonably explain the production of the ion charge
states and their relative abundances for either high or low Z atoms.

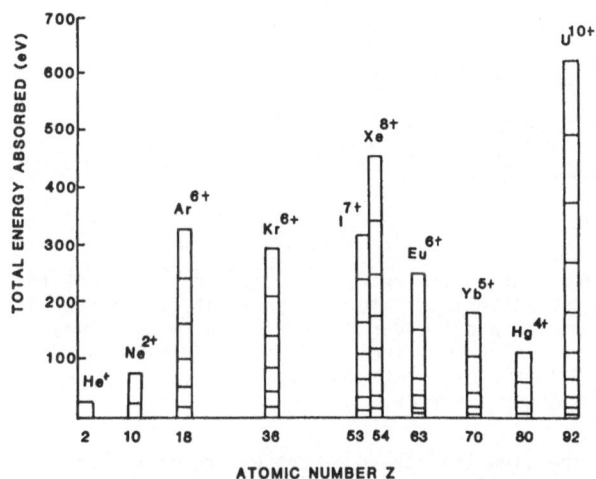

Fig. 1: Data concerning the multiple ionization of atoms produced by irradiation at 193 nm. Plot of total ionization energies of the observed charge
states as a function of atomic number (Z). The coincidence of an H_2O^+ background signal prevented the I^{7+} species from being positively identified.

The confluence of all these features leads to a physical picture of the
interaction that involves an ordered driven response of the electrons in the
outer atomic subshells. Although this picture furnishes a description of
the qualitative features of the experimental findings, much remains to be
done to consolidate this speculative model.

Recently, a magnetic mirror time-of-flight electron energy spectrometer
was used to study the energy spectrum of the emitted electrons. The spectrometer has the noted advantages of a large solid angle collection of 2π,
good discrimination against electrons produced outside the laser focus, and
the capability of recording the entire spectrum in a single laser pulse.
Figure (2) shows the electron spectra of Se at various intensities. In the
Xe spectra, some peaks can be recognized immediately, such as the 0.7 eV
feature corresponding to two-photon ionization, two consecutive peaks of so-
called above-threshold ionization [10] at 7.1 and 13.5 eV, and the stepwise
ionization from Xe$^+$. The rapid decline of the above-threshold ionization
lines and evidence indicating the small probability for stepwise ionization

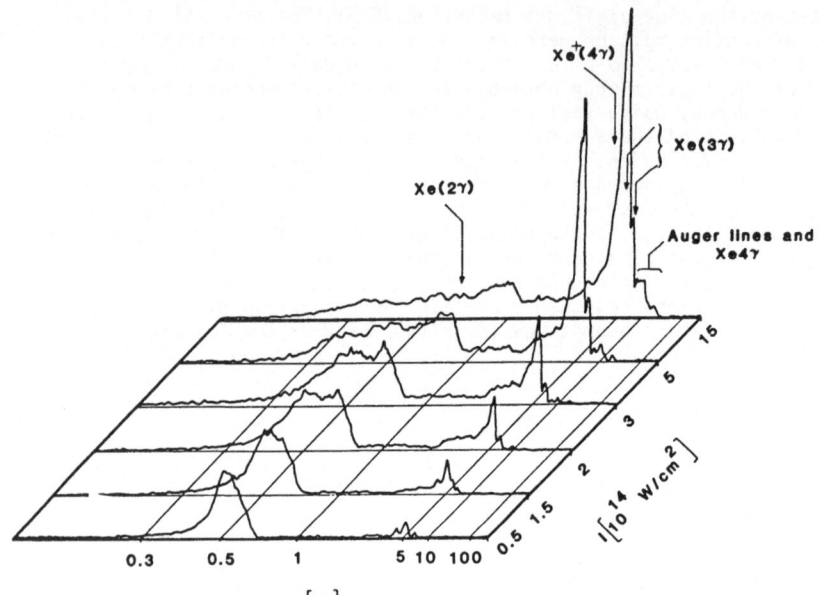

Xe*(4γ)

Xe(3γ)

Xe(2γ)

Auger lines and
Xe4γ

1.5

5

3

2

$I\,[10^{14}\ \text{W/cm}^2]$

1.5

0.5

0.3   0.5   1   5 10 100

Electron Energy [eV]

Fig. 2: Overall xenon time-of-flight photoelectron spectrum from ∿ 0.1 eV to
∿ 100 eV. The uncertainty in the intensity scale is approximately a factor of
two. Irradiation was at 193 nm with a pulse duration of ∿ 5 ps with a lens
with a focal distance of 20.5 cm. Electron lines originating from Xe and Xe$^+$
arising from two (2γ), three (3γ) and (4γ) processes are indicated, along with
a group tentatively assigned as Auger features.

of low-charge states leads to the view that these processes are of relatively
minor importance in the physical coupling generating the high-charge states
observed in the ion experiment. As the laser intensity is increased, the
0.7 eV electron peak exhibits a broadening and suppression similar to that
observed in other experiments using 1.06 μm radiation. [11] This effect can
be understood in the context of the pondermotive potential created by the
radiation field. [10,12] Briefly explained, in order to establish the condition enabling the electron to escape from the region of interaction, the
excess energy acquired by the electron after absorbing a sufficient number
of photons to reach the ionization continuum must be sufficient to account
for the average oscillation energy of a free electron in such a field. This
energy of oscillation is $e^2E^2/4m\omega^2$, in which E, ω, and m are the radiative
electric field, the angular frequency of the laser field, and the electron
mass, respectively. Therefore, the energy condition of multiphoton ionization
can then be written as

$$N\hbar\omega + \varepsilon_x(E) - \varepsilon_x^+ - \bar{\varepsilon}_{osc}(E) > 0 \qquad (2)$$

in which $\varepsilon_x(E) - \varepsilon_x^+(E)$ and $\bar{\varepsilon}_{osc}(E)$ are the ionization and average oscillation
energies in the field, respectively. The intensities at which this effect
begins to appear for 1.06 μm and 193 nm radiation have shown reasonable agreement with experiments.

In addition, in xenon there is an additional series of sharp lines, shown
in Figure (3), which cannot be identified with any stepwise or above-threshold

168

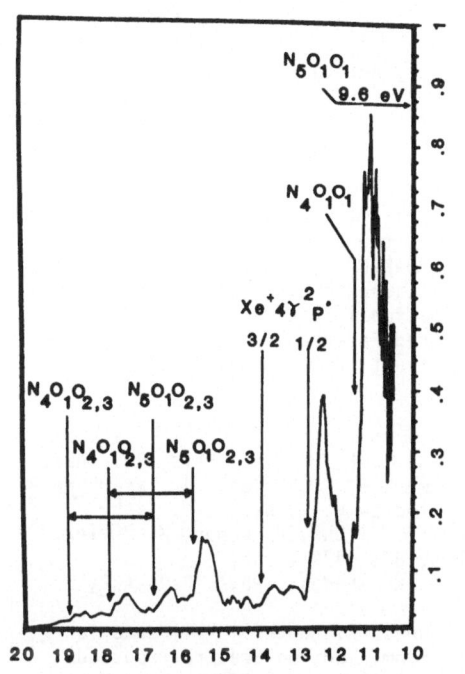

Fig. 3: Prominent transitions observed in the electron spectrum of xenon irradiated with 193 nm at 10^{15} W/cm^2 are shown. Both continuum-continuum ($4\gamma \rightarrow Xe^+_{3/2}$, $Xe^+_{1/2}$) and tentatively assigned Auger ($N_{4,5}OO$) features are apparent. The splittings between the three N_4-N_5 pairs, two of which are shown by horizontal arrows, have the common value of ~ 2 eV, the known $4d_{5/2}$-$4d_{3/2}$ separation in xenon. The vertical arrows indicate the high-energy edges of the observed features which represent the true energies of the lines.

ionization processes. Significantly, however, their relative energies and strengths match well to the known [13] $N_{4,5}OO$ Auger lines observed in single-photon ionization experiments. An absolute energy shift of ~ 1 eV is evident in the spectrum and is believed to be due to the influence of residual stray electrons produced by scattered light. These Auger lines, of which a total of eight have been detected, represent the first direct evidence of inner-shell excitation by multiphoton processes. However, it is important to note that preliminary findings show that the direct multiphoton coupling to the 4d-shell for production of the 4d hole cannot account for the total strength of the Auger lines. Furthermore, the most energetic electrons observed are those that correspond to the 4d Auger lines. It is believed that the electrons responsible for the high charge states could be present mainly as a low-energy continuum in the spectrum, although confirmation of this aspect still awaits more refined measurement. However, if this hypothesis is correct, this would mean that several electrons are released simultaneously and that their energies are distributed statistically. One can then envisage electrons oscillating collectively with the radiation field and unable to escape due to a large oscillation energy. Such a configuration is likely to be highly unstable when the radiation field is turned off and, consequently, would result in releasing the electrons with broad energy distribution.

In conclusion, the results of the electron experiments have shown that a stepwise process cannot account as the main cause of the observed multi-photon multiple ionization. The weakening effect of the lowest order ioni-zation peak with increasing laser intensity is found in agreement with the understanding of the pondermotive effect using 193 nm laser radiation. And, most significantly, the first direct evidence of multiphoton inner-shell ex-citation is observed. These experiments provide insights into the nature of the strong field interaction that involves an ordered coherent motion of the electrons. Quantum mechanically, such states are described as multiply excited levels. These atomic motions could potentially provide a means of producing inner-shell excitations suitable for the generation of coherent short wavelength radiation.

3. Acknowledgments

The authors wish to acknowledge the contribution of I. McIntyre and the expert technical assistance of M. J. Scaggs, J. R. Wright, and P. Slagle. This work was supported by the Office of Naval Research, the Air Force Office of Scientific Research under contract number F49630-83-K-0014, the Department of Energy under grant number DE-AC02-83ER13137, the Lawrence Livermore National Laboratory under contract number 5765705, the National Science Foundation under grant number PHY 81-16626, the Defense Advanced Research Projects Agency, and the Los Alamos National Laboratory under contract number 9-X54-C6096-1.

4. References

H. Pummer, H. Egger, T. S. Luk, T. Srinivasan, C. K. Rhodes: Phys. Rev. A28, 795 (1983); H. Egger, T. S. Luk, H. Pummer, C. K. Rhodes: in Laser Spectroscopy VI, Proc. 6th Intl. Conf., edited by H. P. Weber, W. Luthy (Springer-Verlag, Berlin, 1983) p. 385; T. S. Luk, H. Egger, W. Müller, H. Pummer, C. K. Rhodes: J. Chem. Phys. 82, 4479 (1985).

T. S. Luk, U. Johann, H. Egger, H. Pummer, C. K. Rhodes: Phys. Rev. A32, 1 (1985).

U. Johann, T. S. Luk, H. Egger, H. Pummer, and C. K. Rhodes: "Evidence for Atomic Inner-Shell Excitation in Xenon from Electron Spectra Produced by Collision-Free Multiphoton Processes at 193 nm," Conference on Lasers and Electro-Optics '85, Baltimore, Maryland, to be published.

W. Müller, M. Shahidi, H. Egger, T. S. Luk, H. Pummer, and C. K. Rhodes: "Extreme Ultraviolet Fluorescence Spectra of Krypton and Xenon under High Intensity Ultraviolet Laser Excitation," submitted for publication.

T. S. Luk, H. Pummer, K. Boyer, M. Shahidi, H. Egger, and C. K. Rhodes: Phys. Rev. Lett. 51, 110 (1983).

L. A. Lompré, G. Mainfray: Multiphoton Processes, edited by P. Lambropoulos, S. J. Smith (Springer-Verlag, Berlin, 1984) p. 23.

H. Egger, T. S. Luk, K. Boyer, D. F. Muller, H. Pummer, T. Srinivasan, and C. K. Rhodes: Appl. Phys. Lett. 41, 1032 (1982).

K. Boyer, C. K. Rhodes: Phys. Rev. Lett. 54, 1490 (1985).

P. Kruit, F. H. Reed, J. Phys. E16, 373 (1983).

H. G. Muller, A. Tip: Phys. Rev. A30, 3039 (1984).

P. Kruit, J. Kimman, H. G. Muller, M. J. van der Wiel: Phys. Rev. A28, 248 (1983).

A. Szöke: "Interpretation of Electron Spectra Obtained from Multiphoton Ionization of Atoms in Strong Fields," to be published.

S. Southworth, U. Becker, C. M. Treusdale, P. H. Kobrin, D. W. Luidle, S. Owaki, S. Shirley: Phys. Rev. A28, 261 (1983).

Autoionizing States of Cd$^+$ Determined by Picosecond Soft-X-Ray Ionization Spectroscopy

W.T. Silfvast, O.R. Wood II, J.J. Macklin, and D.Y. Al-Salameh

AT & T Bell Laboratories, Holmdel, NJ 07733, USA

1. Introduction

A new technique, picosecond soft-x-ray ionization spectroscopy, has been used to detect autoionizing levels belonging to the $4d^9 5s6s$ electronic configuration of Cd$^+$, lying approximately 4 eV above the ionization limit, by observing their rapid emission to lower-lying bound levels. The levels in Cd$^+$, which are the first identified autoionizing levels of that ion species, were directly pumped by photoionization with broadband soft-x-ray emission from a laser-produced plasma produced by focusing the 50 mJ, 70 psec output from a Nd:YAG laser onto a metal target inside a Cd vapor cell [1]. The autoionizing time of the $4d^9 5s6s$ $^2D_{5/2}$ state was measured to be 460 +/- 50 psec which is in good agreement with the RCN/RCG multiconfiguration Hartree-Fock atomic physics code [2] prediction of 490 psec.

2. Discussion

The broadband soft-x-ray flux from a laser-produced plasma has been shown to be an efficient photoionization source to produce metastable states [3] and to directly pump lasers [1]. Photoionization involving the removal of a single electron from an atomic gas or vapor typically occurs over a photon energy range of from several eV to hundreds of eV depending upon the binding energy and angular momentum of the removed electron. Maximum photoionization cross-sections for the removal of individual electrons range from less than 1 Mb to several tens of Mb's with larger angular momentum electrons such as d and f-electrons having the larger cross-sections. Photoionization involving two electrons has a much lower probability of occurring, and such events are often masked by the dominant single electron removal. The $4d^9 5s6s$ autoionizing levels of Cd$^+$, detected by the rapid excitation technique, are examples of such two-electron ionization processes. Since these autoionizing states had not previously been identified, the RCN/RCG atomic physics code was used to determine their approximate location, and the oscillator strengths coupling them to lower-lying levels. The first even-parity core-excited states above $4d^9 5s^2$, the eight $4d^9 5s6s$ states, were predicted by the code to range from 156,800 to 169,500 cm^{-1} (the ionization limit for Cd$^+$ is at 135,374 cm^{-1}) and to have autoionizing times ranging from 5×10^{-13} to 10^{-9} sec.

In addition to autoionization, strong radiative emission (predominantly in the 1600 - 2200 Å region) can occur selectively from these states to many of the $4d^9 5s5p$ lower-lying states (there are 20 such bound states, 19 of which have been identified). Because of the two-electron nature of the process, direct photoionization pumping to these $4d^9 5s6s$ states from the $4d^{10} 5s^2$ neutral ground-state would be expected to be relatively weak when compared with the strong one-electron photoionization of the $4d^9 5s^2$ state. However, if the photoionization flux is high enough, excitation and subsequent emission from the longer lived autoionizing states is conceivable.

3. Results

Strong emission from long-lived autoionizing states in Cd+ has been observed for the first time by combining rapid excitation using the soft-x-ray flux from a laser-produced plasma with fast detection, using a 125 psec risetime microchannel plate detector and a fast oscilloscope. A 50 mJ, ~70 psec duration pulse from a Nd:YAG laser can produce a 70 psec duration soft-x-ray flux of 10^{15} photons (assuming 10% conversion efficiency[3]). Using such a source, approximately 15 subnanosecond-duration emission lines were observed in the Cd photoionization spectrum between 2000 and 4000 Å under conditions in which typical Cd^+ bound states were decaying in times of 10 nsec or longer and Cd^{++} states were decaying in times of 2 to 3 nsec. The subnanosecond duration emission lines, necessarily autoionizing transitions due to their fast decay (plasma densities were not high enough to account for such a rapid decay by electron collisions), were computer sorted for common upper levels and using the known $4d^9 5s5p$ levels as lower levels. One of the levels found using this technique led to the identification of the $4d^9 5s6s$ $^2D_{5/2}$ (^3D) state shown in Table I. Five emission lines identified the level to within 2 cm^{-1} and were in good agreement with the predicted oscillator strengths of the atomic physics code.

The decay of the 2055 Å transition was measured with a microchannel plate detector using a time-correlated photon-counting technique [4]. The result, indicating a single exponential decay with a decay time of 460 +/- 50 psec, is shown in Figure 1 together with a measurement of the 5320 Å excitation pulse from a frequency-doubled Nd:YAG laser. The autoionizing

Table I Observed transitions from $4d^9 5s6s$ $(^1D)^2D_{5/2}$ autoionizing state located at 168041.1 +/- 2.0 cm^{-1}.

Lower State $4d^9 5s5p$	Measured Wavelength [Å]	Relative Intensity Observed	Calculated
$(^3D)^2D^o_{5/2}$	2752.5	13	10
$(^1D)^2D^o_{5/2}$	2137.9	80	46
$(^1D)^2F^o_{7/2}$	2054.2	100	100
$(^1D)^2P^o_{3/2}$	2040.7	78	71
$(^1D)^2F^o_{5/2}$	1956.7	17	11

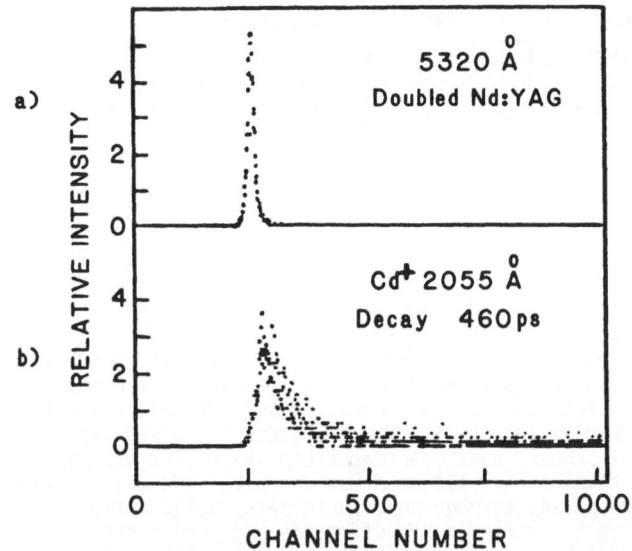

Fig. 1 Typical emission spectra obtained using time-correlated photon-counting technique (a) 5320 Å emission from doubled Nd:YAG excitation source; (b) 2055 Å emission from $4d^9 5s6s\,^2D_{5/2}$ autoionizing level in Cd^+,

time of this level was predicted by the atomic physics code to be 490 psec.

4. Conclusions

The use of high-flux, 70 psec duration soft-x-ray photoionization pulses and fast detection techniques in Cd demonstrates that characteristic autoionizing radiation can easily be observed and can be used to identify continuum states in Cd as well as in other ions. Such techniques are useful not only in obtaining new information about continuum states but may also lead to XUV lasers.

5. References

1. W.T. Silfvast, J.J. Macklin and O.R. Wood, II: Opt. Lett. **8**, 551 (1983).
2. R.D. Cowan: The Theory of Atomic Structure and Spectra (Berkeley: University of California Press, 1981), Secs. 8-1 and 16-1.
3. R.G. Caro, J.C. Wang, R.W. Falcone, J.R. Young and S.E. Harris: Appl. Phys. Lett. **42**, 9 (1983).
4. S.K. Poultney: In Advances in Electronics and Electron Physics L. Marton, ed. (Academic, New York, 1972), Vol. 31.

Coherent VUV and XUV Radiation Tunable to 90 nm, and Spectra of Rare Gas Dimers

B.P. Stoicheff, P.R. Herman, P.E. LaRocque, and R.H. Lipson

Department of Physics, University of Toronto,
Toronto, Ontario, M5S 1A7, Canada

1. VUV and XUV Sources Using Mg, Zn, and Hg Vapors

Progress in the development of coherent, tunable sources for the VUV and XUV has reached a stage where these sources will soon be found in most spectroscopic laboratories. The principal methods used in generating such radiation are frequency tripling and 4-wave sum-mixing (4-WSM) in rare gases and metal vapors [1, 2]. Recently, several novel schemes have been described which overcome the lack of transmitting window materials at $\lambda < 104$ nm. For example, rare gases as well as molecular gases have been pulsed through supersonic jets [3,4], a laminar flow of H_2 has been used with a curtain of Ne buffer gas [5], and a rotating pinhole has been synchronized with pulses of the primary laser [6], all providing optical transmission during XUV generation with minimum gas flow. By these various means, it has been possible to produce XUV radiation which is tunable over limited regions, to wavelengths as short as 50 nm [7].

In our laboratory, VUV and XUV radiation has been generated in vapors of Mg, Zn, and more recently Hg, by 4-WSM with 2-photon resonance enhancement. The method makes use of coherent radiation from two pulsed dye lasers (pumped by N_2, KrF, or XeCl lasers): one (ν_1) is tuned to a 2-photon-allowed resonance of the vapor, and the second (ν_2) is tunable over a broad frequency range such that $2\nu_1 + \nu_2$ corresponds to an autoionizing resonance [8]. The relevant energy levels used for 2-photon resonances in Mg, Zn, and Hg are shown in Fig. 1.

For work with the Mg and Zn vapors, uniform vapor densities resulting in stable VUV emissions are obtained over periods of several hours, with the use of double heat-pipe ovens [9]. The heat-pipe oven for Hg vapor (Fig. 2) is a simple cell of pyrex glass. Liquid Hg and its vapor are confined to the central heated section by water-cooled jackets at each end which are tapered to return condensed Hg back to the hot zone. For generation of VUV radiation from 126 to 104.5 nm, a LiF plate ~ 0.5 mm thick forms the exit window, and a vapor pressure of up to 95 torr is used with an equal pressure of He buffer gas. For XUV generation, $\lambda < 104.5$ nm, the exit window is a glass disc 2 mm thick with a central section ~ 2 mm diam containing an array of 50 μm diam.capillaries. These pores reduce the gas conductance by a factor of ~ 750 over that of a single aperture of equivalent area, at a He pressure of 35 torr, and provide an efficient XUV window with $\sim 50\%$ transmission [10].

Up to the present time, 4-WSM in Mg, Zn, and Hg vapors has been used to generate continuously tunable radiation over the wavelength range 174 to 87.5 nm (or $\sim 57,000$ cm^{-1}); from 174 to 120 nm in Mg, 140 to 106 nm in Zn, and 126 to 87.5 nm in Hg. The emission of VUV and XUV radiation is in 5 ns pulses, at a repetition rate of 20 Hz, with $\sim 10^6$ to 10^{11} photons/pulse (depending on the wavelength region) and linewidths of $0.1 - 0.5$ cm^{-1}. Maximum intensities are obtained for energies near the respective ionization limits of these elements, with significant degradation of intensity at shorter wavelengths as shown for Hg vapor in Fig. 3. These sources have provided sufficient inten-

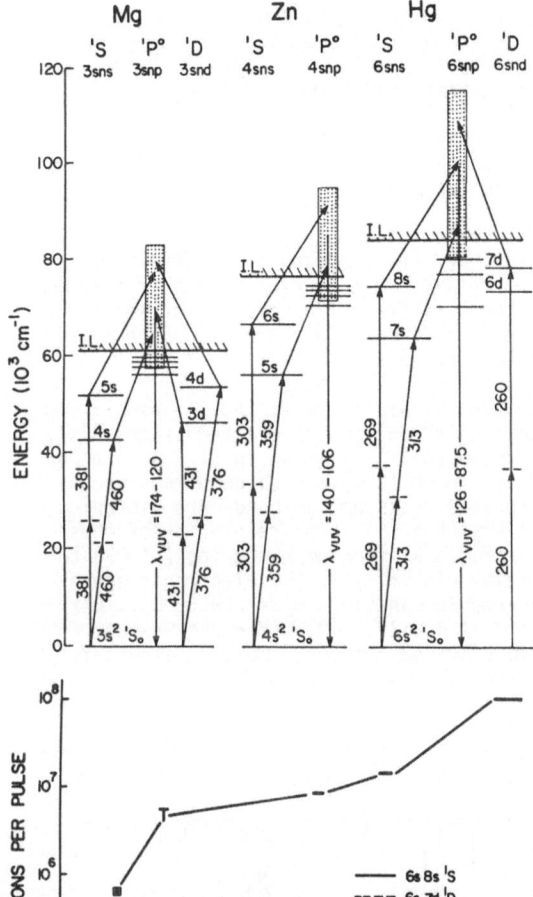

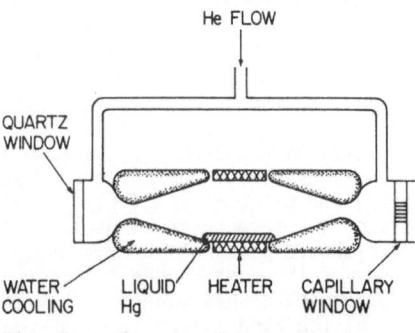

Fig. 1. Partial energy level diagrams for Mg, Zn, and Hg showing levels used for 2-photon resonance enhancement and corresponding wavelengths in nm. Regions of ionizing continua and broad autoionizing levels which provide tunability are shown by hatched areas.

Fig. 2. Diagram of Hg heat-pipe with capillary-array window used for XUV generation.

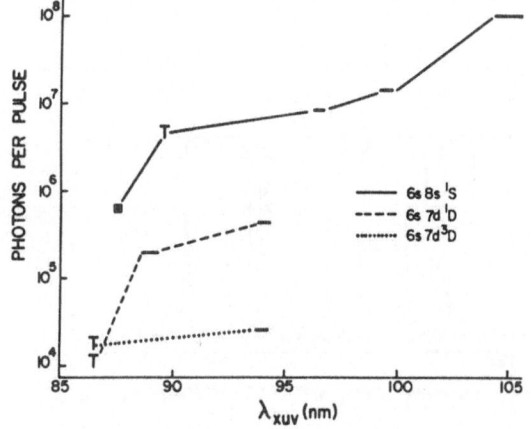

Fig. 3. Measured photons/pulse vs λ for three levels used in resonance enhancement with Hg vapor.

sity, monochromaticity, and tunability to make possible several spectroscopic studies which would have been extremely difficult using other techniques.

2. Spectra of Xe₂, Kr₂, and Ar₂

An example of the unique capabilities of these sources is provided by our recent investigations of the electronic spectra of the rare gas dimers Xe_2, Kr_2, and Ar_2. These molecules are of interest because they are model systems for van der Waals interactions and because of their potential as media for VUV excimer lasers. However, despite the large effort devoted to studies of these molecules, surprizingly little spectroscopic information is available on their excited states. Prior to the present investigations, vibrational constants for the ground states of Xe_2 [11], Kr_2 [12], and Ar_2 [13] and of one excited state of Kr_2 were known, as well as the internuclear separation for Ar_2 [14].

Many of the problems encountered in earlier studies were circumvented by the use of a pulsed supersonic jet to form cold (<10 K) ground state dimers, and by the use of the above tunable, coherent sources to excite fluorescence spectra. This combination of techniques resulted in the observation of vibrationally resolved electronic spectra for all three molecules, including rotational structure for Kr_2 and Ar_2. Spectra involving transitions from the ground state $X^1\Sigma_g^+(v'' = 0)$ to the lowest three (stable) excited states were recorded and analyzed for each of Xe_2, Kr_2, and Ar_2, leading to the first determinations of molecular constants and potential energy curves for their excited states as well as the internuclear distance of Kr_2. The results for Xe_2 have just been published [15] and the analyses of Ar_2 are in progress: here a brief summary of the Kr_2 spectra and results are presented.

Three band systems of Kr_2 were investigated at 125, 124, and 117 nm arising from transitions from the ground state $X^1\Sigma_g^+(v'' = 0)$ to the excited states $A^3\Sigma_u^+$, $B^1\Sigma_u^+$, and $C^1\Sigma_u^+$, respectively. One of these $(B \leftarrow X)$ is shown in Fig. 4. It consists of 9 structured bands separated by ~ 70 cm^{-1}, with each band made up of 8 blue-shaded peaks of ~ 2 cm^{-1} width. These 8 peaks or sub-bands are vibronic bands of the most abundant isotopes of Kr_2 (Fig. 5). The resolution of these isotopic bands has made possible the vibrational quantum numbering (v' = 30 to 38) of the band system, and the evaluation (in cm^{-1}) of the electronic energy $T_e' = 75426.8 \pm 2.0$, vibrational and anharmonic constants $\omega_e' = 219.5 \pm 0.1$, $\omega_e' x_e' = 2.2325 \pm 0.0015$, and potential well-depth $D_e' = 5629.1 \pm 2.5$, for the B state. Similar spectra and their analyses have provided the corresponding constants for the A and C states.

It will be noted that in Figs. 4 and 5 there is an indication of rotational structure in the strongest peaks. When a 10% mixture in He of a single isotope

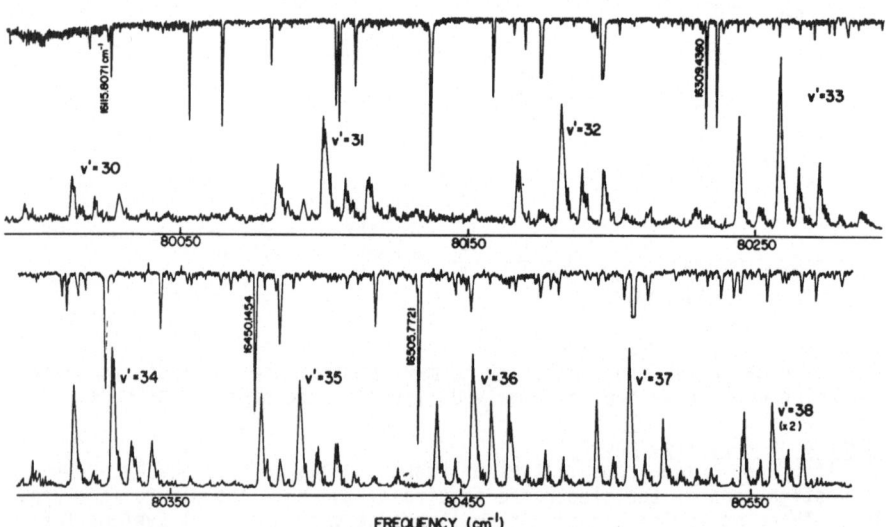

Fig. 4. Fluorescence excitation spectrum of Kr_2 near 124 nm showing 9 different vibronic bands (v' = 30 to 38) with structure due to isotopic molecules. The upper traces are optogalvanic spectra of U used for wavelength calibration.

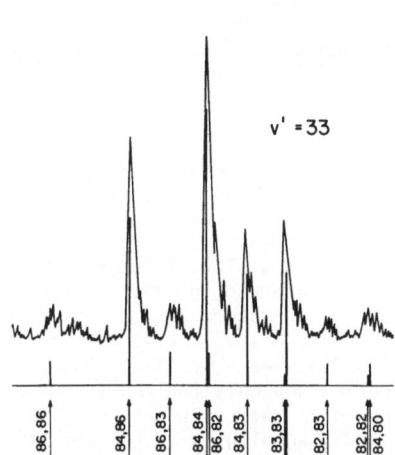

v' = 33

86,86 84,86 86,83 84,84 84,83 83,83 82,83 82,82
                    86,82   84,80

Fig. 5. Comparison of observed
and calculated "stick" spectra
for vibronic bands (v' = 33) of
10 Kr$_2$ isotopes. Note that all
9 bands in Fig. 4 have this
same structure.

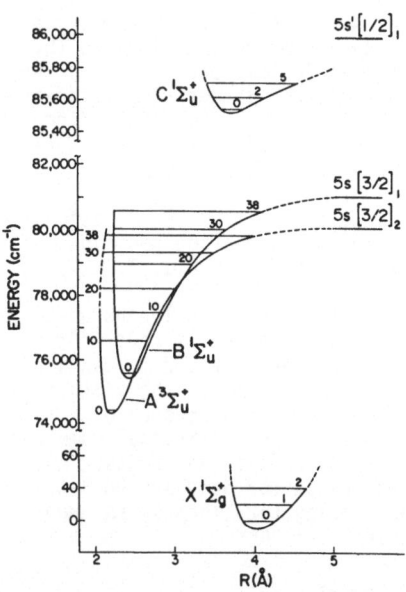

Fig. 7. Potential energy curves
for the ground state and first
three excited states of Kr$_2$.

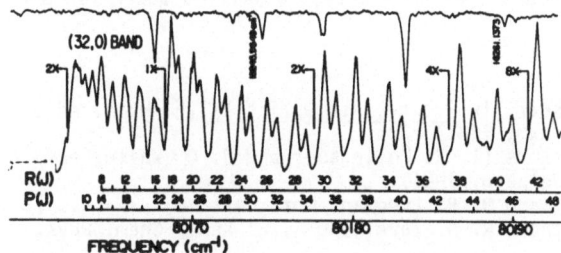

Fig. 6. Spectrum of the
B ← X (32,0) band of
^{86}Kr$_2$ showing resolved
P and R rotational
branches.

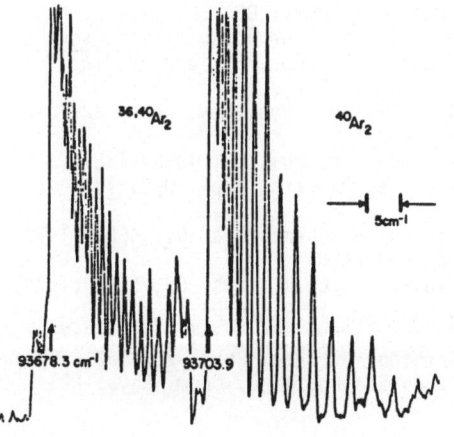

Fig. 8. Rotational structure in
B ← X (27,0) bands of ^{36,40}Ar$_2$
and ^{40}Ar$_2$.

of ^{86}Kr was expanded in the pulsed nozzle and spectra were excited in the relatively warmer region close to the nozzle extensive rotational structure was observed for all band systems. An example is shown in the spectrum of Fig. 6 taken at much larger dispersion and resolution than that of Fig. 5. The usual analysis of these spectra yielded rotational constants of the upper and ground states, and led to the first determination of the internuclear distance in Kr_2, namely $r_e'' = 4.017 \pm 0.012\,Å$.

Finally, the derived rotational and vibrational constants were used to calculate the shapes and positions of the four potential energy curves for Kr_2. These are shown in Fig. 7.

For Ar_2, three analogous band systems were observed at 107, 106, and 105 nm. However, since the isotope ^{40}Ar has a high natural abundance of 99.6%, only a single series of vibronic bands was observed for each system, and this is not sufficient to deduce the vibrational numbering. Thus a small sample of ^{36}Ar was obtained, and spectra of $^{36,40}Ar_2$ as well as $^{40}Ar_2$ were studied, resulting in unambiguous vibrational numbering and molecular constants for the ground and lowest three excited states. An example of the rotational structures of isotopic bands for band system B-X is given in Fig. 8, showing twice the rotational spacing for $^{40}Ar_2$ (with every second level missing) as for the heteronuclear $^{36,40}Ar_2$.

3. Conclusion

The present investigation using tunable, coherent, VUV radiation from 4-WSM in Mg, Zn, and Hg vapors has yielded spectroscopic constants for Xe_2, Kr_2, and Ar_2. Clearly, the experimental techniques used here will be applicable to the whole family of rare gas dimers and their mixtures, as well as other van der Waals molecules.

4. References

1. W. Jamroz and B. P. Stoicheff, in Progress in Optics XX (E. Wolf, ed., North-Holland, Amsterdam, 1983), 325.
2. C. R. Vidal, in Tunable Lasers (I. F. Mollenauer and J. C. White, eds., Springer-Verlag, Heidelberg, 1984).
3. J. Bokor, P. H. Buchsbaum, and R. R. Freeman, Opt. Lett. 8, 217 (1983).
4. E. E. Marinero, C. T. Rettner, R. N. Zare, and A. H. Kung, Chem. Phys. Lett. 95, 486 (1983).
5. T. Srinivasan, H. Egger, H. Pummer, and C. K. Rhodes, IEEE J. Quantum Electron. QE-19, 1270 (1983).
6. K. D. Bonin and T. J. McIlrath, J. Opt. Soc. Amer. B2, 527 (1985).
7. R. Wallenstein, R. Hilbig, G. Hilber, A. Lago, and W. Wolff in Laser Spectroscopy VII (T. W. Hänsch and Y. R. Shen, Springer-Verlag, Heidelberg, 1985).
8. R. T. Hodgson, P. P. Sorokin, and J. J. Wynne, Phys. Rev. Lett. 32, 343 (1974).
9. H. Scheingraber and C. R. Vidal, Rev. Sci. Instrum. 52, 1010 (1981).
10. T. B. Lucatorto, T. J. McIlrath, and J. R. Roberts, Appl. Opt. 18, 2505 (1979).
11. D. E. Freeman, K. Yoshino, and Y. Tanaka, J. Chem. Phys. 61, 4880 (1974): M.-C. Castex, J. Chem. Phys. 74, 759 (1981).
12. Y. Tanaka, K. Yoshino, and D. E. Freeman, J. Chem. Phys. 59, 5160 (1973).
13. Y. Tanaka and K. Yoshino, J. Chem. Phys. 53, 2012 (1970).
14. E. A. Colbourn and A. E. Douglas, J. Chem. Phys. 65, 1741 (1976): Y. Tanaka, W. C. Walker, and K. Yoshino, J. Chem. Phys. 70, 380 (1979).
15. R. H. Lipson, P. E. LaRocque, and B. P. Stoicheff, J. Chem. Phys. 82, 4470 (1985).

Molecular Spectroscopy by Stepwise Two-Photon Ion-Pair Production at 71 nm

A.H. Kung

San Francisco Laser Center, Department of Chemistry,
University of California, Berkeley, CA 94720, USA

R.H. Page, R.J. Larkin, Y.R. Shen, and Y.T. Lee

Materials and Molecular Research Division, Lawrence Berkeley Laboratory,
University of California, Berkeley, CA 94720, USA

The Rydberg states of H_2 have been a continuing subject of intensive study by various research groups.[1-4] However, understanding of the high lying electronic states of this molecule has been inhibited by the lack of spectroscopic data in the region < 75 nm. Experimental studies have been difficult because spectroscopic features are generally buried under an intense absorption or photoionization continuum.[5] Intense, high-resolution excitation sources are not easily available. Recent developments on tunable, narrowband, coherent XUV sources [6] provide new means of studying the spectroscopy in this region with high resolution (± .0005 nm). We have applied the technique of stepwise two-photon excitation to study photoionization of H_2 in a molecular beam using the two lowest excited states of H_2 as the intermediate level. This excitation, coupled with the detection of background-free H^- ions has enabled us to uncover, for the first time, spectroscopic features that are difficult to observe in positive ion detection.[7] These features have been successfully assigned to new Rydberg series converging to the high vibrations of the H_2^+ ground electronic state.

In this study, the first photon at ∿ 96 nm is derived by frequence tripling of a pulsed tunable dye laser in a pulsed jet of CO. The frequency is tuned on resonance with a ro-vibronic level of H_2 in the Lyman (B $^1\Sigma_u^+$) band with v'=12, 13 or the Werner (C $^1\Pi_u$) band with v'=2. The excited H_2 molecule absorbs a second photon that is the UV output from a second dye laser (see Figure 1). The frequency of this second photon is chosen such that the sum of the frequencies of the first and the second photons can be higher than 71.57 nm, the threshold for H^- ion production. Here, the H^- ions are formed from H_2 by photopredissociation through the ion-pair channel. Monitoring the negative ions with a quadrupole mass spectrometer and tuning the second

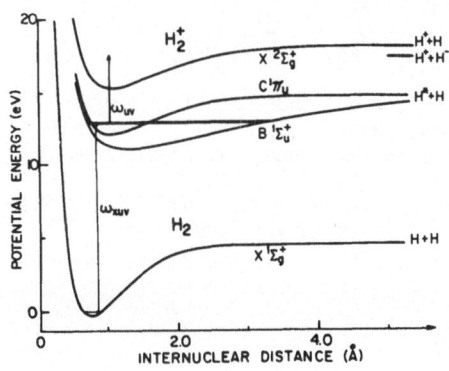

Figure 1: Potential energy diagram of H_2, showing two-photon excitation scheme.

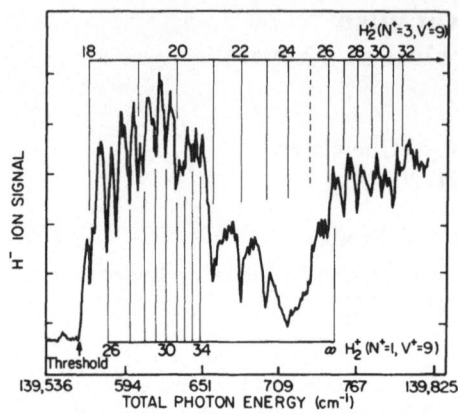

Figure 2: H_2 photoionization spectrum following stepwise two-photon excitation. The first photon is fixed at the $C^1\Pi_u$ (v'=2) Q(1) transition frequency.

laser will show the threshold for ion-pair production followed by a H^- ion continuum. Interference of this ion-pair continuum by the major, molecular ion production channel will result in structure in the continuum.[8]

Figure 2 is an example of the resulting spectrum. This spectrum consists of a structured continuum with sharp dips superimposed on the structures. The origin of the gross features is not understood at this point. However, the dips have been successfully assigned. They correspond to previously unobserved Rydberg series of H_2 that converge to the H_2^+ ($v^+=9$, N=1) and H_2^+($v^+=9$, N=3) vibrations as the series limit. By tuning to different intermediate states, we have also observed several preliminary, new Rydberg series converging to the $v^+=9$ vibration of the H_2^+ ion with the rotational quantum number as high as N=5. These represent the first observation of large Δv (≥ 9) and ΔJ (≥ 3) transitions in the spectroscopy of H_2.

In conclusion, we have described a new technique that couples non-linear spectroscopy with negative ion detection for molecular hydrogen spectroscopic studies in the XUV region. The technique should be easily extendable to studies of other diatomics and polyatomic molecules.

Acknowledgements   This work was supported by the United States Department of Energy under grant DE-AC03-76SF-00098. AHK's work is also supported by the National Science Foundation under grant CHE-8303208.

References
1. G. Herzberg, J. Mol. Spectrosc. 33, 147 (1970).
2. G. Herzberg and Ch. Jungen, J. Mol. Spectrosc. 41, 425 (1972).
3. P.M. Dehmer and W.A. Chupka, J. Chem. Phys. 65, 2243 (1976).
4. K.H. Welge, private communication (1985).
5. J. Berkowitz: Photoabsorption, Photoionization, and Photoelectron Spectroscopy, (Academic Press, New York, 1979).
6. C.T. Rettner, E.E. Marinero, R.N. Zare, and A.H. Kung, J. Phys. Chem. 88, 4459 (1984).
7. W.A. Chupka, P.M. Dehmer, and W.T. Jivery, J. Chem. Phys. 63, 3929 (1975).
8. U. Fano, and J.W. Cooper, Rev. of Mod. Phys. 40, 441, (1968).

Broadly Turnable Coherent VUV Radiation Generated by Frequency Mixing in Gases

R. Hilbig, G. Hilber, A. Lago, A. Timmermann, and R. Wallenstein

Fakultät für Physik, Universität Bielefeld, D-4800 Bielefeld 1,
Fed. Rep. of Germany

We report on the generation of tunable coherent radiation in the vacuum ultraviolet (VUV) by third- and fifth-order sum and difference frequency conversion of intense dye laser radiation in rare gases and metal vapors.

The nonresonant frequency mixing of third order ($\omega_{vuv} = 2\omega_1 \pm \omega_2$) generates coherent light in the spectral range of 72 to 210 nm. Phase matching conditions restrict the tuning range of the sum frequency to wavelength regions of negative mismatch Δk defined as the difference between the wave vectors of the VUV radiation and the driving polarization [1]. Frequency tripling and sum frequency conversion in the rare gases Ne, Ar, Kr and Xe and in Hg-vapor thus provided VUV in all wavelength regions in which these gases are negative dispersive [2-4]. This includes, for example, the continuum of Xe($\lambda_{vuv} < 92$ nm) and Kr($\lambda_{vuv} < 84$ nm).

In contrast to the sum frequency the difference frequency mixing is not restricted by the dispersion of the medium. It is therefore well suited for the generation of VUV light which is continuously tunable in the entire range of $\lambda_{vuv} = 110$-220 nm [5].

At laser pulse powers of a few megawatts the efficiency of the nonresonant frequency conversion is typically 10^{-5} to 10^{-6}. Tuning the laser frequency to a two-photon resonance the resonant enhancement of the induced polarization can provide conversion efficiencies of 10^{-3} to 10^{-4} even at input powers as low as 10-100 kW.

The two-photon resonant frequency mixing has been investigated, for example, in Xe, Kr and Hg [6]. The experimental results provided tunable VUV of kW pulse power tunable in almost the entire range of 72 to 210 nm.

With present dye laser systems third-order frequency conversion generates VUV at wavelengths $\lambda_{vuv} > 72$ nm. Intense tunable VUV at shorter wavelengths ($\lambda_{vuv} = 42$-72 nm) can be produced by fifth-order processes. This is shown by the first results on resonant fifth-order frequency mixing in Ar and Ne which generates continuously tunable radiation at $\lambda_{vuv} = 50$-72 nm.

In the case of Ar the UV light (λ_1 = 319.8 nm) of a frequency doubled dye laser (λ_2 = 639.7 nm) is four-photon resonant with the Ar-transition 3p - 9p (5/2,2). Simultaneously $3\omega_1$ ($\lambda_1/3$ = 106.61 nm) almost coincides with the resonance transition 3p - 4s (3/2,1) (λ_{res} = 106.66 nm). Thus the conversions ω_{vuv} = $5\omega_1$ and ω_{vuv} = $4\omega_1 + \omega_2$ are almost three- and four-photon resonant. Because of the resonant enhancements, these conversions produce intense VUV at 63.9 and 71 nm. Tuning ω_2 in the range λ_2 = 216-800 nm, the mixing ω_{vuv} = $4\omega_1 + \omega_2$ provides VUV at λ_{vuv} = 58-72 nm. In Ne, similar frequency mixing processes generate light in the region of λ_{vuv} = 50-57 nm.

In addition to the conversion of pulsed dye laser radiation single frequency cw coherent VUV light is generated by tripling and sum frequency mixing the outputs of stabilized dye lasers (Spectra Physics Model 380 D) in Mg and Sr vapor [7]. Tuning, for example λ_L to the Mg two-photon resonance 3^1S_0 - 3^1D_2(λ_L = 430.88 nm) a laser power P_L of 0.2 W generated VUV radiation (λ_{vuv} = 143.6 nm) of more than $1.2 \cdot 10^5$ photons/sec (P_{vuv} = $1.8 \cdot 10^{-13}$ W). In a similar way resonant sum frequency mixing of the radiation of a Rhodamine 6G dye laser (tuned to the Sr 4^1S_0 - 4^1D_2 two-photon transition (λ_L = 575.8 nm)), of a single-mode Ar^+ laser and a blue dye ring laser produced in Sr up to $3 \cdot 10^{10}$ photons at wavelengths in the range of λ_{vuv} = 171-192 nm.

An increase of the VUV output should be obtained at higher input power which will be achieved by placing the conversion cell into an external ring resonator. The expected VUV output of 10^{10} to 10^{12} photons/sec will be sufficient for spectroscopic applications. Because of the narrow line width and the very precise frequency control of cw laser systems the generated cw VUV will render possible laser VUV spectroscopy of highest spectral resolution.

References:
1 G.C. Bjorklund, IEEE J. Quantum Electron., QE-11, 287 (1975).
2 R. Wallenstein, Laser u. Optoelektron. 14, 29 (1982), and references therein.
3 W. Zapka, D. Cotter, and U. Brackmann, Opt. Commun. 36, (1981).
4 R. Hilbig, and R. Wallenstein, IEEE J. Quantum Electron., QE-17, 1566 (1981); Opt. Commun. 44, 283 (1983); Opt. Commun. 49, 297 (1984).
5 R. Hilbig, and R. Wallenstein, Appl. Optics, 21, 913 (1982).
6 R. Hilbig, and R. Wallenstein, IEEE J. Quantum Electron., QE-19, 194 (1983); IEEE J. Quantum Electron., QE-19, 1785 (1983); and references therein.
7 A. Timmermann, and R. Wallenstein, Opt. Lett. 8, 517 (1983).

Generation of Coherent UV- and VUV-Radiation by the Anti-Stokes Raman Process

H. Welling, K. Ludewigt, H. Schmidt, R. Dierking, and B. Wellegehausen

Institut für Quantenoptik, Universität Hannover, Welfengarten 1, D-3000 Hannover 1, Fed. Rep. of Germany

Frequency up-conversion by the anti-Stokes Raman laser process offers interesting perspectives for an efficient generation of powerful tunable coherent radiation in the uv and vuv spectral range [1,2]. Candidates for pulsed anti-Stokes lasers in the 100 nm to 200 nm spectral range are the atoms of group VI: O, S, Se and Te, where the high lying metastable ^1S levels can be used as storage levels in the up-conversion process [3].

In this contribution, first realization of Se-anti-Stokes Raman laser oscillation at 158.7 nm and 167.5 nm is reported (Fig. 1). The necessary population inversion between the $4p^1S_0$ metastable starting level and the lower lying $4p^1D_2$ or $4p^3P$ final levels is generated by photodissociation of the gaseous molecule COSe with ArF (193 nm) excimer laser radiation. For a COSe vapor pressure of about 0.5 mbar and a photodissociation energy of typically 10 mJ (15 ns; intensity ~ $1 \cdot 10^7$ Wcm^{-2}) inversion densities of about $3 \cdot 10^{15}$ cm^{-3} are expected [4].

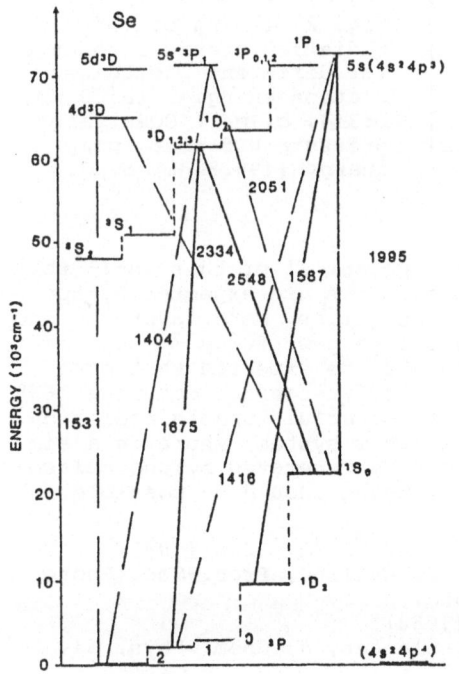

Fig. 1 Se-level scheme.
—— observed pump- and anti--Stokes laser transitions.
--- transitions, where strong anti-Stokes fluorescence has been observed with pump radiation of a few µJ (wavelengths in Å).

183

For the conversion of 254.8 nm into 167.5 nm ($^1S_0 \rightarrow {}^3D_1 \rightarrow {}^3P_1$) obtained data are a threshold pump energy of 20 µJ (10 ns) for a cell length of 14 cm and an inversion density of about $3 \cdot 10^{15}$ cm^{-3}. At a maximum pump energy of 2 mJ (obtained by frequency doubling of dye laser radiation with a cooled ADP crystal) a tuning range of ± 5 cm^{-1} and an output energy of about 5 µJ was achieved.

For the conversion of 199.5 nm into 158.7 nm ($^1S_0 \rightarrow {}^1P_1 \rightarrow {}^1D_2$) threshold pump energies of less than 0.5 µJ have been observed for the given conditions (cell length, inversion density). The required pump radiation at 199.5 nm was generated as the seventh anti-Stokes component of 475 nm dye radiation in molecular hydrogen. For a maximum pump energy of only 5 µJ we could obtain an output energy of more than 1 µJ and a tuning range of ± 10 cm^{-1} (Fig. 2).

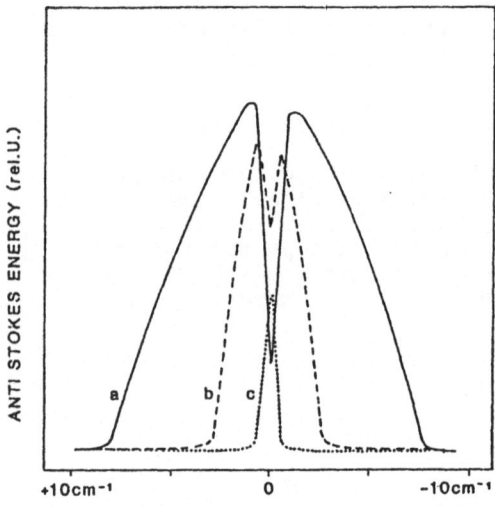

PUMPLASER DETUNING (cm⁻¹)

Fig. 2 Tuning of 158.7 nm radiation around resonance for different photodissociation energies (a: 10 mJ, b: 3mJ, c: 1mJ). COSe vapor pressure 0.5 mbar, pump energy (199.5 nm) 5 µJ.

These first data demonstrate the potential of this anti-Stokes laser process. Further strong improvements are expected by applying longer cells, higher inversion densities, more favorable photodissociation wavelengths [4], and more powerful pump sources. Scaling up present results, it can also be expected that even tunable ArF laser radiation (193 nm; off-resonance detuning 1668 cm^{-1}) may be directly converted into powerful tunable radiation around 154.5 nm. Considering the sulphur system, where in a similar way the necessary inversion can be generated by photodissociation of COS, still shorter wavelengths should be possible.

1 J.C. White, D. Henderson, Phys. Rev. A 25, 1226 (1982)
2 B. Wellegehausen, K. Ludewigt, H. Welling, Proc.-Soc. Photoopt. Instr. Eng. 492, 10 (1985)
3 J.C. White, Opt. Lett. 9, 38 (1984)
4 G. Black, R.L. Sharpless, T.G. Slanger, J. Chem. Phys. 64, 3985 (1976)

Part VII

Nonlinear Optics
and Wave Mixing Spectroscopy

The Relationship Between Collision-Assisted Zeeman and Hanle Resonances and Transverse Optical Pumping

N. Bloembergen and Y.H. Zou*

Division of Applied Sciences, Harvard University,
Cambridge, MA 02138, USA

Four-wave light mixing experiments to study the phenomenon of collision-induced coherence were reported at two previous laser spectroscopy conferences [1,2]. In the meantime a full-length paper has reviewed the collision-induced Zeeman coherences [3]. More recently, the observation of collision-induced Hanle resonances in zero magnetic field has also been reported by us [4]. The connection of these collision-assisted Hanle resonances and transverse optical pumping was pointed out [5]. The language of optical pumping [6] was also employed by KOSTER et al. [7] to describe Hanle resonances in phase conjugate four-wave mixing experiments. It is the purpose of this note to elaborate on this relationship.and to discuss differences in the behavior of the collision-assisted four-wave-mixing signal for tuning near the D_1 and D_2 resonance lines, respectively.

The experimental configuration of the phase conjugate four-wave mixing experiment is shown schematically in Fig. 1. The earth's magnetic field is carefully balanced by three pairs of Helmholz coils. A small field $B_0\hat{z}$ can be applied at right angles to the $\hat{x}$-direction of propagation of the light beams. The incident beams with wave vectors k_1 and k_2 make a small angle θ of about 10^{-2} radians. These linearly polarized beams are orthogonally polarized in the $\hat{z}$- and $\hat{y}$-direction, respectively. Their frequency is tuned with a deviation Δ from D_1- or D_2-resonance, respectively, where Δ lies typically in the range 10-100 GHz. These two beams set up a coherence grating [3] with wave vector $\Delta k = k_1 - k_2$ in the Na-vapor. The grating is probed by the backward wave k_3 in the negative $\hat{x}$-direction, polarized along $\hat{z}$, and leads to the phase-conjugate output beam with wave vector k_4, polarized in the $\hat{y}$-direction. All beams are derived from an argon laser pumped, single mode, frequency-stabilized c.w. dye laser (Coherent 599-21).

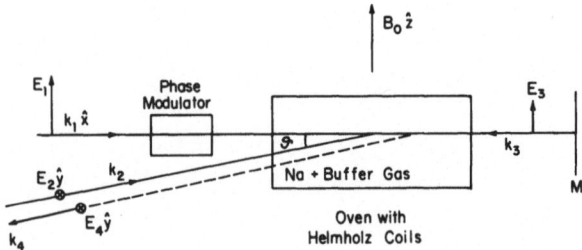

Fig.1 Geometry for the collision-enhanced Zeeman and Hanle resonances in four-wave light mixing

---

*Chinese Visiting Scholar, on leave from Peking University, Beijing, People's Republic of China

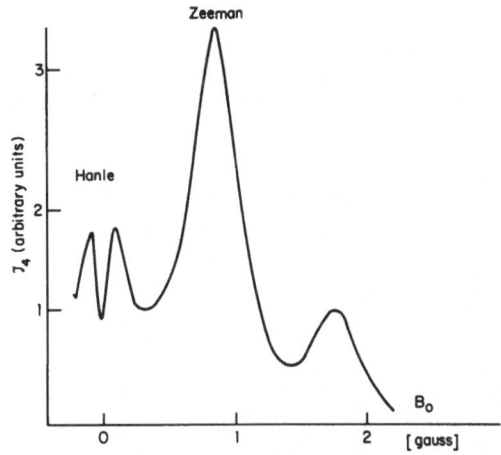

Fig.2 The intensity $\mathscr{I}_4$ of the phase conjugate wave as a function of magnetic field. The phase of the light wave E_1 is modulated at 620 kHz. See text for further experimental details

A new feature is the addition of a phase modulator. The beam E_1 can be supplied with side bands at $\omega_1 = \omega \pm n\omega_m$. The beam E_2 is not modulated, $\omega_2 = \omega$. Thus the coherence grating is modulated at $n\omega_m$. As the magnetic field is varied, one observes Zeeman resonances for $n\hbar\omega_m = g\beta B_0$. For $B_0 = 0$, a Hanle-type resonance is obtained. In Fig. 2 the experimentally observed intensity of the phase-conjugate beam, $\mathscr{I}_4$, is plotted vs B_0. The modulation frequency is $\omega_m/2\pi = 620$ kHz. The power at the carrier frequency is 1.4 mW, or about 4 W/cm^2 in the focal region. The power in the first side band is 2 mW, and in the second side band 0.5 mW. A prominent Zeeman resonance occurs at $B_0 = 0.89$ gauss and a much weaker one at twice this field. The Hanle resonance at $B_0 = 0$ shows a strong saturation dip at this power level. The data were taken with $\Delta = 55$ GHz below $^2P_{1/2}$ at 2010 torr partial pressure of He. The angle between the beams is $\theta = 1°$, and the linewidth is determined by residual Doppler broadening and some power broadening.

Figure 3 shows a narrow Hanle-type resonance, obtained without phase modulation. The FWHM width is about 15 kHz. The detuning is $\Delta = 50$ GHz below the $^3P_{1/2}$ level. In this case the intensity in the interaction region is

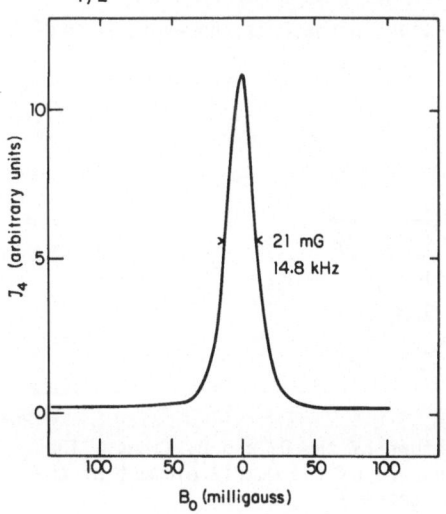

21 mG
14.8 kHz

Fig.3 Experimental recording of a Hanle resonance in four-wave light mixing. The FWHM width is less than 15 kHz. The experimental conditions are described in the text

reduced to 0.2 W/cm^2. The angle θ is reduced to 0.3°. The oven temperature is 294°C, corresponding to Na vapor pressure of about 2 millitorr. A high partial pressure of argon, p_{Ar} = 8400 torr, is used to further reduce the residual Doppler width,

$$\Delta\omega_{res\ D} = \bar{v} \cdot |k_1 - k_2| \approx \left(\frac{kT}{M}\right) |k_1|^2 \theta^2 \tau_c \tag{1}$$

Here $\tau_c \propto p_{Ar}^{-1}$ is the time between collisions of the Na-atom (with mass M) and the buffer gas, and $\bar{v}$ is the diffusive velocity. As shown previously [3,4], a perturbation theory approach results in the intensity of the phase conjugate beam $\mathscr{I}_4$ being proportional to

$$\mathscr{I}_4 \propto \left| \sum_{n,n'} \frac{\mu_{gn}\mu_{ng'}\mu_{g'n'}\mu_{n'g}}{(-\Delta+i\Gamma)(g_F\mu_B B_0 - (k_1-k_2)\cdot\bar{y}+i\Gamma_{gg'})} \left\{ \frac{i\Gamma(\rho_{gg}^{(0)} + \rho_{g'g'}^{(0)})}{\Delta^2 + \Gamma^2} \right\} \right|^2 \tag{2}$$

Here $|n\rangle$ and $|n'\rangle$ are states of the $3^2P_{1/2}$ or $3^2P_{3/2}$ manifold. Γ is the width of the 3S-3P transitions,

$$\Gamma = (1/2\ t_{sp}) + c_1 p \tag{3}$$

The buffer gas pressures used are always sufficiently high that the collisional broadening (10 MHz/torr for Na-Ar collisions) is very large compared to the natural linewidth for spontaneous emission. If nuclear spin is ignored, the ground states $|g\rangle$ and $|g'\rangle$ correspond to the electron spin states $m_s = \pm 1/2$, respectively, of the $3^2S_{1/2}$ configuration. The pertinent energy levels and the square of the pertinent matrix elements are reproduced in Fig. 4. The quantity $\Gamma_{gg'}$ corresponds to the width of a resonance between the two Zeeman levels of the ground state. This width is much narrower than the natural width (1/2 t_{sp}) and is determined by contributions from spin exchange (Na-Na) collisions. Collisions of Na with noble gas atoms are very ineffective in perturbing the ground state, although an increase of $\Gamma_{gg'}$ at high buffer gas pressure of Xe has been observed [5]. Equation (2) describes the observed features of the very narrow Hanle and Zeeman resonances, which have an integrated intensity proportional to $\Gamma^2 \propto p^2$. They are thus "collision-assisted".

The regime of validity of the perturbation approach is, however, severely restricted by the condition,

$$\left| \frac{\hbar^{-2}\mu_{gn}\mu_{ng'}E_1 E_2^* \Gamma}{\Delta^2 + \Gamma^2} \right| << \Gamma_{gg'} \tag{4}$$

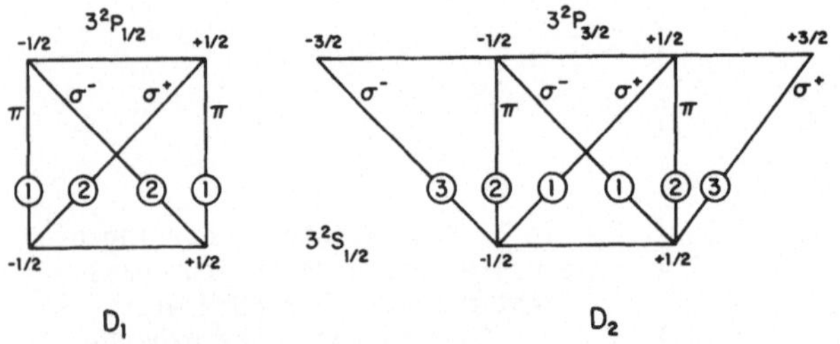

Fig.4 The level diagram and Zeeman structure of the D_1 and D_2 lines. The circled numbers are proportional to the square of the matrix element of the indicated transitions. (Nuclear spin is ignored)

It corresponds to the condition that changes in the diagonal elements of the density matrix, or of the populations, may be ignored in the lowest order calculation of $\rho_{gg}^{(2)}$, starting from $\rho_{gg}^{(o)} = \rho_{g'g'}^{(o)} = 1/2$.

In zero magnetic field there is, of course, no reason to prefer the $\hat{z}$-axis, and it is more logical to take the $\hat{x}$-axis, the direction of the light beams, as the axis of quantization. In the former case, one says that a coherence grating $\rho_{gg}^{(2)} \propto E_1 E_2^* \exp\{i(\underline{k}_1 - \underline{k}_2)\cdot r\}$ is created by the presence of the two light beams. In the latter case the beams cause an $\hat{x}$-directed population grating. In either case a transverse polarization is induced which would precess around the $\hat{z}$-axis with a frequency $\hbar^{-1}g\beta B_0$ in the presence of a magnetic field. It is quite clear that the two beams with orthogonal linear polarization, equal amplitudes $|E_1| = |E_2|$, and a relative phase-shift $(\underline{k}_1 - \underline{k}_2)\cdot r$, give rise to a left circular polarization at locations spaced by $(\underline{k}_1 - \underline{k}_2)\cdot r = 2m\pi$. Half way between these positions the light is circularly polarized in the opposite sense. The high buffer gas pressure prevents the Na-atoms from diffusing rapidly. Therefore these gratings exist for a time determined by the inverse width of the ground state resonances, $\Gamma_{gg'}^{-1}$, or $(\Delta\omega_{res\ D})^{-1}$. The language of transverse optical pumping [5,6,7] is therefore appropriate, and this permits an extension to the domain of stronger optical pumping where condition (4) is violated,

$$(1/2\ t_{sp}) \gg \left| \frac{\hbar^{-2}\mu_{gn}\mu_{ng'}E_1 E_2^*\Gamma}{\Delta^2 + \Gamma^2} \right| \gg \Gamma_{gg'} \qquad (5)$$

Note that the term in the middle of this inequality represents the spatial modulation of the pumping rate to the 3P-states by the presence of the two light beams. The collisions are extremely effective in producing an equal distribution over all 3P-states. The subsequent decay by spontaneous emission repopulates all ground states at equal rates. Since the repopulation rate $(1/2\ t_{sp})$ is fast, but the redistribution rate among the ground states, $\Gamma_{gg'}$, is slow compared to the pumping rates, the $\hat{x}$-directed ground state populations reach a steady state, where the population in each state is inversely proportional to the pumping-out rate. For large detuning Δ this rate is collision-assisted and given by the middle term in eq. (5). For a circularly polarized light near the D_1 line, the diagram in Fig. 4 shows that the total $\hat{x}$-directed population will be in the $m_s = + 1/2$ state, or $\rho_{1/2,1/2}^{(\hat{x})} = 1$ and $\rho_{-1/2,-1/2}^{(\hat{x})} = 0$. This is the maximum saturated polarization. For tuning near the D_2 line, the same transverse polarization will reach a limiting value given by $\rho_{1/2,1/2}^{(\hat{x})} = 1/4$, $\rho_{-1/2,-1/2}^{(\hat{x})} = 3/4$. These results of optical pumping for D_1 and D_2 light with complete spin reorientation in the 3P manifold correspond to one of the cases already discussed by FRANZEN [8].

It is clear by comparison of conditions (4) and (5) that the collision-assisted coherence for $\hat{z}$-directed quantization is equivalent to incipient collision-assisted transverse optical pumping with population changes for $\hat{x}$-directed quantization. When condition (4) is satisfied, the relative intensities of the four-wave mixing signal for equal detuning Δ from the D_1 and D_2 resonance, respectively, should lead to a ratio $\mathscr{I}_4(D_1)/\mathscr{I}_4(D_2) = 1$. This follows from the diagrams in Fig. 4, as in both cases only the product of one π- and one σ-transition matrix element, $\mu_z\mu_\pm$, is involved. In the limit of strong pumping, expressed by condition (5), the ratio should reach the limit $\mathscr{I}_4(D_1)/\mathscr{I}(D_2) = 4$, as the maximum attainable transverse polarization for D_2 is half as large as for D_1 pumping. We have verified this experimentally

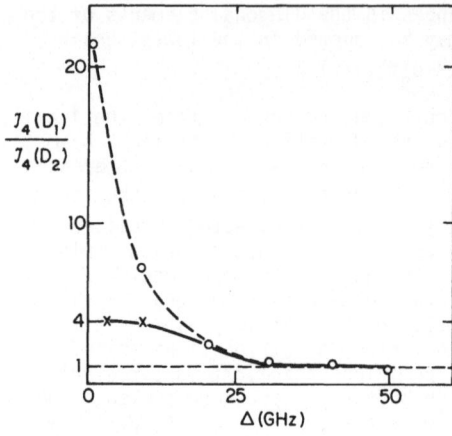

Fig.5 The ratio of the four-wave mixing signal intensity for the same detuning Δ from the D_1 and D_2 resonances, respectively. The following experimental parameters were used: $p_{Xe} = 2000$ torr, $\theta = 0.3°$. The intensity in the focal region is 0.2 W/cm^2. The circles. o, are experimental points for an oven temperature $T_C = 260°C$. The crosses, x, are for $T_C = 160°C$. Consult text for further discussion

by measuring this ratio as a function of the detuning Δ, keeping the light beam intensities and the buffer gas pressure, and consequently Γ, fixed. When the detuning is decreased, the crossover from condition (4) to (5) takes place, as illustrated in Fig. 5.

For small detuning Δ from the D_2 line, it is important to keep the Na-vapor pressure low to avoid attenuation by linear absorption of the circu-larly polarized light. The D_1 light is never absorbed for strong optical pumping. Complete bleaching takes place as the population of one of the ground states which could absorb is emptied. For higher Na vapor pressure and smaller detunings, the observed ratio $\mathscr{I}(D_1)/\mathscr{I}(D_2)$ may exceed 20, due to attenuation of D_2 light. This behavior is also indicated in Fig. 5.

When condition (5) is satisfied, the transverse polarization reaches a limiting value. Since the phase conjugate signal intensity $\mathscr{I}_4$ is propor-tional to the square of this polarization, the intensity should become inde-pendent of the light field amplitudes $|E_1|$ and $|E_2|$ and of Γ. Since the buffer gas pressure dependence of $\mathscr{I}_4$ is contained in Γ, the intensity $\mathscr{I}_4$ should become independent of the buffer gas pressure. This saturation of the signal at high buffer gas pressure has been observed and is shown in Fig. 6.

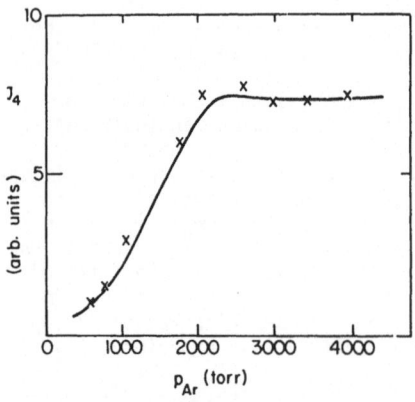

Fig.6 The four-wave mixing signal vs argon pressure. $\Delta = 20$ GHz below D_1 resonance, $\theta = 0.8°$, $T_C = 265°C$. The pump power density in the focal region is 8 W/cm^2

When a magnetic field $B_0\hat{z}$ is applied, the $\hat{x}$-directed polarization rotates around the $\hat{z}$-axis. A Zeeman resonance now occurs, for $\omega_1-\omega_2 = g\beta\hbar^{-1}B_0$. We may consider the pumping field to remain at ω_1, but the state of polarization at a fixed location in space is modulated at the difference frequency. When the polarization has turned through 180° from the $+\hat{x}$ to the $-\hat{x}$ direction, the polarization has switched from σ^- to σ^+ at the resonance. The pumping continues in the rotating frame. This situation of modulated transverse optical pumping in an external magnetic field was first considered by BELL and BLOOM more than 20 years ago [9]. Thus, the collision-induced coherent Zeeman resonances are equivalent to collision-assisted modulated transverse optical polarization pumping. This correspondence further clarifies the initially somewhat paradoxical nature of collision-induced coherence.

This research was supported by the Joint Services Electronics Program of the U.S. Department of Defense under contract N00014-84-K-0465.

References

1. N. Bloembergen, A.R. Bogdan, M.C. Downer: in Laser Spectroscopy V, edited by A.R. W. McKellar, T. Oka, B.P. Stoicheff (Springer, Heidelberg 1981), p. 157
2. L.J. Rothberg, N. Bloembergen: in Laser Spectroscopy VI, edited by H.P. Weber, W. Luthy (Springer, Heidelberg 1983), P. 178
3. L.J. Rothberg, N. Bloembergen: Phys. Rev. A 30, 820 (1984)
4. N. Bloembergen, Y.H. Zou, L.J. Rothberg: Phys. Rev. Lett. 54, 186 (1985)
5. N. Bloembergen: in Proceedings of the "Symposium Kastler", Paris, January 1985, edited by F. Laloë, J. de Physique, to be published
6. See, for example, C. Cohen-Tannoudji, A. Kastler: in Progress in Optics V, edited by E. Wolf (Wiley, New York 1966), p. 1
7. E. Köster, J. Mlynek, W. Lange: Opt. Commun. 53, 53 (1985)
8. W. Franzen, A.G. Emslie: Phys. Rev. 108, 1453 (1957)
9. W.E. Bell, A.L. Bloom: Phys. Rev. Lett. 6, 280 (1961)

Ionization-Raman Double-Resonance Spectroscopy

P. Esherick and A. Owyoung

Sandia National Laboratories*, Division 1124,
Albuquerque, NM 87185, USA

1. Abstract

Ionization-detected stimulated Raman spectroscopy (IDSRS)[1] is a new technique for high-sensitivity, high-resolution Raman studies based on resonant
laser ionization/detection of vibrationally excited molecules. We present
results on the application of IDSRS to the Raman spectroscopy of benzene. Our
spectra concentrate on the region near 1600 cm^{-1} where the overlap of
transitions to the (ν_{16}, $\nu_2+\nu_{18}$) Fermi dyad produces a particularly complex
spectrum. By taking advantage of the double-resonance capabilities of IDSRS,
we are able to cleanly separate transitions to each of the two vibrational
bands, and thereby greatly simplify their spectra.

2. Introduction

The IDSRS excitation/detection process involves three sequential steps as
illustrated in Fig. 1. Molecules are vibrationally excited in the first step
via resonantly pumped stimulated Raman transitions, and then selectively ionized in the second step by a tunable UV source. In the final step, the ionized species are collected by biased electrodes,where they are detected as
current in an external circuit. Raman spectra are obtained by scanning the
frequency of one of the Raman pump sources and monitoring the ionization signal as a function of the frequency difference (Stokes shift) between this
laser and the second, fixed-frequency, visible pump laser. IDSRS is closely

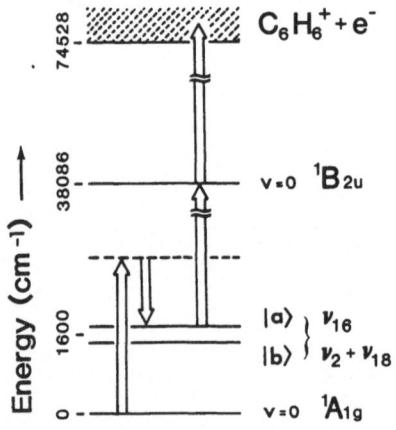

Fig. 1. Energy level diagram
illustrating the IDSRS excitation
scheme employed in our studies of
the ν_{16} and $\nu_2+\nu_{18}$ bands of benzene.

*This work performed at Sandia National Laboratories supported by the
U.S. Department of Energy under Contract number DE-ACO4-76DPO076.

related to several other optical-optical double-resonance techniques involving folded excitation schemes[2-5]. Most similar is the approach of ORR and co-workers[2] in which stimulated Raman pumping is monitored using laser-induced fluorescence rather than resonant ionization.

3. Apparatus

The basic elements of the IDSRS apparatus consist of the UV and Raman excitation laser sources, a simple pulsed molecular jet, and charge collection electrodes and detection electronics. The bandwidths of the Raman excitation sources are of particular importance to this experiment both for efficient pumping of the Raman transitions as well as for resolving the rotational detail within the spectrum. We achieve near-Fourier-transform-limited bandwidths (0.002 cm^{-1}) for these sources by pulse amplifying two single-mode CW sources, a 647.1-nm Kr ion laser and a tunable dye laser operating near 586 nm. Output energies of 5 mJ or more are achievable for each source, however due to the effects of AC Stark shift[6] induced line broadening, the best data are often obtained with lower energies.

A commercial frequency-doubled dye laser is used to generate the 274-nm light required to resonantly ionize the vibrationally excited benzene molecules. Output energies of several millijoules in a 3-4 ns pulse are easily attainable, even with a frequency narrowing intra-cavity etalon installed. The dye oscillator can be pressure scanned over a 25 cm^{-1} interval. Frequency markers during UV scans are provided by both a 1 cm^{-1} free-spectral-range etalon and an I_2 fluorescence cell.

Detection of the ionization signal relies on a fairly simple parallel-plate electrode collection system. With a maximum preamplifier charge-to-voltage sensitivity of 2 volts/picocoulomb, integrating in a boxcar averager gives us a best-case detection limit of the order of 0.001 picocoulombs or 6000 electrons.

The principal limitation on the sensitivity of IDSRS is the background ionization signal that is present even in the absence of any Raman pumping. In both our earlier studies of NO[1] and benzene the primary source of this background signal is the thermal population of the vibrational mode under study. This thermal background can be effectively eliminated by cooling the sample in a supersonic free-expansion jet, which also localizes the gas flow in the region where all three laser beams overlap.

Our free-expansion jet consists of a piezo-electric valve with a 2-mm orifice coupled to a 20-mm-long by 1-mm-i.d. teflon tube (see Fig. 1) which extends the orifice away from the metal face of the valve. This allows us to significantly decrease the distance between the orifice and the point at which we probe the expansion without disturbing the electric fields essential to our ion detection system. Typically, we operate at a distance of 5 mm from the tip of the teflon extension with 80-torr of benzene seeded in argon or helium at a total pressure of 700-750 torr. Assuming that molecules diverge from the orifice with a cone angle of 90°, our measured flow rates imply the benzene density at the point sampled is of the order 3×10^{15} molecules/cm^3, a density equivalent to roughly 0.1-torr pressure at room temperature.

4. IDSRS spectra of Benzene

The e_{2g} vibrational bands of benzene, including the (v_{16}, v_2+v_{18}) Fermi dyad, have recently been studied via spontaneous Raman methods by HOLLINGER and WELSH[7]. Their spectra show evidence of rotational structure, however, the limited resolving power available in spontaneous Raman spectra precluded complete resolution of the rotational structure. The region near 1600 cm^{-1}, which lies almost precisely halfway between the band centers ν_{16} and $\nu_2+\nu_{18}$ Fermi dyad, is extremely congested due to the overlapping of transitions from the upper and lower components of the dyad. However, by taking advantage of the double-resonance capabilities of IDSRS we are able to isolate the two bands, using the UV resonances as a means to pick out for display the rotational peaks of either one or the other vibrational transition. The success of this technique is illustrated in Fig. 2.

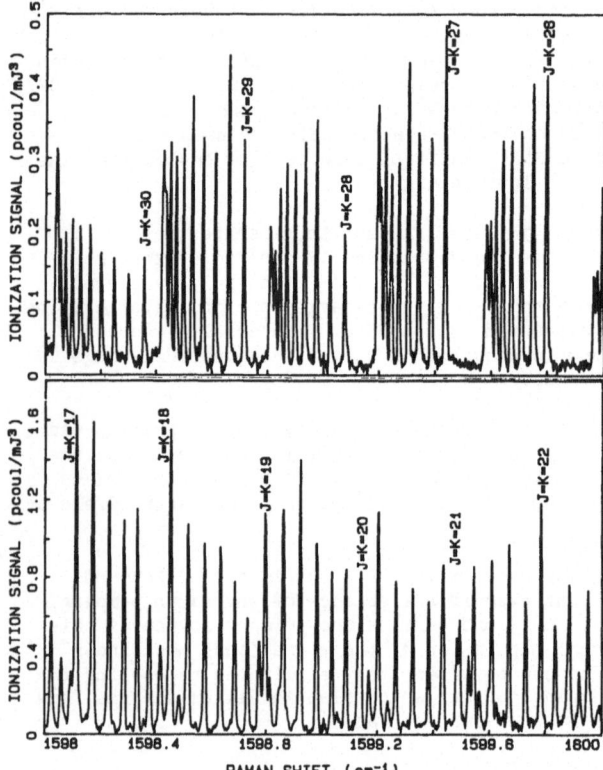

Fig. 2. Ionization detected Raman spectra in the region of overlap between the ^OO branch transitions of the upper (1609 cm^{-1}) vibrational state and the ^SS branch transitions of the lower (1591 cm^{-1}) state. (a) UV wavelength tuned to 36467 cm^{-1} in the ^PP branch of the upper state.

(b) UV wavelength tuned to 36496 cm^{-1} in the ^rR branch of the lower state.

In Fig. 2 (a) we selectively display transitions connected to the upper level of the Fermi dyad simply by setting the UV source to a frequency corresponding to ^PP branch transitions to $^1B_{2u}$ originating from this level. Since this frequency is well below the absorptions to $^1B_{2u}$ from the lower vibrational level, we completely exclude the contributions to the Raman spectrum from the lower level. The reverse is true, however, in Fig. 2b where we have tuned the UV source to be resonant with ^rR branch transitions originating from the lower level.

We have obtained additional spectra of benzene over the spectral region from 1583 to 1617 cm^{-1}. This region covers the majority of the transitions arising from the ν_{16} and $\nu_2+\nu_{18}$ Fermi dyad. To our knowledge this constitutes the first high-quality, rotationally resolved Raman spectrum of a degenerate vibrational mode in a polyatomic molecule. A full analysis of this data, with the intent of detailing the effects of the Fermi interaction of these two close-lying states, is currently underway[8,9] in collaboration with J. Pliva of Pennsylvania State University.

5. Conclusions

The application of resonant laser ionization techniques in conjunction with stimulated Raman pumping has allowed us to obtain high-resolution Raman spectra with unprecedented sensitivities. By using near-Fourier-transform-limited bandwidth lasers in the stimulated Raman pumping step we are able to simultaneously achieve an instrumental spectral resolution of the same order as the Doppler width of the transitions studied. Integration of IDSRS with time-of-flight mass spectrometry and particle counting detection will render the method even more sensitive and selective.

Acknowledgment

The authors would like to acknowledge the expert technical assistance of Mr. A. W. Staton in the development and maintenance of the experimental apparatus used in this effort.

References:

1. P. Esherick and A. Owyoung, Chem. Phys. Lett. 103, 235 (1983).
2. D. A. King, R. Haines, N. R. Isenor and B. J. Orr, Optics Letters 8, 629 (1983).
3. C. Kittrell, E. Abramson, J. L. Kinsey, S. A. McDonald, D. E. Reisner, R. W. Field, and D. H. Katayama, J. Chem. Phys. 75, 2056 (1981).
4. D. E. Reisner, P. H. Vaccaro, C. Kittrell, R. W. Field, J. L. Kinsey, and H.-L. Dai, J. Chem. Phys. 77, 573 (1982).
5. D. E. Cooper, C. M. Klimcak and J. E. Wessel, Phys. Rev. Lett. 46, 324 (1981).
6. L. A. Rahn, R. L. Farrow, M. L. Kozykowski and P. L. Mattern, Phys. Rev. Lett. 45, 620 (1980).
7. A. B. Hollinger and H. L. Welsh, Can. J. Phys. 56, 1513 (1978).
8. P. Esherick, A. Owyoung and J. Pliva, submitted to J. Chem. Phys.
9. J. Pliva, P. Esherick and A. Owyoung, to be published.

Raman-Optical Double Resonance Spectroscopy of Glyoxal Vapor

B.J. Orr, A.B. Duval*, and D.A. King†

School of Chemistry, University of New South Wales, P.O. Box 1, Kensington, Sydney, 2033 Australia

The range of coherent nonlinear-optical techniques available for Raman spectroscopy has recently been enhanced by the development of Raman-optical double resonance (Raman-ODR) methods, in which the effect of coherent Raman excitation of a molecule is monitored by either laser-induced fluorescence (LIF) [1] or multiphoton ionisation [2]. Representative applications [1,2] of these methods have demonstrated a 1000-fold enhancement of Raman spectroscopic sensitivity, relative to that for more conventional techniques such as coherent anti-Stokes Raman spectroscopy (CARS) or stimulated Raman gain (SRG) spectroscopy. The feasibility of this approach is supported by studies of other coherent Raman phenomena, such as photoacoustic Raman spectroscopy (PARS) or CARS saturation, and by predictions of SRG coefficients.

Figure 1 depicts the Raman-LIF double-resonance (Raman-LIFDR) excitation scheme used in our work. The energy levels and wavelengths specified are those of glyoxal ($C_2H_2O_2$), a planar centrosymmetric molecule which is in many

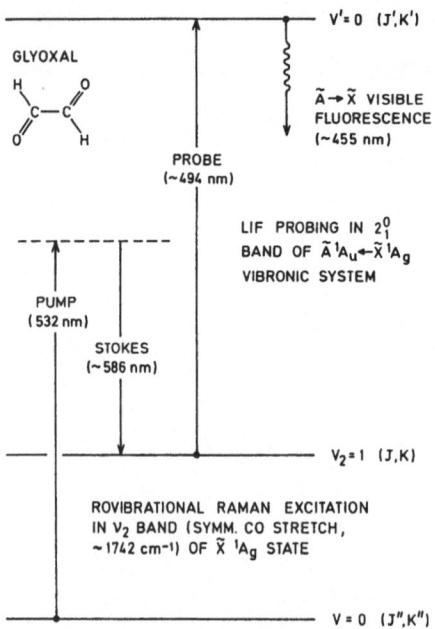

Fig. 1 Optical excitation scheme for time-resolved Raman-LIFDR experiments in glyoxal. Coherent rovibrational Raman excitation in the v_2 (symmetric C=O stretch, ~ 1742 cm^{-1}) band is probed by rovibronic absorption in the A←X 2^0_1 hot band (~ 494 nm) and then fluorescence in the A→X (0-0) band (~ 455 nm). Labels such as (J,K) denote rotational states in the limit of a prolate symmetric rotor, which acts as an adequate approximation in this context

* Now at : BHP Central Research Laboratory, Wallsend, AUSTRALIA 2287
† Now at : Ginzton Laboratory, Stanford University, Stanford, CA 94305, U.S.A.

respects ideal for exploratory Raman–LIFDR experiments. Our technique [1,3] requires three pulsed laser beams, of which two provide the Raman PUMP and STOKES radiation for coherent Raman excitation and the third functions as the PROBE for LIF detection. Signal strength is critically dependent on the temporal overlap of the three laser pulses (which is optimised by monitoring CARS and/or PARS signals) and on the focusing and alignment of the PROBE beam with respect to the Raman-excited zone. The remainder of the detection system (bandpass optical filter, photomultiplier and boxcar detector) derives its relative simplicity from the spectroscopic specificity and timing precision provided by the Raman-pumping/LIF-probing pulse sequence.

Figure 2 illustrates the sensitivity attainable with Raman–LIFDR. The lower trace is a survey LIF excitation spectrum of glyoxal vapor in the vicinity of the $A \leftarrow X\ 2^0_1$ vibronic hot band. The origin of this band is indicated, as is some of its more prominent rotational fine structure, consisting of a series of ^rR (i.e. $\Delta J = \Delta K = +1$) sub-band heads for K=7–13. The upper trace of Fig.2 shows the corresponding spectrum after 230-fold reduction of the glyoxal vapor pressure, with Raman excitation applied at a fixed Stokes-shift frequency ($\omega_P - \omega_S$) set ~2 cm^{-1} below that of the ν_2 band origin, thereby enhancing the intensity of much of the rovibronic band contour by a factor of at least 600. Furthermore, the product of pulse-sequence time delay (~15 ns) and sample pressure (~0.13 Torr) is small (~2 ns Torr) compared to that for hard-sphere gas-kinetic collisions (~130 ns Torr), so that the Raman–LIFDR features in Fig.2 have been generated under effectively collision-free conditions.

The result of tuning the Stokes-shift frequency ($\omega_P - \omega_S$) to the spectrally congested central peak of the ν_2 band, as in Fig.2, is that many rotational states (J,K) in the $\nu_2=1$ vibrational level are simultaneously Raman-excited. This yields a Raman–LIFDR spectral contour which, although greatly enhanced in

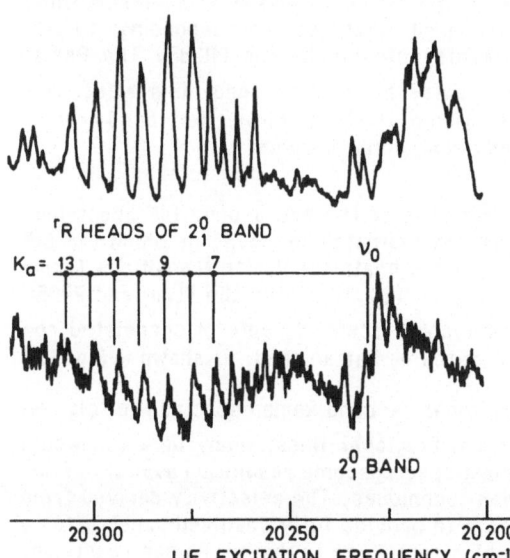

Fig.2 Raman–LIFDR enhancement of rovibronic hot-band absorption in glyoxal vapor at 35°C. A LIF excitation spectrum, recorded at a pressure of 30 Torr and with an optical bandwidth of ~0.7 cm^{-1}, is shown in the lower trace. Upper trace is a Raman–LIFDR spectrum, recorded under the following conditions : glyoxal vapor pressure is reduced to ~0.13 Torr; Raman-excitation (Stokes shift) frequency is fixed at ~1740 cm^{-1}, with ~0.1-cm^{-1} effective bandwidth and ~17 mJ in both Raman PUMP and STOKES pulses; delay between Raman-excitation and PROBE pulses is ~15 ns; other conditions are common to both traces

197

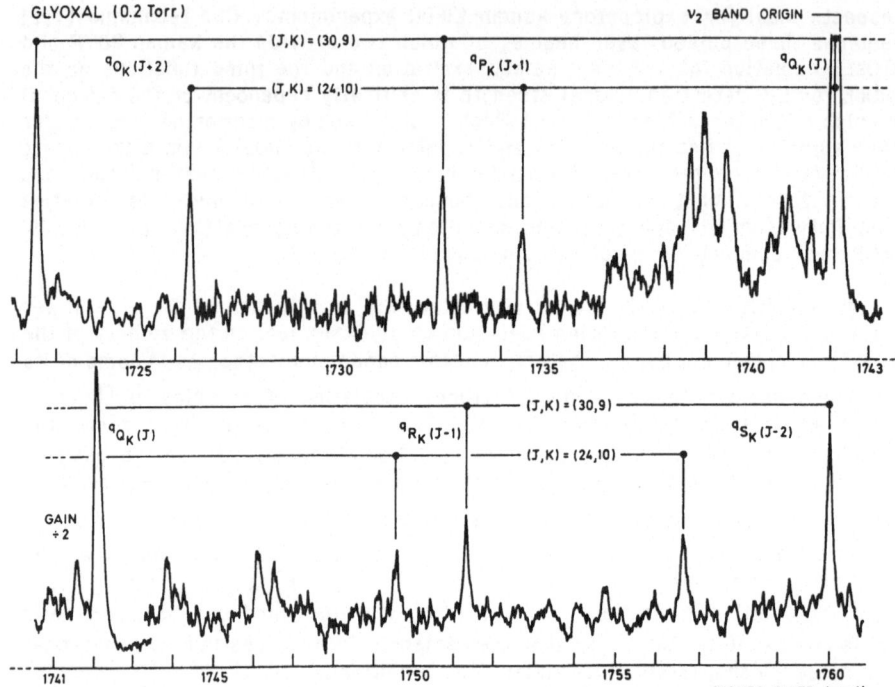

GLYOXAL (0.2 Torr)

v_2 BAND ORIGIN

$(J,K) = (30,9)$

$^qO_K(J+2)$

$(J,K) = (24,10)$

$^qP_K(J+1)$

$^qQ_K(J)$

1725   1730   1735   1740   1743

$^qQ_K(J)$

$(J,K) = (30,9)$

$^qR_K(J-1)$

$^qS_K(J-2)$

$(J,K) = (24,10)$

GAIN ÷2

1741   1745   1750   1755   1760

STOKES SHIFT (cm⁻¹)

Fig.3 State-specific Raman-LIFDR spectrum for glyoxal vapor at 0.2 Torr and 35°C. The PROBE frequency (optical bandwidth ~0.2 cm^{-1}) is held fixed at ~20272 cm^{-1}, which coincides with the $^rQ_9(30)$ and $^rP_{10}(24)$ transitions of the A←X 2^0_1 band. Tunable STOKES (~586.5 nm, 15 mJ/pulse, ~0.05 cm^{-1} bandwidth) and fixed-wavelength Raman PUMP (532 nm, 15 mJ/pulse, ~0.1 cm^{-1} bandwidth) radiation is used to generate a well-resolved Raman spectrum which consists of O-,P-,Q-,R-, and S-branch features corresponding to the intermediate ($v_2=1$) rovibrational levels selected by the PROBE. The Raman excitation pulses precede the PROBE pulse by ~15 ns, providing effectively collision-free conditions. The inset portion of the trace (lower left-hand region) has been recorded with a two-fold reduction in instrumental gain

intensity, differs only marginally from that of the background LIF spectrum. Much greater rotational specificity is demonstrated in Fig.3, in which the LIF PROBE conditions are employed to select rovibrational Raman transitions terminating in $v_2=1$ levels with (J,K)=(30,9) or (24,10). With the PUMP and PROBE wavelengths fixed and the STOKES wavelength scanned, sets of correlated rovibrational features are projected out of the v_2 Raman band, as shown in Fig.3.

It should be noted that the conventional v_2-band Raman spectrum of glyoxal would consist of a great number of rovibrational lines, many of which would remain overlapped even with the highest spectroscopic resolution available from either spontaneous or coherent Raman techniques. The selectivity derived from our double-resonance approach is able to provide fully resolved spectra for a small subset of the quantum states which contribute to the full Raman spectrum.

Moreover, this has been realised with relatively modest bandwidth (~ 0.1 cm^{-1}) for the Raman-excitation process. Figure 3 shows a number of additional Raman-LIFDR spectral features (e.g., those around 1738-1741 cm^{-1}), which are most readily ascribed to contributions from hot bands involving sequences in the low-frequency modes, v_7 and v_{12} ; such features correlate closely with otherwise anomalous v_2-band contours observed in CARS and PARS spectra of glyoxal [3].

A variety of Raman-LIFDR spectra such as that in Fig.3 have been recorded and used to determine new spectroscopic constants for glyoxal. Differences in the rotational constants $\langle B \rangle = \frac{1}{2}(B+C)$ and $(A-\langle B \rangle)$ between the A,v'=0 and X,v_2=1 vibronic states are: $10^3\Delta\langle B \rangle = -3.8 \pm 0.2$ cm^{-1} and $\Delta(A-\langle B \rangle) = +0.126 \pm 0.012$ cm^{-1}; these agree satisfactorily with previous estimates [4]. In addition, the vibrational fundamental v_2 is found to be 1742.3$\pm$0.5 cm^{-1}, which implies that the origin for the A$\leftarrow$X 2^0_1 band is 20231.1$\pm$0.5 cm^{-1}.

In a more mechanistic application of Raman-LIFDR, it has been possible to make direct observations of collision-induced rotational relaxation in the ground electronic manifold of glyoxal. Variation of the time delay between Raman-excitation and LIF-PROBE pulses provides kinetic data which show that the rotational-relaxation rate is ~ 6.5 times gas-kinetic. Such collision-induced rotational relaxation out of X,v_2=1(J,K) rovibrational states of glyoxal is found to be remarkably similar, both in terms of rate and lack of propensity rules, to that out of the excited A,v'=0 (J',K') rovibronic states [5].

Our investigations of glyoxal have been used in this paper to indicate the potential of the Raman-LIFDR technique with regard to sensitivity enhancement, spectroscopic information, and energy-transfer studies. Although efficient Raman-ODR excitation schemes are restricted to a limited range of molecules and vibrational modes, that range is sufficiently extensive to ensure the utility of Raman-LIFDR in a variety of future applications. Its ultimate potential for studies of state-selective molecular dynamics is extremely encouraging.

[1] D.A.King, R.Haines, N.R.Isenor & B.J.Orr : Optics Letters **8**, 629 (1983)
[2] P.Esherick and A.Owyoung : Chem.Phys.Letters **103**, 235 (1983)
[3] A.B.Duval, D.A.King, R.Haines, N.R.Isenor & B.J.Orr : "Fluorescence-detected Raman-optical double resonance spectroscopy of glyoxal vapor", J. Opt. Soc. Amer. B, in press (1985) ; "Coherent Raman spectroscopy of glyoxal vapour", J. Raman Spectrosc., in press (1985)
[4] C.S.Parmenter and B.F.Rordorf : Chem. Phys. **27**, 1 (1978)
[5] B.F.Rordorf, A.E.W.Knight & C.S.Parmenter : Chem. Phys. **27**, 11 (1978)

Selective Vibrational Pumping of a Molecular Beam by Stimulated Raman Process

F. Shimizu

Department of Applied Physics, University of Tokyo,
Bunkyo-ku, Tokyo 113, Japan

K. Shimizu and H. Takuma

Institute for Laser Science, University of Electro-Communications,
Chofu-shi, Tokyo 182, Japan

The purpose of this paper is to demonstrate a general technique of the selective labeling of internal energy levels for the analysis of thermal collision processes between neutral molecules. Laser is undoubtedly one of the most powerful tools for this purpose. Number of studies have been reported on the internal state distribution of molecules after collision using laser-induced fluorescence technique (LIF). However, it is rather difficult to label individual energy level of molecules before the collision. An electronically excited state can be selectively pumped by direct transition using tunable lasers. For vibrationally and rotationally excited states, prohibitive number of lasers are necessary to change large quantum number by direct pumping. An alternative method is to pump out thermal population, which is applicable only for low-lying levels. We have demonstrated folded two photon process via an electronically excited state can be used to pump selectively a vibrationally excited level of a molecular beam [1,2]. This technique can be applied for a wide range of molecules, because the selection rule on vibrational transitions $\Delta v=1$ does not apply if the potential surfaces of the excited and ground electronic states have different shape.

Several schemes of the two-step pumping are possible for the present purpose. Two important factors to affect the result are the relative magnitude between the transit time τ_t and the spontaneous lifetime τ_s of the intermediate level, and the amount of frequency detuning. Three possibilities are shown schematically in Fig. 1. The first laser L_1 is resonant to the transition between the ground level S_1 and the intermediate electronically excited level S_2. The second laser L_2 connects S_2 and a vibrationally excited level S_3. In Fig. 1(a), L_1 is on resonance and τ_t is longer than τ_s. The population of S_1 is pumped by L_1 to S_2, and then decays to a wide range of vibrationally excited levels by spontaneous emission. The application of L_2 does the same role on S_3 as L_1 does on S_1. Therefore, S_3 is marked as a hole among other vibrational levels which are populated by the spontaneous emission from S_2. This is not an efficient method of labelling, because the amount of population in S_3 is only a small fraction of initial population in S_1. When $\tau_s > \tau_t$ (Fig. 1(b)), the initial population in S_1 is shared equally among three levels during the interaction, if the

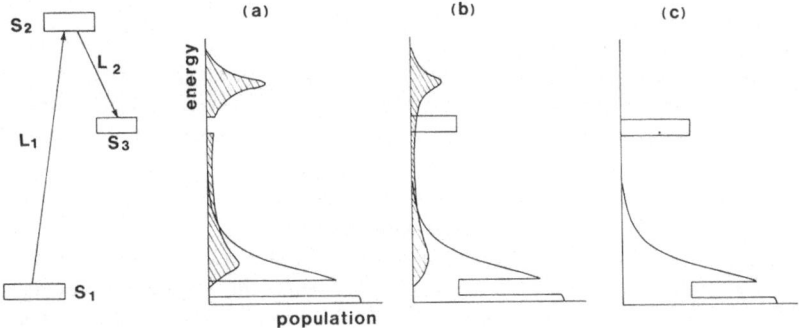

Fig. 1 Schematic drawings of three possible pumping schemes. (a) Lasers are on resonance and $\tau_t > \tau_s$, (b) $\tau_t < \tau_s$ and (c) stimulated Raman pumping.

laser intensities are sufficiently strong. After the molecule passes through the interaction region, the population in S_2 gradually decays to vibrationally excited levels. The final population in S_3 is 1/3 of the initial S_1 population. For the laser diameter of 1mm, typical transit time is 1µs. Therefore, when $S_1 - S_2$ is an allowed transition, we usually expect the case in Fig. 1(a). We may use spin-forbidden or weaker transition to realize the case in Fig.1(b). More attractive scheme is to use stimulated Raman process (Fig. 1(c)), where L_1 and L_2 are slightly off-resonant to respective transitions. Since S_2 is not actually populated, the spontaneous emission is irrelevant. The initial population in S_1 is shared equally between S_1 and S_3. They do not change even after the molecule passes through the interaction region. The population in levels other than S_1 and S_3 is not affected. Although the Raman pumping is a higher order process compared to the direct pumping, the efficiency at the same laser intensity is approximately the same, provided that the molecule has an appropriate electronically excited state. For the direct pumping, the transition moment must be sufficiently small not to depopulate S_3 by spontaneous emission. Therefore, a higher power is necessary. The laser power requirement for both cases is roughly $|E|^2 \approx \hbar^2/(\tau_t \tau_s \mu^2)$, where μ is the electronic transition moment, and E is the field amplitude of the pumping laser. Note that $\tau_s \mu^2$ is independent of the magnitude of the transition moment.

More detailed analysis is done by using familiar density matrix equations for a three-level system interacting two near-resonant monochromatic fields. The effect of the transit time is incorporated by introducing the effective decay rate $1/\tau_t$. Then, the equations can be easily solved. The results of the numerical evaluation are summarized as follows. (i) The number of molecules I_3 in S_3 normalized by the number in S_1 before the interaction is larger than 0.2, if both L_1 and L_2 are sufficiently intense to saturate respective transitions. For many diatomic molecules, this condition is satisfied by relatively low power lasers, such as ion laser pumped dye lasers. (ii) The width of the Raman resonance at the peak of I_3 is only several times of the transit width. This requires a very good control of the laser frequency and limits the angle of the molecular beam to be pumped to a very small value. This occurs, because the frequency detuning at the maximum I_3 increases with increasing laser intensity. One may operate the lasers closer to the resonance to increase the resonance width. (iii) The population in unwanted vibrational levels (I_n) can be suppressed only by increasing the intensity of L_2 and the frequency detuning. At large detuning, the ratio I_n/I_3 is approximately $4\tau_t^2/|x_2|^2$, where $x_2 = \mu_2 E_2/\hbar$ is the Rabi frequency of L_2. Figure 2 shows a typical laser power-dependence of I_3 and I_n/I_3 when the spontaneous width is 10 times larger than the transit width. The branching ratio of the spontaneous emission from S_2 to S_3 is assumed to be 10^{-2}.

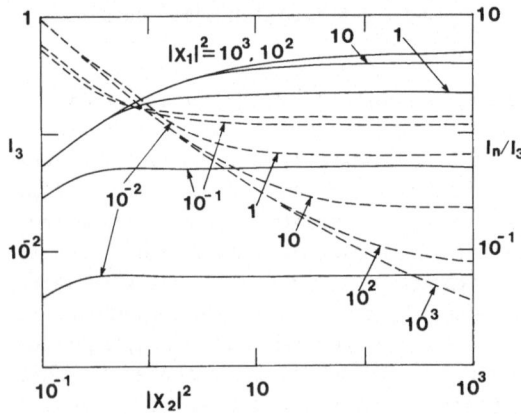

Fig.2 The relative population I_3 in S_3 (solid lines) and the ratio I_n/I_3 (broken lines) as a funcion of laser intensities $|x_1|^2$ and $|x_2|^2$. The subscripts 1 and 2 denote L_1 and L_2, respectively. The laser power is in unit of the square of the transit width, $1/\tau_t^2$.

The experiment was done using a Na_2 supersonic molecular beam. The Na_2 beam was produced from a 70μm pinhole on 50μm thick tantalum foil. It was seeded by Ar gas with the stagnation pressure of 2 to 4 atmosphere. The beam crossed the pumping lasers L_1 and L_2 perpendicularly at 15mm downstream of the nozzle just before the first skimmer. We employed the levels $X^1\Sigma_g^+(0,3)$, $A^1\Sigma_u^+(25,4)$ and $X^1\Sigma_g^+(31,5)$ for S_1, S_2 and S_3, respectively. The population in S_3 was monitored by LIF using the third laser L_3 at 50mm further downstream from the pumping point. The transition used for the monitoring was between S_3 and $A^1\Sigma_u^+(52,4)$ (S_4). The laser diameter at the crossing point was approximately 300μm. S_2 has the spontaneous lifetime of 12ns. Considering the average Na_2 speed of 10^3m/s, the ratio τ_t/τ_s in this experiment was approximately 30. The branching ratio of the transition from S_2 to S_3 was approximately 1%, which was measured by the fluorescence spectrum from S_2. The laser power was 300mW for L_1 and 80mW for L_2 at the pumping point. Figure 3(a) shows LIF intensity from S_4 as a function of the detuning of L_1 and L_2. L_2 is scanned repeatedly around the resonance, while L_1 is scanned slowly in one direction. The peak which run diagonal in the figure shows the stimulated Raman pumping, and the peak along $\omega_1=0$ is the population created by the

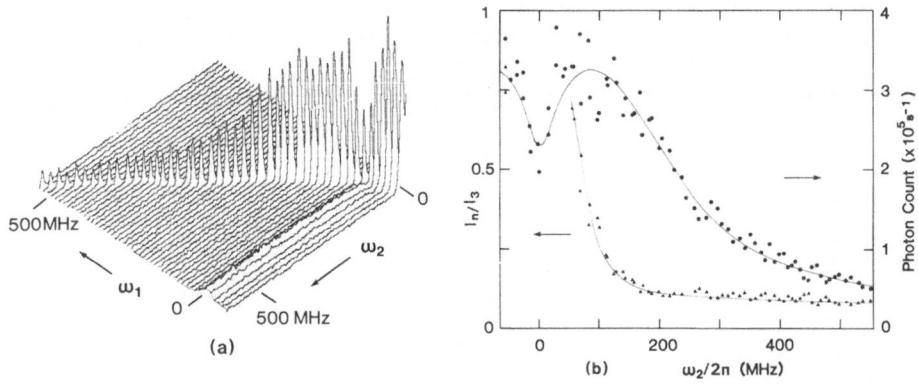

Fig.3 (a) Observed fluorescence intensity from S_4 as a function of detunings ω_1 for L_1 and ω_2 for L_2. (b) The fluorescence intensity along the Raman peak and the ratio I_n/I_3.

spontaneous emission. The Raman peak have the maximum at around $\omega_1 \approx \omega_2 = 100MHz$, and its intensity is 3.3×10^5 photoelectrons/s. Considering the overall detection efficiency of less than 10^{-2}, the beam flux in S_3 is larger than 3×10^7, which is sufficient for many collision experiments. The relative pupulation I_3 is estimated from the hight of the spontaneous peak $(2.6 \times 10^4 s^{-1})$ and the branching ratio from S_2 to S_3 (9.7×10^{-3}). The result is $I_3 = 0.13$ at the maximum. The ratio I_n/I_3 can be also estimated from the width of the spontaneous peak, and is shown in Fig. 3(b) as a function of the detuning $\omega_1 = \omega_2$, together with the fluorescence intensity at the Raman peak. The width of the Raman peak was 18MHz at the peak of I_3. The above results agree within a few factor from the theoretical prediction. The diffecence may be attributed to the spacial inhomogeneity of pumping lasers. Generally, it decreases I_3 and increases the resonance width. It is rather difficult to estimate quantitatively the effect of the inhomogeneity. In the actual application of this technique, it is preferable to analyze beam characteristics after it is pumped, rather than to try to create a prescribed beam. The former can be done easily by LIF using an independent monitor laser.

REFERENCES
1. F. Shimizu, K. Shimizu and H. Takuma; Chem. Phys. Lett. **102**, 375 (1983).
2. F. Shimizu, K. Shimizu and H. Takuma; Phys. Rev. A. **31**, 3132 (1985).

Nonlinear Optical Processes in Micron-Size Droplets

S.-X. Qian*, J.B. Snow†, and R.K. Chang

Yale University, Section of Applied Physics and Center for Laser Diagnostics, New Haven, CT 06520, USA

1. Introduction

The spherical interface of micron-size liquid droplets significantly modifies the spatial distribution of an incident plane wave in the vicinity of the droplets. The liquid-air boundary condition requires a scattered field component which together with the incident wave causes the characteristic angular scattering pattern for an incident wave with wavelength λ_i. Furthermore, the elastic scattering spectrum [1,2] at a fixed observation angle exhibits peaks at specific wavelengths $\lambda_{n,\ell}$. For a droplet of radius a with refractive index m (relative to that of the surrounding medium), these morphology-dependent resonances (MDRs) with mode number n and mode order ℓ are known by the Lorenz-Mie formalism [3] to occur at a specific size parameter $X_{n,\ell}$ with a real value ($\mathrm{Re}X_{n,\ell} = 2\pi a/\lambda_{n,\ell}$) and an imaginary value ($\mathrm{Im}X_{n,\ell} = 2\pi a\Delta\lambda_{n,\ell}/\lambda_{n,\ell}^2$), where $\Delta\lambda_{n,\ell}$ is the linewidth of the peaks in the elastic scattering spectrum [4]. In the literature, MDRs have been referred to as ripple structures, "whispering gallery" modes, or structure resonances [3].

Nonlinear optical emission from micron-size droplets is significantly influenced by the liquid-air interface. The concept of the droplet as an optical cavity and the effect of the interface on the internal field distribution and on phase-matching are discussed. Examples of laser emission, stimulated Raman scattering, and coherent anti-Stokes Raman scattering from droplets are presented.

2. Internal Field Distribution of the Incident Wave

The internal field distribution can be visualized by plotting the results of exact Lorenz-Mie calculations within an equatorial plane [5]. Whether λ_i is off resonance or on resonance with a multitude of MDRs, nonlinear optical generation within the droplets is greatly modified relative to the normal case of extended media for which the plane wave approximation is valid. Inside the droplet, the localized intensity distribution either increases or decreases the effective interaction length relative to 2a, the enhanced intensity increases the nonlinear polarization, and the modified propagation vector k_m alters the phase-matching condition.

For the off-resonance case (i.e., $\lambda_i \neq \lambda_{n,\ell}$), the internal field distribution for large droplets can be estimated with geometric optics (Snell's law) and is localized near the forward portion of the droplet, schemati-

---

*On leave from the Department of Physics, Fudan University, Shanghai, People's Republic of China.
†Present address: Naval Underwater Systems Center, New London Laboratory, New London, CT 06320.

cally shown in Fig. 1(a). The intensity at the "focal volume" can be as much as 10 times larger than that of the incident beam and of other parts of the droplet. The propagation vector within the medium k_m is broadened by an amount Δk [see Fig. 1(a)] even though the incident wave has a unique propagation direction.

For the on-resonance case (i.e., $\lambda_i = \lambda_{n,\ell}$), the internal field distribution has many peaks of comparable intensities that are confined near the droplet circumference. Both the peak and average intensity around the circumference are greatly enhanced (10-10^5 depending on n,ℓ values). Figure 1(b) shows portions of the incident wave which skim the interface, couple into the droplet, and form two counterpropagating waves confined near the interface. At a MDR, these two counterpropagating waves form a standing wave.

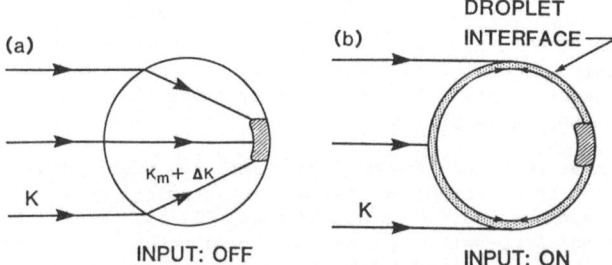

(a)   (b)   DROPLET INTERFACE

$K_m + \Delta K$

K

K

INPUT: OFF   INPUT: ON

Fig. 1.   Schematic of the internal field distribution of a spherical droplet when the incident wavelength (a) does not
correspond to a MDR and (b) does correspond to a MDR.

3. Droplet as an Optical Cavity

The droplet can serve as an optical cavity for specific wavelengths in the inelastic emission spectrum, i.e., for specific shifted wavelengths $\lambda_s = \lambda_{n,\ell}$. Some portions of the internally generated radiation, which have wavelengths commensurate with MDRs, will strike the droplet interface at angles larger than the critical angle and be trapped in the droplet by total internal reflection. After many such reflections and upon completion of a round-trip around the circumference, the phase of this wave is equal to that of the initial wave. The entire droplet therefore acts as a spherical optical cavity. Such a cavity has a large Fresnel number and, consequently, even when the round-trip gain exceeds the round-trip loss, the radiation emerging from the droplet is essentially isotropic.

The Q of the droplet cavity can be estimated from the ratio of $\mathrm{Re}X_{n,\ell}$ to $\mathrm{Im}X_{n,\ell}$. For a lossless medium, the $\mathrm{Im}X_{n,\ell}$ is a measure of the cavity decay rate, the rate at which the internally trapped radiation leaks out through the interface after the radiating source is shut off. This cavity decay time can be in the nanosecond range even though each trip around the circumference of a droplet with a = 10 μm is ∿0.2 psec. Another way to estimate the Q at $\lambda_s = \lambda_{n,\ell}$ is to examine the ratio of $\lambda_{n,\ell}/\Delta\lambda_{n,\ell}$ where $\Delta\lambda_{n,\ell}$ is the linewidth of the MDR. For ethanol or water droplets with a = 10-100 μm, this ratio can be at least as high as 10^3-10^5. Surface irregularities, shape distortions, or optical absorption will lower the ratio.

For spherical droplets with wide gain profiles, the emission spectra consist of a series of peaks with nearly equal wavelength spacing, corresponding to a set of MDRs with the same mode order (i.e., ℓ is fixed) but different mode number (n, n±1, n±2,...). Typically, threshold is first achieved with that set of MDRs having lower $ImX_{n,\ell}$. Upon increasing input pump intensity, an exponential increase for another set of MDRs occurs while the original set exhibits saturation characteristics.

4. Laser Emission

Laser emission from individual ethanol droplets (∿60 µm diam) containing rhodamine 6G with cw argon ion laser pumping (thresholds as low as 10 mW focused to a spot size of ∿200 µm diam) has been reported [6]. The laser intensity of a MDR peak vs input pump intensity was linear at low input, exponential at higher input, and finally saturated at even higher input. The temporal behavior of the laser emission exhibited relaxation oscillations once threshold was reached. Unlike conventional lasers with external mirrors, the laser radiation from the droplets emerged over 4π steradians, consistent with the concept of a droplet as a large Fresnel cavity. No measurable angular pattern or large forward-to-backward asymmetry were noted, indicating that laser emission consists of many modes that are all within the broad fluorescence gain profile.

Photographs of red laser emission from both spherical and highly distorted dye-doped ethanol droplets have been taken. The droplets were pumped by the second-harmonic output (1-10 mJ, 10 nsec, 532 nm, focused to an area of ∿1 mm²) of a Nd:YAG Q-switched laser as the pump source. These near-field photographs clearly illustrate that the optical field within the droplets (regardless of their shape) is confined near the interface where high internal fields and substantial optical feedback exist [7]. The red laser emission highlights the liquid-air interface, even of droplets with highly irregular shapes and of widely varying sizes.

5. Stimulated Raman Scattering (SRS)

For SRS, which is also a threshold process, the droplet interface provides a high Q cavity for selected wavelengths within the Raman gain profile, which is usually broad enough to span one or more MDRs. Since the input radiation is monochromatic (0.5 cm^{-1} for the second harmonic of a Nd:YAG laser), the resonance condition can be fulfilled for only certain droplet radii. The droplet radius can be varied precisely by adjusting the driving frequency of the vibrating orifice droplet generator.

If $\lambda_i \neq \lambda_{n,\ell}$, the input pump intensity is most intense near the forward portion of the droplet [see Fig. 1(a)]. Consequently, the SRS gain is highest in this localized region or "focal volume." For specific wavelengths within the Raman gain profile (i.e., $\lambda_s = \lambda_{n,\ell}$), Stokes waves upon traveling around the circumference will experience gain mainly in this localized region and consequently achieve SRS threshold. For all other wavelengths $\lambda_s \neq \lambda_{n,\ell}$ for which the feedback is much lower, the Stokes radiation will readily emerge from the droplet and the SRS threshold will not be attained.

If $\lambda_i = \lambda_{n,\ell}$, the input pump intensity is more evenly distributed around the circumference [see Fig. 1(b)] and is more intense than the nonresonant case. Consequently, the Raman gain for those $\lambda_s = \lambda_{n,\ell}$ waves can occur throughout their propagation around the circumference. The combined effect of the enhanced pump intensity and the much longer interaction length of ∿2πa in a round-trip provides sufficient gain, even for those $\lambda_s = \lambda_{n,\ell}$ waves which are located in the wings of the Raman gain profile. Regularly

spaced SRS peaks extending far out into the wings of the spontaneous Raman
bandwidth have been observed for ethanol droplets when their radii are
tuned to be commensurate with MDRs at λ_i = 532 nm.

When comparing the input intensity threshold between a droplet and a
much longer optical cell (e.g., ~30 μm diam droplets vs 10 cm cell path
length), it is important to keep in mind that optical feedback exists in
the droplet while the cell involves only a single pass gain. SRS within
the broad Raman bandwidth of the O-H stretching mode of water has been
observed for 40-80 μm diam water droplets with input intensity ~4 GW/cm^2 (at
λ_i = 532 nm and $\lambda_i \neq \lambda_{n,\ell}$) [8]. Although far above the SRS threshold for
droplets, this intensity was not sufficient to achieve SRS threshold from a
10 cm water-filled cell. In fact, for aqueous 0.5 M KNO_3, the SRS of both
the O-H and NO_3^- modes can be observed from individual droplets (see Fig. 2).
The hydrogen-bonding broadened O-H band can support a series of MDRs of
nearly equal wavelength spacing ($\lambda_{n,\ell}$ with different n but fixed ℓ). Surprisingly, the narrow NO_3^- band can be coincident with one of the MDRs so
that $\lambda_s = \lambda_{n,\ell}$. The water-air interface can be thought of as providing the
optical cavity for the SRS associated with the ν_1 mode of the NO_3^- anions.
For ethanol droplets, the SRS threshold is as low as ~100 MW/cm^2.

6. Coherent Anti-Stokes Raman Scattering (CARS)

Since CARS is a four-wave parametric process, it is not a threshold process
as is the case for both laser emission and SRS. It might be anticipated
that the morphology of the droplet should significantly affect the broad-

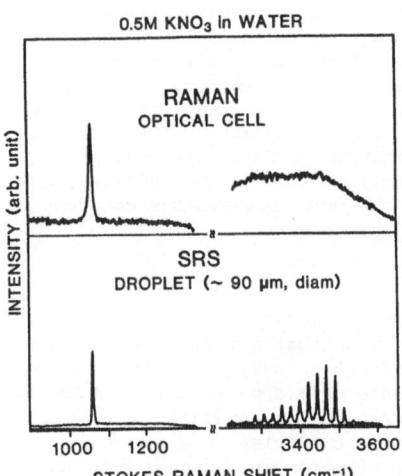

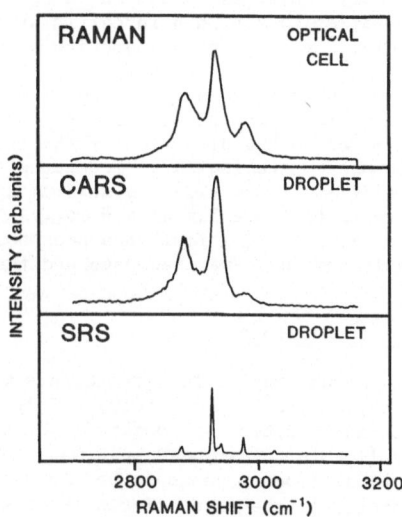

Fig. 2. The single laser pulse
SRS spectrum from a 0.5 M
KNO_3 aqueous droplet compared with the spontane-
ous Raman spectrum from
the same solution in an
optical cell. The NO_3^-
and OH^- modes are clearly
observable in the droplet
spectrum.

Fig. 3. SRS and CARS spectra from
an ethanol droplet com-
pared with the spontane-
ous Raman spectrum of ethanol in a cell. The SRS
spectrum shows MDR peaks.
The CARS spectrum does not
show any MDR peaks and is
spectrally limited by the
bandwidth of the broadband
Stokes beam $<\lambda_s>$.

band CARS generation since the internal fields of the pump wave at λ_i and the broadband Stokes wave ($<\lambda_s>$) can be greatly enhanced when these input waves satisfy the MDR conditions. The Stokes wave, being broadband, is bound to be in resonance with a multitude of $\lambda_{n,\ell}$ while the pump wave, being monochromatic, can be in resonance provided the droplet radius is adjusted. The resultant broadband anti-Stokes wave $<\lambda_{AS}>$ should also span a multitude of $\lambda_{n,\ell}$. Figure 3 shows the broadband CARS spectrum from an ethanol droplet ∿40 μm in diameter as well as the corresponding SRS spectrum. Also shown is the spontaneous Raman spectrum from an optical cell filled with ethanol and irradiated with an Ar laser beam at $\lambda_i = 514.5$ nm.

Contrary to our initial expectations and in contrast to the distinct MDR peaks present in the SRS spectrum, no MDR peaks are observable in the CARS spectrum. Whereas the phase-matching curve for ethanol in an optical cell is narrow (∿±0.2°) and symmetrical, that for an ethanol droplet is broad (∿±2°) and asymmetric, i.e., the CARS intensity for the collinear configuration ($\phi = 0°$) is higher than the intensity at $\phi = 2\phi_{pm}$, where the external phase-matching angle is $\phi_{pm} = 3.4°$.

The insensitivity of the CARS spectra to MDRs of the two input beams at λ_i and $<\lambda_s>$ and the broadened asymmetric phase-matching angular dependence can be explained by noting the internal field distributions and their propagation vectors inside the droplet. Since there are always some wavelengths within $<\lambda_s>$ that are on resonance, we need to consider only whether the λ_i is on resonance or off resonance. In the case of $\lambda_i = \lambda_{n,\ell}$, even though the internal fields are large and more uniformly distributed around the circumference, the phase-matching condition between the four waves cannot be satisfied for path integrals around the circumference. For $\lambda_i \neq \lambda_{n,\ell}$ and $<\lambda_s> \neq \lambda_{n,\ell}$, phase-matching can be fulfilled if the angle between λ_i and $<\lambda_s>$ beams is adjusted. The broadened phase-matching curve results from a short interaction length and a spread of the $\vec{k}_m$ inside the droplet due to the spherical interface. For $\phi = \phi_{pm}$, the path integral along the diameter in the direction of the incident waves is a maximum. The asymmetric phase-matching curve indicates that the triple product of the electric fields at λ_i and $<\lambda_s>$ has maximum spatial overlap at $\phi = 0$ and considerably less spatial overlap at $\phi = 2\phi_{pm}$. The absence of MDRs in the CARS spectrum when $<\lambda_{AS}> = \lambda_{n,\ell}$ implies a poor coupling efficiency between the coherently generated phase-matched wave and the MDRs.

7. Conclusions

The morphology of liquid droplets provides an optical cavity in which substantial feedback can occur. This system offers a novel means of studying nonlinear optical phenomena. The normal plane wave approximation, which is applicable to extended bulk media and the propagating waveguide modes for optical fibers, needs to be reconsidered for the droplet case since the spherical interface affects the spatial distribution and intensity of the inner fields, the propagation direction, and the optical feedback for the inelastically generated radiation.

Acknowledgments

We gratefully acknowledge the partial support of this work by the Army Research Office (Contract No. DAAG29-85-K-0063) and the Donors of the Petroleum Research Fund, administered by the American Chemical Society.

References

1. A. Ashkin, Science 210, 1081 (1980).
2. J.F. Owen, P.W. Barber, B.J. Messinger, and R.K. Chang, Opt. Lett. 6, 272 (1981).
3. C.F. Bohren and D.R. Huffman, Absorption and Scattering of Light by Small Particles (Wiley, New York, 1983).
4. P.R. Conwell, P.W. Barber, and C.K. Rushforth, J. Opt. Soc. Am. A 1, 62 (1984).
5. J.F. Owen, R.K. Chang, and P.W. Barber, Opt. Lett. 6, 540 (1981); A.B. Pluchino, Appl. Opt. 20, 2986 (1981).
6. H.-M. Tzeng, K.F. Wall, M.B. Long, and R.K. Chang, Opt. Lett. 9, 499 (1984).
7. S.-X. Qian, J.B. Snow, and R.K. Chang, "Laser Droplets: Highlighting the Liquid-Air Interface by Laser Emission," submitted to Science.
8. J.B. Snow, S.-X. Qian, and R.K. Chang, Opt. Lett. 10, 37 (1985).

CARS Studies of NaH$_2$ Collision Pairs

S.L. Cunha, P. Hering, and K.L. Kompa

Max-Planck-Institut für Quantenoptik,
D-8046 Garching, Fed. Rep. of Germany

The spectroscopy of collision pairs is of considerable importance for theoretical reasons and for the understanding of many experiments in laser chemistry /1/. The absorption of laser light during collision and the distribution of the absorbed energy among the separating collision partners is our present interest.

The Na + H$_2$ system has been chosen both for theoretical and experimental reasons. Ab initio potential energy surfaces are available and E-VRT energy-transfer experiments have been performed /2/. The so called bond-stretch attraction model has been applied to explain the quenching process of Na(3^2P) by H$_2$. The energy-transfer from Na(3^2P) to rovibronic states of H$_2$ arises from crossings of the ground and excited states. The positions of these crossings are therefore crucial. No precise determination of vibrational and rotational state population of H$_2$ after the energy-transfer process has been done so far. In particular the role of rotation could not be determined.

Coherent Anti-Stokes Raman Spectroscopy (CARS) has been used successfully in many fields /3/. We apply the CARS method to measure the rovibronic state population of H$_2$ before and after collision with a Na atom dressed in a laser field of
known frequency and strength. The wavelength of the dressing field for our first experiments is centered around the Na D lines.

The experimental setup (Fig. 1) consists of a flashlamp-pumped dye laser system (FLPL) with variable pulselength which irradiates the Na-H$_2$

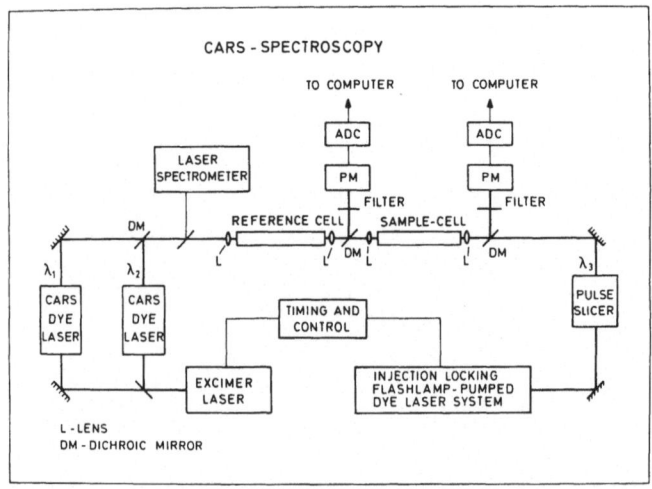

Fig. 1   Experimental setup

gas mixture. A CARS laser system detects the population of individual rovibronic states of H_2. With an adjustable time control the CARS laser can monitor the requested population before, during and after the irradiation of the FLP-laser with a time resolution of 10 ns.

Without the FLP-laser only v = 0 levels of H_2 are populated. With the FLP-laser we can detect strong depopulation of v = 0 and population of rovibronic levels up to v = 3.

Figure 2 shows as an example the population of individual rotational levels of the v = 3 band.

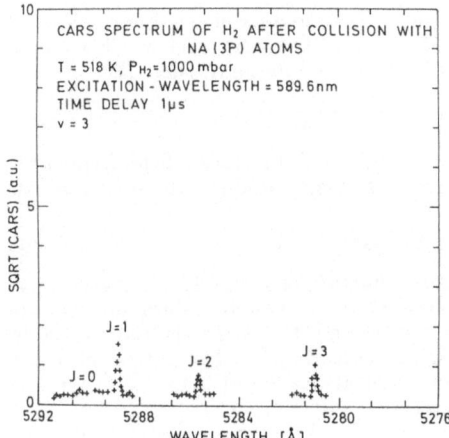

Fig. 2
CARS spectrum of v = 3 band

With this method we can determine the time-dependence of the population and depopulation of single rovibronic states of H_2 after collision with sodium dressed in a laser field. Thus it is possible to detect all states which are involved in the quenching process, including the rotational levels /4/.

References
1   P. Hering, Y. Rabin: Chem. Phys. Lett. 77, 506 (1981)
     P. Hering, P.R. Brooks, R.F. Curl jr., R.R. Judson, R.S. Lowe:
        Phys. Rev. Lett. 44, 687 (1980)
     R.C. Maguire, P.R. Brooks, R.F. Curl jr.: Phys. Rev. Lett. 50,
        1918 (1983)
2   I.V. Hertel: in "Dynamics of the Excited State", edited by K. Lawley
        (Wiley, New York, 1982)
     P. Botschwina, W. Meyer, I.V. Hertel, W. Reiland: J. Chem. Phys. 75,
        5438 (1981)
3   S.A.J. Druet, J.P.E. Taran: Prog. Quant. Electr. 7, 1 (1981)
4   S.L. Cunha, P. Hering, K.L. Kompa: Opt. Comm. to be published

Two-Photon Absorption from a Phase-Diffusing Field

D.S. Elliott[†], M.W. Hamilton*, K. Arnett, and S.J. Smith

Joint Institute for Laboratory Astrophysics of the University of Colorado and the National Bureau of Standards, Boulder, CO 80309, USA

We report here experimental results on the effect of random frequency fluctuations of a laser field on the process of two-photon absorption. The laser field which we produce experimentally realizes the properties of the phase diffusion model. This field is characterized by

$$E(t) = E_0 \exp\{-i[\omega_0 t + \int \omega(t)dt]\} \tag{1}$$

where E_0 and ω_0 are the constant field amplitude and the mean field frequency, respectively. The frequency, $\omega(t)$, is a random Gaussian variable, whose correlation function is given by

$$<\omega(t)\omega(t+\tau)> = b\beta e^{-\beta|\tau|} \tag{2}$$

where b is the spectral density of the frequency fluctuations, and $1/\beta$ is the correlation time of the fluctuations. These two parameters completely determine the statistical properties of the laser field, and correspond to experimentally variable features of the noise power spectrum applied to the laser field. The power spectrum of this laser can be calculated analytically for arbitrary values of b and β. Two limiting cases of the power spectrum are of particular interest, one corresponding to a nearly Gaussian laser power spectrum and the other to a nearly Lorentzian laser power spectrum. The former results when the frequency fluctuations are slow, but large in magnitude ($\beta \ll$ b), such that the power spectrum is essentially identical to the distribution function of the frequency fluctuations. The nearly Lorentzian laser power spectrum (of width b) results when the frequency fluctuations are rapid, but small in magnitude ($\beta \gg$ b).

In 1968, B. R. Mollow [1] calculated that, for the case of the Lorentzian laser power spectrum, the two-photon absorption linewidth should increase as four times the laser width. Similar calculations in that same report using a field model with amplitude fluctuations but having the same Lorentzian power spectrum yielded an absorption width which increased as twice the laser width, showing the importance of the nature of the field fluctuations. Knowledge of only the laser power spectrum is in general insufficient for predicting laser bandwidth effects. We have extended Mollow's calculations to the case of two-photon absorption for the laser with the Gaussian power spectrum. Since absorption of the light can take place when these slow fluctuations "tune" the laser frequency to one half the transition frequency, the width of the absorption spectrum is twice the width of the laser.

In addition to effects resulting from the finite correlation time of the fluctuations described in the preceding paragraph, this experiment presents an opportunity for observing effects due to the partial correlation of two different fields. We have extended Mollow's work to include this effect also. The corresponding experiment

[†]Present address: School of Electrical Engineering, Purdue University, West Lafayette, IN 47907 USA

*Present address Sektion Physik der Universität München, Am Coulombwall 1, 8046 Garching, Fed. Rep. of Germany

makes use of the standard counter-propagating laser beam technique used for suppressing the Doppler broadening of the two-photon absorption line shape. The time necessary for the light to travel from the vapor cell to the reflector and back to the vapor cell can result in partial decorrelation of the field fluctuations. For the Lorentzian laser power spectrum, a decorrelation occurs for delay times longer than one half the inverse of the laser band width, and the absorption spectrum is reduced to twice the laser bandwidth.

We have measured [2] the dependence of the width of the two-photon absorption line on the width of the laser power spectrum using the 3^2S F$=2 \rightarrow 5^2S$ F$=2$ transition in atomic sodium. The cw laser beam was tuned through the frequency range around one-half the atomic transition frequency, and the absorption lineshape determined by detecting the 330 nm fluorescent radiation corresponding to the second step of the $5S \rightarrow 4P \rightarrow 3S$ decay of the excited state sodium. The dependence of the absorption linewidth on the laser linewidth is shown in Fig. 1. Curves a and b correspond to a nearly Lorentzian laser lineshape, with a different degree of correlation between the counterpropagating beams. The laser beams are strongly correlated for curve a, due to a delay time of only 1 nsec, and the slope of the curve is nearly four, as predicted by Mollow. The curvature for large laser linewidths is due to the increasing Gaussian nature of the laser spectrum. Curve b corresponds to a delay time of 17 nsec, and the effect of decorrelation of the fields is evident. Data corresponding to a 24 nsec delay time, not shown, indicate a further reduction in absorption bandwidth. Curve c represents the absorption linewidth for a laser whose power spectrum is nearly Gaussian in shape. The data is expected to fall on a line of slope two for the true Gaussian lineshape. The solid lines in each case represent the calculated dependence of the absorption linewidth on the laser linewidth.

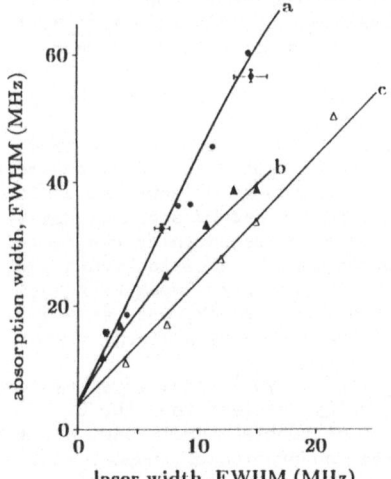

Fig. 1. Two-photon absorption linewidth versus laser linewidth for a phase diffusion field. The data in curves a (•) and b (▲) correspond to a nearly Lorentzian lineshape, while the laser spectrum for the data in curve c (Δ) is nearly Gaussian. The correlation between the laser fields is nearly complete for the data in curves a and c, while a 17 nsec delay has allowed partial decorrelation of the fields for curve b.

This work was supported by the U.S. Department of Energy, Office of Basic Energy Sciences.

1. B. R. Mollow: Phys. Rev. *175*, 1555 (1968)
2. D. S. Elliott, M. W. Hamilton, K. Arnett, and S. J. Smith: Phys. Rev. Lett. *53*, 439 (1984); and to be published, Phys. Rev. A

Optical Heterodyne Three-Level Spectroscopy

S. Le Boiteux, D. Bloch, and M. Ducloy

Laboratoire de Physique des Lasers, LA 282, Université Paris-Nord,
Av. J.B. Clément, F-93430 Villetaneuse, France

In the recent years, the sensitivity of Doppler-free non linear spectroscopy - saturated absorption (SA), two-photon spectroscopy, Raman spectroscopy... - has been remarkably improved by the development of high-frequency (HF) modulation and heterodyne detection techniques. In particular, shot-noise limited sensitivity is now routinely obtained in various types of experiments, because the amplitude noise of lasers is essentially a low-frequency noise. Here, we discuss some of the possibilities opened by the extension of these HF modulation techniques to three-level spectroscopy, when two different lasers are used. We report on a saturation spectroscopy experiment in the UV range [1] and we also propose a new technique to monitor pressure-induced extra-resonances (PIER).

A typical experiment of optical heterodyne spectroscopy on a three-level system (1-0-2) consists in applying an amplitude modulation at frequency δ on a saturating beam ω_1, quasi-resonant for the 0-1 transition, and in detecting the modulation induced on a probe beam ω_2, nearly resonant for the 0-2 transition. Two geometries are considered : the pump and probe beam can be either counter-propagating (as in SA experiments), or co-propagating ; in both cases, the resonance conditions for the appearance of the probe modulation are Doppler-free, due to velocity selection.

By applying this technique to combined visible -UV three-level spectroscopy, we were able to observe <u>weak</u> UV transitions of neon with a Doppler-free resolution. Thanks to the recent development of tunable cw UV sources, it is nowadays possible to perform high- resolution Doppler-free UV spectroscopy by standard SA techniques : however, such experiments are essentially restricted to intense resonance lines because the UV sources must be able to saturate the transition, in spite of their limited power. On the contrary, weak UV transitions can be monitored by three-level saturation spectroscopy, because the saturation is thus produced on the coupled 0-1 transition by a strong visible laser.

As an example we have studied the $4p_9$-$1s_5$-$2p_9$ (V-type) three-level system of Ne. The sensitivity of our method is outlined by the fact that the UV transition $1s_5$-$4p_9$ (λ = 298 nm), in the same experimental conditions, was not observable by standard linear techniques (linear absorption or induced fluorescence). The interest of using HF modulation appears obviously on Fig.1. Various UV transitions of Ne could be observed, like $1s_5$-$4p_6$, $1s_5$-$4p_{10}$, $1s_4$-$4p_3$, and it appeared that the most favorable three-level systems were those for which the common level population could be completely depleted by optical pumping via the 0-1 transition (e.g. $1s_5$-$2p_8$ Ne transition). We have measured the resonance pressure broadening for several Ne systems. In the case of the $4p_6$-$1s_5$-$2p_8$ system, the observed broadening (of the order of 57 MHz/Torr) must be ascribed mainly to the $1s_5$-$2p_8$ transition. The experiments were performed both in co-and counter-propagating geometry. In a counter-propagating geometry, the transmission change signal on the UV probe beam is only due to population changes induced in the common level by the saturating beam, while in co-propagating geometry, a coherent Raman process yields an additional con-

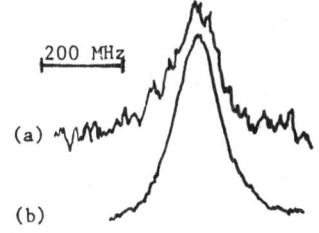

Fig. 1 *UV transmission lineshape when the visible frequency ω_1 is scanned : (a) δ = 37 kHz, time constant τ = 1 s ; (b) δ = 1.7 MHz, τ = 0.1 s*

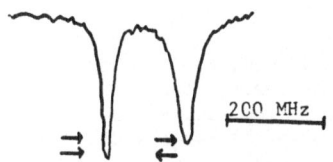

Fig. 2 *UV signal lineshape when ω_1 is scanned : comparison between co-propagating ($\rightrightarrows$) and counter-propagating ($\rightleftarrows$) geometry*

tribution to the signal, increasing its amplitude and narrowing its lineshape. This effect is evidenced in Fig. 2.

As is well known in HF heterodyne spectroscopy, the spectrum of an atomic transition appears generally as a pair of resonances when the modulation frequency is larger than the optical width. Indeed, the spectrum of the modulated pump beam being composed of frequencies $\omega_1 + \delta/2$ and $\omega_1 - \delta/2$, the interaction with the probe beam at ω_2 can lead to the emission of e.m. fields at $\omega_2 + \delta$ or $\omega_2 - \delta$ (according to the ordering of interactions), which are detected through the heterodyne beating with ω_2, and for which the respective conditions of resonance are different. In the co-propagating geometry, the resonance conditions for the standard population process and for the Raman coherence are not the same. The Raman contribution yields a quadruplet or a sextuplet structure, according to whether one has $\omega_1 < \omega_2$ or $\omega_1 > \omega_2$. Inside this multiplet Raman structure, one doublet is always located on the same position as the population doublet, leading to an interference effect between population and Raman coherence processes.

In a recent theoretical analysis, we have shown that, when cascade effects induced by spontaneous emission can be neglected, and in the approximation of weak incident intensities, the interference between population and coherence effects is totally destructive, in the <u>absence of dephasing collisions</u>. Hence, the resonance doublet associated to the population term should appear only in presence of collisions, and represents a new kind of pressure-induced extra-resonance (PIER), whose origin is comparable to the ones observed by BLOEMBERGEN et al. However, since these new resonances are predicted for resonant laser irradiations, they should be observed with <u>low gas pressures</u> and <u>low power</u> cw lasers. They could yield additional informations on collisional processes in the impact regime, in a pressure-range different from the one usually analyzed with PIER resonances.

1. S. Le Boiteux <u>et al</u>, Opt. Commun. <u>52</u>, 274 (1984).

Competition Between Photoionization and Two-Photon Raman Coupling

G. Leuchs [1], G. Alber, and S.J. Smith [2]

Joint Institute for Laboratory Astrophysics, University of Colorado and National Bureau of Standards, Boulder, CO 80309, USA

In transitions between bound states, multiphoton processes become important at higher laser intensities when the Rabi-frequency is comparable to or larger than the width of the transition line. This has been shown, e.g., in the study of resonance fluorescence of a two-level atom [1]. Bound-free transitions, however, cannot be saturated owing to the large width of the continuum. Consequently, multiphoton processes should not be important. The only effects expected at higher laser intensities are depletion of the bound state population and possibly continuum-continuum transitions, the latter resulting in multiple peaks in the photo-electron energy spectrum [2].

The experiment described in the following focusses on the intensity dependence of the photoionization process leading to the first, lowest energy electron peak, which can be reached by one-photon absorption. According to the simple picture discussed above, this bound-free transition should not show any intensity dependence apart from depletion of the bound state. In contrast to this expectation we demonstrate that a third-order process involving Raman coupling to a nearby nearly degenerate state may effectively compete with the one-photon absorption process.

In the experiment sodium atoms of a thermal beam have been excited via the $3^2P_{1/2}$ to the $n^2D_{3/2}$ state, using pulsed dye lasers pumped by the second and third harmonic of a Nd:YAG laser. The fundamental beam of the Nd:YAG laser at $\lambda=1.06\mu$ was optically delayed and also directed onto the atomic beam. The delay was adjusted such that the 1.06μ laser pulse interacted with the atoms only after the exciting dye laser pulses had left the interaction region. With this scheme we studied one-photon ionization out of the excited $n^2D_{3/2}$ state. From measurements of total ionization signal as a function of laser intensity, a linear dependence of the signal on the intensity was found at low laser intensities. At higher intensities ($\sim10^8$ W/cm^2 for n=8) the signal leveled off. This is consistent with the bound-state depletion mentioned above. However, the total ionization signal is not very sensitive to the characteristics of the intermediate state. A more stringent test is provided by measurement of the angular distribution of the photoelectrons [2] by simultaneously rotating the linear polarizations of all nearly collinearly propagating lasers, and by detecting electrons emitted at right angles to the laser beams. Based on the simple argument made above, this distribution should not change at higher intensity of the Nd:YAG laser. The only effect of increasing the laser intensity should be that more and more atoms are ionized during the early part of the laser

---

[1] Heisenberg Fellow of the Deutsche Forschungsgemeinschaft, present address: Max-Planck-Institut für Quantenoptik, 8046 Garching, FRG.
[2] Quantum Physics Division, National Bureau of Standards.

pulse. The kinetic energy of the photoelectron was measured using the time-of-flight technique, proving that no continuum-continuum transitions were involved in this experiment.

The Figure shows polar diagrams of angular distributions of photoelectrons measured for two different Nd:YAG laser intensities and for two different intermediate states, $7^2D_{3/2}$ and $8^2D_{3/2}$. The data are represented by the points and the solid lines result from a least squares fit. At the lower laser intensity, the observed angular distribution has the shape predicted by lowest order perturbation theory, which is the same for n=7 and 8. At the high laser intensity, however, the distributions change drastically, and, what is more, they then differ drastically for n=7 and 8. We explain this effect by two-photon Raman coupling from the $n^2D_{3/2}$ to the $n^2D_{5/2}$ state [3]. Similar Raman coupling between nearly degenerate states has been taken into account in the theoretical description of, for example, quantum beats in photoionization [4,5]. The experiment discussed here has the advantage to isolate the effects of Raman coupling.

For the $8^2D_{3/2}$ state the effect was found to be especially strong owing to the near resonance with the 4p state. The difference in shape at high intensity for n=7 and 8 is due to the fact that $E(n^2D_{3/2}) - h\nu$ is slightly below (above) $E(4p)$ for the n=7 (n=8) state, so the sign of the coherent admixture of the $n^2D_{5/2}$ state is different in the two cases.

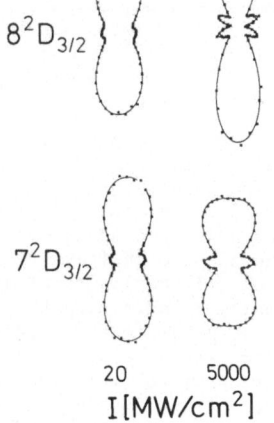

The photoelectron angular distribution can be described by a linear combination of $|Y_{30}|^2$ and $|Y_{31}|^2$ [2]. The Raman coupling leads to a mixing of the two fine structure states $|J=3/2,M=1/2\rangle$ and $|J=5/2,M=1/2\rangle$. If the Raman coupling term has a negative sign, as in the case of the 8d-state, this results in an enhancement of the $|Y_{30}|^2$ contribution to the angular distribution at higher laser intensities as shown in the Figure. Similarly, the positive sign of the Raman term in the case of the 7d state enhances the $|Y_{31}|^2$ contribution. The measurement of the electron angular distributions at different laser intensities together with a more detailed theoretical analysis allows one to study the dynamics of the competition between the one-photon and the three-photon process.

$8^2D_{3/2}$

$7^2D_{3/2}$

20   5000

$I[MW/cm^2]$

References

1. J.D. Cresser, J. Häger, G. Leuchs, M. Rateike and H. Walther: Topics in Current Physics Vol. 27, ed. by R. Bonifacio (Springer-Verlag, Berlin, Heidelberg, New York 1982), p.21, and references therein.
2. Chin and Lambropoulos, Multiphoton Ionization of Atoms, (Academic Press, Toronto, Orlando 1984).
3. G. Leuchs: in Multiphoton Processes, ed. by P. Lambropoulos and S.J. Smith (Springer-Verlag, Berlin, Heidelberg 1984).
4. A.T. Georges and P. Lambropoulos, Phys. Rev. A 18, 1072 (1978).
5. P.L. Knight, Opt. Commun. 32, 261 (1980).

Propagation Effects
in Strong Pump-Weak Probe Interactions

*F.P. Mattar**

Physics Department, New York University and Regional Laser Center**,
Massachusetts Institute of Technology, Cambridge, MA 02139, USA

Analytic and numerical calculations for the propagational dependence
of Rabi split resulting from a strong resonant pump and a weak probe are
presented. A semi-classical formulism was adopted for both the perturba-
tional (using stationary phase and steepest descent asymptotic techniques)
and the rigorous computational treatments. This calculation can be considered as a double self-induced transparency where the strong pump (2π on-
axis area) experiences coherent pulse break up and depletion as the weak
probe (0.02π area) cooperatively builds up. Different transition gain
ratios (i.e., oscillator strength: $\mu^2\omega$) are considered to insure that the
weak probe does not get (a) delayed with respect to the pump or (b) get out
of synchronization, and cease to overlap. The probe detuning can be as
large as the input on-axis pump Rabi frequency. The interplay of nonlinear
Raman gain action, dispersion and diffraction give rise concomittantly to a
number of effects previously studied independently such as self-phase modulation, wave front encoding, transverse ring formation, self-focusing,
quasi-trapping (when pump detuning is also allowed), and asymptotic three-
level solitary waves.

With the goal of understanding various aspects of Raman propagation,
we have analyzed the effect of pump strength, namely (a) Rabi-splitting
(side-band in $\rho_{33}(\infty)$); (b) coherent pulse break-up "a la S.I.T."; (c) diffraction related effects (self-focusing, multiple foci); and (d) pump detuning - leading to quasi-trapping since lensing effects associated with
the phase ϕ that evolves due to dispersion can balance the ϕ originating
from diffraction coupling. We have also studied the effects of probe detuning by searching to elucidate which probe detuning will lead to larger
probe magnification for a given pump strength. We assessed the dependence
of relative delay between pump and probe as a function of: (a) the pump
strength; (b) the probe detuning; (c) the beam profile; and (d) the gain
ratio (that is, the oscillator's strength ratio). We have proceeded in the
understanding of the effect of optical thickness "$\alpha\ell$" in conjugation of
pump depletion for the following physical situations: (a) for both resonant and off-resonant pump; and (b) for plane-wave and non-uniform plane-
wave.

To illustrate the physics, we present two cases: the first one, on-
resonance Fresnel dependence illustrating the pump depletion and the probe
build-up. The larger the Fresnel number the smaller is the probe build-up
equivalently. This can be seen from the radially integrated output power
(OPower laser is the pump coherences laser B is the probe). The smaller
Fresnel number, which is defined by the larger reciprocal radial width
TBRHON, experiences a larger buld-up and a smaller delay, whereas case two

* Partially supported by ONR (N000-14-77-C-0553), AFOSR and ARO under the
contract DAAG-29-84-K-0137 and NSF-PHY-84-06107 as well as Batelle/CRDC.

**This is an NSF Regional Instrumentation facility.

deals with the dispersion in the case of plane and non-plane wave, the on-resonance gain is the largest as the detuning increases, the magnification of the probe is reduced. Further calculations are to be published in JOSA-B.

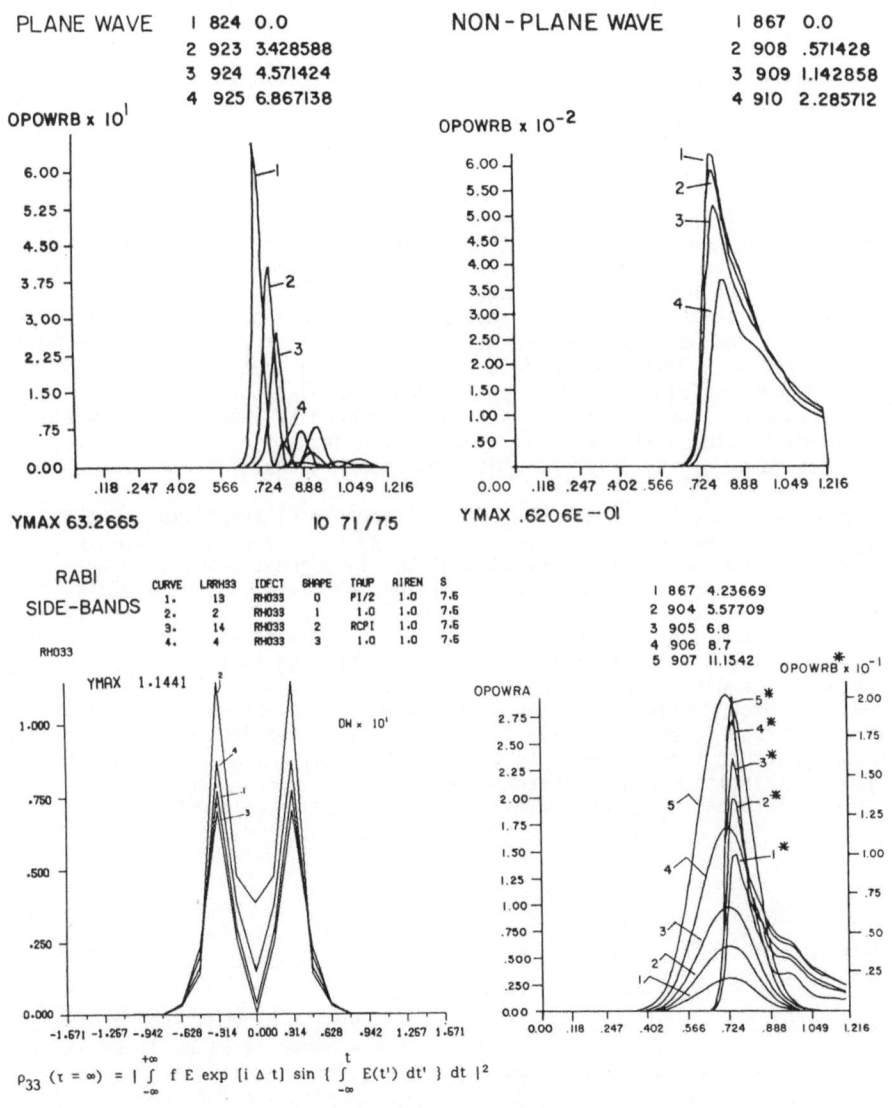

PLANE WAVE

1	824	0.0
2	923	3.428588
3	924	4.571424
4	925	6.867138

OPOWRB x 10^1

YMAX 63.2665

IO 71/75

NON-PLANE WAVE

1	867	0.0
2	908	.571428
3	909	1.142858
4	910	2.285712

OPOWRB x 10^{-2}

YMAX .6206E−01

RABI
SIDE-BANDS

RH033

YMAX 1.1441

CURVE	LRRH33	IDFCT	SHAPE	TAUP	RIREN	S
1.	13	RH033	0	PI/2	1.0	7.6
2.	2	RH033	1	1.0	1.0	7.6
3.	14	RH033	2	RCPI	1.0	7.6
4.	4	RH033	3	1.0	1.0	7.6

DW x 10^1

1	867	4.23669
2	904	5.57709
3	905	6.8
4	906	8.7
5	907	11.1542

OPOWRA

OPOWRB x 10^{-1}

$$\rho_{33} (\tau = \infty) = | \int_{-\infty}^{+\infty} f\, E \exp [i\, \Delta\, t]\, \sin \{ \int_{-\infty}^{t} E(t')\, dt' \}\, dt |^2$$

with f the fractional pump to probe Rabi frequency ratio (f << 1) and Δ the probe detuning; $\theta = s\pi$; shape sech, sech^2, Gaussian & Lorentzian

219

Direct Observation of Incoherent Population Effect and Ground-State Coherence Components in Nonlinear Hanle Effect

R.J. McLean, D.S. Gough, and P. Hannaford

CSIRO Division of Chemical Physics, P.O. Box 160, Clayton, Victoria, 3168 Australia

The generation of Zeeman coherence in the lower level of a linear-σ polarized laser excited $J_{lower}=1$ to $J_{upper}=0$ transition may lead to an intensity minimum in fluorescence from the upper level as an applied magnetic field is scanned through zero. The intensity variation, resulting from the coupling of the lower level coherence to the upper level population, is known as a nonabsorption resonance [1], and has a width determined by the rate at which the lower level Zeeman coherence is destroyed. A detailed theoretical treatment [2] has shown that the magnetic field dependence of the upper level fluorescence from a Doppler broadened atomic ensemble is also influenced by incoherent population transfer amongst the states of the laser excited transition that leads to a broader resonance characterized by the homogeneous transition width.

By performing the experiment on the 570.7 nm $^7F_1 \rightarrow {}^7F_0^0$ transition from the ground-state of samarium it has been possible to achieve conditions which not only have allowed the first observation of the incoherent population transfer resonance in a $J_{lower}=1$ to $J_{upper}=0$ transition, but also cause the widths of the two resonances to differ by such a large factor that they are readily distinguished in a single magnetic field scan of a few gauss (Fig. 1). Although the samarium vapour is generated by cathodic sputtering in a rare-gas discharge, resonances whose widths correspond to a rate of collisional disalignment in the samarium ground-state as low as 50 kHz have been observed. The interpretation of the experimental profiles is verified by independent experiments that separately measure the ground-state collisional disalignment rate and the homogeneous transition width. Furthermore, the behaviour of the magnetic field profiles is in broad agreement with the predictions of the theoretical model of [2].

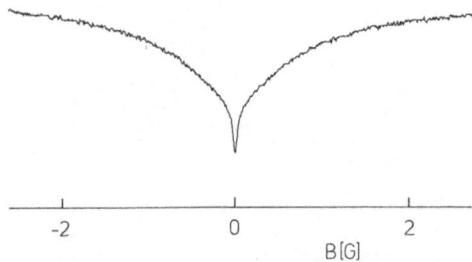

-2   0   2
B[G]

Figure 1. Fluorescence profile from upper level of laser excited 570.7 nm samarium transition. Neon pressure 0.7 Torr

References

1. H.R. Gray, R.M. Whitley and C.R. Stroud: Opt. Lett. _3_, 218 (1978); G. Orriols: Nuovo. Cim. B _53_, 1 (1979)
2. R.J. McLean, R.J. Ballagh and D.M. Warrington: J. Phys. B (in press)

Lower-Level Zeeman Coherence Effects
in Competitive σ^+ - σ^- Mode Interactions

C. Parigger, R.J. Ballagh, and W.J. Sandle
Physics Department, University of Otago, Dunedin, New Zealand

P. Hannaford
Division of Chemical Physics, CSIRO, Melbourne, Australia

A collection of atoms in an optical cavity may mediate a competitive interaction between σ^+ and σ^- modes, and so transform a symmetric input (equal amplitudes of σ^+ and σ^-) to an asymmetric output (unequal amplitudes) [1] . This phenomenon (called polarization switching), has a strong dependence on the particular choice of atomic transition and while predicted and observed to occur for a J = 1/2 to J = 1/2 transition [2], nevertheless it will not occur (in the absence of a magnetic field) for a J_{lower} = 1 to J_{upper} = 0 transition [3]. The difference is due to the Zeeman coherence developed in the lower level, which dominates the electric dipole response and is the cause of the well known phenomenon of population trapping [4]and a variety of non-linear behaviour (e.g.[5]).

In a J = 1/2 to J = 1/2 transition, the electric dipole depends only on the populations of the individual states, and provided redistribution amongst the lower states is slow, longitudinal optical pumping allows the weaker of the σ^+ or σ^- modes to be more strongly absorbed. Thus an imbalance of σ^+ and σ^- generated by fluctuations in the cavity will be enhanced, leading to polarization switching. In the J_{lower} = 1 to J_{upper} = 0 transition, the population contribution to the dipole response will again, through longitudinal optical pumping, tend to enhance imbalance between σ^+ and σ^- modes. However the Zeeman (alignment) coherence, originating from transverse optical pumping, has a contribution to the electric dipole which opposes imbalance and tends to restore equality between σ^+ and σ^- . With a physical choice of lower-level relaxation rates, the coherence contribution is sufficient to prevent polarization switching occurring.

Application of a longitudinal magnetic field reduces the Zeeman coherence and, for appropriate input power, stable asymmetric output appears together with polarization sensitive switching. We have experimentally explored these effects using the $^7F_1$-$^7F_0^0$ (570.68 nm) transition of Sm I in a Fabry-Perot cavity, applying a range of longitudinal fields, and using linearly polarized (symmetric) input, resonant, or near-resonant to both the atomic transition and a cavity mode [6]. The behaviour of σ^+ and σ^- output intensities shows a variety of phenomena. In the absence of a magnetic field (± 4 µT) we find, in accordance with predictions, that the cavity output displays simple bistable switching, but no asymmetry. In Figure 1(a) we show one of the observed behaviours of an input-output intensity scan (of both σ^+ and σ^-), for small longitudinal magnetic field (55 µT), and with the laser detuned by approximately 750 MHz on the low frequency side of the ^{154}Sm transition. The switching is complex, with first one then the other polarization being dominant. We call this effect magnetically induced polarization switching. Related phenomena have been observed for the D, transition of sodium [7,8].

Using standard techniques (e.g. [2]) we have obtained an analytic expression for the behaviour of the system (plane wave, ring cavity,

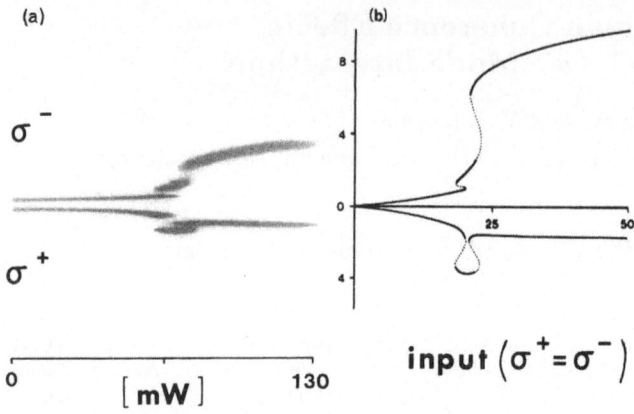

(a)

(b)

σ^-

σ^+

0   [mW]   130

input $\left(\sigma^+ = \sigma^-\right)$

Figure 1. Transmitted versus input intensities for the J = 1 to J = 0 transition with a 55 µT longitudinal magnetic field (a) multiple up-down experimental scans; (b) theory (scaled as in [6])

homogeneous broadening) in the mean field approximation. In Figure 1(b) we plot the theoretical results for scaled intensities, using parameter values which approximate the experimental conditions of Fig.1(a). Good qualitative agreement between theory and experiment is seen [6].

1. M.Kitano, T.Yabuzaki, T.Ogawa: Phys.Rev.Lett. 46, 926 (1981)
2. M.W.Hamilton, R.J.Ballagh, W.J.Sandle: Z.Phys.B 49, 263 (1982)
   M.W.Hamilton, W.J.Sandle, J.T.Chilwell, J.S.Satchell, D.M.Warrington: Opt.Commun. 48, 190 (1983)
3. R.J.Ballagh, V.Jain: Phil.Trans.R.Soc. London A 313, 445 (1984)
4. G.Orriols: Nuovo Cimento 53 B, 1 (1979)
   R.J.McLean, R.J.Ballagh, D.M.Warrington: J.Phys.B (in press)
5. D.F.Walls, P.Zoller: Opt.Commun. 34, 260 (1980)
6. C.Parigger, P.Hannaford, W.J.Sandle, R.J.Ballagh: Phys.Rev.A 31, (1985)
7. F.Mitschke, J.Mlynek, W.Lange: Phys.Rev.Lett. 50, 1660 (1983)
8. G.Giusfredi, P.Salieri, S.Cecchi, F.T.Arecchi: Opt.Commun. 54, 39 (1985)

Four-Wave Mixing as a Means
of Studying Intracollision Dynamics - A Theoretical Study

Y. Prior

Chemical Physics Department, The Weizmann Institute of Science,
Rehovot 76100, Israel

A. Ben-Reuven

School of Chemistry, Tel-Aviv University, Ramat-Aviv,
Tel-Aviv 69978, Israel

Collision dynamics in gases can be studied by a variety of spectroscopic
approaches involving resonance scattering of coherent radiation. These
approaches are generally classified as *incoherent* or *coherent* scattering
experiments. An outstanding example of the first class is collisional
redistribution in resonance fluorescence [1], and of the second class is
(collision-induced) Four-Wave Mixing (FWM) [2].

In this work we present a nonimpact theory of 4-wave mixing and intracollisional dynamics, the details of which will be published elsewhere [3].
The work consists of two major parts:

(1) A unified *microscopic* derivation (starting from the Liouville-von
Neumann equation for the gas-plus-radiation system) is provided for both
the incoherent and coherent scattering. The two-point nature of the
coherent scattering, involving the product of scattering amplitudes at two
distinct points in space, is emphasized, in contrast to the one point
nature of incoherent scattering.

(2) The theory is extended to the *nonimpact* domain, including effects of
intracollisional dynamics. This extension is based on methods developed
recently in the theory of spectral line shapes [1]. The multimode character
of FWM is used in order to introduce effects of multi-photon transitions
occuring within a single collision.

The theoretical disposition starts from a general expression for resonance
scattering of the radiation obtained from the Liouville-von Neumann equation.
The expression is reduced to a description of two atoms (or molecules) in
the gas, their surrounding medium (molecules and radiation) acting as a
thermal bath, and the input and output radiation modes. The familiar impact-
limit expressions for redistribution and FWM are derived as special cases.
The analysis then extends to nonimpact phenomena, providing explicit expressions for the scattering rates in the quasi-static limit of semiclassical
(vertical) Franck-Condon transitions during the collision.

Possible applications of the nonimpact theory include:

(a) *Intracollisional dynamical phenomena* (rainbow effects and satellites,
quasibound states tunneling etc.) including two-photon effects with or
without intermittent time delay.
(b) *Bound-continuum transitions* including photodissociation and photoexci-
merization dynamics.
(c) *Short-lived molecular complexes* (van der Waals molecules, dimers, free
radicals) and their formation or dissociation dynamics.

(d) *Chemical reaction dynamics* (long-lived transition states, tunneling and effects of quasibound states, etc.) including time-delayed two-photon resonances involving transitions on opposite sides of the transition state. Such studies can augment recent work done on single-photon spectra [4].

Possible repercussions of the two-point nature of FWM are also discussed, including:

(i) Effects of *long-range correlations* between internal and translational degrees of freedom in dense fluids, such as extended (nonlocalized) electronic excitations.

(ii) *Spatial effects in laser noise* requiring departure from the treatment of noise by modifications of the (one-point) Bloch equations.

Experimental work on several aspects of the theory presented here is in progress [5].

References
1. K. Burnett: Phys. Reports, 118, 339 (1985) and references therein
2. Y. Prior, A.R. Bogdan, M. Dagenais and N. Bloembergen: Phys. Rev. Lett. 46, 111 (1981)
3. Y. Prior and A. Ben-Reuven: Nonimpact Theory of Four-Wave Mixing and Intracollisional Dynamics, Phys. Rev. A, (1985) submitted
4. J.C. Polanyi: Faraday Discuss. Chem. Soc., 67, 129 (1979); P. Arrowsmith, S.H.P. Bly, P.E. Charters and J.C. Polanyi: J. Chem. Phys. 79, 283 (1983)
5. M. Rosenbluh, Y. Shevi and Y. Prior: (to be published); N. Horesh and Y. Prior: (to be published).

Stimulated Emission via Unequal Frequency Two-Step Hybrid Resonance in K Vapor*

L.J. Qin, Z.G. Wang, K.C. Zhang, L.S. Ma, Y.Q. Lin, and I.S. Cheng

Department of Phyics, East China Normal University,
Shanghai 200062, People's Republic of China

We had observed the equal frequency two-photon hybrid resonance laser in potassium vapor pumped by a dye laser operated on 6911 Å or 6939 Å [1]. Recently more stimulated radiation in potassium vapor via unequal frequency two-step hybrid resonance were detected. When the dye laser 1 pumped by a N_2 laser was tuned within the wavelength region of 6200-7000 Å, the potassium dimers were excited to the state $B^1\Pi_u$ from ground state $X^1\Sigma_g$. Through K_2^*-K collision, K atoms were excited to the 4P state from ground-state 4S. When the dye laser 2 pumped by the same N_2 laser was used to excite the K atoms to 7S or 5D state from 4P, the stimulated radiation at 7.84 μm (7S-6P$_{1/2}$), 7.89 μm (7S-6P$_{3/2}$), 8.45 μm (5D-6P$_{1/2}$), 8.51 μm (5D-6P$_{3/2}$), 4.86 μm (5D-4F) and cascade-stimulated radiation at 6.46 μm (6P$_{1/2}$-6S), 6.43 μm (6P$_{3/2}$-6S), 6.20 μm (6P$_{1/2}$-4D), 6.24 μm (6P$_{3/2}$-4D), 3.64 μm (6S-5P$_{1/2}$), 3.66 μm (6S-5P$_{3/2}$), 3.71 μm (4D-5P$_{1/2}$), 3.73 μm (4D-5P$_{3/2}$), 3.14 μm (5P$_{1/2}$-3D), 3.16 μm (5P$_{3/2}$-3D) and 2.71 μm (5P$_{3/2}$-5S) were then detected.

Laser 1 with dye DCM and laser 2 with dye Rh690 could be tuned to cover 6200-7000 Å and 5700-6000 Å respectively. The output energy of the two-dye lasers were about 40 μJ with the linewidth about 0.1 Å and the pulse duration about 5 nS. The beam from laser 2 was delayed by an optical delay system and then coincided with the beam from laser 1. The two beams were focused onto the center of a heat-pipe oven with length of 70 cm. About 10 gm of potassium were put into the stainless steel oven with the entrance and the exit windows of quartz and KRS-5 crystal respectively. The heating region of the oven was about 22 cm in length with temperature controlled around 430°C. Other parts of the set-up was the same as described in reference [1].

When we simultaneously set the laser 1 on any wavelength between 6200-7000 Å, corresponding to the $X^1\Sigma_g$ —$B^1\Pi_u$ transition of K_2, and set the laser 2 on 5782.3 Å (4P$_{1/2}$-7S), 5801.8 Å (4P$_{3/2}$-7S), 5812.2 Å (4P$_{1/2}$-5D) or 5831.9 Å (4P$_{3/2}$-5D), we detected the strong forward directional radiation from 7S or 5D state and their cascade directional radiation as well. The stimulated radiation energy changed slightly when the laser 1 scanned from 6200 to 7000 Å. But once the laser 1 was blocked, all of the stimulated radiation signals disappeared almost. If we blocked the laser 2 instead of blocking the laser 1, the stimulated ra-

* Project Supported by the Science Fund of Chinese Academy of Sciences

diation signals almost disappeared too. All this means that the population on 4P resulted from the following pumping and energy-transfer processes:

$$K_2(X^1\Sigma_g) + h\nu \ (6200\text{-}7000 \ \mathring{A}) \xrightarrow{\text{(absorption)}} K_2^*(B^1\Pi_u) \quad \ldots\ldots(1)$$

$$K_2^*(B^1\Pi_u) + K(4S) \xrightarrow{\text{(collision)}} K_2(X^1\Sigma_g) + K^*(4P_{1/2,3/2}) \quad \ldots\ldots(2)$$

The energy difference of the left from the right side in equation (2) is smaller than the interval between neibouring vibrational levels of the K_2 ground state (about 90 cm^{-1}). So the energy-transfer from $K_2^*(B^1\Pi_u)$ to K(4S) is very efficient.

Furthermore we varied the delay time T for the pulse of laser 2 from that of the laser 1. The experimental results showed that the contribution of the laser 2 in generating the stimulated radiation was not the same under different T. The stimulated radiation signals were maximum at T = 8 nS. At T = 14 nS or T = 2 nS, the signals were down to about half of maximum. By T = 0 or T = 17 nS, the signals were down to 1/10 of maximum. The pulse duration of the laser 2 was also about 5 nS, which seems long enough to partly populate 4P state by energy-transfer from K_2^* to K with the delay time T about zero or even slightly in advance of the pulse of the laser 1. Considering the lifetime of 4P state in potassium is 28 nS [2,3], the experimental results and the analysis should be reasonable.

References:

1. Z.G.Wang, L.J.Qin, L.S.Ma, Y.G.Lin and I.S.Cheng, Optics Communications, 51, 151 (1984)

2. G.Copley and K.Krause, Can. J. Phys., 47, 533 (1969)

3. J.K.Link, J.O.S.A., 56, 1195 (1966)

Rayleigh-Brillouin Gain Spectroscopy in Gases

C.Y. She*, H. Moosmüller, G.C. Herring, S.Y. Tang, and S.A. Lee

Physics Department, Colorado State University, Ft. Collins, CO 80523, USA

Over the past decade, we have seen extensive research on and application with Raman gain and other types of coherent Raman spectroscopies. These activities have made clear the advantages of coherent spectroscopy:high signal-to-noise, excellent discrimination from stray light and fluorescence backgrounds, and high spectral resolution. While Raman spectroscopy probes frequency shifts typically 10 cm^{-1} to 3000 cm^{-1} from the exciting laser line, information on thermal properties of the medium, which include thermal conductivity, shear and bulk viscosities, and lower-frequency excitations may be obtained from quasi-elastic Rayleigh-Brillouin scattering.

We have performed the first Rayleigh-Brillouin gain spectroscopy in gases [1]. This was done with two mildly focused, counter-propagating laser beams crossed at a small angle of ≈1° in a Brewster-windowed gas cell. As the pump beam is scanned across the probe beam in frequency, a high resolution stimulated Rayleigh-Brillouin (gain/loss) spectrum of the gas in the cell is obtained. Analysis of this experimental spectrum provides a method for determining thermal dynamic parameters and for testing the collision models of the medium. As an example, Rayleigh-Brillouin gain spectra of Ar at 296K were taken. As can be seen in Fig. 1, their spectral shapes depend on the gas pressure used. For monatomic gas, the state of the medium is specified by a dimensionless parameter y which is proportional to the ratio of wavelength, $2\pi/K$, to the mean free-path, l, where K indicates momentum transfer. Different authors have defined y differently; two popular versions for Maxwell potential have been used: y=0.89/Kl (this work, Yip and Nelkin [2] Tenti et al [3] and She et al [1]), and y=1.33/Kl (Sugawara et al [4] and Clark [5]). Taken the pump laser linewidth into account, theoretical spectra based on different models are also shown in the figure for comparison. For low pressures (y=0.3, 0.5), the early simple kinetic model [1] (solid curve) yields results in agreement with the experimental spectra. At these pressures, the Brillouin doublet is highly suppressed and one sees only a combined scattering feature commonly referred to as Rayleigh spectrum. In the transition region, y=1.5, the hydrodynamic model (dashed curve) still fails, although it begins to work for y=2.5, where sound wave propagation begins to take place. The minor but discernible discrepancy with the simple kinetic model can be reconciled by a more exact numerical solution to the Boltzman equation [4]. The numerically calculated points, indicated by triangles, are in agreement with the measured spectra. Since the signal is proportional to exp(G)-1, the spectral function is expected to be asymmetric when the gain, G, is greater than 0.1. This is already noticeable for Ar with y=2.5. A more dramatic case of gain asymmetry can be seen in the Rayleigh-Brillouin gain spectrum of SF$_6$ with y=7.2. This anomaly

---

* On sabbatical at the University of Maryland, College Park, MD 20742

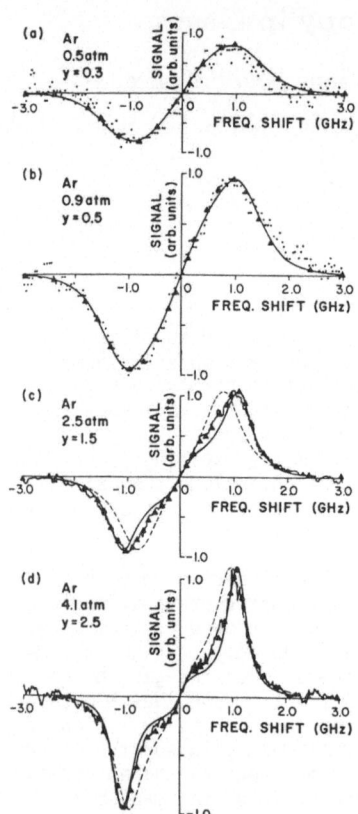

Fig. 1. Rayleigh-Brillouin gain spectra of Ar at different y values compared to the hydrodynamic model (dashed lines), the simple kinetic model (smooth solid lines) and the numerically exact kinetic model (triangles). The experimental curves are shown as dots in (a) and (b.) and as wiggly solid lines in (c) and (d).

between the gain and loss portions of the spectrum provides a method for direct determination of the peak gain, G, which is found to be linearly proportional to the pump power [1].

The gain spectroscopy reported here is only one branch of a larger family of coherent Rayleigh-Brillouin spectroscopies which should also be pursued. With improved lasers, e.g., a cw single-frequency pump laser, spectra with beams crossing at arbitrary angles and higher spectral resolution should be possible. Application of this technique to other gaseous systems is obvious. With this technique, preliminary study of concentration fluctuations in a gas mixture [1] and measurement of high-pressure supersonic flow have already been made [6].

References

1. C. Y. She, G. C. Herring, H. Moosmuller and S. A. Lee, Phys. Rev. A31, 3733 (1985).
2. S. Yip and Nelkin, Phys. Rev. 135, A1241 (1964).
3. G. Tenti, C. D. Boley and R. C. Desai, Can. J. Phys. 52, 285 (1974).
4. A. Sugawara, S. Yip and L. Sirovich, Phy. Fluids 11, 925 (1968).
5. N. A. Clark, Phys. Rev. A12, 232 (1975).
6. G. C. Herring, H. Moosmuller, S. A. Lee and C. Y. She, Opt. Lett. 8,602 (1983).

Excited State Raman and Resonance Enhanced Three-Photon Scattering in Sodium

Y. Shevy, M. Rosenbluh, and H. Friedmann

Bar Ilan University, Physics Department, Ramat-Gan, Israel

We describe the observation of spontaneous emission in the fluorescence spectra of Na vapor excited near the Na D lines by a focused, pulsed dye laser. The fluorescence peaks have been shown to originate from two simultaneously acting physical processes; excited state Raman scattering (ESRS) and resonantly enhanced three-photon scattering (RETPS).

The physical origin of the fluorescence peaks can be understood by considering Fig. I. Shown here are the Na ground state ($3S_{1/2}$) and two excited states ($3P_{1/2}$ and $3P_{3/2}$) interacting with a laser at frequency ω_L, tuned for this particular case to be between the two 3P states. Shown as dashed lines are the virtual states created by the laser-atom interaction. For the case of Fig. Ia and Ib, the scattered photons at ω_a and ω_b are produced by an ESRS process which is the time-reversed analogue (emission followed by absorption) of the commonly observed Raman scattering (absorption followed by emission). The creation of populations in the excited states, which is necessary for this process is possible through RETPS, discussed below, off-resonant excitation and Na-Na collisions. All of these processes are represented schematically by the double arrows in Fig. Ia and Ib.

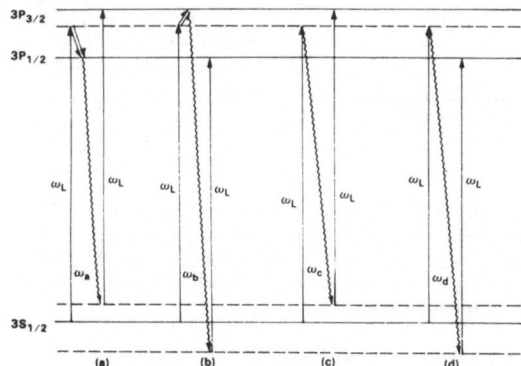

Fig. I
Diagrammatic representation of the interaction of a strong laser interacting with the three-level Na system. Diagrams a and b depict excited state Raman scattering and c and d show three-photon scattering processes.

In Fig. Ic and Id, two possible modes of RETPS are shown, each terminating in a different excited Na state. For all four scattering processes the scattered frequencies are distinct and resolvable. Thus, one would expect to see seven peaks in the fluorescence spectra of Na. One due to Rayleigh scattering, the two Na D line resonances and the four shown in Fig. I. Our observations have indeed shown such fluorescence peaks for detunings as large as 20cm⁻¹ on either side of the D lines. In Figs. IIa and IIb we show two typical spectra, obtained under conditions described

229

previously [1] for a detuning of the laser between the D lines and to the red side of D_1 respectively. In Fig. IIb all four resonances, corresponding to those shown in Fig. I can be identified. In Fig. IIa only three peaks are shown since the resonance enhancement for the three-photon scattering peak at ω_c is weak and although it has been observed it is not shown in the figure.

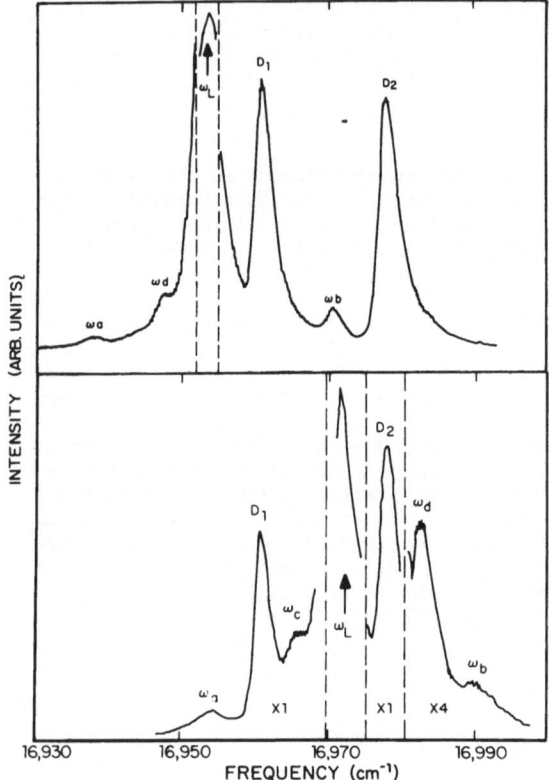

Fig. II
(a) Spontaneous fluorescence observed with the laser tuned to the red of the D_1 line

(b) Similar spectrum for a laser detuning between the D lines. For both spectra the fluorescence intensity scale at the laser frequency has been reduced by many orders of magnitude.

These observations suggest that identical physical mechanisms could operate to yield stimulated emission, provided that appropriate population differences are present. We have recently observed such stimulated emission [2] as part of the co-propagating radiation and conical emission spectra observed in Na-vapor-strong laser interactions [3].

1. Y. Shevy, M. Rosenbluh, H. Friedmann, Phys. Rev. A **31**, 1209 (1985).
2. Y. Shevy, M. Rosenbluh, to be published.
3. E.A. Chauchard, Y.H. Meyer, Opt. Comm. **52**, 141 (1984) and references therein.

Frequency-Domain Nonlinear-Optical Measurement of Femtosecond Relaxation

R. Trebino, C.E. Barker, and A.E. Siegman

Stanford University, Stanford, CA 94305, USA

Femtosecond optical studies of molecular motion in liquids are providing new insight into the fundamental properties of the liquid state. On this timescale, numerous physical processes, e.g., local density fluctuations, intermolecular interactions, and energy relaxation, occur, and, as a result, theoretical and experimental relaxation curves can be quite complex [1]. High-precision and high-resolution experimental techniques that access this fundamental regime are thus important for interpreting dynamical models of liquid physics. We have developed a novel frequency-domain very-high-temporal-resolution nonlinear-optical technique for measuring femtosecond events [2-4] and have employed it to study ultra-fast processes in liquids.

Specifically, we measure the dispersion of the third-order nonlinear-optical susceptibility, $\chi_{ijkl}^{(3)}(\omega_1-\omega_2+\omega_3)$ with $\omega_1-\omega_2 \approx 0$. The frequency-dependence of the third-order susceptibility under this resonance condition is the Fourier transform of the material response [3]. We have developed a three-laser frequency domain technique (nondegenerate four-wave mixing) that allows the measurement (with a resolution of ~1 fsec) of any optically induced femtosecond effect in any material phase. This technique provides high signal-to-noise ratio and background-free operation [4]. Additionally, any element of the third-order susceptibility tensor can be studied, and much freedom in parameter choice is available. Such versatility is essential in probing the dynamics of molecular interactions in liquids.

We have performed various studies of materials in the liquid phase. Studies of internal conversion in triphenylmethane dyes in solution have revealed sum-of-exponential decays with very fast femtosecond components (see Fig. 1; the data are fit to the Fourier transform of a two-exponential decay: $A \exp(-t/\tau_1) + (1-A)\exp(-t/\tau_2)$). We have also measured

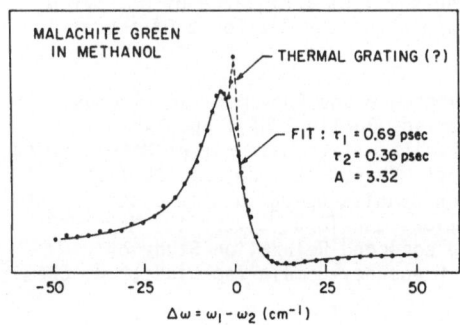

Fig. 1. Malachite green in methanol measured with ortho-gonally polarized excitation beams (to eliminate thermal gratings).

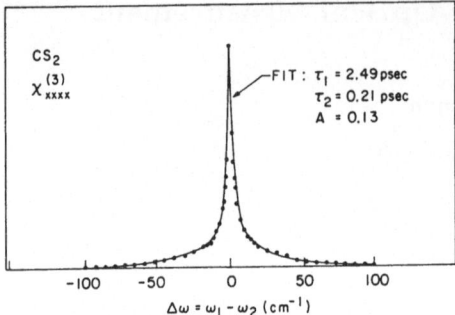

Fig. 2. The xxxx element of the third-order susceptibility of carbon disulfide

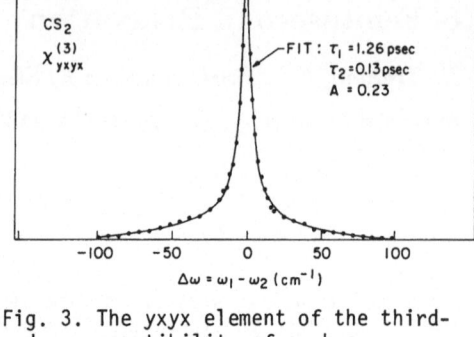

Fig. 3. The yxyx element of the third-order susceptibility of carbon disulfide

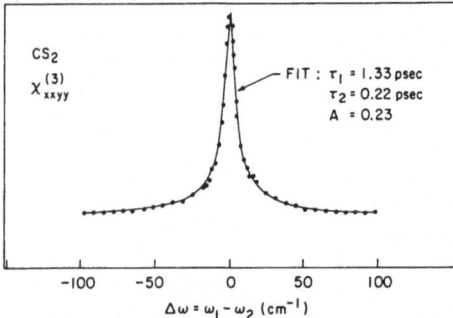

Fig. 4. The xxyy element of the third-order susceptibility of carbon disulfide

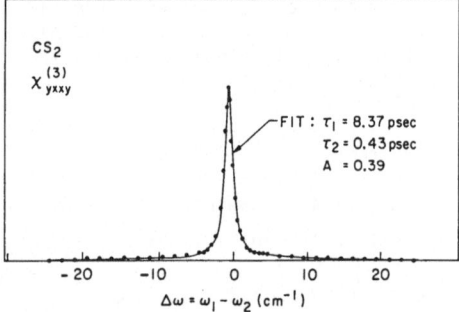

Fig. 5. The yxxy element of the third-order susceptibility of carbon disulfide

orientational relaxation in carbon disulfide, obtaining high signal-to-noise ratio lineshapes for the various components of the susceptibility (see Figs. 2-5; these data are also fit to a two-exponential decay model). We easily observe the subpicosecond interaction-induced effects, but more importantly, residuals rests on the fits reveal the lack of a good fit to a simple sum-of-exponentials decay. This result is in agreement with molecular dynamics simulations that indicate that the Kerr transient is a complex combination of electronic (negligible here) terms, reorientational terms, interaction-induced terms, and cross terms [1].

The authors gratefully acknowledge the skillful assistance of Mr. David Arnone and Mr. Sung-Joo Yoo, and the support of the Air Force Office of Scientific Research.

1. P.A. Madden: "Interaction-Induced Subpicosecond Phenomena in Liquids," Ultrafast Phenomena IV, (Springer-Verlag, Berlin 1984), pp. 245-251.
2. A.E. Siegman, "Proposed Picosecond Excited-State Measurement Method using a Tunable-Laser-Induced Grating," Appl. Phys. Lett. 30, 21 (1977).
3. R. Trebino, "Subpicosecond-Relaxation Studies using Tunable-Laser-Induced Grating Techniques," Ph.D. Dissertation, Stanford University, May 1983.
4. Rick Trebino and A.E. Siegman, "Subpicosecond Relaxation Study of Malachite Green using a Three-Laser Frequency-Domain Technique," J. Chem. Phys. 79, 3621 (1983).

Spectroscopy of Atoms Dressed by Optical Photons in Nearly Degenerate Four-Wave Mixing

P. Verkerk, M. Pinard, and G. Grynberg

Laboratoire de Spectroscopie Hertzienne de l'Ecole Normale Supérieure, Université Pierre et Marie Curie, F-75230 Paris Cedex 05, France

The concept of dressed-atom [1] is often useful when an atom interacts with an intense electromagnetic field. In the case of degenerate four-wave mixing in Doppler-broadened systems, it has been shown that the saturation effects can be clarified using the dressed-atom approach [2]. In the case of nearly degenerate four-wave mixing, we show thereafter that narrow natural-width limited resonances are observed which correspond to an enhancement of the four-wave mixing process due to transitions between energy levels of the dressed-atom. The fact that Rabi sidebands could be observed in nearly degenerate four-wave mixing was first pointed out by Harter and Boyd [3] who use a model of homogeneous broadened medium. Later on, Steel and Lind [4] show that the lineshape in nearly degenerate four-wave mixing is strongly modified by a saturating beam and that some indication of Rabi sidebands is observed experimentally. The understanding of the effect of an inhomogeneous broadening has permitted to improve the experimental conditions and to obtain clearer results.

The experiment is performed in a neon discharge on the 607 nm line ($J=1 \to J=0$ transition) with three incident beams of same polarization. The pump beams E_1 and E_2 have different frequencies ω_L and ω_L'. The probe beam E_3 makes a slight angle with E_2 and has the same frequency as E_1. The beam E_1 has a much higher intensity ($P_1 \sim 10$ W/cm^2) than the two other beams. The experimental curves are obtained by recording the intensity of the generated wave for several values of the frequency detuning ($\omega_0 - \omega_L$) when the frequency ω_L' of the second laser is scanned. An example of recording corresponding to $\omega_0 - \omega_L = 250$ MHz is shown on Fig. 1a. Two narrow resonances are clearly observed which correspond to the degenerate resonance ($\omega_L = \omega_L'$) and to a Rabi sideband (Δ_1). The width of these resonances is smaller than the Doppler width and than the resonance Rabi frequency. We have reported on Fig. 1b the experimental position of the resonances obtained for several values of ($\omega_0 - \omega_L$). The comparison with theoretical curves shows a good agreement. Other details about this experiment are given in [5].

We have studied several other configurations of polarization and frequencies. For example, the exchange of the probe and the weak pump beam in the preceeding experiment gives other kinds of lineshape where the Rabi sidebands are no longer observed. The case of cross-polarized pump beams has also been studied in detail. For this situation, it has been shown [6] that narrow resonances can be observed in degenerate four-wave mixing when a magnetic field is applied. Resonances having closely connected characteristics are observed without external field in nearly degenerate four-wave mixing.

In conclusion, we believe that these experimental observations clearly demonstrate the interest of the dressed-atom approach in four-wave mixing.

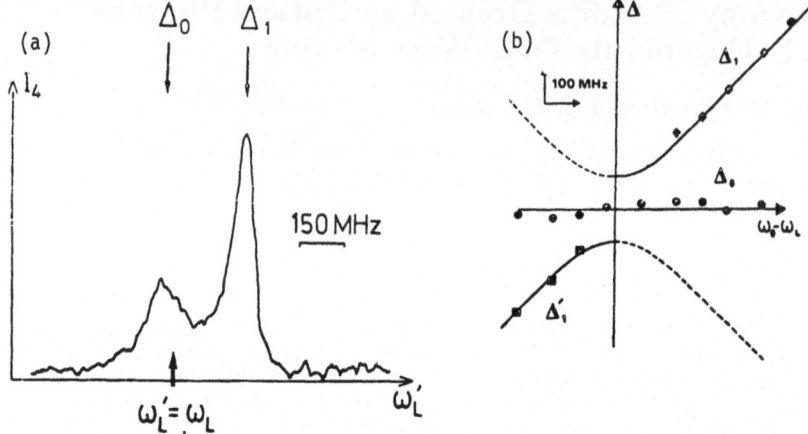

Fig. 1 : a) Experimental recording of the intensity of the phase conjugate beam versus the frequency of the weak pump beam ω_L'. The frequency ω_L of the intense pump beam and of the probe beam is detuned from the atomic resonance ($\omega_0 - \omega_L$ = 250 MHz). The resonance Rabi frequency is 100 MHz. All the beams have the same polarization. - b) Position of the resonances for several values of the frequency detuning ($\omega_0 - \omega_L$). Δ_1 and Δ_1' correspond to the two Rabi sidebands and Δ_0 to the degenerate resonance. The dashed part of the curve corresponds to a domain where the intensity of the Rabi sidebands is vanishing at the secular limit.

It should be emphasized that the dressed-atom method does not only predict the position of the resonances, but also their heights and widths. The comparison with the experimental results has shown a good agreement. We believe that the effect of one saturating beam in four-wave mixing is now well understood both in the cases of homogeneous and Doppler-broadened inhomogeneous widths.

1 C. Cohen-Tannoudji in Frontiers in Laser Spectroscopy, edited by R. Balian S. Haroche and S. Liberman, p. 3 (North Holland 1977)
2 G. Grynberg, M. Pinard and P. Verkerk : Opt. Comm. 50, 261 (1984)
3 D.J. Harter and R.W. Boyd : IEEE J. Quant. El. QE16, 1126 (1980)
4 D.G. Steel and R.C. Lind : Opt. Lett. 6, 587 (1981)
5 P. Verkerk, M. Pinard and G. Grynberg : Opt. Comm. (to be published)
6 M. Pinard, P. Verkerk and G. Grynberg : Opt. Lett. 9, 399 (1984)

Measurement of Relaxation Time of Liquid Crystal by Bifurcation in Optical Bistability

Hong-jun Zhang, Jian-hua Dai, and Peng-ye Wang

Institute of Physics, Academia Sinica, P.O. Box 603,
Beijing, People's Republic of China

In the hybrid optical bistable devices, the Ikeda instability /1,2/ exists when the feed-back time delay (t_R) is much longer than the relaxation time (τ) of the nonlinear medium. The oscillation period (P2) after the first bifurcation point is $T_0=2t_R$. In fact, the relaxation time of the medium is still affect the oscillation period.

The system can be described by the relaxation equation:

$$\tau\frac{dx(t)}{dt} = -x(t)+f(x(t-t_R);\mu) \tag{1}$$

where $f(x:\mu)$ is a nonlinear function which describes the optical bistable system; μ is the control parameter; $x(t)$ is the output intensity. To investigate the stability of a steady-state, we follow the linearized analysis. Let x^* be the steady-state solution, i.e.

$$x^* = f(x^*;\mu)$$

By expanding the Eq.(1) around x^* to the linear term of $\varepsilon(t)=x(t)-x^*$, where $\varepsilon(t)$ is small, we are led to an equation:

$$\tau\frac{d\varepsilon(t)}{dt} = -\varepsilon(t)+f'(x^*;\mu)\varepsilon(t-t_R).$$

The eigen equation of it is

$$(\lambda+1)e^{\frac{\lambda}{\tau}t_R} = f'(x^*;\mu).$$

Let the complex eigenvalues $\lambda=\alpha+i\beta$, we can get

$$(\alpha+1)^2+\beta^2 = e^{-\frac{2t_R}{\tau}\alpha}f'(x^*:\mu) \tag{2}$$

$$\alpha+1+\beta\,\text{ctg}(\frac{t_R}{\tau}\beta) = 0. \tag{3}$$

The stability criterion is provided by the transcendental equation (3).If $\alpha<0$ the system is stable and $\alpha>0$ instable. The imaginary part represents the angle frequency $\omega=\beta/\tau$. Near the instable boundary, i.e.$\alpha=0$, from Eq.(3)

$$\beta = -\text{tg}(\frac{t_R}{\tau}\beta). \tag{4}$$

Let $\beta_1 = \frac{t_R}{\tau}\beta$, Eq.(4) becomes

$$\frac{\tau}{t_R}\beta_1 = -\text{tg}\beta_1. \tag{5}$$

When $\tau\neq0$ and $\tau/t_R\ll1$, take the first order approximation, the non-zero solution of Eq.(5)

$$\beta_1'= \frac{t_R}{t_R+\tau}\pi.$$

So that the period of oscillation is

$$T_0 = 2t_R\pi/\beta_1 = 2(t_R+\tau). \tag{6}$$

We expect to measure the relaxation time of the nonlinear medium by means of the result of Eq.(6). The approximation degree can be given by

$$\Delta T_0/\tau = \frac{2}{\pi}\frac{t_R}{\tau}\left|\beta_1^* - \beta_1'\right|$$

where β_1^* is the exact solution of Eq.(5). This is related to the ratio t_R/τ and it can be found that $\Delta T_0/\tau < 5\%$ when $t_R/\tau > 10$ by the numerical method.

In the liquid crystal hybrid optical bistable device /3/, we measured the relaxation time of the $90°$ twisted nematic liquid crystal.

The precision of measurement can be estimated as follows:

The response time of the optic-eletro detector and the amplifier are both shorter than 1ms. The sampling time of the micro-processor is about 2ms. It is much shorter than the relaxation time of the liquid crystal (of the order of 10^2 milisecond). The measurement error of T_0 is about 5ms. In our condition, $t_R/\tau > 15$ and the temperature is $24°C$. We measured that the relaxation time of the $90°$ twisted nematic liquid crystal is $\tau = 142$ms. We must notice that in the normal method /4/ of measurement of the relaxation time of liquid crystals, the relaxation time is defined by the duration of the contrast of 90% to 10%. But in our method it is defined by the duration of which the output intensity decays from 1 to e^{-1} after the applied voltage is removed suddenly.

In conclusion, a new method of measurement of the relaxation time of liquid crystals by means of the period-doubling bifurcation in optical bistability is suggested. This method is not only suitable to the $90°$ twisted nematic liquid crystal but also to other molecular alignments. The function of $f(x; \mu)$ may be of different form to different liquid crystal alignments. But we can also control the parameter of the system to the P2 oscillation state.

REFERENCES

1. K.Ikeda, Opt. Commun. 30,257(1979)
2. K.Ikeda, H.Daido and O.Akimoto, Phys. Rev. Lett. 45, 709(1980)
3. Zhang Hong-jun, Dai Jian-hua,Wang Peng-ye and Jin Chao-ding, JOSA, (to be published)
4. M.Schadt, W.Helfrich, Appl. Phys. Lett. 18, 127(1971)

Part VIII

Quantum Optics,
Squeezed States, and Chaos

Test of Photon Statistics by Atomic Beam Deflection

Y.Z. Wang, W.G. Huang, Y.D. Cheng, and L. Liu

Shanghai Institute of Optics and Fine Mechanics, Academia Sinica,
Shanghai, People's Republic of China

1. Introduction

There has been growing interest recently in optical phenomena that exhibits
purely quantum-mechanical feature of radiation field[1-5]. Mandel studied the
distribution of the number of photons emitted in a given time by a two-level
atom in a resonant coherent exciting field[1]. The theory indicates that photon statistics of fluorescence of the two-level atom is sub-Poisson photon
statistics. As a measure of the departure of photon statistics from Poisson
law, Mandel introduced a Q-parameter

$$Q=[\langle(\Delta n)^2\rangle-\langle n\rangle]/\langle n\rangle, \tag{1}$$

where $\langle n\rangle$ is the mean number of photons emitted in a given time by the two-
level atom, and $\langle(\Delta n)^2\rangle$ is the variance of the emitted photon number. For the
sub-Poissonian statistics, Q is less than zero.

Cook pointed out that there exists a simple relationship between the photon
statistics in resonance fluorescence and the statistics of the momentum transferred to an atom by a plane travelling wave[2]. From this relation he derived
the expressions for mean photon number $\langle n\rangle$ and $\langle(\Delta n)^2\rangle$, and obtained Q as

$$Q=-[\Omega^2(3\beta^2-\Delta^2)]/[2(\Delta^2+\beta^2+\tfrac{1}{2}\Omega^2)^2], \tag{2}$$

where Ω is the Rabi frequency of the two-level atom in the travelling wave,
β is half the Einstein A-coefficient, and $\Delta=\omega-\omega_o$ is the detuning frequency
between the field frequency ω and the atomic resonant frequency ω_o. This expression indicates that, in addition to the sub-Poissonian statistics, super-
Poissonian photon statistics [$\langle(\Delta n)^2\rangle$ greater than $\langle n\rangle$] occurs in resonance
fluorescence for certain off-resonance cases. This result is of interest be-
cause it shows that sub-Poissonian photon statistics are not a necessary consequence of photon anti-bunching in time, which is always present in the
radiation from a single two-level atom. Anti-bunching and sub-Poissonian statistics are distinct effects that need not necessarily occur together[5]. It
is important that measurement of Q should be made to verify the theoretical
predictions.

On the basis of the relation between the photon statistics in resonance
fluorescence and the statistics of the momentum transferred to an atom by the
plane travelling wave, Cook first suggested an atomic beam deflection experiment to demonstrate the non-Poisson statistics[2]. An atomic beam is transversely illuminated by a laser beam. In this case the Q is expressed as a
function of the deflection angle ,

$$Q=\frac{M\langle 1/v^2\rangle}{\hbar k\langle 1/v\rangle}\cdot\frac{[\langle(\Delta\theta)^2\rangle-\langle(\Delta\theta)^2\rangle-S^2\langle\theta\rangle^2]}{\langle\theta\rangle}-\frac{7}{5} \tag{3}$$

where M is the mass of atom, v is the velocity of the atom, $\hbar k$ is the photon momentum. $\langle\theta\rangle$ is the mean deflection angle, $\langle(\Delta\theta)^2\rangle$ is the atomic beam spreading under laser beam illumination. $\langle(\Delta\theta)_0^2\rangle$ describes the initial beam divergence and $S^2\langle\theta\rangle^2$ is the contribution to the beam spreading resulting from the distribution of the atom's velocity. S^2 is expressed as $S^2=[\langle 1/v^4\rangle-\langle 1/v^2\rangle^2]/\langle 1/v^2\rangle^2$. The quantities $\langle(\Delta\theta)_0^2\rangle$, $\langle(\Delta\theta)^2\rangle$ and $\langle\theta\rangle$ are directly measured in the experiment. For a thermal atomic beam $S^2\langle\theta\rangle^2$ is quite large, the part of $\langle(\Delta\theta)^2\rangle$ associated with the photon statistics is easily lost in the uncertainty of the measured value of $\langle(\Delta\theta)^2\rangle$. To overcome this problem, S must be decreased by velocity selection, meanwhile the useful density of atomic beam decreased strongly. For a significant measurement $\langle(\Delta v)^2\rangle^{\frac{1}{2}}/\langle v\rangle$ must be less than 1/40[2]. Another way this problem is to use a multiple laser beam to defelct the atomic beam. In this method the velocity distribution does not introduce any additional spreading of the atomic beam. It allows us to accomplish the task more easily.

2. Test of Photon Statistics by Atomic Beam Deflection

We suggest an experiment of atomic beam deflection by a multiple laser beam to test photon statistics. The principle of the experiment is shown in Fig.1. The atomic beam travels across the multiple laser beam which propagates forward and backward between the reflectors, the mean light pressure of laser beams cannot change the motion of the atoms, but due to the quantum fluctuation a diffusion of atomic momentum occurs and the atomic beam is spread.

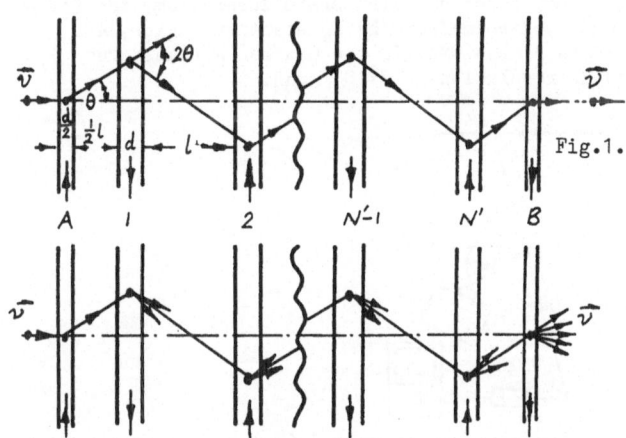

Fig.1. (a) Atomic Beam Deflection by Mean Light Pressure d-width of laser beam; l-distance between laser beams; N=N'+1

(b) Atomic Beam Spread by Momentum Diffusion d-width of laser beam; l-distance between laser beams; N=N'+1

In the case of multiple laser beam, following Cook[2] the Q-parameter has been derived as

$$Q = \frac{2M^2(\Delta^2+\beta^2+\frac{1}{4}\Omega^2)\langle v^3\rangle}{Nd(\hbar k)^2\beta\Omega^2}[\langle(\Delta\theta)^2\rangle - \langle(\Delta\theta)_0^2\rangle] - \frac{7}{5} \qquad (4)$$

where N is the number of laser beams, d is the diameter of the beams. All quantities in (4) are directly measured in the experiment.

This method has two advantages: First, to compare (4) with (3) shows that the velocity distribution does not introduce any additional spreading of the atomic beam. Second, because of the perpendicular interaction of the laser beams with the atomic beam, the signal to noise ratio is much greater than that in the experiment using a velocity selected atomic beam.

3. Experiment and Result

An experiment is being carried out on a precisely collimated sodium beam. The scheme of the experiment is a little different from that mentioned above, as

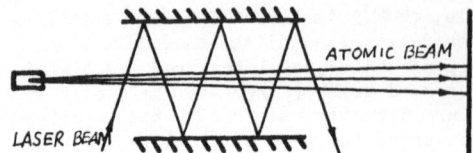

Fig.2 Another schematic diagram
of the experiment

shown in Fig.2. The laser beam is incident at an oblique angle ϕ with the atomic beam and the frequency is tuned to be nearly resonant with the $3^2S_\frac{1}{2}$(F=2) ---$3^2P_\frac{3}{2}$(F=3) transition of sodium. Due to the Doppler frequency shift, only part of the atoms in the beam is resonant with the laser beam. If $\phi=84°$, the velocity distribution of the resonant atoms is about $(\Delta v^2)^{\frac{1}{2}}/v_0 \sim 5 \times 10^{-2}$. In this case Q-parameter is expressed as

$$ Q = \frac{M^2 v_0^3 (1+G)}{(\hbar k)^2 N d \beta G \sin^2\phi} \cdot [\langle(\Delta\theta)^2\rangle - \langle(\Delta\theta)_0^2\rangle] - \frac{7}{5} . \tag{5}$$

Figure 3 shows the experimental setup, which was described in [6]. In this experiment an OMA (Optical Multi-Channel Analyzer OMA-II) is used to detect the fluorescence of the atomic beam induced by the probe laser beam, the fluorescence spot is located 68 cm downstream from the interaction region, and imaged by a camera on the surface of the OMA. If the OMA has a transverse resolution of 25 μm, the angular resolution is 3×10^{-5} rad.

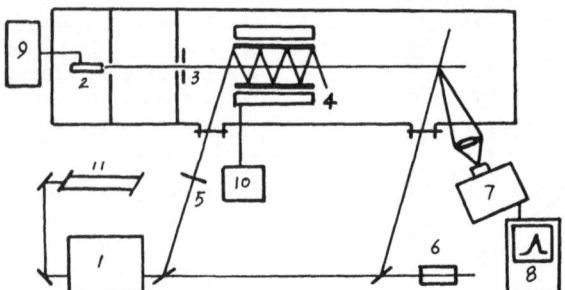

Fig.3. The schematic diagram of the experiment setup 1-CW laser; 2-oven; 3-collimator; 4-D.C. magnet; 5-λ/4 plate; 6-Na absorption cell; 7-OMA; 8-oscilloscope; 9-temperature control device; 10-D.C. coil; 11- Ar$^+$ laser

When the laser beam incident to the mirrors is interrupted, the OMA records an intensity distribution of the fluorescence, as shown in Fig.4(a), which indicates the initial divergence of the atomic beam. When the laser beam is incident to the mirrors, the atomic beam is spreading by the momentum diffusion, as shown in Fig.4(b). If the amplitude of the distribution is expressed by J, $\langle(\Delta x)^2\rangle$ can be obtained as,

$$ \langle(\Delta x)^2\rangle = \sum_{i=1}^{q} (\Delta x_i)^2 J_i / \sum_{i=1}^{q} J_i , \tag{6}$$

where q is the number of the OMA detector channels. Using the experimental data and (6), we have obtained $\langle(\Delta\theta)_0^2\rangle = 1.1 \times 10^{-7}$ and $\langle(\Delta\theta)^2\rangle = 8.3 \times 10^{-8}$. The experimental parameters used are the following: the laser power P=3 mW, the diameter of the laser beam d=2.0 mm, the saturation factor G=1.27, the number of the laser beams N=28, the reflectivity of the mirrors r=99.8%, the temperature of sodium source T=710K, the mean velocity of the atoms $\langle v\rangle$=895 m/s.

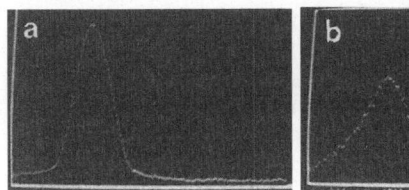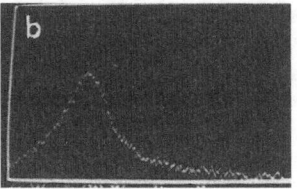

Fig.4. The result of the experiment (a)-undiffused; (b)-diffused

Using (6), we have $Q = -0.79 \pm 0.14$.   From (2) we caculated the theoretical
$Q = -0.73$. The difference between experimental and theoretical results is mainly due to measurement error of laser power inside the vacuum chamber. Other
errors are small, for example, laser beam damp due to the loss of reflection
between the mirrors introduces a small deflection angle of the atomic beam,
which can be expressed as $S^2\langle\theta\rangle^2$[2]. In our experiment $S^2\langle\theta\rangle^2 \sim 10^{-10}$ rad^2, it
is much smaller than $\langle(\Delta\theta)^2\rangle$.

4. Conclusion
We have done the atomic beam deflection experiment to test the photon statistics. The preliminary experiment shows that the photon statistics in resonance
fluorescence of the two-level atom is sub-Poissonian, $\langle(\Delta n)^2\rangle$ is less than
$\langle n\rangle$. Our results are consistent with those of QED theory of resonance fluorescence and the quantum theory of the motion of atom.

Reference
1. L.Mandel, Opt. Lett., 4, 205 (1979)
2. R.J.Cook, Opt. Commun. 35, 347 (1980)
3. R.J.Cook, Phys. Rev., A22, 1078 (1980)
4. S.Stenholm, Phys. Rev. A27, 2513 (1983)
5. R.Short and L.Mandel, Phys. Rev. Lett., 51, 384 (1983)
6. Y.Z.Wang et.al., Scientia Sinica, A27, 881 (1984)

Chaotic Dynamical Behavior in Lasers

N.B. Abraham, D.K. Bandy, and M.F.H. Tarroja
Dept. of Physics, Bryn Mawr College, Bryn Mawr, PA 19010, USA

R.S. Gioggia and S.P. Adams
Dept. of Physics, Widener University, Chester, PA 19013, USA

L.M. Narducci
Dept. of Physics, Drexel University, Philadelphia, PA 19104, USA

L.A. Lugiato
Dipartimento di Fisica dell' Università, I-20133 Milano, Italy

We describe new uses of the spectrum of the output of a laser system to determine details of the changes in the dynamical behavior of lasers exhibiting highly nonlinear dynamics including deterministic chaos.

1. Chaotic Laser Systems

Chaos is by now a relatively popular subject seeming to be almost a fad or a cult with its somewhat cryptic technical jargon. Yet scanning the last few years of Physical Review Letters, Physical Review A, Physics Letters A, and Physica D, one quickly learns that this subject is of wide-ranging and serious interest to many physicists. The term "chaos" has gained an increasingly refined technical meaning which often leads us to use the apparently contradictory combination "deterministic chaos". Briefly we should note two different types of nonlinear dynamical systems where chaos can be found and studied. In Hamiltonian systems, nonlinear deterministic motion can be obtained on the constant energy surfaces of the phase space of the system. In these cases there are interesting connections between chaotic and ergodic motion. Somewhat in contrast, dissipation in systems leads to an evolution from initial conditions in phase space to an attracting subset which may be a familiar type of solution such as a constant solution (a single point in phase space) or a periodic solution (a closed loop in phase space) or may be an exotic type called a "strange (or chaotic) attractor". Such exotic objects have the properties of sensitive dependence on initial conditions (resulting in exponential divergence of neighboring trajectories in phase space), stretching and folding in various directions near the attractor, and irregular (nonperiodic) evolution of any one of the nonlinearly coupled variables. The attracting set in these cases can be shown to be self-similar on different scales corresponding to its having a fractal dimensionality. Such chaotic solutions may occur in systems described by coupled nonlinear ordinary differential equations so long as there are at least three variables. The resulting chaotic behavior has many of the features of what is usually called noise. The aperiodic time evolution, the broadband power spectrum, and a degree of unpredictability after short times suggest a kind of pink noise. Instead, this is a subtle and exotic form of deterministic evolution and does not have its origins in stochastic noise perturbations in either the equations or the experiments. A type of chaotic behavior is also found in systems which can be described by maps in which previous values predict the next value in a sequence. Much of the recently popularized theoretical and numerical work has dealt with one-dimensional maps where only the previous value is needed to predict the successive value.

Several good reviews are now available [1-2] and there is a variety of particular papers on measuring and characterizing chaos [3]. From the point of view of studies of laser systems we usually begin from stable or simply periodic behavior and look experimentally or theoretically for more complex time dependent behavior. Thus the field is generally defined as the study of nonlinear dynamics and instabilities in lasers and other nonlinear optical systems. Several useful reviews, collections of articles and monographs have appeared [4-6]. The kinds of laser problems that are being addressed include the stability of single mode lasers, the response of systems to modulation, the pulse to pulse variations in a modelocked multimode laser, the stability of injection-locked lasers, the interaction of modes in a bidirectional ring lasers, and the stability of lasers with intracavity absorbers, among others.

As pointed out in many recent papers, chaotic behavior of systems is not news, what is new is the understanding that the behavior may have its roots in dynamical processes and not in external perturbations or thermodynamic fluctuations of internal parameters.

The study of laser chaos began in the late 1950's and early 1960's [see the introduction in Ref. 5] but the connections between the dynamical behavior found in theoretical models and the "dirty" experimental results of that period could not be found. As nonlinear dynamics has emerged as a specialized interdisciplinary concentration in science, certain signatures have been identified which virtually guarantee that the underlying processes are deterministic. Among those signatures are the changing spectral patterns that show, for example, a succession of period doubling bifurcations indicating that the system must make increasingly many loops in its phase space before exactly repeating. This route is often known as the "Feigenbaum route to chaos" [7] as he was among the first to regularize this description and to find its unversal characteristics. Another route to chaos is the successive appearance of incommensurate frequencies in the spectrum, or chaos by way of the breaking of toroidal motion in phase space (typically no more than two incommensurate frequencies can coexist before chaos sets in). Details of these routes (and others) are presented systematically by Eckmann [1]. Unfortunately not all systems fall into simply classifiable univeral categories and laser systems are usually too complex to fit the simple sequences though some agreements have been found betweeen experimental results and both the simple and complex theories.

Much of recent work [3] has focussed on quantitatively analyzing the time evolution of chaotic systems, thus going beyond the use of qualitative measures such as broad band spectra and routes to chaos. Many of these have been dramatically successful and these tests make it possible to say with assurance that the broadband, "noisy" spectra that occur for some lasers result from the low-dimensional deterministic evolution. However, these number-crunching oriented techniques are of little interest to spectroscopists and are relatively hard to comprehend physically. For our discussion of chaos in this setting, we wish to return to a view of the spectrum with a different eye, one that is even more sensitive to subtle spectral variations and demonstrate that in both experiments and models there is more dynamical information than was previously believed or perceived.

2. Different Kinds of Time Dependent Solutions

Stable laser behavior refers to what is often called cw laser action, the existence of a constant amplitude for the monochromatic optical carrier wave. It is of interest to examine several of the time-dependent-amplitude solutions found in laser systems. Their intensity power spectra as they have been seen experimentally are shown in Figure 1, where they correspond to weak modulation (a), strong modulation (b), and clearly aperiodic modulation (c). We have found it extremely useful to examine not only the intensity power spectra, but also the optical spectrum [8]. Familiar techniques for doing this are the use of monochromators or Fabry-Perot interferometers when the optical spectrum is spread widely in wavelength. In our case we have used heterodyne techniques with a stabilized reference laser to observe the relatively closely spaced frequencies as are also shown in Figure 1. By suitable choice of reference laser frequency, the heterodyne signals are shifted from zero to be centered at about 70 MHz so that each spectrum in Figure 1 is a combination of a low-frequency part which is the homodyne spectrum (intensity power spectrum) and a high-frequency part with is the heterodyne spectrum. The spectra are shown for studies of a unidirectional, single-mode, ring laser using the 3.51 μm transition in xenon for the lasing action. This is a well-studied inhomogeneously broadened and high gain system that is particularly susceptible to chaotic behavior.

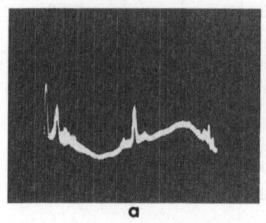 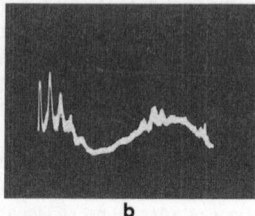 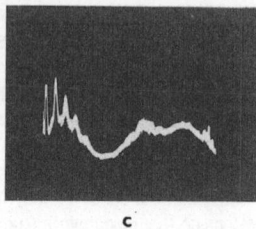

Figure 1. Sample laser homodyne and heterodyne spectra. Vertical scale is logarithmic while the horizontal scale covers 0-100 MHz linearly.

The additional spectral information helps us to see that in the first case the instability is a weak modulation of the amplitude of the previously stable solution resulting in sidebands to the optical carrier. In the second case the system is strongly modulated and we can note particularly that by the zero power at the previous optical carrier frequency we can conclude that there is no remaining constant amplitude term, but instead a fully symmetric (positive and negative) amplitude modulation. The third case shows the chaos in the broadband nature of the optical spectrum.

3. Symmetry Breaking versus Period Doublings

Subtle changes in the processes are also revealed by the heterodyne spectra. For example, Figure 2 shows an intensity power spectrum that has a small peak at one half of the principal pulsing frequency. This is consistent with the evidence that the intensity pulses alternate in peak heights. Examination of the heterodyne spectrum shows that there is a small peak at the optical carrier frequency suggesting a large modulation which has a residual dc component suggesting an asymmetric modulation. Thus the heterodyne spectrum helps us to distinguish between popular interpretations that these were period doubling transitions and symmetry-breaking transitions [9] as have occurred in this case.

We should note that heterodyning is not only a useful experimental technique, but can also be an effective technique for the interpretation of numerical results. We have heterodyned reference signals and numerically generated results for the complex electric field amplitude in order to better interpret the results of simulations [10]. Several such spectra are shown in Figure 3 where both periodic and aperiodic signals can be found and identified.

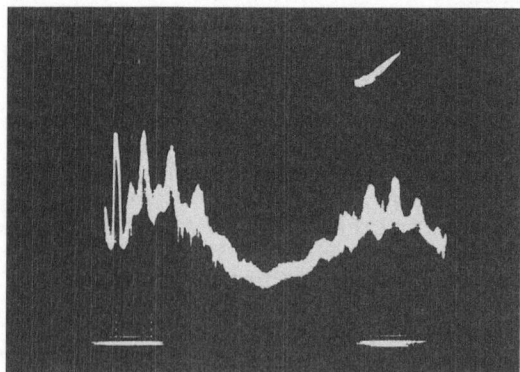

Figure 2. Homodyne and heterodyne spectra showing a case of an asymmetric attractor that might be misinterpreted as a period-doubled case.

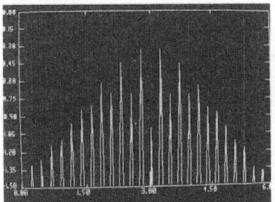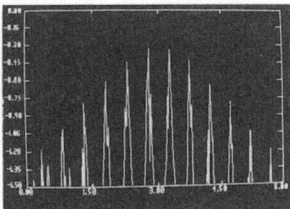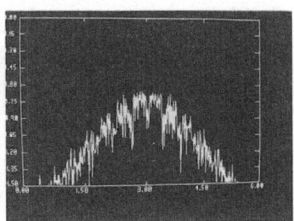

Figure 3. Examples of heterodyne spectra for numerical solutions of the appropriate model for the experiments described above.

4. Distinguishing Chaos

Several issues arise regarding the identification of chaos. First, we wish to determine if the aperiodic behavior arises from deterministic chaos or from stochastic noise. Techniques have been proposed and applied with success in making the distinctions between experimental results coming from chaos and the predictions for random numbers [8]. This is particularly helpful in cases of "weak" chaos, which results in only a slight blurring of the periodic spectra. As a further extension we have recently compared broadband signals from laser chaos with broadband signals from amplified spontaneous emission and have found clear distinctions that indicate that one has its origins in low-dimensional deterministic evolution and that the other is either high dimensional or stochastic (as might be expected for a large number of independently emitting atoms).

The reasons for wanting such a distinction are clear. Broadband sources of high power and relatively narrow spectral density have various applications. Chaotic laser sources may provide this type of pseudo-incoherent source. Where chaos exists, other forms of more desirable dynamical time dependent behavior may also exist including stable pulsing or periodic behavior which can be sought by changes in laser parameters. Where the noise is undesirable it is very important to know if it arises from chaos or stochastic processes or perturbations as the solutions one attempts depend critically on the origin and type of noise.

References:
1. J. P. Eckmann, Rev. Mod. Phys., 53, 643 (1981).
2. Universality in Chaos, ed. P. Cvitanovic (A. Hilger, London, 1984).
3. N. B. Abraham, J. P. Gollub and H. L. Swinney, Physica, 11D, 252-264 (1984).
4. N. B. Abraham, Laser Focus, May 1983, pp. 73-81.
5. Special Issue of the Journal of the Optical Society of America B, 2, January 1985.
6. Instabilities and Chaos in Quantum Optical Systems, ed. F. T. Arecchi and R. G. Harrison, (Springer-Verlag, Heidelberg, to be published).
7. M. J. Feigenbaum, J. Stat. Phys., 19, 25 (1978).
8. N. B. Abraham, A. M. Albano, T. H. Chyba, L. M. Hoffer, M. F. H. Tarroja, S. P. Adams, and R. S. Gioggia in ref. 6.
9. P. Coullet and C. Vanneste, Hel. Phys. Acta, 56, 813-823 (1983); Y. Kuramoto and S. Koga, Phys. Lett., 92A, 1-4 (1982).
10. M.F.H. Tarroja, N. B. Abraham, D. K. Bandy, T Isaacs, R. S. Gioggia, S. P. Adams, L. M. Narducci and L. A. Lugiato, in Proceedings of International Meeting on Instabilities and Dynamics of Lasers and Nonlinear Optical Systems, R. Boyd, L. Narducci, and M. Raymer, eds., (Cambridge U. Press, to be published).

Self-Pulsing and Chaos
in Optically Bistable and Tristable Systems

T. Yabuzaki, M. Kitano, and T. Ogawa

Radio Atmospheric Science Center, Kyoto University,
Uji, Kyoto 611, Japan

Introduction

Self-pulsing and chaos in optically bistable systems have attracted a great deal of attention from the theoretical interests of dynamical behavior of nonequilibrium systems. The systems studied so far are mostly those exhibiting bistability with hysteresis, which appears as an intensity change of output light as the incident light intensity is varied. In this paper, we report on the phenomena observable in the systems exhibiting the new type optical bistabity and tristability, which were proposed by us[1,2] and experimentally realized by Cecchi et al.,[3] Hamilton et al.,[4] and Giusfredi et al.[5] These bistability and tristability bring about symmetry breaking (i.e. pitchfork bifurcation), different from ordinary bistability (multistability) with hysteresis. We mention first the self-pulsing caused by the self-sustained spin precession, giving particular attention to the involved phase transitions. Secondly, we mention chaotic behaviors caused by the delayed feedback, which are also considerably different from those in ordinary bistable systems.

Optical Systems

As shown in Fig. 1, both systems are all-optical and contain atoms with spin in the ground—state which plays a role of a nonlinear dispersive medium. The incident light is linearly polarized and tuned on a wing of the absorption line. It can be shown that each system has a positive feedback loop with respect to the intensity difference of σ_+ circularly polarized components and to the spin-orientation through competitive optical pumping.[1,2] When a small amount of intensity difference ΔI_+ between σ_+ components exists, spin-orientation is produced by optical pumping, which produces the difference of refractive indeces Δn_+ for σ_+ light. In case of the bistable system, Δn_+ causes the rotation of polarization of the incident light, which results in the change of relative angle between the directions of polarization and optical axis of the $\lambda/8$ plate, and increases ΔI_+. While, in the case of the tristable system, Δn_+ changes the resonant condition for σ_+ light and increases ΔI_+. As shown in Fig. 2, when the incident light intensity I_0

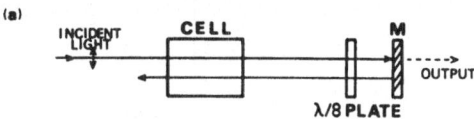

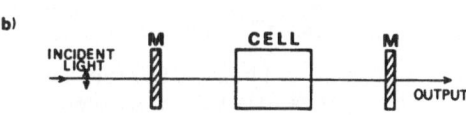

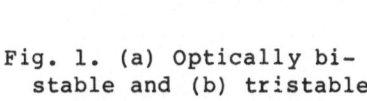

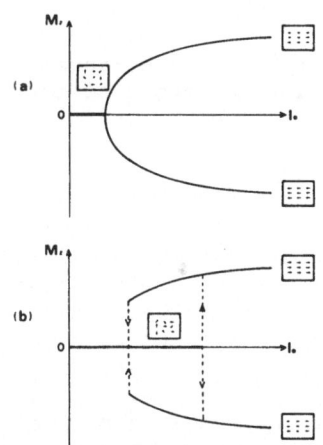

Fig. 1. (a) Optically bi-
stable and (b) tristable
systems behaving with
symmetry-breaking.

Fig. 2. Steady-state magnetization M_z as a function
of incident light intensity
I_0, for (a) bistable and
(b) tristabe systems.

exceeds critical values and the loop-gain exceeds unity,
symmetry breaking takes place. Namely, even if the incident
light is linearly polarized, the polarization of light inside
the systems becomes σ_+- or σ_--dominant, which produces
spontaneously the magnetization $\vec{M}$ to the direction parallel or
antiparallel to the incident beam.

Self-Pulsing by Self-Sustained Spin-Precession

We found that, when a transverse magnetic field H_0 is
applied to the atoms in the tristable system (Fig.1 (b)), the
spin-orientation produced by the symmetry-breaking precesses
around H_0 without decay, because of the alternative switching
of σ_+ components in the cavity.[6] This phenomenon was observed
by Mitschke et al.[7], in the Fabry-Perot cavity filled with
sodium vapor. It can be shown that the similar self-pulsing
takes place in the optically bistable system shown in Fig.
1(a), when a transverse magnetic field H_0 is applied. In the
presence of H_0, the Bloch equation describing the motion of the
magnetizatio M_z becomes

$$d^2m_z/dt^2 + f(m_z)dm_z/dt + g(m_z) = 0, \qquad (1)$$

with

$$f(m_z) = \Gamma(2 + 2I_0 - kL\cos 2kLm_x), \qquad (2)$$

$$g(m_z) = \Omega_0{}^2 m_z + \Gamma^2(I_0+1)m_z$$
$$- (\Gamma^2/2)I_0(I_0+1)\sin 2kLm_z . \qquad (3)$$

247

where Γ is the spin-relaxation
rate, k is the linear wavenumber,
L is the length of the cell, and
Ω_0 is the Larmor frequency. One
may find that when $\sin(2kLm_z)$ is
expanded with respect to m_z up to
the second order, Eq. (1) becomes
the van der Pol equation. We
have numerically calculated the
steady state magnetization, and
found that there exist roughly
three types of solutions:
monostable solution at $\vec{M}=0$,
bistable solutions, and
precession around H_0 along a
limit cycle. Figure 3 shows the
phase diagram, i.e. the
conditions of I_0 and Ω_0 to get
three types of solutions.

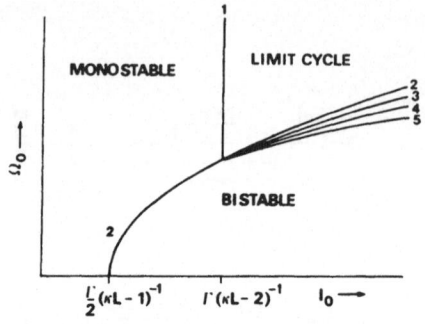

Fig. 3. Phase diagram for
three distinct states
of the system.

Chaotic Behaviors of Tristable System

The chaotic behavior of optically bistable systems has
extensively studed so far, since the first work of Ikeda.[8] With
respect to the bistable system shown in Fig. 1(a), we have
recently studied the effects of the delayed feedback[9] and shown
that the asymmetric chaos induced at a bifurcation is symmetry-
recovered at another bifurcation, which can be viewed as a
"crisis" of chaos.[10]

Here we present the delay-induced chaos and related
phenomena in the tristable system shown in Fig. 1(b).
Numerical calculation has been done in the case that the delay
time is much longer than the system response time, for which
the behavior of the system can be described by a difference
equation. Figure 4 shows an example of calculated bifurcation
diagrams, i.e. the plots of orbit of the intensity difference
of intracavity $\sigma_\pm$ components normalized by the incident light
intensity I_0 for given values of I_0. In Fig. 4, symmetric
periodic oscillation can be seen for $I_0 \lesssim 4$ (quasi-periodic
oscillation at $I_0 \sim 3$). In the region $4 \lesssim I_0 \lesssim 8$ we see the
existence of three stable states: two being stationary states
and a chaotic state which disappears for $I_0 \gtrsim 8$. After the
bifurcation at $I_0 \sim 13$, shows asymmetric oscillation takes place
between sub-branches of the upper or lower state. It must be
interesting to note that, in the high intensity region $I_0 \gtrsim 8$,
the choice of the system to take the upper or lower state is
almost at random, not determined by the sense of the initial
condition of $\Delta I_\pm$, if $\Delta I_\pm \lesssim 0.4$ in the present case. This is
because the system varies chaotically from the initial
condition within the vestige of strange attractor, from which

it escapes by chance after many, typically 50, random walks and
approaches the upper or lower state. We have calculated the
initial condition dependence on the choice of states, and found
that it has structures as shown in Fig. 5, where dots represent
the cases that the system chooses the upper state shown in Fig
4. For the negative initial conditions, the structures becomes
antisymmetric to those in Fig. 5.

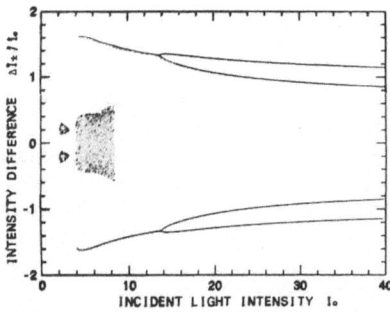

Fig. 4. Typical Bifurcation   Fig 5. Initial condition
        diagram.   dependence on the choice
                                           to take the upper state.

This work is supported by the Ministry of Education, Science
and Culture in Japan, under a Grant-in-Aid for Scientific
Research.

References
1. T. Yabuzaki, M. Kitano, and T. Ogawa, Phys. Rev. A 29,
   1964 (1984).
2. M. Kitano, T. Yabuzaki, and T. Ogawa, Phys. Rev. Lett. 46,
   826 (1981).
3. S. Cecchi, G. Giusfredi, E. Petriella, and P. Salieri,
   Phys. Rev. Lett. 49, 1928 (1982).
4. M. W. Hamilton, W. J. Sandle, J. T. Chilwell, J. S.
   Satchell and D. M. Warrington, Opt. Comm. 48, 190 (1983).
5. G. Guisfredi, P. Salieri, S. Cecchi, and F. T. Arrechi,
   Opt. Comm. 54, 39 (1985).
6. M. Kitano, T. Yabuzaki, and T. Ogawa, Phys. Rev. A 24,
   3156 (1981).
7. F. Mitschke, J. Mlynek, and W. Lange, Phys. Rev. Lett.
   50, 1660 (1983).
8. K. Ikeda, H. Daido, and O. Akimoto, Phys. Rev. Lett. 45,
   709 (1980).
9. M. Kitano, T. Yabuzaki, and T. Ogawa, Phys. Rev. A 29,
   1228 (1984).
10. C. Grebogi, E. Otto, and J. A. York, Physica 7D, 181
    (1983).

Experimentalists' Difficulties in Optical Squeezed State Generation

M.D. Levenson and R.M. Shelby

IBM Research Division K32/281, 5600 Cottle Rd., San Jose, CA 95193, USA

Squeezed states of light are theoretically predicted states of the radiation field which have phase-dependent quantum fluctuations, with the fluctuations corresponding to one phase being less than those of a coherent state [1]. These states have eluded experimental demonstration, at least so far. From an experimentalist's point of view, squeezed state research can best be described as a series of difficulties that must somehow be overcome.

First Difficulty: Figuring out what the Theorists are Talking About

To an experimentalist, a light wave is a sinusoidal electric field: $E = E_1 \cos\omega t + E_2 \sin\omega t$ where the quantities E_1 and E_2 are called quadrature amplitudes. These quadrature amplitudes can be portrayed axes of a two-dimensional "map" or "phasor diagram," and any state of the field can be specified on such a map. A classical state would correspond to a single point because both quadratures can be perfectly specified. In quantum mechanics, the quadrature amplitudes are conjugate variables subject to an uncertainty principle. Thus quantum states must be specified by probability contours drawn on the phasor diagram. A coherent state would be represented by circular contours centered at the average value of the field. A squeezed state would have "elliptical" contours, with minor axes smaller than the radius of the corresponding contour of the coherent state. Thus one field quadrature amplitude of a squeezed state is less uncertain than that of a coherent state. Measurements in that quadrature are more precise than the "ordinary quantum limit" [2].

The uncertainty in the electric field amplitude is usually attributed to "vacuum fluctuations." Thus, if one is to believe in squeezed states, one must also believe that nonlinear optical interactions can alter the vacuum fluctuations. Also, explicit calculation of the mean square fluctuation yields the shape in Fig. 1, which does not correspond to an experimentalist's expectations for an ellipse. Thus after one has figured out what the theorists are talking about, one is confronted with the additional difficulty of believing it.

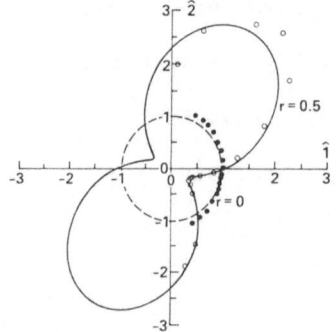

Figure 1. Polar plots of the normalized noise level as a function of phase angle for two different pump powers. The solid circles are for 180 mW pump power and fit a curve for a squeezed state with r=0.50. The open circles for 16 μW pump fit a curve which simulates a coherent state. The deviations of the points from the theoretical curves mostly reflect servo malfunctions.

Second Difficulty: Measuring a Quadrature Amplitude

Simple light detection measures the energy flux which is proportional to the sum of the squares of the quadrature amplitudes. Optical heterodyne and homodyne detection can measure single quadratures, but special procedures – such as the Yuen-Shapiro test for squeezing – are necessary to observe the predicted reduction in quantum noise [3,4]. In the Yuen-Shapiro test, the light wave to be measured and a strong, coherent local oscillator wave are combined on a low reflection beam splitter. The reflectivity of the beam splitter is so low that all of the signal beam passes through to the photodetector, but the local oscillator is so strong that the largest amplitude reaching the detector is the local oscillator amplitude. However, quantum noise theory predicts that the largest fluctuations are those of the signal, and the dominant term in the detector noise results from the interference of these signal quantum fluctuations with the local oscillator.

The photodetector produces a current with an average proportional to the square of the local oscillator amplitude and a mean square deviation proportional both to the average current and to the mean square fluctuations of the signal quadrature in phase with the local oscillator. This mean square quadrature fluctuation depends upon phase as shown in Fig. 1. For a coherent state, it is independent of the local oscillator phase. A squeezed state would show a sinusoidal dependence, with a minimum below the coherent state value. The Yuen Shapiro test is to vary the local oscillator phase and search for that dependence.

Third Difficulty: Laser Fluctuations

Lasers are very noisy, and their characteristic amplitude and phase variations are typically eight orders of magnitude larger than quantum noise. To detect the suppression of quantum noise, one must avoid laser noise. The best trick for doing that is to use an electronic spectrum analyzer to detect high-frequency fluctuations of the photodetector current. Most laser noise ends at 4 MHz, and thus quantum limited detection is possible at high frequencies. There is a squeezing test similar to the Yuen Shapiro method that applies to this heterodyne detected case, but it is more complex. Quantum noise theory also requires that the squeezing interaction couple frequencies both above and below the local oscillator frequency if the noise reduction is to be greater than 3dB. Otherwise, the vacuum fluctuations in the unsqueezed frequency band mix with the local oscillator and dominate the noise [3].

Fourth Difficulty: Squeezing Vacuum States

The interaction Hamiltonians that produce squeezed states are proportional to the square of a creation operator or the product of two creation operators. This is different from the Hamiltonians for coherent states which are linear in the photon creation and destruction operators. The most promising squeezing interactions are 3 and 4 wave mixing. These parametric interactions couple the input fluctuations of the signal and idler waves (which may be vacuum states) in a way that amplifies one linear combination and attenuates another. It is the attenuated combination that is detected as the squeezed quadrature.

Fifth Difficulty: "No Nonabsorbing Material is Nonlinear Enough to Produce Detectable Squeezing with c.w. Laser Pumping in the Length of a Single Beam Waist"

Absorption ruins squeezing by randomly destroying photons that had been emitted in highly correlated pairs. Detectable squeezing implies a factor of 10 reduction in the noise, which can be detected unequivocally. Pulsed lasers are simply too unreproducible for such subtle measurements, and really big c.w. lasers are very expensive.

One solution of this problem is to enclose the nonlinear medium in a "single-port" optical cavity, thus allowing many beam waists worth of interaction [5]. That idea seems likely to work, but my favored solution is to use a single mode optical fiber, where the light is tightly confined in a core and the length can be equivalent to millions of beam waists. In fact a 4 micron core fused silica fiber will give detectable squeezing with 1W of pump power and 50m of interaction length [6].

Sixth Difficulty: One Cannot Get 1 W of Light through a Fiber

The stimulated Brillouin effect limits the single mode power that can be transmitted through a fiber. The threshold for SBS in 50m of fiber is about 100 mW. Only the threshold power can be transmitted, the rest is reflected. The solution is to impose a temperature gradient along the fiber, thus broadening the SBS gain line. The fiber doesn't last long, however.

Seventh Difficulty: Thermal Noise in Fibers

We were surprised to find that fibers themselves generate noise by means of spontaneous forward Brillouin scattering from the acoustic eigenmodes of the fiber. This Guided Acoustic Wave Brillouin Scattering (GAWBS) produces a complex structured spectrum of noise well above the quantum noise level [7]. Figure 2 shows a better-than-typical GAWBS spectrum. The widths of the peaks depend on the acoustical properties of the jacket material used to protect the silica fiber. The fiber used for Fig. 2 has an aluminum jacket and the peaks are narrow. There even seems to be a region where the noise level approaches the quantum limit.

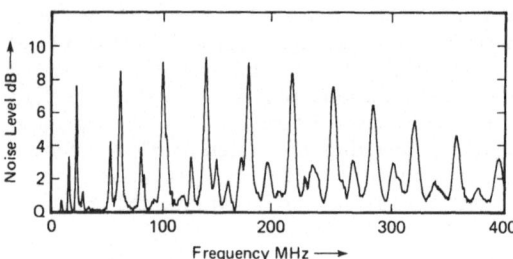

Figure 2. GAWBS Spectrum of an aluminum jacketed fiber. The baseline is the quantum noise level. A small region around 40 MHz shows no excess noise.

Eighth Difficulty: Obtaining More Such Fiber

Unsolved.

Ninth Difficulty: Proving That We Aren't Crazy

The nonlinear interactions that suppress quantum noise also should reduce classical noise on the input beam. We have used the apparatus of Fig. 3 to test this hypothesis and prove that our proposed squeezed state generation and detection system is likely to work [8]. The noise is generated by electro-optical amplitude and phase modulators driven by an r.f. noise source. The modulator drives are carefully balanced to simulate the phase independence of a coherent state, but the noise level is 100 times larger. The noise and a strong pump wave are coupled into a fiber, and the fiber output is reflected from an interferometer into a photodiode. The interferometer phase shifts and attenuates the pump, and the reflected pump frequency wave acts as a local oscillator at the detector.

The interferometer resonance is locked at some offset from the pump frequency, and the phase is varied by changing this offset. The average power at the detector and the noise level at 50 MHz are both measured as a function of the interferometer offset. With a short fiber, the noise is proportional to the detector power and independent of phase. With 180 mW pump in a 100m fiber, the noise becomes phase dependent, with a minimum on one side of the minimum detected power. The points plotted in Fig. 1 are the ratios of the noise level to detector power plotted as a function of phase. The high power in the fiber clearly "squeezes" the externally generated noise by a factor of 3, just as the theory would predict. The noise minimum is at 30 degrees of phase shift; the calculated squeeze parameter is r=0.55. If GAWBS can somehow be suppressed, we should be able to demonstrate a true squeezed state of light, in spite of all nine difficulties. Meanwhile, we remain entangled in optical fiber technology.

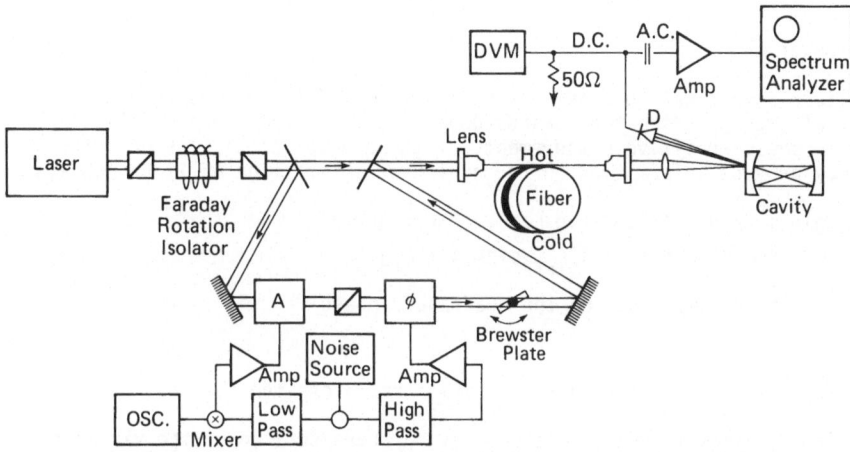

Figure 3. Apparatus for demonstrating "squeezing" of classical noise. The amplitude and phase modulators are marked A and ϕ, respectively. The digital volt meter (DVM) measures the average detector current. Other details are in the text.

REFERENCES

1. D. F. Walls: Nature 306, 141 (1983) and references therein.

2. C. M. Caves: Phys. Rev. D 23, 1693 (1981).

3. H. P. Yuen and J. H. Shapiro: Opt. Lett. 4, 334 (1979).

4. B. L. Schumaker: Opt. Lett. 9, 189 (1984).

5. B. Yurke: Phys. Rev A 29, 408 (1984).

6. M. D. Levenson, R. M. Shelby, M. Reid, D. F. Walls and A. Aspect: Phys Rev A (submitted).

7. R. M. Shelby, M. D. Levenson and P. W. Bayer: Phys. Rev. Lett. 54, 939 (1985).

8. M. D. Levenson, R. M. Shelby and S. H. Perlmutter: Optics Letters (submitted).

Phase-Sensitive Quantum Spectroscopy

D.F. Walls and M.D. Reid
Department of Physics, University of Waikato, Hamilton, New Zealand

P. Zoller
Institute for Theoretical Physics, University of Innsbruck,
Innsbruck, Austria

M.J. Collett
Department of Physics, University of Essex, Colchester, U.K.

1 Introduction

A new field investigating phase-dependent quantum fluctuations is developing.
Present efforts are directed towards generating squeezed states of light [1].
In future phase-sensitive measurements of the quantum statistics of emitted
light may provide information on the nonlinear interaction of light and atoms.

A single mode electric field may be expressed as

$$E(t) = \lambda[X_\theta\cos(\omega t + \theta) + X_{\theta-\pi/2}\sin(\omega t + \theta)] \tag{1}$$

where X_θ is the amplitude of the quadrature phase

$$X_\theta = ae^{i\theta} + a^+e^{-i\theta} \tag{2}$$

(a and a^+ are the annihilation and creation operators for the mode).

The fluctuations in one quadrature may be characterized by the variance
$V(X_\theta)$. For a coherent state $V(X_\theta) = 1$. A squeezed state has less fluctuations in one quadrature than a coherent state thus $V(X_\theta) < 1$ ($V_{(X_\theta - \pi/2)} > 1$).

For a multimode field one may define a squeezing spectrum

$$V(X_\theta,\omega) = \int d\tau e^{-i\omega\tau}[<X_\theta(\tau)X_\theta(0)> - <X_\theta(\tau)><X_\theta(0)>] \tag{3}$$

$V(X_\theta,\omega)$ is a measure of the fluctuations in the X_θ quadrature at frequency ω.
For a coherent state $V(X_\theta,\omega) = 1$ fluctuations below the quantum noise limit
or squeezing at frequency ω is characterized by $V(X_\theta,\omega) < 1$.

The squeezing spectrum may be measured by a heterodyne detection scheme
[2,3]. The signal field is combined with a local oscillator on a beam splitter
before being detected on a photodiode. The spectrum of fluctuations in the
photocurrent $V(i,\omega)$ is then measured with a spectrum analyser. $V(i,\omega)$ is
directly proportional to $V(X_\theta,\omega)$ of the signal where θ is determined by the
phase of the local oscillator with respect to the signal. We shall now
consider schemes to reduce $V(X_\theta,\omega)$ below the quantum noise limit of a
coherent state.

2 Generation of Squeezed States

a Resonance Fluorescence from a Two-Level Atom

The spectrum of squeezing in the fluorescent light from a coherently driven
two-level atom has been calculated by COLLETT ET.AL. [4]. The squeezing

254

spectrum is plotted in Fig. 1a for the Rabi frequency Ω chosen to give maximum squeezing. The maximum squeezing occuring at $\omega = 0$ corresponds to a reduction in fluctuations of only 28%. In Fig. 1b the variance of fluctuations in the two quadratures are plotted for the Rabi frequency much greater than the natural linewidth γ. While there is no squeezing this shows how the central peak and the two sidebands in the fluorescent triplet [5] may be separated with phase sensitive spectroscopy.

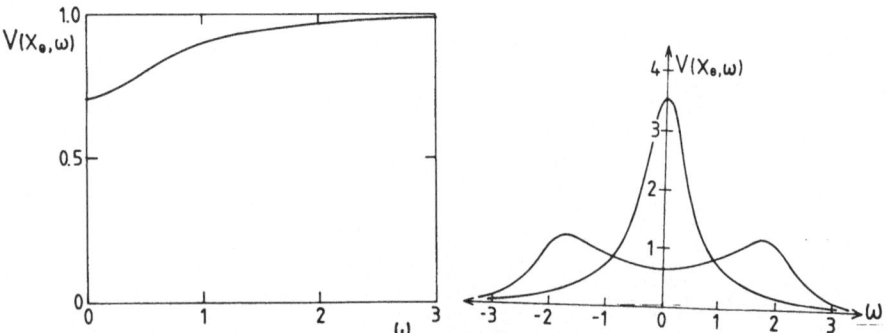

Fig. 1 $V(X_\theta, \omega)$ for resonance fluorescence

(a) $\dfrac{\Omega}{\gamma} = \dfrac{1}{2\sqrt{3}}$ (b) $\dfrac{\Omega}{\gamma} = 2$

b Parametric Amplification and Intracavity Four-Wave Mixing

Squeezed light may be produced via the nonlinear optical processes of parametric amplification and four-wave mixing inside an optical cavity. The medium may be modelled by a classical nonlinear susceptibility χ. The maximum squeezing occurs at the oscillation threshold. This corresponds to phase sensitive critical fluctuations where $V(X_\theta) \to 0$ but $V(X_{\theta-\pi/2}) \to \infty$. The squeezing spectrum at threshold [6,7] is plotted in Fig. 3 for (a) a cavity with two equally transmitting mirrors, (b) a cavity with one mirror perfectly reflecting. The best squeezing results are from a single ended cavity (YURKE [8]).

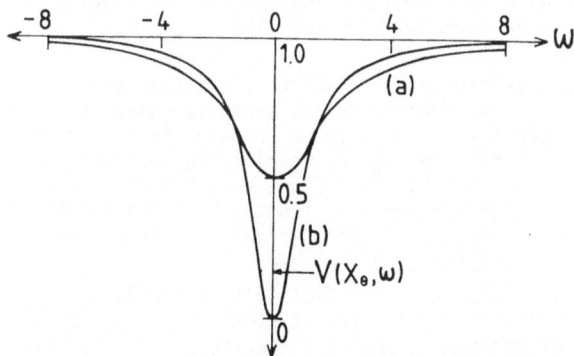

Fig. 2 $V(X_\theta, \omega)$ for a parametric oscillator at threshold
(a) double-ended cavity
(b) single-ended cavity

c Second Harmonic Generation

Second harmonic generation inside an optical cavity will also produce squeezed
light [9,10]. The squeezing spectrum of the output field [11] is shown in
Fig. 3 for (a) the fundamental field with $\gamma_f \gg \gamma_{SH}$, (b) the second harmonic
field with $\gamma_{SH} \gg \gamma_f$. (γ_f and γ_{SH} are the cavity loss rates for the fundamental and second harmonic modes respectively). The spectra plotted are
for the threshold of self oscillations [12] where maximum squeezing occurs.

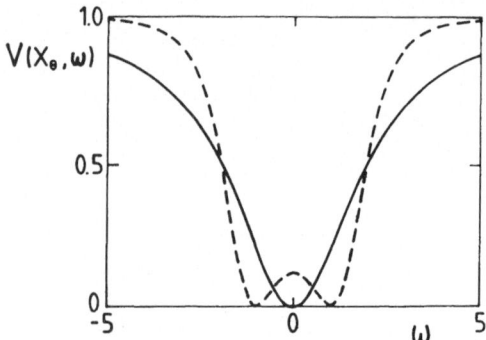

Fig. 3 Second harmonic generation
(a) $V(X_\theta,\omega)$ for the fundamental ($\gamma_f \gg \gamma_{SH}$)
(b) $V(X_\theta,\omega)$ for the second harmonic ($\gamma_{SH} \gg \gamma_f$)

3 Spontaneous Emission Limits to Squeezing

The above calculations are all based on a classical susceptibility χ for the
medium. When the medium is quantised spontaneous emission from the atomic
energy levels will place limits on the squeezing attainable. A quantised
treatment of four—wave mixing using a two—level model for the medium has been
given by REID and WALLS [13]. In order to suppress the noise due to spontaneous emission one is required to operate in the dispersive limit and far
from atomic saturation. This imposes the conditions, $\Delta \gg 1$, $I_p \ll \Delta^2$,
$10I_p{}^2 \ll \Delta^3$ where Δ is the atomic detuning in units of the atomic linewidth
and I_p is the pump intensity normalised by the resonant saturation intensity
of the atom. In order for the gain to exceed the loss we further require
$I_p/\Delta \gg 1$. These conditions may be satisfied for Δ in the range $10^3 - 10^4$
and $I_p > 10^3 - 10^4$.

In a cavity in order to reach threshold one requires $(2C/\Delta)(2I_p/\Delta^2) \sim 1$,
where C is the cavity cooperativity parameter. Hence a minimum value of C is
required, otherwise the pump intensity I_p needs to be so large as to induce
unwanted spontaneous emission effects. In Fig. 4, we plot the maximum
squeezing for $\Delta = 10^4$, for two values of C. In fact, $C = \alpha L/1-R$ where α
is the line-centre weak field attenuation per unit length, L is the length of
the medium and R is the mirror reflectivity. Also $\alpha = 3\lambda^2\rho/2\pi$ where λ is the
field wavelength and ρ the atomic density.

A value of $2C = 10^6$ in a cavity with mirror reflectivity $R = 0.99$ and
$L = 50$ cm corresponds to an atomic density for sodium atoms of $\rho = 10^{11}/cm^3$.
At these densities the effect of collisions on the squeezing are negligible.

A three—level atomic medium in a lamda configuration offers an alternative
mechanism to suppress spontaneous emission. The ground—state coherences
created by the coherent pump field provides a mechanism for nonlinearity

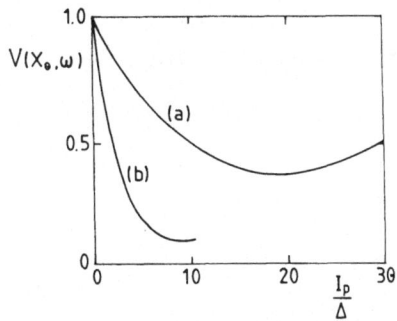

Fig. 4 $V(X_\theta,0)$ for intracavity four-wave mixing, as a function of pump intensity for $\Delta = 10^4$
(a) $2C = 10^6$ (b) $2C = 5.10^6$

without absorption. This enables spontaneous emission to be suppressed at lower detunings with consequently lower requirements on the pump power, the cavity finesse and the atomic density. [14]

Acknowledgements

This work was supported by the New Zealand Universities Grants Committee and the Österreichische Forschungsgemeinschaft.

References

1 For a review see D.F. Walls: Nature 306, 141 (1983)

2 H.P. Yuen and J.H. Shapiro: IEEE Trans.Inf.Theory, 24, 657 (1978)

3 M.D. Levenson, R. Shelby, M.D. Reid, D.F. Walls and A. Aspect: (in press)

4 M.J. Collett, D.F. Walls and P. Zoller: Optics Comm., 52, 145 (1984)

5 B. Mollow: Phys.Rev., 188, 1969 (1969)

6 M.J. Collett and C.W. Gardiner: Phys.Rev., 30A, 1386 (1984)

7 C.W. Gardiner and C.M. Savage: Optics Comm., 50, 173 (1984)

8 B. Yurke: Phys.Rev., A29, 408 (1984)

9 L.A. Lugiato, G. Strini and F. de Martini: Opt.Lett., 8, 256 (1983)

10 L. Mandel: Optics Comm., 42, 437 (1982)

11 M.J. Collett and D.F. Walls: Phys.Rev., (in press)

12 P.D. Drummond, K.J. McNeil and D.F. Walls: Optics Comm., 28, 255 (1979)

13 M.D. Reid and D.F. Walls: Phys.Rev.,31A, 1622 (1985)

14 M.D. Reid, D.F. Walls and B.J. Dalton: (to be published)

Demonstration of Self-Pulsing Instabilities in a Single-Mode Homogeneously Broadened Raman Laser

R.G. Harrison and D.J. Biswas

Physics Department, Heriot-Watt University, Edinburgh EH14 4AS, U.K.

Here we report the first experimental observation of instabilities leading to chaos in a single-mode homogeneously broadened Raman laser system. These effects have been obtained over a wide range of operating conditions including those for optimum lasing suggesting that this behaviour is indeed general for this broad class of near-resonantly enhanced two-photon lasers. The route to chaos in this system is characterized by period doubling.

The laser, which uses NH_3 as the active medium near-resonantly pumped on the aR(6,0) transition by a CO_2 laser, emits on aP(8,0) transition at $\sim$ 12.8 μm. This lasing transition has recently been clearly identified as Raman in origin[1] for NH_3 pressures 1 - 20 torr and pump intensity $\sim$ 600kW/cm^2 (operating conditions for our experiment).

A transversely excited atmospheric (TEA) CO_2 laser was used as the pump source and operated on a single transverse and axial mode to generate temporally smooth long pulses (2 μsec FWHM with 250 KW peak power). The optically pumped Fabry-Perot cavity of length $\sim$ 25 cm was provided with a piezoelectric tuning (PZT) facility and contained a KBr Brewster terminated NH_3 cell of length 22 cm. The cavity consisted of a concave (2 m radius of curvature) gold mirror and a Ge mirror of 64% R at 12.8 μm.

Transverse scans of the NH_3 emission taken by a pyroelectric array detector confirmed an essentially Gaussian spatial profile; indicating operation on the lowest order longitudinal mode. For a pump intensity of $\sim$ 500 kW/cm^2 lasing was obtained up to a pressure broadened gain-bandwidth of $\sim$ 187 MHz, substantially smaller than the free spectral range (FSR) of the laser cavity viz, 600 MHz, thus ensuring single-mode condition. This was further established from determining the cavity tuning range as a function of pressure over which lasing was obtained. This is consistent with the predicted dependence of gain bandwidth on pressure viz, a FWHM value of 17.049 MHz/torr. This therefore indicates a gain which is little over double the lasing threshold.

Chaotic and periodic pulsation behaviour in the NH_3 emission was sensitive to cavity length tuning and occurred over NH_3 pressures of 5-9 torr, smaller than the total range 3-11 torr for lasing emission, most pronounced effects occurring at a pressure $\sim$ 8 torr. At this pressure homogeneous broadening was 136 MHz while cavity linewidth was estimated to be $\sim$ 170 MHz thus establishing a bad cavity condition. At higher pressure homogeneous broadening got too large and finally this bad cavity instability was killed. A single period doubling route to chaos was observed with the fine cavity length tuning as a control parameter. The base period in the top trace is $\sim$ 18 nsec compared to the cavity round trip time of 1.6 nsec. The pulsations appear to be steady while the pumping is uniform and this therefore suggests a clear case of single-mode instabilities at relatively low excitation above threshold.

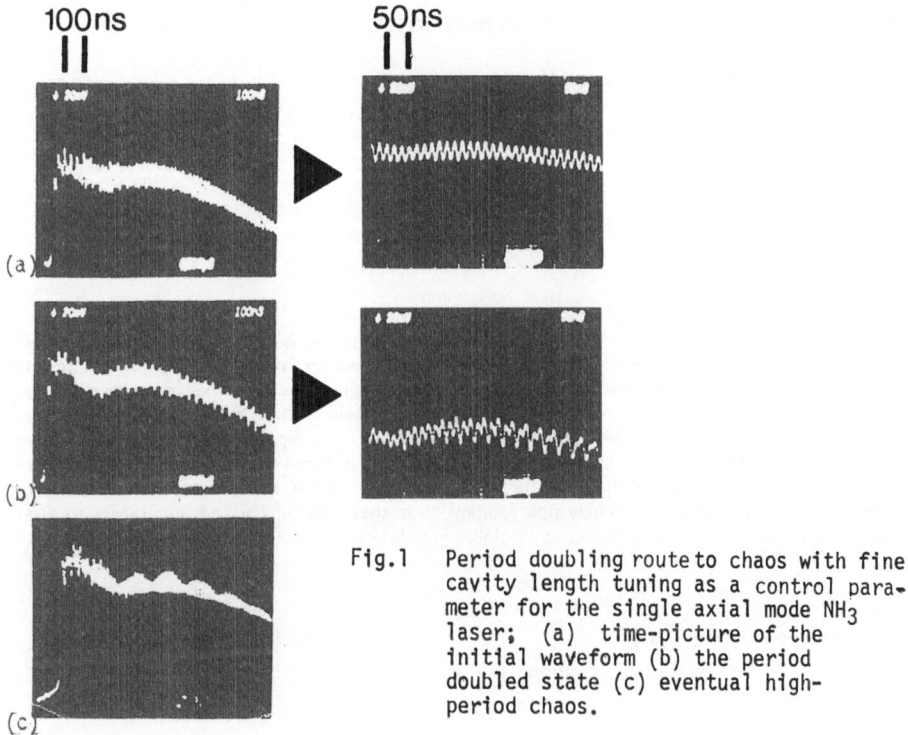

100ns

50ns

(a)

(b)

(c)

Fig.1   Period doubling route to chaos with fine
cavity length tuning as a control parameter for the single axial mode NH_3
laser;   (a)   time-picture of the
initial waveform (b) the period
doubled state (c) eventual high-
period chaos.

In conclusion, we have demonstrated huge pulsating instabilities with
identifiable routes to chaos in the emission from a mid-infrared single mode
homogeneously broadened two-photon Raman laser at an excitation level much
less than that considered necessary for 2-level single mode laser systems.
We note that these phenomenon are obtained over a wide range of operating
conditions representative of those common to conventional laser systems of
this class.   Although quantitative understanding of two-photon Raman lasers,
extensively used for mid to far infrared generation, is in a somewhat
embryonic state, our observations have identified them as specially suitable
systems for the manifestation of instability phenomenon.   We therefore hope
our results will stimulate further theoretical interest towards elucidating
the dynamic behaviour of these systems in regard to such effects.

H.D. Morrison, B.K. Garside and J. Reid: IEEE J.Q.E., _20_, 1060 (1984).

Theory of Two-Photon Resonance Fluorescence

D.A. Holm and M. Sargent III

Optical Sciences Center, University of Arizona, Tucson, AZ 85721, USA

Spontaneous emission between two states of the same parity is forbidden by electric dipole selection rules. However by means of a two-photon transition such a process can occur. In this work we calculate the substantially enhanced spontaneous emission spectrum from a two-photon two-level medium under the influence of a pump field at half the transition frequency. We use the two-photon two-level model[1] shown in Fig. 1. The dipole matrix element between levels a and b is zero, and the pump field at frequency ν_2 is approximately one half of the frequency difference $\omega = \omega_a - \omega_b$. Dipole transitions from states a and b to the intermediate levels j are allowed and sufficiently nonresonant that they can be treated accurately to first order. Hence the j levels acquire no appreciable population.

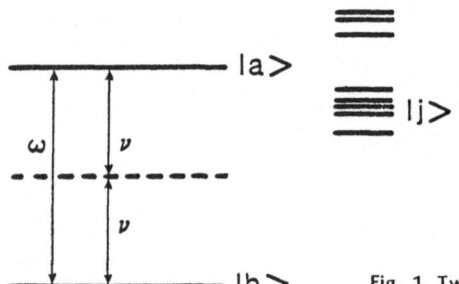

Fig. 1 Two-photon two-level model

We consider spontaneous emission from the upper level a to the intermediate level j or from the intermediate level j to the lower level b. The strong pump field is taken to be classical and monochromatic. As for one-photon resonance fluorescence,[2] only the scattered mode need be quantized. The two-photon resonance fluorescence spectrum formula[3] is considerably more complicated than the one-photon result. This is due in part to spontaneous emission originating from both the upper level a as in the one-photon case, and also from the off-resonance levels j. It is also due to the Stark shift of the two-photon transition frequency. The physical origin of this shift comes from the frequency shifts of levels a and b induced by virtual transitions to the off-resonant j levels. In the one-photon case this shift can be neglected since such non-resonant interactions are small compared to resonant ones, but for the two-photon two-level model, the shift is of the same magnitude as the other parameters.

In the absence of the Stark shift, the two-photon spectrum has the same shape as the one-photon case, even preserving the 1:3:1 sidebands to central peak height ratio. Other one-photon features are present as well, such as the spectrum remaining symmetric when the pump field is detuned from the atomic resonance. However, in the one-photon case the scattered radiation initially increases linearly with the dimensionless pump field intensity I_2 and then approaches a constant as $I_2 \to \infty$. For the two-photon model, the total scattered radiation increases as the square of I_2 initially, but then continues to increase linearly with I_2 for $I_2 \gg 1$, i.e., pump intensity $\gg$ the saturation intensity.

Nonzero Stark shifts can yield radically different spectra from the familiar three-peaked spectrum. The Stark shift introduces dispersive-like features leading to pronounced asymme-

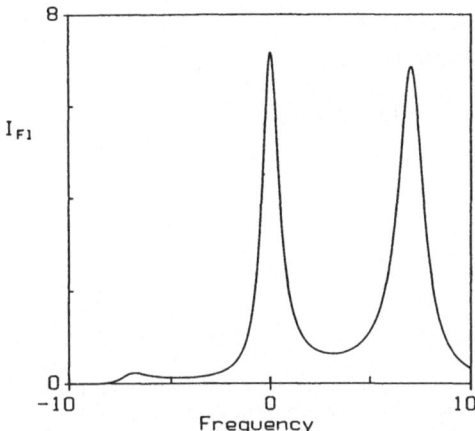

Fig. 2 Two-peaked spectrum versus frequency for an intensity $I_2 = 10$, $T_2 = 2T_1$, a Stark shift $\omega_s I_2 = 5/T_1$, and detuning $\omega - 2\nu_2 = -5/T_1$

tries and ordinarily to a smaller central peak. Figure 2 shows the spectrum for $\omega_s I_2 = 5/T_1$, and a detuning $\omega - 2\nu_2 = -5/T_1$. Here by "balancing" the detuning against the Stark shift, we maintain the central peak height at the $\omega_s = 0$ centrally tuned value, while amplifying one side-band and nearly eliminating the other.

The two-photon two-level resonance fluorescence spectrum also has a significantly more complex expression for the elastic portion of the spectrum, sometimes called the Rayleigh peak. The Rayleigh scattering from this model may be calculated either directly from the quantum mechanical calculation[3], or by determining the semiclassical complex polarization amplitude for this level scheme and finding its squared modulus. The Rayleigh scattering consists of three contributions, 1) the off-resonant dipoles, 2) the two-photon two-level coherence ρ_{ab}, and 3) the interference between the dipoles and ρ_{ab}. Because of this third term, the elastic scattering is asymmetric with respect to detuning, totally unlike the one-photon resonance fluorescence case. Depending upon the values of the off-resonant dipoles, this part of the spectrum may either dominate the total emission, or, for an appropriate detuning and intensity I_2, be very small.[3]

REFERENCES

1. M. Takatsuji, Phys. Rev. A4, 808, (1971).
2. B. R. Mollow, Phys. Rev. 188, 1969 (1969).
3. D. A. Holm and M. Sargent, Opt. Lett. 10, August (1985).

Squeezing Light Noise in a Cavity
Near the Vacuum Fluctuation Level

R.E. Slusher, L. Hollberg, B. Yurke, and J.C. Mertz

AT & T Bell Laboratories, Murray Hill, NJ 07974, USA

We have observed phase-sensitive amplification and attenuation of random optical noise using the nonlinear four-wave-mixing interaction [1]. This "squeezing" of the noise into a particular quadrature of the optical phase is homodyne detected at power levels near the vacuum fluctuation limit. It has been predicted [2,3] that the four-wave-mixing process in a cavity [2] can produce squeezed states [3] (or squeezed coherent states [2]) of the optical field whose quantum fluctuations in one-phase quadrature are less than the vacuum fluctuation level (interpreted classically as the shot noise level).

The nonlinear medium used in these experiments [4] consists of a sodium atomic beam (density $\sim 2\times10^{11}$ cm^{-3}) which is crossed by counterpropagating pump beams which are tuned ~ 1 GHz away from the Na transitions at 589.0 nm. Also crossing at right angles to the atomic beam is a buildup cavity which collects and enhances the squeezed fields generated spontaneously by four-wave-mixing. The buildup cavity (free spectral range of 140 MHz) is controlled to be resonant with the pump laser frequency by locking it to a weak beam which is frequency-shifted from the pump by 420 MHz.

The nondegenerate four-wave-mixing process generates pairs of photons symmetrically located in frequency above and below and pump laser frequency. Cavity fields coherent with these photon pairs build up during successive passes in the frequency-locked cavity. The resulting light transmitted through a 2% output mirror (the other mirror reflectivity is ~ 0.995) is then detected by a balanced homodyne receiver [4,5]. It is the phase-dependent noise characteristics of this detected signal which show "squeezing" near the vacuum fluctuation level. In this system, the phase reference is either the four-wave-mixing pump phase or the local oscillator phase.

Of particular interest are those frequency components of the noise shifted from the pump frequency by integral multiples of the cavity mode spacing [2]. An example of this phase-sensitive noise is shown in Fig. 1. Here the oscillatory behavior of the signal exhibits the phase-sensitive amplification and attenuation of noise about an intermediate level V_T. The noise level V_T is obtained with no four-wave-mixing and is a combination of nearly equal components of fundamental vacuum fluctuation noise and spontaneous emission noise generated by the pump beam. At the maximum attenuation points, the noise is well below the combined (vacuum plus spontaneous) random noise level V_T, yet it is still ~ 0.7 dB above the vacuum noise level.

By simultaneously detecting the noise at 140 MHz and 280 MHz, we have experimentally verified predictions [2] of a relative 180° phase-shift for the squeezing at adjacent cavity modes.

Though the squeezing which we observe has not yet reduced the total noise below the vacuum fluctuation level [6], it is sufficiently close to be of

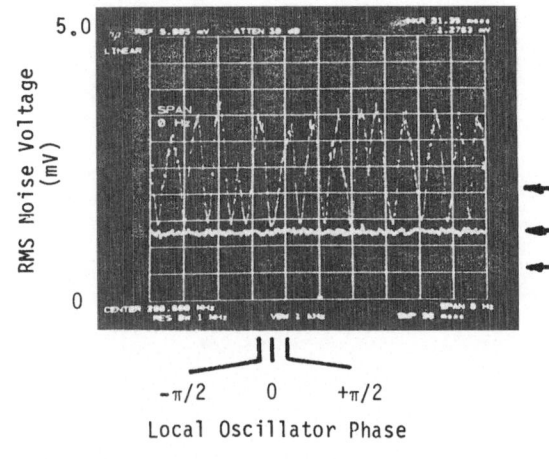

RMS Noise Voltage (mV)

5.0

0

−π/2   0   +π/2

Local Oscillator Phase

Fig. 1. Phase-sensitive amplification and attenuation of random optical noise by four-wave-mixing in a cavity. The upper oscillatory trace is the rms noise voltage at 280 MHz as a function of local oscillator phase. V_T This noise is amplified V_{LO} and attenuated relative V_A to the level V_T. The lower trace

$$V_{LO} = \left(V_V^2 + V_A^2\right)^{\frac{1}{2}}$$

is obtained with the pump off and local oscillator on. This is the rms sum of the vacuum fluctuation noise level V_V plus a small contribution from amplifier noise V_A.

interest in the quantum limit. For example, for the data in Fig. 1, the magnitude of the noise which we detect must be viewed as being composed of a combination of ∿0.8 photon per mode of spontaneous emission noise and ½ photon per mode of vacuum fluctuation noise. The measured squeezing of this combined noise is 3 dB.

All noise sources, gains, losses and efficiencies can be measured in our experiment. These measured parameters allow quantitative comparisons with the theory. Extrapolation to accessible parameter regimes indicates that squeezed light levels well below the vacuum fluctuation level can be obtained and eventually applied to sensitive interferometry.

References

1. H. P. Yuen and J. H. Shapiro, Opt. Lett. 4, 334 (1979).
2. B. Yurke, Phys. Rev. A29, 408 (1984) and Phys. Rev. A (to be published).
3. D. F. Walls, Nature 306, 141 (1983).
4. R. E. Slusher, L. Hollberg, B. Yurke, J. C. Mertz and J. F. Valley, Phys. Rev. A31, 3512 (1985).
5. H. P. Yuen and V. W. S. Chan, Opt. Lett. 8, 177 (1983).
6. Related results are recently obtained in different systems; M. D. Levenson, R. M. Shelby, S. H. Perlmutter, CLEO (1985), also in this volume; and M. W. Maeda, P. Kumar and J. H. Shapiro, Proceedings of Optical Instabilities Conference, Rochester (June 1985).

Part IX

Coherent Transient Effects

Nonlinear Optical Phenomena Beyond Optical Bloch Equation

E. Hanamura and H. Tsunetsugu

Department of Applied Physics, The University of Tokyo,
Bunkyo-ku, Tokyo 113, Japan

1. Introduction

Almost all nonlinear optical phenomena have been successfully described by solving the Maxwell equations coupled with the optical
Bloch equations[1]. In these descriptions, first the effects of
the reservoir on the electronic system were taken into account
as the longitudinal and transverse relaxation constants on the
diagonal and off-diagonal components of the density operators,
respectively. Then,the optical Bloch equations for the density
operators are solved under the external pump-fields, in terms of
which the nonlinear polarization is evaluated. With these polarizations as the source term, the Maxwell equation for the signal
field is solved both in the frequency and time regions. These
procedures are justified only under limited conditions.

We present the general theory of nonlinear optical responses
without using the Maxwell and optical Bloch equations, and several new phenomena which the usual descriptions could not predict.
Here,the effect of the reservoir on the system is represented as
the frequency modulation $\delta\omega(t)$ of the excited state, which has
the finite correlation time τ_c. First, it is noted that the way
of relaxation depends on the observing phenomena and the magnitude of the pump field,while a single transverse relaxation time
T_2 could describe all the optical phenomena in the conventional
treatments. As an example, the optical nutation and free induc-
tion-decay behavior are discussed in Section 2. Second, the
spatial as well as temporal correlation of the frequency modulations is shown in Section 3 to be reflected clearly on the transient response,due to the degenerate four wave mixing.

2. Coherent Optical Transients

The Hamiltonian of the electronic system S and the reservoir B
is represented as

$$H(t) = H_S + H_B + H_{SB} + V(t), \quad \overline{H}(t) = H(t) + V_{ob}(t), \quad (1)$$

where $V(t)$ describes the interaction of the electronic system
with pump fields and $V_{ob}(t)$ that with the observing field. The
electronic system is considered as an assembly of two levels $|a>$
and $|b>$ with the energies $\hbar\omega_a$ and $\hbar\omega_b$, and the reservoir
Hamiltonian H_B is eliminated so that H_{SB} is represented as the
frequency modulation $\delta\omega(t)$ of the excited state $|b>$ [2]. This
system is pumped by a single or several external fields through
$V(t)$. Let us observe the emitted photon intensity with (ω_0, k_0)

per second, which is denoted by

$$B = \frac{d}{dt}(a_0^\dagger a_0) = i[V_{ob}(t), a_0^\dagger a_0] \equiv i v_{ob}(t) a_0^\dagger a_0. \tag{2}$$

Then the observed quantity can be represented in terms of the wave function $\Psi(t)$ or the density operator $\bar{\rho}(t) \equiv |\Psi(t)><\Psi(t)|$ of the system under the pump and the observed fields as

$$<\Psi(t)|B|\Psi(t)> = \mathrm{Tr} B\bar{\rho}(t) \equiv <<B|\bar{\rho}(t)>>, \tag{3}$$

where $\bar{\rho}(t) = \exp[-i\int_{-\infty}^{t}(L+v_{ob})d\tau]\rho_0$, and $L(t) \equiv [H(t),\ ]$.

In the optical nutation (ON) measurement, the electronic and reservoir systems are in the thermal equilibrium $\rho_0 \equiv |a>\rho_B<a|$ at $t \leq 0$, where $|a>$ denotes the electronic ground state. The pump field $E(t) = E\cos(kr-\omega t)$ is applied and the emission of the photon ω_0 is observed at $t > 0$. The observed field is weak, so that the interaction $V_{ob}(t)$ may be treated perturbationally. Therefore to the first order,

$$<<B|\bar{\rho}(t)>> = (-i)^2 \int_0^t dt' <<a_0^\dagger a_0 | v_{ob}(t)\exp[-i\int_{t'}^{t}L(\tau)d\tau]v_{ob}(t')$$

$$\cdot \exp[-i\int_0^{t'}L(\tau)d\tau\ |\rho_0>>. \tag{4}$$

In the ON measurement, we are not interested in the spectrum but the time-dependence of the emitted intensity. Therefore the observing quantity is

$$I_{ON}(t) \equiv \int d\omega_0 <<B|\bar{\rho}(t)>>. \tag{5}$$

$v_{ob}(t)$ and $v_{ob}(t')$ in eq.(4) contain the factors $\exp[\pm i\omega_0(t\pm t')]$, and the combination $\exp[\pm i\omega_0(t-t')]$ gives a finite contribution. Then, from the integral over ω_0 in eq.(5), we have a factor $2\pi\delta(t-t')$ for eq.(4). As a result,

$$I_{ON}(t) = 2\pi|V_{ab}|^2 \sum_{i,j}^{N} <<b_i b_j | \exp[-i\int_0^t L(\tau)d\tau]|a_i a_j>>$$

$$\cdot \exp[-ik_0(r_i-r_j)]. \tag{6}$$

Here V_{ab} is the matrix element denoting the electronic transition from $|b>$ to $|a>$ emitting a photon ω_0. Taking into account the spatial correlation, we have

$$I_{ON}(t) = 2\pi|V_{ab}|^2[N^2\delta_{k,k_0} +N]<<bb|\exp[-i\int_0^t L(\tau)d\tau]|aa>>. \tag{7}$$

The density matrix $\rho_{bb}(t) \equiv <<bb|\exp[-i\int_0^t L(\tau)d\tau]|aa>>$ can be obtained by the inverse Laplace transform of $\rho[s]$ derived in [3]. It is noted here that the ON signal can be represented by the single Green function $G(t) \equiv \exp[-i\int_0^t L(\tau)d\tau]$ of the system under the pump field. We evaluated $\rho_{bb}(t)$ already in [4]. As to the frequency modulation $\delta\omega(t)$, we took the Gaussian model: $<\delta\omega(t_1)\delta\omega(t_2)>_B = \Delta^2 \exp(-\gamma_c|t_1-t_2|)$. Under such nearly resonant condition or strong pumping as $|\omega_{ba}-\omega|<<\mu E$, the nutation signal is shown to decay with $\Gamma = 3\gamma/2 + \Delta^2\tau_c/\{1+(\chi\tau_c)^2\}$, oscillating with the Rabi frequency $\chi=\mu E$, where $\tau_c\equiv\gamma_c^{-1}$. Note the power-dependent relaxation Γ.

The observation of the free-induction decay (FID) has been an useful method to observe the transverse relaxation time T_2. However, the large deviation from the result of the optical ^{2}Bloch equation with the constant T_2 was observed [5] and was explained in terms of the frequency modulation model by the present author [6]. Furthermore in this section, the vertex correction effects [7] are shown to play an important role in the relaxation process of the FID in contrast with the optical nutation. The pumping condition of the FID is quite opposite to that of the ON. The pump field E(t) is applied until t=0 and after it is switched off at t≥0, the decaying of the emitted photon intensity is measured. Therefore the FID signal is expressed as the product of two Green functions:

$$I_{FID}(t) = 2\pi|NV_{ab}|^2 <<bb|G_0(t,0)G(0,-\infty)|aa>>_B. \qquad (8)$$

Here $G_0(t,0)\equiv\exp[-i\int_0^t L_0(t')dt']$ and $G(0,-\infty)\equiv\exp[-i\int_{-\infty}^0 L(t')dt']$ denote the propagations under the Hamiltonian H_0 and $H=H_0+V(t)$, respectively, without and with the pumping field. The double Laplace transform of $G_0(t,\tau)G(\tau,0)$ was obtained [7] and the FID behavior was evaluated in terms of this [8]. The electronic states are in the excited state |b> both just before and after the switching-off time t=0 so that the frequency modulations work extending over two time regions t≤0 and t≥0 due to the finite correlation time τ_c. This effect is called the vertex correction. This is in contrast with the ON where no vertex correction works. As a result, the relaxation behavior differs by the vertex correction for the FID and the ON measurements. The vertex corrections on the FID are the more effective for the slower modulation $\gamma_c<<\Delta$ and the weaker pump field $\mu E<<\Delta$ or γ_c. This is because it takes an order of the correlation time τ_c before the frequency modulation gives a finite effect and because the effect of the frequency modulation is reduced under such a strong pumping as $\mu E>>\Delta$ or γ_c. This comes from the fact that the reservoir elementary excitations become unable to follow the motion of the electronic system [2,6].

3. Four Wave Mixing [9]

Four Wave Mixing (4WM) is the observation of the photon emission under the three pump fields. Expanding the density operator $\bar{\rho}(t)$ in the interactions of the electronic system with these fields, and considering the pumping by two pulse fields with time delay T: $E(t)=E_1\exp[i(\omega_1 t-k_1 r)]\delta(t)+E_2\exp[i(\omega_2 t-k_2 r)]\delta(t-T)+c.c.$, we observe $I_{4WM}(t)$ in the direction of $k=2k_2-k_1$. We consider that the transition of frequency ω_{ba}^i at the i-th site has the static imhomogeneous broadening σ in addition to the dynamical frequency modulaion $\delta\omega_i(t)$ of the excited state |b> at the i-th atom. After we average over ω_{ba}, we have the following expression [9]:

$$I_{4WM}(t) = A \sum_{i,j} \exp[i\Delta k(r_i-r_j)]\exp[-2\gamma t-2(t-2T)^2\sigma^2]$$
$$<\exp[i\int_T^t \{\delta\omega_i(\tau)-\delta\omega_j(\tau)\}d\tau-i\int_0^T \{\delta\omega_i(\tau)-\delta\omega_j(\tau)\}d\tau]>_B, \qquad (9)$$

where $2\gamma\equiv1/T_1$ is the decay rate of the excited state $|b>$ and $\Delta k= k+k_1-2k_2$. This gives the photon echo at t=2T. Note that the second pulse reverses the phase of the system as in the photon echo, so that the frequency modulation $\delta\omega_i(\tau)$ contribute with the opposite sign for the first and second time intervals $0\leq t<T$ and $T\leq t<2T$. The contributions from the i-th and the j-th sites have also the opposite signs. Therefore this 4WM and the photon echo give the information on the temporal and spatial correlations very effectively. We take into account these correlations of the frequency modulation in the following form:

$$<\delta\omega_i(t_1)\delta\omega_j(t_2)>_B = \Delta^2 g(r_{ij})\exp\{-\gamma_c|t_1-t_2|\}, \qquad (10)$$

where $g(r_{ij}) = 1$ for $r_{ij}<R_c$, and 0 otherwise,

and R_c is the characteristic length of the spatial correlation. In the direction of the complete phase matching $\Delta k=0$, i.e., $k=2k_2 -k_1$, we observe the signal decaying with 2γ and $2(\gamma+\gamma')$ with the intensity ratio R_c^3 to $R^3-R_c^3$. $4\pi R^3/3$ is a measure of the volume in which 4WM is actively induced and $\gamma'\equiv\Delta^2\tau_c$. When the frequency modulation has the long-ranged correlation $R_c\simeq R$, e.g., near the second order phase-transition, only the signal with the longitudinal decay $2\gamma=1/T_1$ remains finite. In the direction of such an incomplete phase-matching as $R^{-1}<<\Delta k\lesssim<R_c^{-1}$, the signal vanishes at $2T\rightarrow0$, shows the peak at $2T=(2\gamma')^{-1}\log\{(\gamma+\gamma')/\gamma\}$ and decays with 2γ afterwards.

We may conclude that our present treatment is inevitable when we want to discuss the case of the temporally or spatially long-ranged correlation in comparison with the characteristic time χ^{-1} or T_2 or the characteristic length λ of the involving wavelength.

References

1. Y.L. Shen: The Principles of Nonlinear Optics (John Wiley & Sons, 1984).
2. E. Hanamura: J. Phys Soc. Jpn. 52, 2258 (1983).
3. H. Tsunetsugu, T. Taniguchi and E. Hanamura: Solid State Commun. 52, 663 (1984).
4. E. Hanamura: J. Phys. Soc. Jpn 52, 3265 (1983).
5. R.G. DeVoe and R.G. Brewer: Phys. Rev. Lett. 50, 1269 (1983).
6. E. Hanamura: J. Phys. Soc. Jpn. 52, 3678 (1983).
7. H. Tsunetsugu and E. Hanamura: Solid State Commun. 55, 397 (1985).
8. H. Tsunetsugu and E. Hanamura: to be submitted to Phys. Rev.
9. E. Hanamura: Solid State Commun. 51, 697 (1984).

Modified Bloch Equations for Solids

P.R. Berman
Physics Department, New York University, 4 Washington Place,
New York, NY 10003, USA

R.G. Brewer
IBM Research Laboratory, San Jose, CA 95193, USA

In a recent experiment of DEVOE and BREWER [1], it was concluded that the optical Bloch equations are incapable of describing the saturation phenomena observed. Optical free-induction decay (FID) measurements of the impurity ion crystal Pr^{3+}:LaF_3 were conducted,where the Pr^{3+} ions are coherently prepared by a laser field under steady-state conditions,and then freely precess when the driving field is removed. At low optical fields, the observed Pr^{3+} optical line width is dominated by magnetic fluctuations,arising from pairs of fluorine nuclear flip-flops. At high optical fields, this nuclear broadening mechanism is quenched,and the Bloch equations are seriously violated. On physical grounds, this failure is due to a time-averaging of the magnetic interaction as the optical nutation frequency increases [2]. The phenomenological dipole dephasing time T_2 of the Bloch equations is therefore not a true constant,but lengthens with increasing field strength.

Several theories [2] - [8] have been proposed to explain the DeVoe-Brewer results. Most of these theories are an extension of a method developed by REDFIELD [9] for modifying the Bloch equations to include the effects of local, magnetically-induced fluctuations of the transition frequency of the radiating spins.

In this report, we present theoretical results for the optical FID decay rate obtained by assuming specific models for the local field fluctuations imposed on the ions' optical resonance frequency, and compare the results with "diffusion-type" theories [2] - [5] of frequency fluctuations. Whereas all the theories explain the overall qualitative structure of the experimental results, especially in strong fields, the predictions in the low-field regime differ sufficiently to allow for the possibility of an experimental test of one theory over another. The low-field data of DeVoe and Brewer are not yet precise enough to provide a critical test of any of the theories, but future experiments may be more conclusive.

Without going into the details of the calculation, a good qualitative understanding of the FID decay rate can still be obtained using simple physical considerations. The FID signal results when an ensemble of ions such as Pr^{3+} is initially prepared by an external optical field and then freely radiates when the field is suddenly removed. A large-scale inhomogeneous width Δ^* characterizes the sample, leading to a first-order FID signal that decays rapidly in a time $(\Delta^*)^{-1}$ - an effect that we do not consider here. The contribution to the FID signal which *does* concern us arises from a nonlinear interaction with the external field. As in the case of an inhomogeneously broadened vapor, the external field "burns" a hole in the ground state population and creates a bump in the excited state population, corresponding to that subgroup of ions with transition frequencies that resonantly interact with the external field. The FID decay rate is determined by the power-broadened line width of the hole resulting from the preparative phase *plus* the free-precession decay when the field is turned off.

In the absence of frequency fluctuations for a two-state system, the FID decay rate given by the Bloch equations is

$$\gamma_F = [(1/T_2^0)^2 + \chi^2(T_1^0/T_2^0)]^{1/2} + 1/T_2^0 , \tag{1}$$

where T_1^0 and T_2^0 are relaxation times in the absence of any frequency fluctuations and χ is the Rabi frequency of the external field. The bracketed term in (1) is the power-broadened hole width and the second term is the contribution of the free-decay period.

Frequency fluctuations modify the FID signal by affecting both the width of the hole and the free-decay process. We adopt a simple model, often applicable to solids, in which local field fluctuations produce frequency shifts $\delta\epsilon(t)$ in the ion resonance frequency. Only Markoffian processes are considered; i.e., the frequency following a fluctuation depends at most only on the frequency before the fluctuation. As such, the frequency jump processes can be characterized by the three parameters:

Γ = rate at which fluctuations occur ("jump rate"),

$\delta\epsilon$ = rms frequency shift per fluctuation,

ϵ_0 = $\sqrt{2}$ times rms frequency associated with the frequency fluctuation distribution.

The quantity ϵ_0, in effect, is a measure of the maximum frequency displacement produced by the local field fluctuations. It is assumed that $\Gamma, \delta\epsilon$ and ϵ_0 are much smaller than the large inhomogeneous width Δ^*.

In this work, we restrict the discussion to three types of Markoffian processes: (1) a Gaussian-Markoffian or diffusion-type model; (2) a "Difference" model in which the probability of finding a frequency ϵ following a fluctuation when the frequency was ϵ' before the fluctuation is a function of $(\epsilon-\epsilon')$ only; and (3) a "Strong" model in which the frequency distribution following a fluctuation is the equilibrium distribution characterized by width ϵ_0. In the diffusion model, $\Gamma, \delta\epsilon$ and ϵ_0 appear only in the combinations [10]:

$\beta = \Gamma\delta\epsilon/\epsilon_0$ = effective jump rate,

$q = \beta\epsilon_0^2/2$ = diffusion coefficient.

We now proceed to give the FID decay rates calculated for each model. Only the weak field regime is considered since, for strong fields, all models predict the same decay rate $\gamma_F \sim \chi(T_1^0/T_2^0)^{\frac{1}{2}}$ owing to the fact that, for large χ, frequency fluctuations become unimportant on the coherence time scale associated with the Rabi oscillations.[1] As has been noted [1], this result

---

[1] The zero-field intercept extrapolated from the strong field regime differs for the various models and could be used in a consistency check of the theories. Moreover, values of γ_F in the intermediate field regime can also be used to distinguish the theories.

disagrees with the conventional Bloch prediction $\gamma_F \sim \chi(T_1/T_2)^{1/2}$, where the low-field values T_1 and T_2 include the effects of frequency fluctuations.

Diffusion Model. Two limits, $\varepsilon_0 \ll \beta$ and $\varepsilon_0 \gg \beta$, can be distinguished. If $\varepsilon_0 \ll \beta$, one finds [2] – [5] an FID decay rate

$$\gamma_F \sim 2/T_2 \quad ; \quad (1/T_2) = (1/T_2^0) + \varepsilon_0^2/2\beta, \tag{2}$$

which implies that the T_2 associated with the weak-field regime can be much smaller than T_2^0 if $\varepsilon_0^2/2\beta \gg 1/T_2^0$. When $\varepsilon_0 \gg \beta$ the problem is best solved by the "Difference" model (discussion below) and, for $\varepsilon_0(\beta T_1^0)^{1/2} \gg 1/T_2^0$, we find a decay rate of order

$$\gamma_F \sim 2/T_2 \quad ; \quad (1/T_2) = \tfrac{1}{2}(\beta T_1^0/2)^{1/2}\varepsilon_0 . \tag{3}$$

Difference Model. The difference model [10] does not obey detailed balancing and cannot be used in the limit $\varepsilon_0 \ll \Gamma$ since frequency fluctuations *do* restore equilibrium in this limit. Thus, use of the difference model, which is characterized by a "kernel" $W(\varepsilon' \to \varepsilon)$ [$W(\varepsilon' \to \varepsilon)$ is the probability density per unit time for fluctuations to change the frequency from ε' to ε] given by

$$W(\varepsilon' \to \varepsilon) = \Gamma(2\pi\delta\varepsilon^2)^{-1/2} \exp[-(\varepsilon-\varepsilon')^2/2\delta\varepsilon^2] , \tag{4}$$

is limited to the case $\varepsilon_0 \gg \Gamma$. Within this restriction, one can distinguish two sub-cases, $\delta\varepsilon < (1/T_2^0) + \Gamma$, $\delta\varepsilon > (1/T_2^0) + \Gamma$.

If $\delta\varepsilon < (1/T_2^0) + \Gamma$, one recovers the diffusion limit. The broadening of the hole is of order $\sqrt{n}\delta\varepsilon$, where $n = \Gamma T_1^0$ is approximately the number of frequency jumps occurring in time T_1^0. Considering only the interesting case when this width is considerably larger than the natural width $1/T_2^0$, i.e. when

$$(\Gamma T_1^0)^{1/2}\delta\varepsilon T_2^0 = (\beta T_1^0)^{1/2}\varepsilon_0 T_2^0 \gg 1, \tag{5}$$

we find an FID signal which decays as

$$\text{FID signal} \propto \tfrac{\pi}{2} - \tan^{-1}[(\beta T_1^0/2)^{1/2}\varepsilon_0 t] . \tag{6}$$

The decay rate (3) agrees qualitatively with the numerical work of JAVANAINEN [5] using a diffusion model.

If $\delta\varepsilon > (1/T_2^0) + \Gamma$, fluctuations take the ion frequency outside the hole burned by the field and simply increase the decay rate. One finds

$$\gamma_F \sim 2/T_2 \quad ; \quad 1/T_2 = 1/T_2^0 + \Gamma . \tag{7}$$

Strong Model. In the strong model, the kernel is given by

$$W(\varepsilon' \to \varepsilon) = \Gamma(\pi\varepsilon_0^2)^{-1/2} \exp(-\varepsilon^2/\varepsilon_0^2), \tag{8}$$

and is independent of the initial frequency ε'. If $\varepsilon_0 \ll \Gamma$, we find [8] a decay rate

$$\gamma_F \sim 2/T_2 \quad ; \quad 1/T_2 = 1/T_2^0 + \varepsilon_0^2/2\Gamma, \tag{9}$$

which is similar to (2). For large fluctuation rates, it is unimportant whether a few strong or many weak fluctuations redistribute the frequency. If $\varepsilon_0 \gg \Gamma$, $1/T_2^0$, χ, fluctuations shift the frequency out of the hole burned by the field and simply increase the decay rate as

$$\gamma_F \sim 2/T_2 \quad ; \quad 1/T_2 = 1/T_2^0 + \Gamma \; . \tag{10}$$

The physical mechanism is the same as that leading to (7).

All the weak field results for the FID decay rate gives values of $\gamma_F \equiv 2/T_2 \neq 2/T_2^0$. Since the form of T_2 is model-dependent, the different models lead to different predictions for the weak-field regime. However, with the exception of (6) which indicates a non-exponential decay, one cannot distinguish the various theories in the weak-field regime unless one has an independent means for experimentally varying the parameters $\delta\varepsilon, \Gamma$ and ε_0. This appears feasible in molecular vapors by varying the perturber-to-active-atom mass ratio or the perturber pressure, but may prove difficult in solids. If an independent variation of the parameters is not possible, one must compare the various theories over the range from weak to strong external fields to obtain a best fit to the data, a procedure that may prove useful when new experimental data become available.

This work is partially supported by the U.S. Office of Naval Research.

References

1. R.G. DeVoe and R.G. Brewer: Phys. Rev. Lett. 50, 1269 (1983); 52, 1354 (1984)
2. A. Schenzle, M. Mitsunaga, R.G. DeVoe and R.G. Brewer: Phys. Rev. A30, 325 (1984); R.G. Brewer and R.G. DeVoe: Phys. Rev. Lett. 52, 1354 (1984)
3. M. Yamanoi and J.H. Eberly: Phys. Rev. Lett. 52, 1353 (1984): J. Opt. Soc. Am. B1, 751 (1984)
4. E. Hanamura: J. Phys. Soc. Jpn. 52, 2258 (1983); 52, 3265 (1983); 52, 3678 (1983)
5. J. Javanainen: Opt. Comm. 50, 26 (1984)
6. K. Wodkiewicz, B.W. Shore, and J.H. Eberly: Phys. Rev. A30, 2390 (1984); K. Wodkiewicz and J.H. Eberly: Phys. Rev. A32 (in press)
7. P.A. Apanasevich, S. Ya. Kilin, A.P. Nizovtsev and N.S. Onishchenko: Opt. Comm. 52, 279 (1984)
8. P.R. Berman and R.G. Brewer: submitted to Phys. Rev. A
9. A.G. Redfield: Phys. Rev. 98, 1787 (1955)
10. P.R. Berman: Phys. Rev. A9, 2170 (1974); P.R. Berman, J.M. Levy and R.G. Brewer: Phys. Rev. A11, 1668 (1975)

Spin-Spin Reservoir Cross-Relaxation of Pr^{+3}: LaF_3 via Optical Pumping

F.W. Otto, M. Lukac, and E.L. Hahn

University of California, Physics Department,
Berkeley, CA 94720, USA

Abstract: The Zeeman quadrupole dipole-coupled Hamiltonian spectrum of hyper-fine optical ground states of Pr^{+3}:LaF_3 is measured by novel magnetic field sweep and rf resonance techniques. Spin reservoir cross-relaxation between Pr^{+3} and F ensembles is detected. A scheme is presented for achieving polari-zation of the F spin reservoir in times shorter than the spin-lattice relaxa-tion time.

1) Introduction: Critical changes of laser hole-burning or resonant laser transmission in a solid which depend upon the effects of nuclear spin-spin res-ervoir level-crossings are observed in Pr^{+3}:LaF_3 for the $^3H_4 \leftrightarrow ^1D_2$ inhomogeneous-ly broadened (10 GHz fwhm) optical transition [1,2]. An externally applied DC magnetic field is swept at an optimum rate dB/dt (500 G/sec) through values where the Zeeman-split nuclear quadrupole hyperfine optical ground states of Pr^{+3} (see Fig. 1) cross with fluorine nuclear spin levels. Therefore, single spin-flips of the rarer (0.5% abundant) Pr spins occur with opposite single and less probable double spin-flips of fluorine nuclei. If B is held fixed at any level-crossing, no changes in laser transmission are observed; and if the sweep rate dB/dt is too slow or too rapid, the effect is reduced. Similarly, for any B field (>50 G) held constant, externally applied continuous-wave rf excitation of NMR transitions among any pair of Pr ground state levels (which simulates

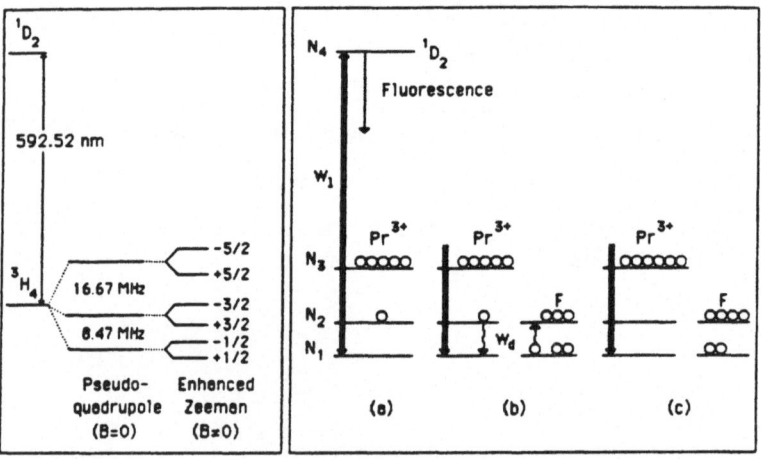

Fig.1. The optical ground state splittings of Pr^{+3}

Fig.2. The corss-relaxation process. n_3 represents the four levels not parti-cipating in the cross-relaxation with F. Because of inhomogeneous broadening, two other cases (not shown here) occur simultaneously: laser resonant with n_2 (producing reverse F polarization); and laser resonant with any of the n_3 levels (not affecting the fluorines)

cross-relaxation with F nuclei) does not measureably alter the transmitted laser intensity. However, when the NMR rf is turned on suddenly, the transmission is observed to undergo a transient reduction and to recover towards the
previous steady-state transmission. Prior to the level-crossing imposed by
field-sweep dB/dt (or to sudden onset of NMR for B held constant at any value)
the laser-excited levels are bleached (Fig. 2a) for sufficient laser intensity.
Because of the fluorescence mechanism (lifetime τ_F = 0.5 msec), particles which
occupied the empty levels in the past have transferred to the remaining five
out of six nuclear levels, all of which have spin-lattice relaxation times, T_1,
of at least one minute. The sudden onset of Pr↔F spin-spin coupling during
field sweep, or the sudden application of NMR, allows the transient transfer of
population back into the empty laser-bleached levels, thus momentarily restoring some degree of laser absorption (Fig. 2b). After these processes two out
of the six levels are empty (Fig. 2c), and the steady-state transmission is the
same as before. As is the case with the sweep rate, the effect is reduced if
the laser intensity is too low or too high. (The laser intensity applied was
2mW/mm², and the laser linewidth was 1 MHz.) A numerical model based on rate
equations was used to confirm this critical dependence on the sweep rate and
laser intensity [3].

The transient field sweep and NMR effects noted above proved to be a very
good means by which we have mapped out the Pr^{+3} ground state hyperfine spin
level spectrum, as well as confirmed the onset of level-crossing with the F
spin system. No crossings with the more complex lanthanum (I = 7/2) quadrupole levels were identified.

2) Rf Spectroscopy: The Hamiltonian for the fluorine spins in a magnetic
field is

$$H_F = -\hbar\gamma^F\Sigma_i B_i S_i \text{ , where } \gamma^F = 4.0055 \text{ kHz/G.}$$

The spin Hamiltonian of Pr^{+3} was given by Teplov [4];

$$H_{Pr} = -\hbar\Sigma_i\gamma_i^{Pr}B_i I_i + D[I_z^2 + I(I+1)/3] + E[I_x^2 - I_y^2]$$

where D = 4.1795 ± 0.0013 MHz and E = 0.154 ± 0.004 MHz [1].

A study of the enhanced nuclear Zeeman tensor γ^{Pr} has been made in low magnetic fields (B = 100 G) [5]. Our work extends the study to higher values of B
where the enhanced Zeeman interaction becomes comparable to the pseudo-quadrupole interaction. Taking Erickson's values for the γ^{Pr} tensor and fitting our
data to the numerically diagonalized Hamiltonian, we determined (adopting the
convention that E>0) that the x-axis of the enhanced quadrupole tensor must be
chosen along the crystal C_2 symmetry axis (which is opposite to Erickson's
choice of x and y axes for the gyromagnetic tensor, γ^{Pr}). Figure 3a displays
the rf data and the numerical fit for an orientation of B of α = 1.5 degrees
with respect to one C_2 axis and perpendicular to the C_3 axis. While our fit
was insensitive to the value of $\gamma_y^{Pr}(\gamma_y^{Pr}/2\pi$ = 4.98 kHz/G) [1], we obtained
$\gamma_z^{Pr}/2\pi$ = 10/16 ± 0.05 kHz/G and $\gamma_x^{Pr}/2\pi$ = 3.45 ± 0.05 kHz/G.

Of particular interest in carrying out spectroscopy by use of NMR is the
level repulsion noted in Fig. 3b. As the F and Pr levels approach one another, there is a slight deflection from the smooth eigenvalue plot of ω vs B on
either side of the point where exact level-crossings occur. The level repulsion, which is seen as a discontinuity of the NMR spectrum, resembles the
effect of a Pake doublet, except that a distribution of fluorine spin orientations interacts with an isolated Pr spin. Imagine two spins, one of which
possesses an arbitrarily adjustable gyromagnetic ratio, γ. Identical γ's provide a Pake doublet spectrum for I = 1/2. For γ_1 nearly equal to γ_2, the on-

set of mutual spin-flips would produce an additional "coherent splitting" superimposed on the static broadening or shift always present due to longitudinal field-broadening. The case of Pr↔F coupling is in fact much more complicated, since the pseudo-quadrupole Hamiltonian for I = 5/2 must be included, and it is not the γ's which are equal but the corresponding transition frequencies. This off-resonant effect is numerically calculated and plotted in Fig. 3b.

3) Magnetic Field Sweep: Figure 5c displays the observed level-crossing responses during the magnetic field sweep, where the magnetic field B was oriented along one of the crystal C_2 axes. For a laser field E_ℓ polarized perpendicular to B, one obtains a response only from the sites with a quadrupole x-axis at an angle of $\alpha = 60°$ with respect to B. When the laser field E_ℓ is

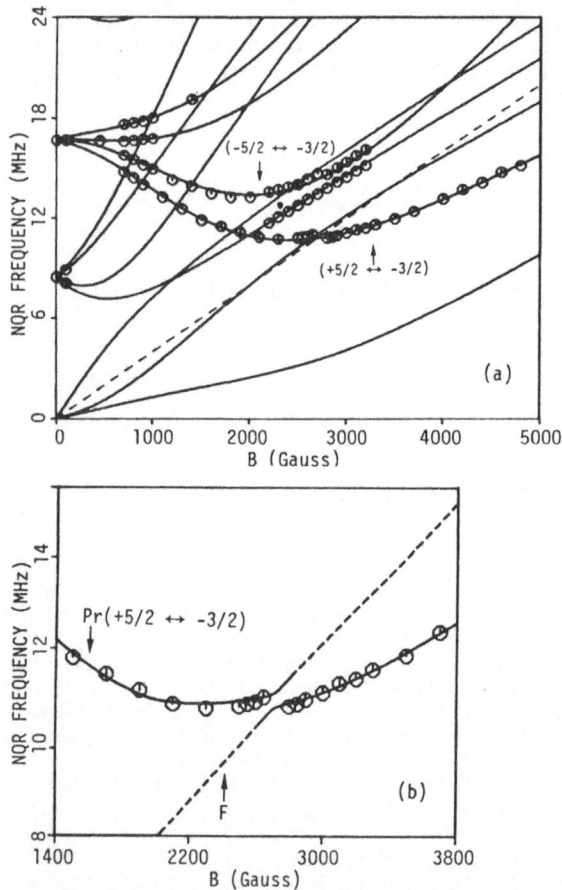

Fig. 3. a) The rf data (circles) and numerical fit (solid line) for $\alpha = 1.5°$, $\gamma_x/2\pi = 3.45$ kHz/G, $\gamma_y/2\pi = 4.98$ kHz/G, and $\gamma_z/2\pi = 10.16$ kHz/G. The dashed line is the fluorine transition. b) Expanded view of the observed level-repulsion (circles) where the condition for cross-relaxation with the fluorine spin reservoir is satisfied. The numerical fit was obtained by diagonalizing a 12 x 12 F-Pr^{+3} coupled Hamiltonian. Only the two transitions of interest are shown

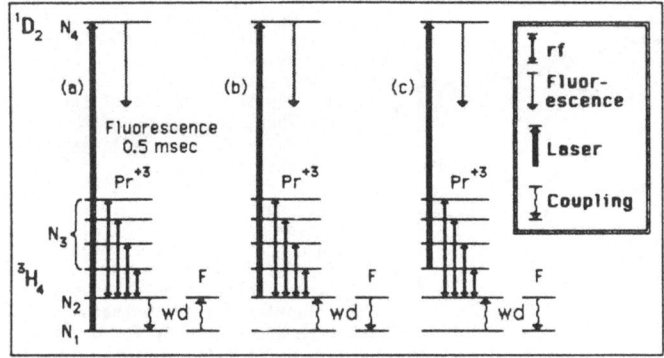

Fig.4. The proposed scheme for achieving fluorine spinreservoir polarization. Inhomogenous broadening causes the pumping cycles shown in a), b) and c) to occur simultaneously

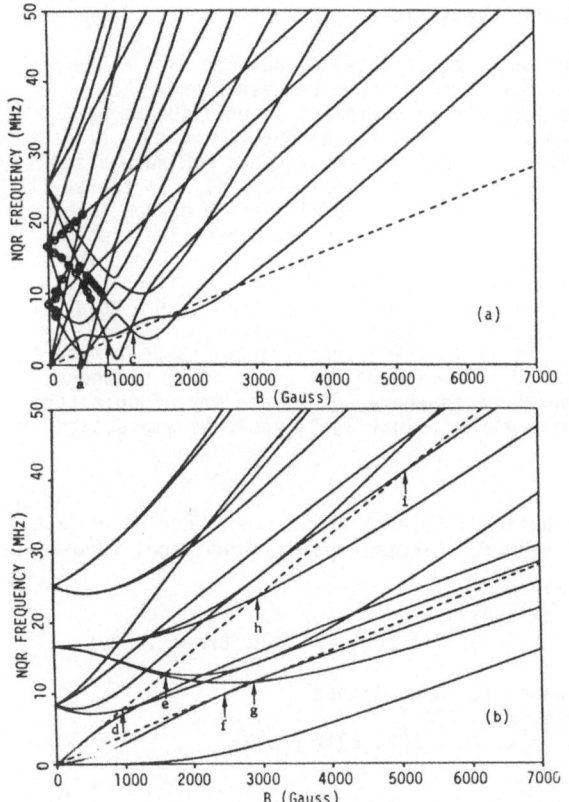

(a)

(b)

Fig. 5a. Rf data (circles) and numerical calculation (solid line) for $\alpha = 60°$, $\gamma_x/2\pi = 3.45$ kHz/G, $\gamma_y/2\pi = 4.98$ kHz/G, and $\gamma_z/2\pi = 10.16$ kHz/G. The dashed line is the fluorine transition. Arrows show where energy-conserving cross-relaxation occurs. Note that only the first crossing point of a given Pr^{+3} line is indicated (see text)

Fig. 5b. Same as a), but with $\alpha = 0°$. Double fluorine spin-flip transitions are also shown (upper dashed line)

polarized parallel to B, an additional set of sites with quadrupole x-axes parallel to B is excited. The eigenlevel diagram for each set of sites is displayed with corresponding level-crossing and enhanced absorption noted.

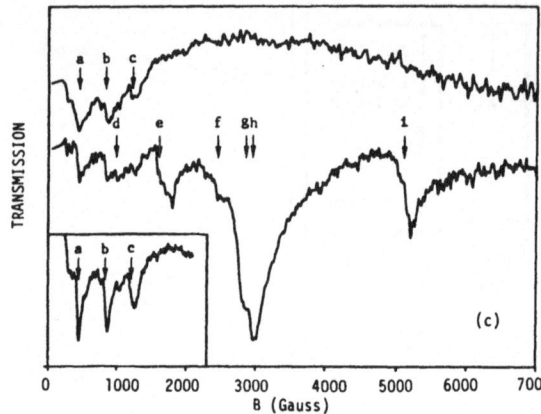

Fig. 5c. Optical transmission during magnetic field-sweep (500 G/sec. The upper curve represents the average of twenty traces, where E_L was perpendicular to B and one of the C_2 axes. Thus, only sites with $\alpha = 60°$ were excited. The lower curve represents the average of 75 traces under the same conditions. The middle curve shows transmission for the case where E_L was parallel to B and one of the C_2 axes. In this case, sites with $\alpha = 60°$ and $\alpha = 0°$ were excited simultaneously. Double fluorine spin-flip transitions were observed only for $\alpha = 0°$

Since the first crossing of a given Pr line with F line depopulates the levels involved, subsequent crossings of the same line are less effective and are not observed. Single and double spin-flip transitions of F nuclei are to be noted.

4) Discussion: From a model based on a set of detailed-balance rate equations which expresses the two-reservoir coupling rate between F and Pr systems, we have confirmed that it is possible to establish an equilibrium net population of F spins in times much shorter than T_1, either at negative or positive temperature. This takes place when four simultaneous NMR saturation transitions couple into one of the two Pr^{+3} transitions (Fig. 4). In the absence of rf, cross-relaxation coupling causes only local heating or cooling of the reservoir spins, although their temperature, averaged over the sample, does not change. The effect of rf coupling is to prevent complete bleaching of the n_1 and n_2 levels and to introduce an asymmetry in the optical pumping. Together with cross-relaxation, this produces a net non-zero rate of F polarization. A rate-equation model predicts that a net polarization of 1% can be achieved in about 1000 seconds, if T_1 is sufficiently long. A Boltzmann population difference can be achieved in a time of the order of 10 seconds. This technique may be of advantage in cases where quick recovery of spin population is required for NMR studies where ordinarily T_1 would be impractically long.

This work was supported by the National Science Foundation. One of us (M.L.) acknowledges the support of the Center for Electro-optics, Ljubljana, Yugoslavia.

References
1. L.E. Erickson, Op. Comm. 21, 147 (1977)
2. J. Mylnek, N.C. Wong, R.G. DeVoe, E.S. Kintzer, and R.G. Brewer, Phys. Rev. Lett. 50, 993 (1983)
3. F.W. Otto, M. Lukac and E.L. Hahn, to be published
4. M.A. Teplov, Sov. Phys. JETP 26, 872 (1968)
5. B.R. Reddy and L.E. Erickson, Phys. Rev. B27, 5217 (1983)

Quantum Evolution of 2-Level Atoms in the Presence of a State Selective Reservoir*

J.F. Lam, D.G. Steel[+], and R.A. McFarlane

Hughes Research Laboratories, Malibu, CA 90265, USA

Recently, we have shown that the nearly degenerate four-wave mixing (NDFWM) process can provide a direct and simultaneous measurement of the longitudinal ($1/T_1$) and transverse ($1/T_2$) relaxation rates of a resonant two-level atom. Furthermore, we observed that in the presence of buffer gases, the linewidth of the resonance describing the longitudinal relaxation process experiences a significant narrowing, below that of the natural width. Using a theoretical model, the narrowing was accounted for as arising from collisional effects in the ground state population[1]. These observations were made for the case where the radiation fields are co-polarized, and we concluded that NDFWM offers a unique technique for the measurement of the ground state velocity changing cross-section[2].

This paper discusses our further studies of the effects of collisions on the spectral and temporal response of NDFWM. Our results indicate for the case of stationary 2-level atoms and for moving atoms characterized by magnetic degeneracy that the addition of buffer gases also gives rise to spectral features having sub-natural linewidths. We also show that, in the presence of buffer gases, the duration of stimulated photon echoes is increased by the ratio of the ground-state collisional lifetime to the spontaneous decay life time of the excited state.

In the case of moving atoms with magnetic degeneracy, we consider the NDFWM spectral response of systems exhibiting Zeeman coherences. The production of such coherences is achieved by having the pump fields 90° cross-polarized with res-pect to the probe field. This scheme provides both phase and polarization conjugation in a resonant medium[3]. In the Doppler limit and in the absence of buffer gases, the spectrum of the generated signal shows two resonances having equal linewidths (given by the spontaneous emission-rate). The first resonance is centered at the pump frequency ω and reflects the fact that the forward pump and probe excite the same velocity group. The second resonance is centered at $3\omega-2\omega_o$, ω_o being the transition frequency. This resonance reflects the fact that the forward pump and the generated signal share the same velocity group and describes the nonlinear response of the optical coherence.

---

*Work supported in part by the U.S. Army Research Office under contract No. DAAG29-81-C-008.

+Present address: Departments of Electrical Engineering and Physics, University of Michigan, Ann Arbor, MI

The equality in the linewidths is a consequence of the resonant atom behaving as closed quantum system. In the presence of buffer gases, the two resonances remain, with the distinction that the linewidths are no longer equal in magnitude. The linewidth of the first resonance is influenced by collisionally-induced changes in the Zeeman coherence of the ground state. Its magnitude is determined by angular momentum-changing collisions. The linewidth of the second resonance is broadened as a function of buffer gas pressure. This is a manifestation of the effects of phase-interrupting collisions on the optical coherence. The predominance of ground state dynamics in determining the details of the first resonance is a consequence of the fact that the resonant atom behaves like an open quantum system in the presence of foreign perturbers.

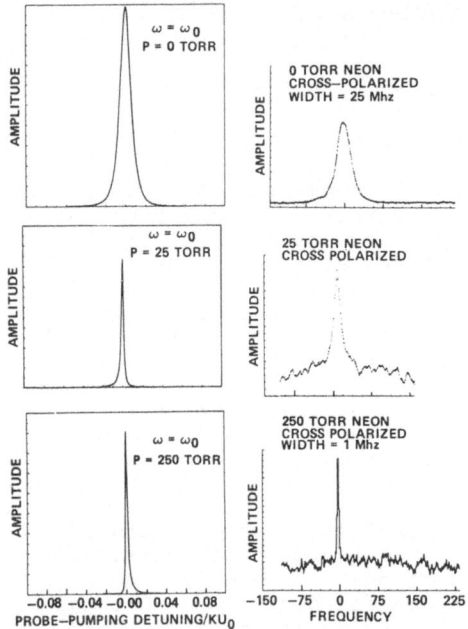

Fig. 1 Spectra of NDFWM for different values of pressure of neon.

The theoretical lineshape was obtained using a perturbative solution of the density matrix equations, taking into account the random motion of the atoms. In the Doppler regime and with the pump frequency set equal to ω_O, the results of the computation agree well with our measurements.

Figure 1 shows the theoretical (left-hand column) and experimental (right-hand column) results for the case when the pump is resonant with the transition $3S_{1/2}(F=2)-3P_{3/2}(F=3)$ of sodium and the pressure of the neon buffer gas is changed from o torr to 250 torr. The measured linewidth at o torr is 25 MHz which is the sum of the spontaneous decay rate (20 MHz), laser jitter (2 MHz) and residual Doppler width (3 MHz). At a neon pressure of 250 torr, the linewidth is measured to be of the order of 2 MHz. This value is a manifestation of the influence of the laser jitter alone. One can deduce an upper bound for the depolarizing collision cross-section for ground state atoms from the spectral width of the four-wave mixing signal. A direct calculation shows that the upper bound for this cross-section is of the order of 10^{-18} cm^2.

In the case of stationary 2-level atoms, we consider the NDFWM spectral response of systems exhibiting population effects due to gratings. The production of population gratings is achieved by having all input fields to be linearly copolarized. The theoretical spectral lineshape is obtained using a perturbation solution of the density matrix equations,and the intensity of the NDFWM signal is

$$I(\delta, \Delta) \alpha \quad \frac{1}{\gamma_c^2 + (\Delta + \delta)^2} \times \frac{1}{\gamma_c^2 + (\Delta - \delta)^2} \times \frac{1}{\gamma_c^2 + \Delta^2}$$

$$\times \frac{(2\gamma_c)^2 + \delta^2}{(\gamma + \Gamma_2)^2 + \delta^2} \times \frac{(\Gamma_1 + \Gamma_2)^2 + (2\delta)^2}{\Gamma_1^2 + \delta^2} \qquad (1)$$

where δ is the detuning between the probe and pump waves, $\Delta = \omega_o - \omega$ is the pump detuning from resonance, $\gamma_c = \gamma/2 + \gamma_{pic}$, γ is the spontaneous decay rate, γ_{pic} is the phase-interrupting collision-rate, Γ_α is the collisionally-induced decay of state α. ($\alpha = 1, 2$).

Two interesting features can be derived from expression (1). First, in the absence of collisions and for $\Delta = 0$, there exists a resonance at $\delta = 0$, i.e. at the transition frequency. The origin of this resonance is attributed to the maximum degree of coherent interaction between the pump and probe waves. The FWHM of this resonance is given by $(\sqrt{2} - 1)\gamma$, which reflects the fact that the lineshape is of the form of a Lorentzian raised to the second power. However, in the presence of buffer gases, the resonance at $\delta = 0$ acquires a width determined by the inverse ground-state lifetime, which is much smaller than the spontaneous emission rate γ. This behavior is shown in Figure 2, where the limiting width of the ground state is taken to be the inverse of the transit time (~ 1 microsecond). Second, we see that in the absence of collisions and for $\Delta = \gamma$, there exists 2 resonances located at $\delta \simeq \pm \Delta$. The resonance at $\delta = \Delta$ is a direct consequence of the coherent interference of the pump and probe waves. The other resonance at $\delta = -\Delta$ arises from the coherent emission generated during the NDFWM process. The shape of each resonance is determined by the relation between two Lorentzians. However, in the presence of buffer gases, a very narrow resonance appears at $\delta = 0$. Its width is determined by the inverse of the ground-state lifetime. This novel behavior is shown in Figure 3. The resonance at $\delta = 0$ is a direct manifestation of the ground-state contribution to the NDFWM signal in the presence of buffer gases.

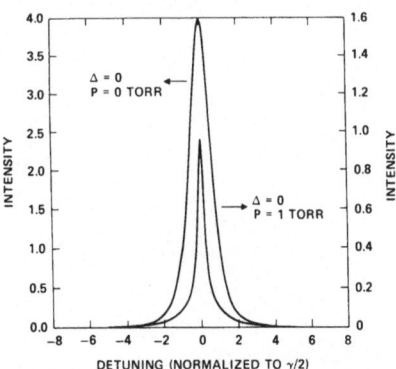

Fig. 2 Spectra of NDFWM for stationary 2-level atoms and $\Delta = 0$.

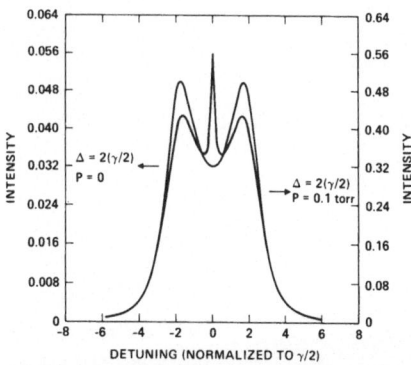

Fig. 3 Spectra of NDFWM for stationary 2-level atoms and $\Delta = \gamma$.

281

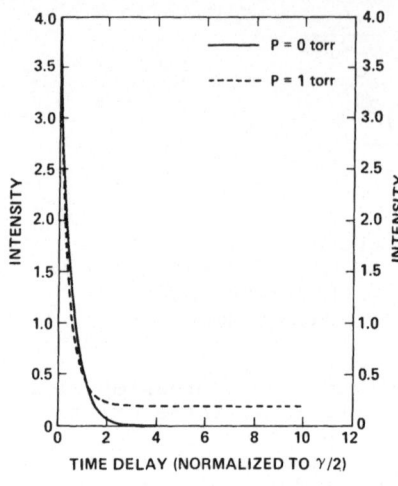

Fig.4 Echo intensity as a function of time delay.

In the case of the stimulated photon echo process, Figure 4 shows the effect of buffer gases on the time evolution of the echo intensity. In order to simulate the effects of collisions on Zeeman coherences, we chose the collisional decay rate for the ground state to be one thousand Hz per torr. The dotted curve in Figure 4 indicates a significant improvement in the duration of the echo process upon the introduction of a small amount of buffer gases.

The authors wish to thank Dr. B.D. Guenther of ARO for his continued interest and support and Prof. P.R. Berman for many stimulating discussions.

References
[1] J.F. Lam, D.G. Steel and R.A. McFarlane, Phys. Rev. Lett. **49**, 1628 (1982).

[2] J.F. Lam, D.G. Steel and R.A. McFarlane, in Laser Spectroscopy VI, H.P. Weber and W. Luthy, editors, (Springer Verlag 1983) pp.315-317.

[3] J.F. Lam and R.L. Abrams, Phys. Rev. A **26**, 1539 (1982).

282

Studies of Two-Level Atoms Identically Prepared by a Phase- and Amplitude-Controlled Excitation Field

Y.S. Bai, A.G. Yodh, and T.W. Mossberg

Department of Physics, Harvard University, Cambridge, MA 02138, USA

While the response of two-level atoms to nearly resonant driving fields is in principle well known, experimental studies of such responses are generally limited to statistical systems where signals from atoms in widely different states relative to the driving field are observed. We have performed a series of experiments in a collimated beam of atomic ^{174}Yb (which constitutes a nearly-ideal two-level system), designed to probe the response of a small sample of nearly-identically prepared two-level atoms to resonant excitation.

Of special interest in our experiment was the study of stationary atom-field states. In the vector model [1], these states correspond to the alignment of the pseudo-spin vector, $\vec{\rho}$, parallel or anti-parallel to the driving field vector $\vec{\omega}$ [2]. In these states, the atom's energy is unchanged by the laser field. We point out that these stationary states are the dressed states familiar from resonance fluorescence theory [3]. In our experiment, we generate these states by first exciting ground-state atoms with a resonant excitation pulse of area $\pi/2$, and then exposing the atoms to a field-shifted in *phase* by 90° (see Fig. 1).

We observe the atomic response to the excitation field by monitoring fluorescence at right angles to both the atomic and laser beams (see Fig. 2).

In the absence of a phase-shift, the atomic fluorescence displays nutations. After the phase-shift, nutations disappear (see Fig. 3).

The capability of studying atoms prepared identically by a nearly resonant excitation field should make it possible to study many basic aspects of the atomic response which have hitherto been taken for granted but gone untested. Interesting new effects may also be observed. For example, atoms excited to specific dressed states as described above should display a novel transient *two*-peak resonance fluorescence spectrum. Of course, this spectrum should eventually evolve back to the usual three-peaked spectrum characteristic of steady-state measurements. Measurements of these effects are underway.

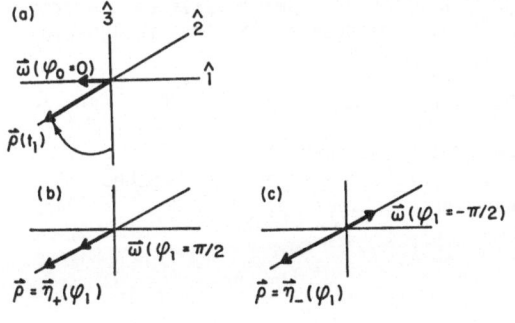

Figure 1: Generation of stationary (i.e. dressed) atom-field states. a) Response of atoms to a resonant pulse of phase $\varphi_0 = 0$ and area $\pi/2$. $\vec{\rho}$ and $\vec{\omega}(\varphi_0)$ are, respectively, the atomic pseudo-spin vector and effective field vector. b) Configuration of $\vec{\rho}$ and $\vec{\omega}$ after the phase of the driving field is shifted by $\pi/2$ radians. c) Same as b after a phase-shift of $-\pi/2$ radians. In both b and c, the pseudo-spin remains stationary despite the presence of the driving field, since $\dot{\vec{\rho}} = \vec{\omega} \times \vec{\rho} = 0$.

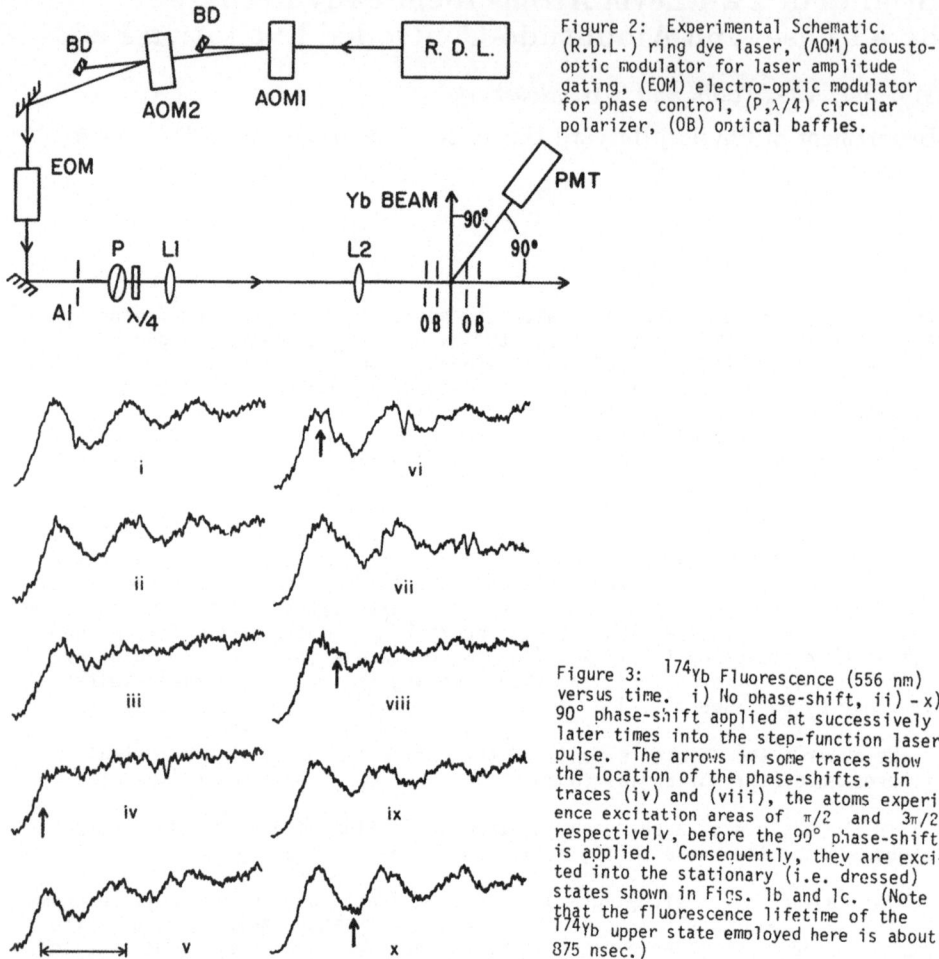

Figure 2: Experimental Schematic. (R.D.L.) ring dye laser, (AOM) acousto-optic modulator for laser amplitude gating, (EOM) electro-optic modulator for phase control, (P,λ/4) circular polarizer, (OB) optical baffles.

Figure 3: ^{174}Yb Fluorescence (556 nm) versus time. i) No phase-shift, ii) - x) 90° phase-shift applied at successively later times into the step-function laser pulse. The arrows in some traces show the location of the phase-shifts. In traces (iv) and (viii), the atoms experience excitation areas of $\pi/2$ and $3\pi/2$, respectively, before the 90° phase-shift is applied. Consequently, they are excited into the stationary (i.e. dressed) states shown in Figs. 1b and 1c. (Note that the fluorescence lifetime of the ^{174}Yb upper state employed here is about 875 nsec.)

We are happy to acknowledge conversations with P.R. Berman. This work was supported by the U.S. NSF (PHY-8207080) and the U.S. Joint Services Electronics Program (N00014-84-K-0465). A.G.Y. acknowledges a U.S. Army fellowship.

References
1. R.P. Feynman, F.L. Vernon Jr., and R.W. Hellwarth, J. Appl. Phys. 28, 49 (1957).
2. S.R. Hartmann and E.L. Hahn, Phys. Rev. 128, 2042 (1962).
3. C. Cohen-Tannoudji in Frontiers in Laser Spectroscopy (Les Houches Session XXVII), R. Balian, S. Haroche, and S. Liberman, Eds., North-Holland, Amsterdam (1977).

Two-Photon Hanle Effect in Ramsey Interrogation

N. Beverini, M. Galli, M. Inguscio, and F. Strumia

Gruppo Nazionale di Struttura della Materia del CNR,
Dipartimento di Fisica dell'Università di Pisa, Piazza Torricelli 2,
I-56100 Pisa, Italy

We have recently reported [1] a homogeneous linewidth version of the nonlinear Hanle effect. The two-photon interaction creates a coherence between Zeeman sublevels, when polarized laser radiation excites components of an optical transition with $\Delta M = \pm 1$ selection rules. When the degeneracy between the M sublevels is removed by an external field B, the coherence is cancelled, causing a decrease in the saturation, and it is possible to record an increase in the absorption. The shape of the effect is directly determined by the level in which the degeneracy is removed, and is insensitive to the laser bandwidth effects (the saturating optical radiation only acts as a coherence coupler between the sublevels).

We have performed experiments on a Ca atomic beam, inducing the transitions from the metastable $^3P_1$ and $^3P_2$ levels to the $^3S_1$ level (natural radiative linewidth 10.5 MHz). The widths of the observed resonances were ultimately determined by the transit time of the atoms crossing the expanded laser beam, and by the residual magnetic field inhomogeneities. Further narrowing of the signals can be achieved by using two interaction regions separated by a length L. The coherence, induced by the laser radiation in the first region, evolves in the dark at a frequency given by the Zeeman separation of the levels. The second laser probes the phase of the evolving coherence and causes oscillations in the recorded fluorescence signal. Typical results are shown in Fig. 1. In this case, phase-sensitive detection of the fluorescence was applied. The width of the fringes was observed to scale almost proportionally to the inverse of L. A width of 6 kHz was observed, for L=3.8 cm. The residual magnetic field inhomogeneity (about 5 mG over 4 cm) was much less effective than for the single-beam experiment.

Instead of the three interactions usually necessary for the extension of the Ramsey technique to the optical region, in our case only two are sufficient. In each interaction region Raman-type transitions occur, creating and probing the coherence. On the other hand, differently from the conventional microwave Ramsey scheme, the number of oscillations is not determined by the velocity monochromaticity. The limitation is rather intrinsic in the system. In fact our measurements are performed by scanning the magnetic field, hence changing not only the frequency of evolution of the coherence, but also its amplitude. In particular the coherence is no longer created when the degeneracy is removed. Similarly, the zero-field signal is not properly a Ramsey oscillation, because at fields near zero the coherence vanishes.

If the laser radiation is polarized parallel to B, so that $\Delta M = 0$ transitions are excited, population effects are evidenced. In fact only transitions

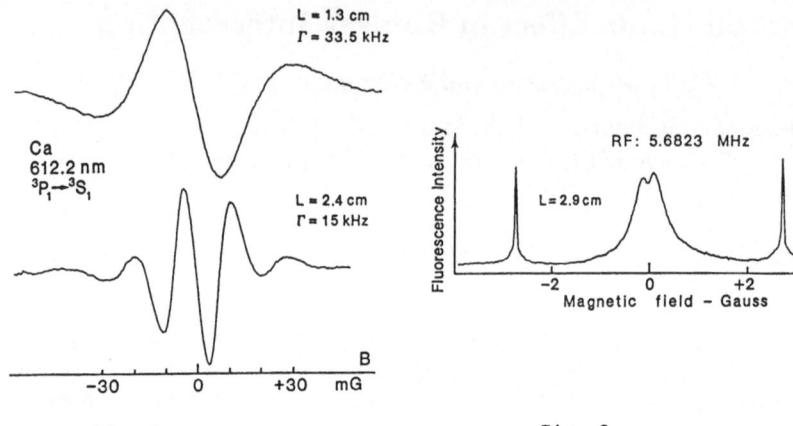

Fig. 1   Fig. 2

from ΛM=-1 and ΔM=+1 sublevels of the $^3P_1$ state can be induced, the M: 0 → 0 transition being forbidden. The optical pumping can be evidenced by scanning B. In fact at zero-field the residual transverse component destroys the Zeeman optical pumping and an increase in the fluorescence can be detected. The effect is enhanced when two separated interaction regions are used. The first interaction region creates the pumping,and in the second one the analysis is performed by detecting the fluorescence. In this scheme the signal is also narrower, because of the increased interrogation time. The optical pumping signal can be observed also for magnetic fields different from zero, if radiofrequency transitions are induced between Zeeman sublevels. Typical results are shown in Fig.2. The signal at zero magnetic fields is broader, because it is also caused by the residual transverse component. At values different from zero,the width is determined only by the transit time. By increasing L to 3.5 cm, we have measured for the rf-optical pumping signals, linewidths of about 30 kHz, unaffectded by the laser intensity and easily saturated by the rf power. This technique, which is another optical version of the atomic beam magnetic resonance, provides now a powerful tool for precise measurements of gyromagnetic factors.

Reference
[1] G. Bertuccelli, N. Beverini, M. Galli, M. Inguscio, F. Strumia , G. Giusfredi: Opt. Lett., 10 , n.6 (1985)

NMR Measurement of Rare Earth Ion
in an Excited State
by Indirect Optical Detection

L.E. Erickson

National Research Council, Ottawa, Canada, K1A OR8

The study of the hyperfine splittings of excited states of ions
in solids is usually restricted to levels which match available
light sources [1]. In this paper, an indirect optical-nuclear
double resonance detection scheme is used to study the hyperfine
interaction of trivalent Praseodymium in an electronic level at
9700 cm^{-1} using a narrowband laser source to pump another electronic level at 16740 cm^{-1} and by observing broadband luminescence
from the pumped level.

The experiment is described with the aid of Fig. 1. The van
Vleck paramagnet, trivalent Praseodymium dilute in Lithium Yttrium
Fluoride was cooled to 2 K. A stabilized cw ring dye laser (16741
cm^{-1}) pumps an individual hyperfine level for each excited ion,
producing a large polarization of the hyperfine levels. This polarization is preserved as the excitation cascades via crystal field
split levels to the ground state. A rf magnetic field induces
transitions between the hf states of the longest lived intermediate
level (the lowest $^1G_4$, lifetime < 100 microsec.) so that on
returning to the ground state, the ion is in a different hyperfine
state than when it was excited. This state is inaccessible to the

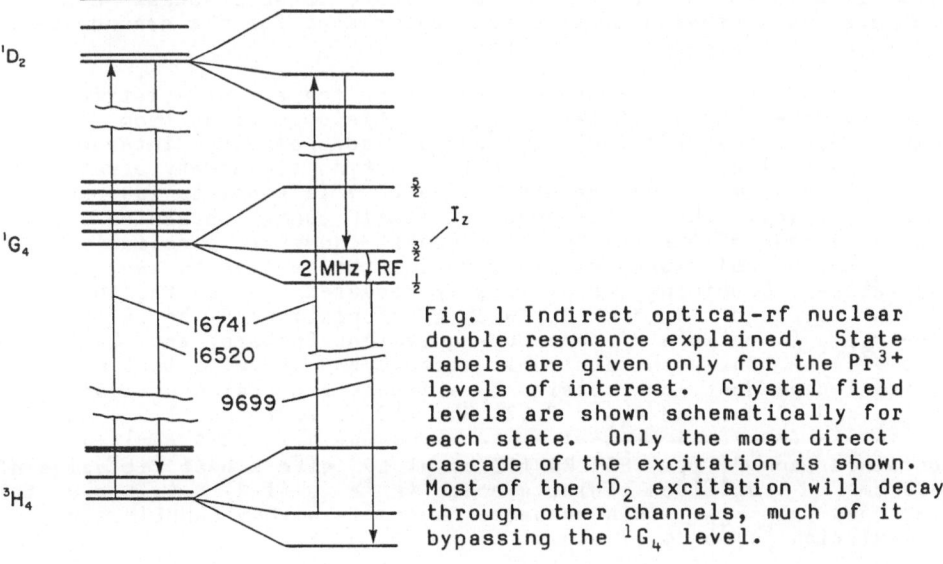

Fig. 1 Indirect optical-rf nuclear
double resonance explained. State
labels are given only for the Pr^{3+}
levels of interest. Crystal field
levels are shown schematically for
each state. Only the most direct
cascade of the excitation is shown.
Most of the $^1D_2$ excitation will decay
through other channels, much of it
bypassing the $^1G_4$ level.

287

pump. As a result the optical absorption at the pump frequency decreases at a rf resonance. The absorption is measured by moni- toring the fluorescence from the lowest $^1D_2$ level to the ground $^3H_4$ multiplet. For the $I_z = 3/2$ to $I_z = 5/2$ transition, the absorption line is shown in Fig. 2. In the absence of an external static magnetic field, the resonance frequencies are 2000 kHz and 4000 kHz with linewidths 95 kHz and 48 kHz respectively. In a 40 G. static magnetic field along the c-axis of the crystal, the 4000 kHz line splits into two components, separated by 210 kHz each 34 kHz wide.

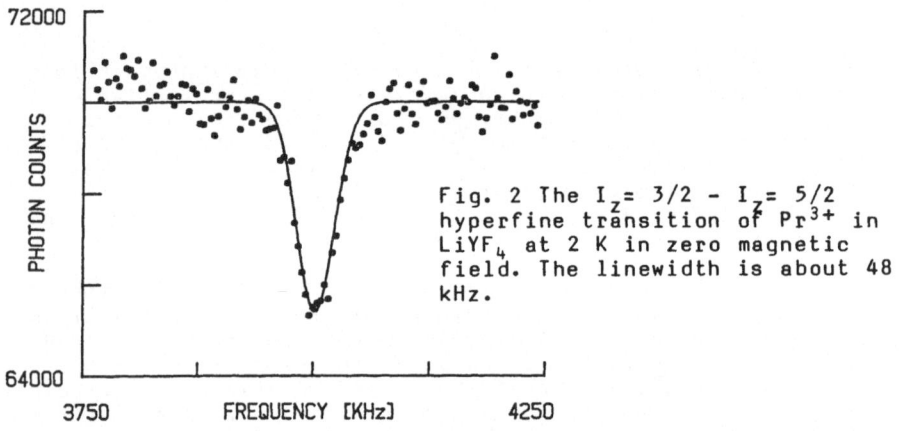

Fig. 2 The $I_z = 3/2 - I_z = 5/2$ hyperfine transition of Pr^{3+} in LiYF$_4$ at 2 K in zero magnetic field. The linewidth is about 48 kHz.

The hyperfine splittings are described by an axial Hamiltonian for I = 5/2. The magnetic field splitting factor ($\gamma_z = 2.64$ kHz/G.) yields the hyperfine constant $A_J = 726$ MHz. A_J for this level is calculated to be 912 MHz using expressions from WYBOURNE [2]. In the calculations, mixing of levels of J≠4 are ignored, as are relativistic corrections. This difference is substantial when compared to a previously reported measurement for the ground state, where exact agreement is obtained.

The quadrupole parameter D consists of the sum of a lattice electric field gradient term, P_{latt}; a field gradient from the 4f electrons of the Pr^{3+} ion, P_{4f}; and a quadrupole-like interaction from the nuclear magnetic dipole interacting via nearby crystal field levels, D_a. The measured value of D is equal to 1000 kHz for the $^1D_2$ level. This measurement by itself cannot determine the contributions of the various components. However, a reliable value of D_a may be calculated easily, reducing the problem to two variables. Combining the previously measured D's [1] for the lowest levels of $^3H_4$ and $^1D_2$, with a calculation of the ratio of the P_{4f}'s (= ratio's of operator-equivalent factors, and $\langle 0| J_z^2 - J(J+1))|0\rangle$ for each electronic state.) and assuming that P_{latt} is state-independent, one obtains by a linear regression analysis $P_{latt} = 0.60$ Mhz and $P_{4f} = 0.028$ Mhz ($^1G_4$), 2.027 MHz ($^1D_2$) and 0.814 Mhz ($^3H_4$). A crystal field parameter $B_{20} = 335$ cm^{-1} is obtained from P_{latt}. ESTEROWITZ et al [3] give a best fit value of 488 cm^{-1}. A shielded radial average $\langle r^{-3}\rangle_{4f}$ (1-R) = 4.78 a.u. is obtained from P_{4f}. This compares favorably with STERNHEIMER's calculation [4] of 4.67 a.u.

References

1. K.K. Sharma and L.E. Erickson: J. Phys. C: Solid State Phys., 14, 1329 (1981).
2. B.G. Wybourne: Spectroscopic Properties of Rare Earths, Interscience (1965), p 115.
3. L. Esterowitz: F.J. Bartoli, R.E. Allen, D.E. Wortman, C.A. Morrison and R.P. Leavitt, Phys. Rev. B19, 6442 (1979).
4. R.M. Sterheimer: Phys. Rev. 146, 140 (1966).

Coherent Ringing in Superfluorescence

D.J. Heinzen, J.E. Thomas, and M.S. Feld

George R. Harrison Spectroscopy Laboratory and Department of Physics, Massachusetts Institute of Technology, Cambridge, MA 02139, USA

Superfluorescence (SF), the cooperative radiation damping of an inverted assembly of two-level atoms, is by now fairly well understood on the basis of both semiclassical and quantized field treatments [1]. However, an important remaining question has been under what conditions, if any, SF can exhibit ringing, and what the nature of this ringing is. In the initial SF experiment, in HF gas [2], ringing was observed and attributed to coherent Rabi-type oscillations. However, in a subsequent series of experiments, in atomic Cs [3], a regime of "single-pulse" SF was identified in which ringing never occurred. We have performed new SF experiments to study this question [4].

Ringing is expected on the basis of uniform-plane-wave (UPW) solutions to the coupled Maxwell-Schroedinger equations. These results show that SF ringing is a coherent Rabi-type oscillation induced by the strong field propagating along the high gain axis of the sample. Two factors may reduce ringing. First, radial ("transverse") variation of the inversion density may lead to SF pulse profiles with ringing periods which vary radially; a smooth total SF pulse may result from spatial averaging of these profiles. Second, non-uniform quantum initiation could, conceivably, cause different transverse regions of the sample to superradiate independently; in this case plane-wave behavior and ringing would not exist.

To test this we have performed experiments examining the emission from small regions of the output face of an SF sample. A cell of length L = 4 cm containing atomic Rb is placed in a magnetic field of 5 kG, and SF is produced on the $6\,^2P_{3/2}$ ($M_J = -3/2$, $M_I = +3/2$) $\rightarrow$ $6\,^2S_{1/2}$ ($-1/2$, $+3/2$) transition of atomic ^{87}Rb at λ = 2.73 μm, inverted using radiation from an E/0 modulated cw laser tuned to the $5\,^2S_{1/2}$ ($-1/2,+3/2$) $\rightarrow$ $6\,^2P_{3/2}$ ($-3/2$, $+3/2$) transition at 420 nm. The dye laser radiation is focussed to a diameter (FW 1/e intensity) of 375 μm, and has a pulse duration of approximately 8 ns and intensity of 60 mW, producing a well-defined cylinder of inverted atoms of radius r_o = 245 μm and Fresnel number $\pi r_o^2 /\lambda L$ = 1.7.

By using single mode laser radiation to pump a non-degenerate system, pure two-level SF can be studied with precise, reproducible conditions. The narrow bandwidth of the pump pulse also inverts a sub-doppler velocity distribution, resulting in long dephasing times (T_2^* = 20 ns) without the use of an atomic beam.

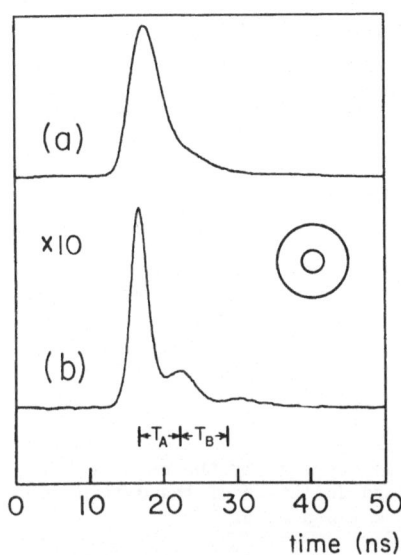

Fig. 1. Pinhole experiment results. (a) SF signal, no pinhole. (b) SF signal, 150 μm diameter pinhole. The pump pulse is 8 ns long and begins approximately at t = 5 ns. The inset circles, drawn to scale, represent the relative diameters of pinhole and sample. The total number of inverted atoms is 5.0 x 10⁸.

The SF emitted into the forward direction is collected by an f/10 lens system and imaged onto a plane P into which a pinhole may be inserted. The radiation is filtered by a Ge flat and imaged onto a fast InAs photodiode, with a response time of 1.6 ns.

As can be seen, (Fig. 1), without a pinhole no ringing is observed, and the pulse shapes appear similar to the Cs experiments [3]. But when the pinhole is inserted into the center of the beam the tail of the pulse is reduced dramatically and ringing is clearly observed. Also shown are the times T_A and T_B calculated from UPW theory for the separations between the first and second pairs of SF lobes, respectively, in excellent agreement with the data.

We conclude that coherent ringing is an intrinsic property of SF, and that its absence in experiments is primarily a spatial averaging effect. The Cs experiment [3] showed no ringing because the entire SF output was observed, whereas it is likely that the HF experiment [2] showed ringing because the detector viewed only a small area in the near field of the beam.

References

1. E.A. Watson, H.M. Gibbs, F. P. Mattar, M. Cormier, Y. Claude, S.L. McCall, and M.S. Feld: Phy. Rev. A 27, 1427 (1983), and references therein.
2. N. Skribanowitz, I.P. Herman, J.C. MacGillivray, and M.S. Feld: Phys. Rev. Lett. 30, 309 (1973).
3. H.M. Gibbs, Q.H.F. Vrehen, and H.M.J. Hikspoors: Phys. Rev. Lett. 39, 547 (1977).
4. D.J. Heinzen, J.E. Thomas, and M.S. Feld: Phys. Rev. Lett. 54, 677 (1985).

Self-Induced Transparency and Photon Echo in Dressed Atoms

Youhong Huang and Fucheng Lin

Shanghai Institute of Optics and Fine Mechanics, Academia Sinica, P.O. Box 8211, Shanghai, People's Republic of China

With the dressed atom approaches, the interaction of two-level atoms in an intense field E_L near resonance at ω_L with two pulses at the two Rabi sidebands $\omega_L \pm \omega_{12}$

$$E_i = E_{i0}(z,t)\cos(\omega_i t - k_i z) \ , \qquad i = 2,1$$

is studied. The physical model and symbols used are exactly the same as [1] and [2]. We assume that E_{10} and E_{20} are real, and depletion of the driving field is negligible. Three waves are in phase-matched case.

The Bloch equation describing the change of the dressed atom state is

$$d\vec{R}/dt = \vec{\Omega} \times \vec{R} \tag{1}$$

The vector $\vec{R} = (R_1, R_2, R_3)$ is defined as

$$R_1 = \sum_n \left(\sigma_{(1,n)(2,n)}e^{-i(\delta t+s)} + \sigma_{(2,n)(1,n)}e^{i(\delta t+s)} \right)$$

$$R_2 = i\sum_n \left(\sigma_{(1,n)(2,n)}e^{-i(\delta t+s)} - \sigma_{(2,n)(1,n)}e^{i(\delta t+s)} \right)$$

$$R_3 = \sum_n \left(\sigma_{(1,n)(1,n)} - \sigma_{(2,n)(2,n)} \right)$$

where $\sigma_{(i,n)(j,n)}$ is the element of the density matrix of the dressed atom and $s=(k_2-k_L)z=(k_L-k_1)z$. The vector $\vec{\Omega}$ is defined as

$$\vec{\Omega} = [-(\beta/\hbar)(E_{20}\cos^2\phi - E_{10}\sin^2\phi),0,\delta \ ]$$

The sign δ is the shift of the sidebands induced by inhomogeneous broadening described by $g(\delta)$.

The motion equations for E_{10} and E_{20} are

$$\left(\frac{\partial}{\partial z}+\frac{\eta}{c}\frac{\partial}{\partial t}\right)\begin{bmatrix}E_{10}(z,t)\\E_{20}(z,t)\end{bmatrix} = \begin{bmatrix}\omega_1\sin^2\phi\\\omega_2\cos^2\phi\end{bmatrix}\frac{c\mu_o N_o \beta}{2\eta}\int_{-\infty}^{\infty}R_2 g(\delta)d\delta \tag{2}$$

From (1) and (2), the area theorem for E_{10} and E_{20} is obtained as

$$dA_i(z)/dt = \alpha_i \sin[A(z)] \tag{3}$$

where

$$\begin{Bmatrix}\alpha_1\\\alpha_2\end{Bmatrix} = \begin{bmatrix}\omega_1\sin^2\phi\\\omega_2\cos^2\phi\end{bmatrix}\frac{\mu_o N_o c\pi g(0)\beta^2}{2\eta\hbar}\frac{(\sin^2\phi - \cos^2\phi)}{(\sin^4\phi + \cos^4\phi)}$$

The effective area $A(z)$ is defined as

$$A(z) = A_2(z)\cos^2\phi - A_1(z)\sin^2\phi$$

For the detuning of the driving field from two-level atom above zero, the coefficients α_1 and α_2 are below zero that represent the amplification of E_1 and the attenuation of E_2, the detuning below zero, the attenuation of E_1 and the amplification of E_2.

From (1) and (2), the E_{10} and E_{20} are deduced

$$\begin{Bmatrix} E_{10}(z,t) \\ E_{20}(z,t) \end{Bmatrix} = \begin{Bmatrix} \omega_1\sin^2\phi \\ \omega_2\cos^2\phi \end{Bmatrix} \frac{E_0(z,t)}{\omega_2\cos^4\phi - \omega_1\sin^4\phi}$$
$$+ \begin{bmatrix} \cos^2\phi \\ \sin^2\phi \end{bmatrix} \frac{\omega_2\cos^2\phi\,\mathcal{E}_{10}(t-\eta z/c) - \omega_1\sin^2\phi\,\mathcal{E}_{20}(t-\eta z/c)}{\omega_2\cos^4\phi - \omega_1\sin^4\phi} \qquad (4)$$

where E_0 respect to the effective area $A(z)$ is defined as

$$E_0 = E_{20}\cos^2\phi - E_{10}\sin^2\phi$$

which evolution on time is completely the same as that in normal SIT [2] and so all the process related is clear. The second term in (4) represents a special four-wave mixing process and will be discussed in more detail. Equation (4) shows that after propagating some distance each pulse will split into two stable modes: one at a speed much slower than the light speed and the other at the light speed. However, under two kinds of input condition

$$\omega_2\cos^2\phi\,\mathcal{E}_{10}(z,t) = \omega_1\sin^2\phi\,\mathcal{E}_{20}(z,t) \qquad \text{and}$$

$$\sin^2\phi\,\mathcal{E}_{10}(z,t) = \cos^2\phi\,\mathcal{E}_{20}(z,t)$$

there may be two kinds of single stable mode: no four-wave mixing process and only it respectively.

If the dressed atoms are excited first on $A(z) = \pi/2$ and then on π after time spacing T, the polarization will be

$$\begin{Bmatrix} P(\omega_2,t) \\ P(\omega_1,t) \end{Bmatrix} = \pm N_0\beta \begin{Bmatrix} \sin^2\phi\sin(\omega_2 t) \\ \cos^2\phi\sin(\omega_1 t) \end{Bmatrix} \exp[- \frac{\pi(t-t_3-t_2+t_1-\hbar/\beta\,|E_0|)}{4T_2^*}]$$

and the direction is along with the z, where

$$T_2^* \doteq \omega_{12}/g(0)$$

is the width of the echo. The exciting pulse may be both a single pulse at one sideband and simultaneous two pulses at two sidebands respectively. Neverthless, the echo pulse is always at two sidebands and there is beat phenomenon in it. In fact, the echo phenomenon means exactly the transient response of the resonant fluorescence. There is no echo in the central one because of no inhomogeneous broadening effect on the centre.

1. C. Cohen-Tannoudji, S. Reynaud.   J. Phys., B10, 345, 2311(1977)

2. L. Allen, J. H. Eberly.   "Optical Resonance and Two-level Atoms"
                              (by John Wiley and Sons Inc 1975)

Photon Echoes on Singlet-Triplet Transitions in Organic Mixed Crystals

G. Wäckerle
Max-Planck-Institut für Medizinische Forschung,
Abteilung Molekulare Physik, Jahnstrasse 29,
D-6900 Heidelberg, Fed. Rep. of Germany

K.P. Dinse
Institut für Physik, Universität Dortmund, Otto-Hahn-Strasse,
D-4600 Dortmund 50, Fed. Rep. of Germany

Since the first observation of two-pulse photon echoes by KURNIT, ABELLA, and HARTMANN in 1964 [1] , a great variety of echo phenomena have been predicted and observed. The application of this type of coherent spectroscopy for the investigation of optical centers in solid matrices was proven to be very attractive, owing to predominant inhomogeneous line broadening which is negated by the echo process.

Mainly by two-pulse (HAHN-) echoes it was shown in mixed organic crystals that contributions from temperature-activated optical dephasing processes can be reduced to values well below the radiative lifetime limits of appr. 1o MHz for an allowed optical transition, if the sample is cooled to about 1.2 K [2]. Further experiments, utilizing parity-forbidden transitions of rare-earth optical centers in anorganic crystals, showed that nuclear spin-dependent interactions with the matrix constitute the low temperature dephasing limit leading to homogeneous optical transitions in the kHz range for selected systems [3,4].

Spin-forbidden transitions in organic molecules with their transition moment in the $1o^{-4} - 1o^{-5}$ Debye range are also characterized by a negligible lifetime width of < 1 kHz. Recently we have shown that low-temperature homogeneous optical linewidths in the 1oo kHz range can be observed, again being controlled by magnetic dipole-dipole interaction of guest and host nuclear spins [5].

Using a 3-pulse (stimulated echo) sequence instead of the familiar HAHN-echo sequence, hyperfine structure in the two electronic states, being connected by the coherent optical field, can be investigated. This possibility results from coherent superpositions of nuclear sublevels, which are created by the optical excitation pulses. In complete analogy to multiple pulse NMR, optical coherence is transferred to the nuclear spin sublevels. This spin coherence manifests itself in a modulated echo amplitude decay, which after Fourier transformation reveals the hyperfine splittings. As the nuclear spin coherence decays with its particular decay time T_{2n}, being controlled by the inhomogeneous width of the NMR transition, hyperfine splittings can be extracted from the optical echo experiment with a resolution, which is no longer controlled by the optical dephasing time, but only by the decay time of the stimulated echo. This decay constant is not even limited by the excited state lifetime [6]. However, spectroscopic information about the excited state is clearly lifetime-controlled.

In the case of di-bromo-benzophenone-d_{1o} (DBBPh) doped into di-bromo-di-phenyl-ether-d_{1o} (DDE) we observed an optical dephasing time

T_2^{opt} = 4 μs, whereas we found a stimulated echo decay time $T_{stim}^{opt} \approx 30$ μs at 1.2 K in zero external magnetic field. Owing to this order-of-magnitude difference, we observed hyperfine transitions with a linewidth much less than the homogeneous width of the optical transition. The figure shows the modulated stimulated echo decay together with its Fourier transform. The transition at 137 kHz can be attributed to a pure quadrupole transition between the zero-field levels of the deuteron nuclear spin interacting with the molecular electric field gradient. For the I = 1 spin transitions at $\nu_+ = e^2qQ(3+\eta)/4h$; $\nu_- = e^2qQ(3-\eta)/4h$; $\nu_0 = e^2qQ\eta/2h$ are expected. Values for deuteron e^2qQ/h in planar aromatic hydrocarbons are found in the narrow range $e^2qQ/h = (180+4)$kHz with $\eta < 0.1$ [7], leading to $(\nu_+ + \nu_-)/2 = (135\pm3)$kHz, in good agreement with the measured value (137 ± 5)kHz.

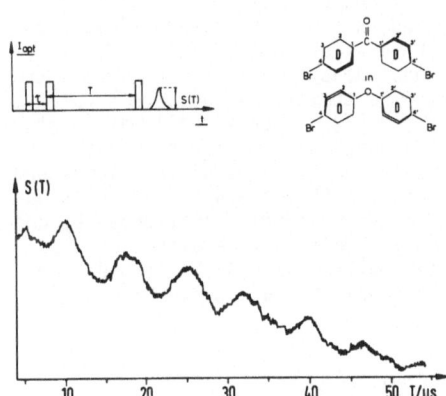

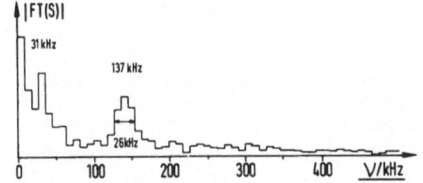

Fig. 1

Modulated stimulated echo decay of perdeuterated di-bromo-benzophenone at 1.2 K using a separation of the first two pulses of 3.8 μs.

References:
1. N.A. Kurnitt, I.D. Abella, and S.R. Hartmann, Phys.Rev.Lett. 13, 567 (1964)
2. W.H. Hesselink and D.A. Wiersma, J.Chem.Phys. 73, 648 (1980)
3. R.M. Macfarlane and R.M. Shelby, Opt.Commun. 39, 169 (1981)
4. S.C. Rand, A. Wokaun, R.G. DeVoe, and R.G. Brewer, Phys.Rev.Lett. 43, 1868 (1979)
5. G. Wäckerle, H. Zimmermann, and K.P. Dinse, Chem.Phys.Lett. 110, 107 (1984)
6. J.B.W. Morsink and D.A. Wiersma, Chem.Phys.Lett. 65, 105 (1979)
7. C. Müller, W. Schajor, H. Zimmermann, and U. Haeberlen, J.Mag.Res. 56, 235 (1984)

Colliding Without Relaxing: The Suppression of Collisional Dephasing with Strong Optical Fields

A.G. Yodh, J. Golub, and T.W. Mossberg

Department of Physics, Harvard University, Cambridge, MA 02138, USA

Recently, considerable interest has arisen in the effect of relaxation on systems which are being strongly driven by a resonant electromagnetic field [1-6]. While this problem was originally encountered in studies of nuclear magnetic resonance, recent interest was sparked by experiments conducted in solid-state materials [5,6] which indicated that strong driving fields can act to suppress optical relaxation. We have experimentally studied the relaxation of optical coherences (off-diagonal density matrix elements) introduced by gas-phase atomic collisions, and find that it can also be suppressed (at least in part) by a strong driving field [7]. Superficially at least, our results are surprising, because the "strong" driving field employed in our experiment was far too weak to influence the dynamics of the picosecond duration collisions and the phase-randomization they produce. It turns out that our driving field suppresses the relaxation normally introduced by collisionally-induced Doppler shifts (velocity changes) whose effect accrues over the relatively long intervals between collisions. As described elsewhere [8], velocity changes occur in distant, weakly-phase-perturbing collisions.

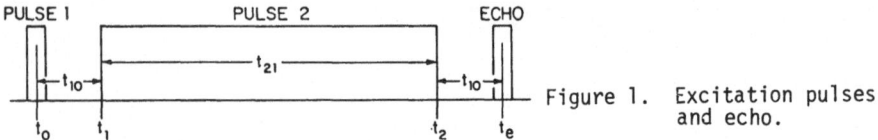

Figure 1. Excitation pulses and echo.

Our modified photon-echo type experiment employed two excitation pulses (see Fig. 1) tuned to resonance with the 555.6 nm $(6s^2)$ $^1S_0$ - $(6s6p)$ $^3P_1$ transition of atomic Yb in a vapor cell. When pulse 2 is sufficiently intense, this excitation scheme produces an echo at the time shown with a duration roughly equal to that of the first pulse. For fixed excitation pulse times, the echo intensity, I_e, decays exponentially with argon perturber gas pressure P, i.e. $I_e = I_0 \exp(-\beta P)$. As shown in Fig. 2 (solid circles), we have measured β as a function of t_{21} while keeping the total interval between pulse 1 and the echo fixed at $t_{eo} = 1200$ nsec. We find that β decreases as pulse 2 fills more and more of the interval between pulse 1 and the echo.

We are interested in the effect of the driving field on the relaxation of the $^1S_0$ - $^3P_1$ optical coherence, but, as recently pointed out [9], the echo information is stored in the level populations as well as the coherence during pulse 2. As a result atoms experiencing strong collisions (and hence large random optical phase perturbations) may in principle still contribute to the echo signal through population-mediated information. It turns out, however, that the sign of this population contribution oscillates with the area of pulse 2 [9]. As a result, in our experiment, where the area of pulse

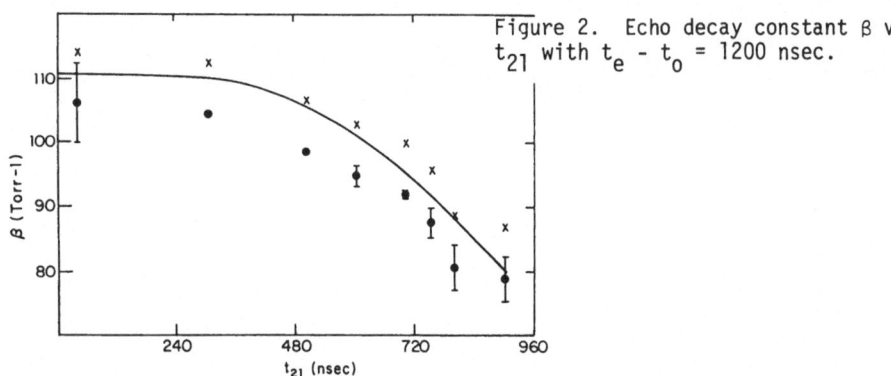

Figure 2. Echo decay constant β vs t_{21} with $t_e - t_o$ = 1200 nsec.

2 varies by ≈10π over the laser beam profile, the net population contribution from atoms having experienced large collisional phase-changes is expected to be very small, and the echo relaxation should very closely reflect the relaxation of the optical coherence.

The solid line in Fig. 2 was computed using known collisional decay parameters for this system [10], and assuming that the driving field completely suspends intercollisional velocity-change mediated phase relaxation, but leaves intracollisional optical phase randomization unaffected. The excellent agreement between our corrected data (X) and calculation support these assumptions.

Our results, characteristic of a completely different relaxation process than that found in solids, should provide an interesting test for theoretical treatments of relaxation in the presence of a strong driving field.

Financial support for this work was provided by the U.S. Joint Services Electronics Program and the National Science Foundation. It was performed at the MIT Laser Research Center. AGY acknowledges a U.S. Army Fellowship, and we thank P. Berman for discussions.

References
1. E. Hanamura, J. Phys. Soc. Jap. 52, 2258, 2267, 3265, and 3678 (1983)
2. M. Yamanoi and J. H. Eberly, J. Opt. Soc. Am. B 1, 751 (1984)
3. P. R. Berman and R. G. Brewer, Phys. Rev. A (in press as of 6/3/85)
4. A. Schenzle, M. Mitsunaga, R. G. DeVoe, and R. G. Brewer, Phys. Rev. A
        30, 325 (1984)
5. R. G. DeVoe and R. G. Brewer, Phys. Rev. Lett. 50, 1269 (1983)
6. T. Endo, T. Muromoto, and T. Hashi, Opt. Commun. 51, 163 (1984)
7. A. G. Yodh, J. Golub, N. W. Carlson, and T. W. Mossberg, Phys. Rev. Lett.
        53, 659 (1984)
8. P. R. Berman, T. W. Mossberg, and S. R. Hartmann, Phys. Rev. A 25, 2550
        (1982)
9. A. V. Durrant and J. Manners, Optica Acta 31, 1167 (1984)
10. R. A. Forber, L. Spinelli, J. E. Thomas, and M. S. Feld, Phys. Rev. Lett.
        50, 331 (1983)

Part X

Surfaces and Clusters

Infrared Laser Photodesorption Spectroscopy of Adsorbed Molecules

T.J. Chuang and I. Hussla

IBM Research Laboratory, San Jose, CA 95193, USA

Vibrational lifetimes in both gaseous and condensed phases can be quite long and in some cases photochemical reactions can take place due to vibrational excitation by infrared radiation. On metal and semiconductor surfaces, vibrational decay rates are generally expected to be very fast because of the existence of efficient energy decay channels: electron-hole pairs, mechanical couplings and phonons. Even so, there is substantial experimental evidence to show that a number of interesting surface chemical processes can still be initiated by IR lasers [1]. One of such IR-stimulated processes involves the excitation of an internal vibrational mode of adsorbed molecules followed by the desorption of the species into the gas phase, *i.e.*, IR photodesorption. The phenomenon has been investigated experimentally in a variety of adsorbate-surface systems [1-3] and attracted considerable theoretical attention [4-7]. Pulsed CO_2 lasers were extensively used in these prior studies. In order to better understand the molecular desorption dynamics, we have extended our investigation to include relatively simple molecules such as NH_3 and ND_3 adsorbed on both metal and dielectric surfaces excited with a tunable IR laser.

The experimental apparatus has been described in detail elsewhere [8,9]. Briefly, a Q-switched Nd:YAG laser is frequency doubled to 532 nm and used to pump a dye laser generating light pulses in 740-840 nm region. Tunable IR pulses in the 2.5-4.2 μm range are generated from the frequency difference between the 1.064 μm radiation and the dye laser in a $LiNbO_3$ crystal. The p-polarized IR beam (6 nsec pulse width, laser linewidth ≤ 1 cm^{-1}) is partially focused and incident at 60-75° from the surface normal covering a surface area of about 5 mm^2. The laser beam enters and leaves an ultrahigh vacuum chamber through sapphire windows. The vacuum chamber is equipped with an ESCA/Auger spectrometer, an ion gun and a quadrupole mass spectrometer for surface characterization and desorption studies. The sample can be cooled to 90K with liquid N_2 or to 10K with a He refrigerating unit. The mass spectrometer is operated in time-of-flight (TOF) mode and placed in a line-of-sight arrangement along the surface normal about 80 mm from the sample. The TOF signal is recorded and signal-averaged with a transient recorder.

The results of NH_3 and ND_3 photodesorption from Cu(100) excited by the tunable IR laser involving N-H stretching modes [8] and by a pulsed CO_2 involving a bending mode [10] have been reported recently. Here, we present mainly the experiments with NH_3, ND_3 and Xe adsorbed on Ag (film) surfaces. Neutral molecular desorption due to resonant absorption of IR photons by adsorbed NH_3 is observed at a substrate temperature of about $T_s=10K$. A typical TOF signal for NH_3 on the Ag surface with $\theta=2$ excited at $\nu=3390$ cm^{-1} and $I=3$ mJ/cm^2 is shown in Fig. 1. The translational temperature (T_d) determined from the maximum of TOF signal (t_m) is about 80K. Thermal contribution from the *direct* laser substrate heating is not negligible because laser-induced thermal desorption is detectable when the laser fluence is raised by a factor of 2.5 or higher. The desorption yield per pulse as a function of laser frequency under such excitation condition is shown in Fig. 2, clearly showing two characteristic IR

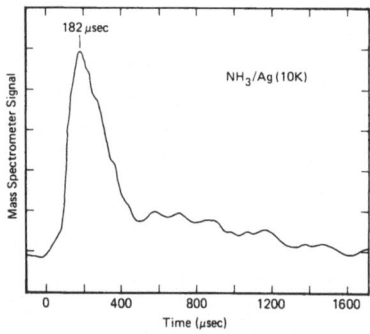

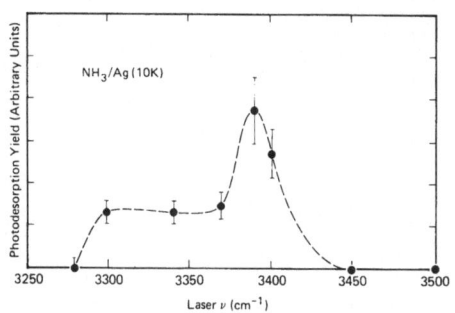

Fig. 1 (left). TOF signal of NH_3 photodesorbed from an Ag film at 10K with $\theta=2$, $\nu=3390$ cm^{-1} and I=3 mJ/cm^2

Fig. 2 (right). Photodesorption yields of NH_3 adsorbed on an Ag film at 10K as a function of laser frequency: $\theta=2$, I=3 mJ/cm^2

absorption-desorption bands. The lower frequency component centered around 3320 cm^{-1} is due to chemisorbed species and is most likely associated with the symmetric N-H stretching mode (ν_s). The high frequency band peaked at 3390 cm^{-1} is due to physisorbed molecules adsorbed on top of the first (chemisorbed) monolayer. This band is most likely associated with the antisymmetric stretching vibration (ν_a). It should be noted that at a monolayer coverage, NH_3 is likely adsorbed with the N atom on top of the Ag surface so that ν_s is along the surface normal and is IR active. On the other hand, the dipole moment associated with ν_a should be perpendicular to the molecular symmetric axis and thus parallel to the surface. Image effects can render this ν_a mode more or less IR inactive. At a multilayer coverage ($\theta>1$), molecular orientation in the physisorbed layer can be randomized and the surface image effects diminish making ν_a mode excitable by the incident IR beam. This interpretation is consistent with the experimental observation that the 3320 cm^{-1} band is more pronounced at $\theta=1$, whereas the 3390 cm^{-1} band is dominant at a multilayer coverage. The spectral assignment is also consistent with the vibrational spectrum of NH_3 adsorbed on Ag(110) obtained by GLAND *et al.* [11] using high-resolution electron energy loss spectroscopy (EELS). The photodesorption yield as a function of laser fluence at a multilayer coverage excited at $\nu=3400$ cm^{-1} is shown in Fig. 3. The desorption quantum yields (*i.e.*, number of molecules desorbed per absorbed IR photon) at both $\theta=1$ and $\theta>1$ are estimated to be less than 5×10^{-4}.

To study the isotope effect, we prepare a NH_3 and ND_3 (1:1) mixture and dose the gas onto the Ag film at 10K with a surface coverage of about 3-4 monolayers. The relative NH_3 and ND_3 surface concentrations are determined directly by laser-induced thermal desorption of the mixture with high intensity (I≥20 mJ/cm^2) IR laser pulses at $\nu=3500$ cm^{-1} not in resonance with any vibrational bands of the adsorbed molecules. When the adsorbed NH_3 is excited at $\nu=3400$ cm^{-1} and I=6 mJ/cm^2, desorption of both NH_3 and ND_3 molecules are detected with almost equal signals by the mass spectrometer. Within the relatively large experimental uncertainty ($\pm25\%$), no significant isotope selectivity in the photodesorption is observed. In a separate experiment, we adsorb 2 monolayers of NH_3 on the Ag film at 12K and then further deposit about 2 atomic layers of Xe on top of the molecular layer. The adsorbate is then excited by IR pulses at $\nu=3400$ cm^{-1} or 3380 cm^{-1} and I=6 mJ/cm^2. Xe desorption is clearly detected. A typical TOF signal is shown in Fig. 4. From the observed t_m, the translational temperature of the desorbed Xe is estimated to be about 70K±25K. From conventional thermal desorption spectrum, it is shown that physisorbed Xe indeed can desorb at 70K.

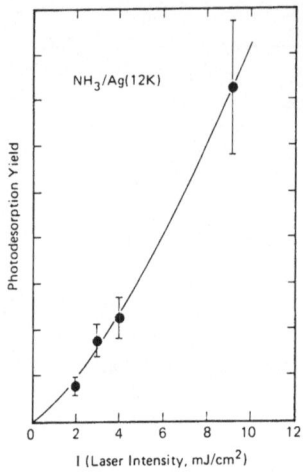

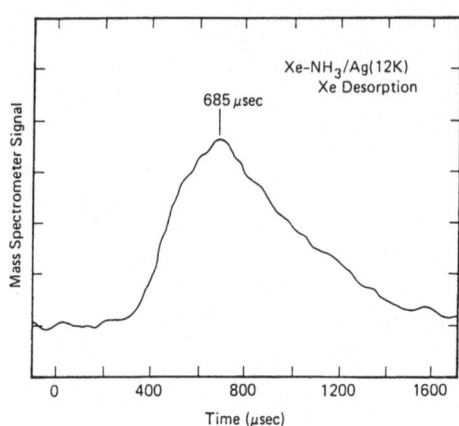

Fig. 3 (left). Photodesorption yields of NH_3/Ag film at 12K as a function of laser intensity: $\theta=4$ and $\nu=3400$ cm^{-1}

Fig. 4 (right). TOF signal of Xe desorbed from Xe ($\theta=2$) condensed on a NH_3 layer ($\theta=2$) adsorbed on an Ag film at 12K excited with an IR laser at $\nu=3400$ cm^{-1} and $I=6$ mJ/cm^2

Under the laser excitation at $\nu=3400$ cm^{-1}, a very small amount of NH_3 desorption is also detected by the mass spectrometer. T_d for the observed NH_3 is roughly 75K±25K, practically the same as that of Xe. The desorption for NH_3 is, however, much less than that without the Xe overlayer. When the laser is tuned off resonance from NH_3 vibrational bands, neither Xe or NH_3 desorption is detected at this laser fluence.

From these and prior experimental studies [3,8-10], we can summarize the major aspects of the IR photodesorption phenomenon as follows: (1) Desorption of molecules due to resonant absorption of IR photons by the adsorbate ("resonant photodesorption") can take place from both dielectric and metal surfaces for a surface coverage as low as a monolayer. Indeed, IR-induced photodesorption (IRPD) can be considered as absorption-desorption spectroscopy quite similar to surface IR absorption-reflection spectroscopy (IRARS). This is due to the fact that there is essentially no mode selectivity in IRPD and the desorption yield depends mainly on the IR absorption cross-section associated with the vibrational mode. Thus, the vibrational structure can be resolved from the photodesorption spectrum. Because of the inherently high spectral resolution (≤ 1 cm^{-1}) provided by the laser light source, IRPD can be developed as a spectroscopic technique for probing weakly-bound surface species with a resolution not attainable by conventional IRARS or EELS. It should be pointed out, however, IRPD can be due to either single-photon or multi-photon absorption. Namely, the photodesorption yield is not necessarily linearly dependent on the incident laser intensity. For molecular systems such as physisorbed NH_3 at more than one monolayer coverages, photodesorption can be induced by single-photon absorption and the desorption spectrum is very similar to the linear IR absorption spectrum. But for a multi-photon excited photodesorption, the desorption spectrum can be substantially narrower than the absorption spectrum as illustrated theoretically by WU *et al.* [7]. (2) IRPD can also provide important insight into molecular interaction dynamics on surfaces. Our studies show that resonantly excited desorption is observable only in a relatively narrow range of laser intensities and the effect of *direct* surface heating due to absorption of laser photons by the substrate is always important. This is true even for highly IR transparent substrates and for highly IR reflective metals. From the results involving NH_3-ND_3 co-adsorbates, it is clear that once the photon energy is absorbed by NH_3 molecules, it is quickly shared with neighboring molecules, even among different isotope species. This rapid energy relaxation could be accomplished *via* ultrafast intermolecular energy transfer process. The NH_3-Xe results, however, indicate that efficient energy transfer does not require strong dipole-dipole coupling. The transfer of energy from molecules to atoms can also be very effective. It is very likely that the rapid energy decay results in the localized heating of the absorbate layer and the substrate surfaces. This *indirect* [1,3] (or "resonant" [5]) heating in combination with the *direct* substrate heating can lead to desorption of some very weakly bound molecules during the laser pulse width. It is concluded that thermal processes are very important in IRPD although these processes alone cannot account for all the major observations of the desorption phenomenon, in particular, the small desorption yields. We also conclude that pure quantum processes such as the elastic and inelastic tunneling processes as discussed by GORTEL *et al.* [5], leading directly to molecular dissociation from the solid surface cannot be by themselves very important. Instead, we suggest that once the photon energy is absorbed by an internal molecular vibrational mode, it is rapidly channeled *via* ultrafast intramolecular relaxation into the lower frequency modes in the molecule-surface potential. This relaxation can be mediated by electron and/or phonon interactions resulting in thermal excitation of the surface potential. This thermal activation can enhance the desorption probability when the molecule is also internally excited.

We wish to acknowledge that this work is supported in part by San Francisco Laser Center, a National Science Foundation Regional Instrumentation Facility, NSF Grant No. CHE 79-16250, awarded to University of California at Berkeley in collaboration with Stanford University.

[1] See the review by T. J. Chuang: Surf. Sci. Reports 3, 1 (1983).
[2] J. Heidberg, H. Stein, E. Riehl and A. Nestman: Z. Physik. Chem. (NF) 121, 145 (1980); J. Heidberg, H. Stein and E. Riehl: Phys. Rev. Lett. 49, 666 (1982).
[3] T. J. Chuang and H. Seki: Phys. Rev. Lett. 49, 382 (1982); H. Seki and T. J. Chuang: Solid State Comm. 44, 473 (1982); T. J. Chuang: J. Electr. Spectr. Relat. Phenom. 29, 125 (1983).
[4] J. Lin and T. F. George: Surf. Sci. 100, 381 (1980); *ibid.* 115, 569 (1982).
[5] Z. W. Gortel, H. J. Kreuzer, P. Piercy and R. Teshima: Phys. Rev. B27, 5066 (1983); *ibid.* B28, 2119 (1983).
[6] F. G. Celii, M. P. Casassa and K. C. Janda: Surf. Sci. 141, 169 (1984).
[7] G. S. Wu, B. Fain, A. R. Ziv and S. H. Lin: Surf. Sci. 147, 537 (1984).
[8] T. J. Chuang and I. Hussla: Phys. Rev. Lett. 52, 2045 (1984); also, in Dynamics on Surfaces, ed. by B. Pullman, J. Jortner, A. Nitzan and B. Gerber (Reidl, Dortrecht, Holland, 1984), p. 313.
[9] T. J. Chuang, H. Seki and I. Hussla: Surf. Sci. 158, 525 (1985).
[10] I. Hussla and T. J. Chuang: Ber. Bunsenges. Phys. Chem. 89, 294 (1985).
[11] J. L. Gland, B. A. Sexton and G. E. Mitchell: Surf. Sci. 115, 623 (1982).

Hydrogen Desorption from Polycrystalline Palladium

H. Zacharias
Fakultät für Physik, Universität Bielefeld,
D-4800 Bielefeld 1, Fed. Rep. of Germany

R. David
Institut für Grenzflächenphysik und Vakuumforschung,
Kernforschungsanlage, D-5170 Jülich, Fed. Rep. of Germany

The investigation of the internal energy distribution of molecules after the interaction with a surface by tunable laser spectroscopy has attracted much interest recently. While several research groups have studied the scattering of molecules from surfaces [1, 2], only a few experiments have been carried out on the desorption of molecules. A major difficulty arises from the fact that the measurement time during molecular desorption from a previously cold surface is relatively short, lasting only minutes. A complete rotational population distribution can thus only be obtained from several successive adsorption - desorption cycles. In this way Cavanagh and King investigated the desorption of NO from Ru(001) [3]. This experimental constraint has been removed recently by studying the desorption of hydrogen which can permeate through metal, allowing thus a continuous desorption experiment at controlled surface temperature [4, 5]. In addition to these experimental advantages detailed information is available about the recombinative desorption of hydrogen from metal surfaces. Angular distributions sharply peaked in the forward direction, proportional to $\cos^n\theta$ with $3 < n < 10$ have been observed [6, 7]. Also fast velocity distributions with mean kinetic energy up to four times the surface thermal energy have been measured [7, 8]. These deviations from Knudsens cosine law and from the Maxwellian kinetic energy distribution suggest dynamical constraints to this process, which should also be reflected in the population distribution of the molecular quantum states. In this contribution we report results on the recombinative desorption of H_2[5] and D_2 molecules from a polycrystalline palladium surface.

The experimental system has been described in detail previously [5]. It consisted of a double wall chamber with the inner wall cooled to liquid nitrogen temperature. In this inner chamber silver sheets partly covered with charcoal pellets were cooled by a closed-cycle refrigerator pump to about 20 K providing thus a hydrocarbon free vacuum. Hydrogen atoms were supplied to the surface by bulk permeation through the 1 mm thick palladium polycrystal at hydrogen pressures between 100 and 2000 mbar and temperatures between 500 and 1050 K.

Rovibrational state distribution of the desorbing H_2 and D_2 molecules were determined by VUV laser induced fluorescence. State selectivity was achieved by single-photon excitation in the B $^1\Sigma_u^+ \leftarrow$ X $^1\Sigma_g^+$ Lyman bands around $\lambda \sim 106$ nm (H_2) and $\lambda \sim 110$ nm (H_2 and D_2). This VUV radiation was generated by frequency tripling near UV radiation focused (f = 70 mm) into a cell containing, respectively, Xe(p $\sim$ 2-30 mbar) and Kr (p $\sim$ 250-400 mbar). Since the tuning range in the latter case is relatively large ($\lambda \sim 110$ to 116 nm), we used this range to probe rotational lines of deuterium in the (0-0), (1-0), (2-0), (3-1), and (4-1) Lyman bands. The VUV radiation was recollimated by a LiF lens and intersected the desorption flux at a distance of about 20 mm from the Pd surface. The VUV intensity was about 2×10^9 photons per pulse with a pulse duration of 5 ns. Hydrogen fluorescence was detected

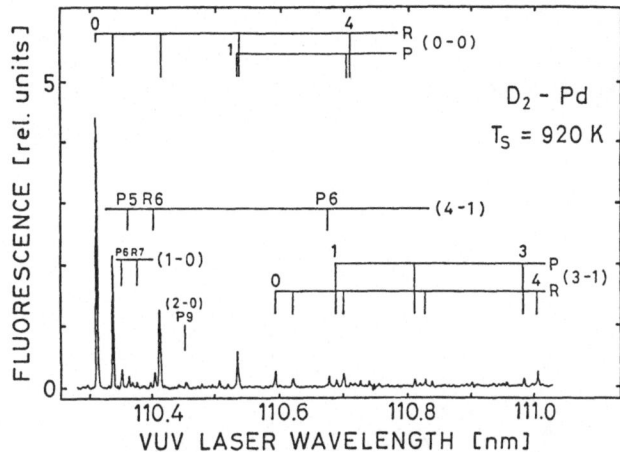

Fig. 1: VUV laser-induced fluorescence excitation spectrum of D_2 desorbing from a polycrystalline Pd surface

by a solar blind photomultiplier placed about 50 mm from the interaction region.

In Fig. 1 the D_2 fluorescence intensity is shown as the wavelength is changed from $\lambda = 110.2$ to $\lambda = 111.0$ nm. The temperature of the Pd surface was $T_S = 945$ K and the deuterium supply pressure behind the Pd disk $P_{D_2} = 2000$ mbar. The spectrum is normalized to the VUV laser intensity. A line identification is given atop the spectrum. The spectrum shown includes also contributions from background D_2 molecules which reappeared in the interaction region after colliding with the cold walls of the chamber. Their contribution to the line heights, mainly to $J'' = 0$ and 1, was determined separately by permitting D_2 molecules to enter the inner chamber through a different gas inlet and was taken into account in the further analysis of the data.

Fig. 2 presents the rotational population distribution of D_2 molecules in $v''=0$ and $v''=1$ after desorption from Pd at $T_S = 920$ K. It is plotted

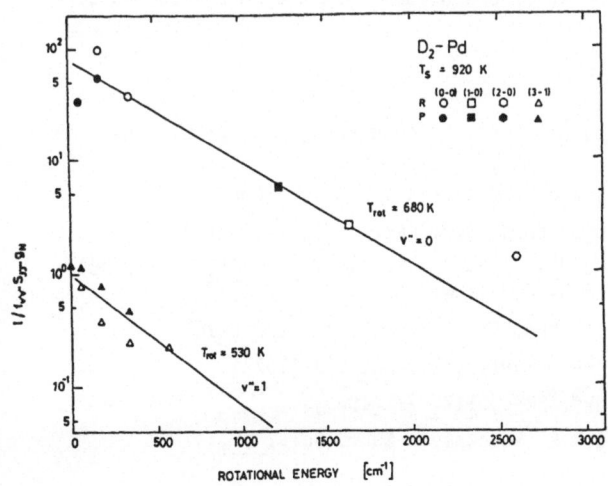

Fig. 2:
Rotational state population of the desorbing D_2 molecules in $v''=0$ and $v''=1$ versus the rotational energy

$\ln[I/(f_{v'v''} \, g_N \, S_{j'j''})]$ versus the rotational energy E_J where g_N denotes the nuclear spin degeneracy, $S_{J'J''}$ the Hönl-London factor, $f_{v'v''}$ the oscillator strength of different vibrational branches and I the normalized line intensity. A slight deviation is observed from an equilibrium Boltzmann distribution which would appear as straight line in this plot.

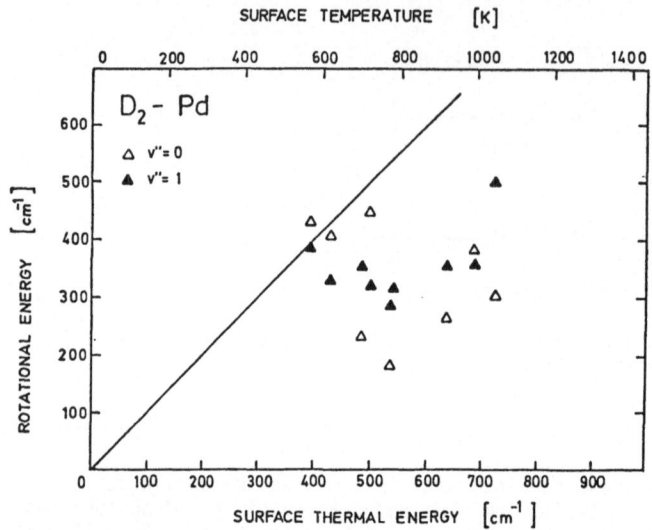

Fig. 3: The average rotational energy in v"=0 and v"=1 is plotted versus the surface thermal energy. The straight line indicates the rotational energy expected for full accommodation

In Fig. 3 average rotational energies $<E_{rot}> = \sum_{J''} N(J'') \cdot E_{rot}(J'')$ for D_2 molecules recombinatively desorbing from the Pd surface are plotted versus the surface thermal energy T_S. It can be seen that the rotational energy is less than the surface thermal energy in the temperature range measured. As can already be noticed in Figures 1 and 2 we observed vibrationally excited D_2 molecules leaving the surface. A preliminary analysis shows that the deduced vibrational temperatures are higher than the corresponding surface temperatures.

References

1 J.A. Barker and D.J. Auerbach: Surf. Science Rep., 4, 1 (1985) and references therein

2 M.M.T. Loy, J.A. Misewich, and H. Zacharias: this volume

3 R.R. Cavanagh and D.S. King: Phys. Rev. Lett., 47, 1829 (1981)

4 G.D. Kubiak, G.O. Sitz, and R.N. Zare: J. Chem. Phys., 81, 6397 (1984); J. Vacuum Sci. Technol., to be published

5 H. Zacharias and R. David: Chem. Phys. Lett., 115, 205 (1985)

6 M. Balooch and R.E. Stickney: Surf. Sci., 38, 313 (1973)

7 G. Comsa and R. David: Surf. Sci., 117, 77 (1982)

8 A.E. Dabiri, T.J. Lee, and R.E. Stickney: Surf. Sci., 26,

Studies of Molecular Monolayers at Air-Liquid Interfaces by Second Harmonic Generation: Question of Orientational Phase Transition

Th. Rasing and Y.R. Shen

Department of Physics, University of California, Berkeley, CA 94720, USA and Center for Advance Materials, Lawrence Berkeley Laboratory, Berkeley, CA 94720, USA

M.W. Kim, S. Grubb, and J. Bock
Exxon Research and Engineering Co., Annandale, NJ 08801, USA

Insoluble molecular monolayers at gas-liquid interfaces provide an insight to the understanding of surfactants, wetting, microemulsions and membrane structures,and offer a possibility to study the rich world of 2-dimensional phase-transitions. In the interpretation of the observed properties of these systems various assumptions about the molecular orientation are often made, but so far few clear experimental data exist [1]. In this paper we will show how optical second harmonic generation (SHG) can be used to measure the molecular orientation of monolayers of surfactant molecules at water-air interfaces. By simultaneously measuring the surface pressure versus surface molecular area we can show for the first time that the observed liquid condensed-liquid expanded transition is an orientational phase-transition.

The SHG radiation from a medium arises from the induced second-order polarization

$$P(2\omega) = \chi^{(2)}(2\omega):E(\omega)E(\omega) \tag{1}$$

in the medium. When the latter has inversion symmetry $\chi^{(2)}$ vanishes in the electric-dipole approximation. This makes SHG an effective surface probe at any interface between two centrosymmetric media, because there the inversion symmetry is necessarily broken. Apart from the intrinsic high spectral and time resolution, optical SHG is a unique surface probe because of its versatility: it can be used at solid-vacuum, solid-solid, solid-air, solid-liquid and liquid-air interfaces, as has been shown recently [2-6].

The surface nonlinear susceptibility $\chi_s^{(2)}$ arising from a monolayer of adsorbates can be written as

$$\chi_s^{(2)} = N_s\langle\alpha^{(2)}\rangle \tag{2}$$

where N_s is the surface density of the molecules and $\langle\alpha^{(2)}\rangle$ is the nonlinear polarizability averaged over the molecular orientational distribution. If $\alpha^{(2)}$ is dominated by a single component $\alpha_{\xi\xi\xi}$ along the molecular axis ξ and the latter is randomly distributed in the azimuthal plane, the nonvanishing components of $\chi_s^{(2)}$ can be written as:[3]

$$\chi_{s,\perp\perp\perp}^{(2)} = N_s\langle\cos^3\theta\rangle\alpha_{\xi\xi\xi}$$

$$\chi_{s,\|\perp\|} = \chi_{s,\perp\|\|} = 1/2(N_s)\langle\sin^2\theta\,\cos\theta\rangle\alpha_{\xi\xi\xi} \tag{3}$$

where θ is the polar angle between the molecular axis and the surface normal and the subindices $\perp$ and $\|$ refer to directions perpendicular and parallel to the surface, respectively. From (3) it follows that a measurement of the ratio of any two linear combinations of $\chi_{s,\perp\perp\perp}$ and $\chi_{s,\|\perp\|}$ can yield a weighted average of θ.

The monolayers were prepared by spreading a solution on a thoroughly cleaned water surface. The trough was made out of glass with the edges coated with paraffin. The density of molecules was controlled by a teflon barrier and the surface tension was measured by a Wilhelmy plate [7]. For the SHG measurements we used the frequency-doubled output of a Q-switched Nd^{3+}:YAG laser at 532 nm with a 7 nsec pulsewidth as the pump beam.

Due to higher-order contributions, there was a non-negligible signal arising from the bare water proportional to $|\chi_w^{(2)}|^2$, while the signal from the surfactant covered surface was proportional to $|\chi_s^{(2)} + \chi_w^{(2)}|^2$. Both were measured separately, so that $\chi_s^{(2)}$ could be deduced.

Using a pump energy of ~ 50 mJ/cm^2/pulse, we found a SHG signal of 2-3 photon/pulse for the bare water and typically 0.1-0.5 photon/pulse for the adsorbates.

As a first example, we have applied this technique to a monolayer of sodium-dodecylnaphthalene-sulfonate (SDNS) floating on a water surface [6]. Figure 1 shows the measured surface pressure π as a function of the surface area per molecule (A). The π-A diagram does not exhibit any discontinuous phase-transition (the rapid increase in π close to 50 Å^2 indicates the formation of a full close-packed monolayer). Figure 2 gives the result of θ as a function of π, showing a smoothly decreasing θ with a limiting inclination angle of ~ 30° at a full monolayer. The nonlinear polarizability of SDNS is dominated by the naphthalene part which is tilted at ~ 30° from the alkyl chain. The final value of $\theta \approx 30°$ then supports the commonly accepted, but never verified, picture that compressing molecules on a liquid surface would force them to stand up.

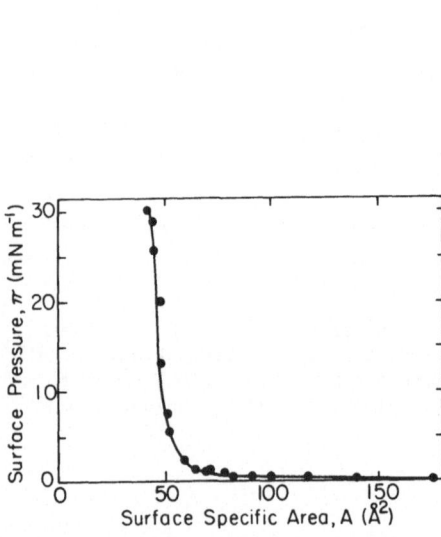

Fig. 1. Surface pressure π of SDNS as a function of the area per molecule A on a water surface

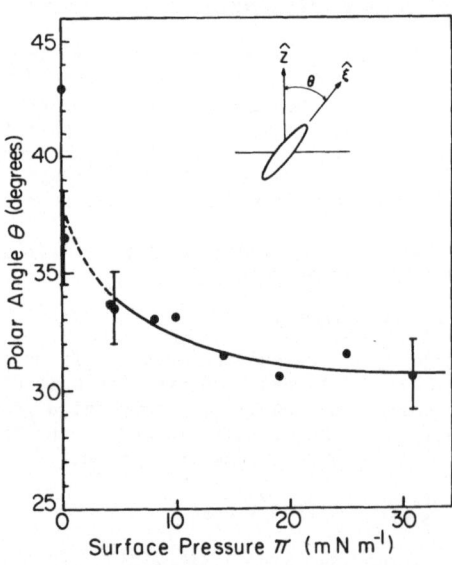

Fig. 2. Tilt angle θ between the molecular axis and the surface normal as a function of the surface pressure π for SDNS on water

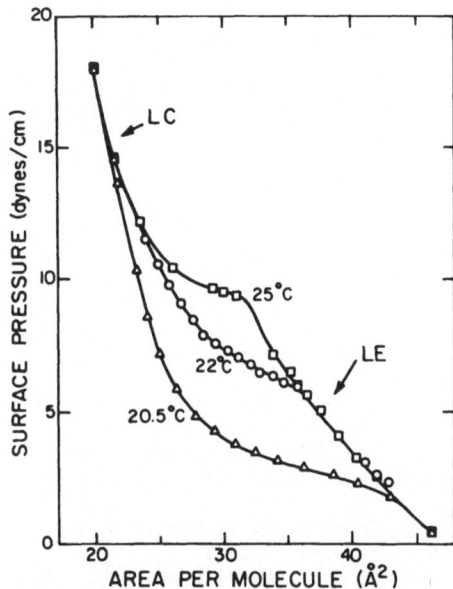

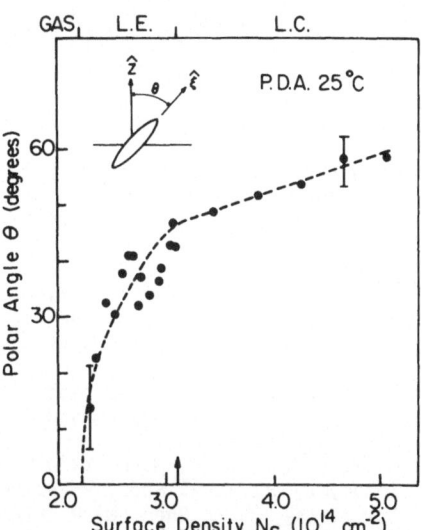

Fig. 3. π–A diagram for PDA at various temperatures. The sharp kink in each curve signals the onset of the LE-LC transition

Fig. 4. Tilt angle θ between the molecular axis and the surface normal as a function of the surface density for PDA on water at 25°C

As a second example, we have studied pentadecanoic acid (PDA) on water. Figure 3 shows the π–A diagram of PDA at various temperatures. The sharp kinks in the middle of the π–A diagrams indicate the transition between the so-called liquid-expanded (LE) and liquid-condensed (LC) phases, which appears to be strongly temperature-dependent. Though known for a long time, the nature of this liquid-liquid transition is still controversial and little understood [1].

It is suspected that the transition is governed by the molecular orientation. Figure 4 shows the results of our orientational measurements for PDA at 25°C. In the LE phase, θ rapidly increases with increasing density until the LE-LC transition is reached, whereafter it changes more slowly and linearly with N_s.

In order to relate θ to the molecular orientation we must determine the dominant SHG contributor on the molecule. We found that the nonlinear polarizability of PDA is dominated by the C-OH bond. Then Fig. 4 shows the C-OH orientation as a function of the molecular density. Physically, we expect that this polar bond would like to stick normally into the water, and hence the molecules would tilt away from the surface normal. Indeed, when we extrapolate the experimental results to lower densities, we find θ approaching 0° at $N_s = 2.2 \times 10^{14}$ cm^{-2} = gas-liquid (LE) transition point. In the LE phase, with increasing N_s, the steric interaction of the hydrocarbon chains of neighboring molecules tends to align the molecules towards, and consequently forces the C-OH orientation away from, the surface normal. At $N_s = 3.1 \times 10^{14}$ cm^{-2} a phase-transition to an oriented liquid occurs.

By measuring the orientation just below the LE-LC transition (in the LE phase) we found θ = 45° ± 3° for all temperatures, though the transition point itself is very temperature-dependent (see Fig. 3). This supports the observation that the LE-LC transition is indeed an orientational phase-transition.

In conclusion, we have shown how optical SHG can be used as a very effective and versatile surface probe. Using this technique we have been able for the first time to follow the molecular orientation of monolayers of molecules on a water-air interface as a function of their surface density, and have shown that the observed LE-LC transition is an orientational phase-transition.

This work was supported by the Director, Office of Energy Research, Office of Basic Energy Sciences, Materials Sciences Division of the U.S. Department of Energy under Contract No. DE-AC03-76SF00098.

References
1. For a review, see, e.g., G. M. Bell, L. L. Coobs, and L. J. Dunne, Chem. Rev. 81, 15 (1981), and references therein.
2. C. K. Chen, T. F. Heinz, D. Ricard, and Y. R. Shen, Phys. Rev. Lett. 46, 15 (1981).
3. T. F. Heinz, H. W. K. Tom, and Y. R. Shen, Phys. Rev. A 28, 1883 (1983).
4. H. W. K. Tom, T. F. Heinz, and Y. R. Shen, Phys. Rev. Lett. 51, 1983 (1983).
5. H. W. K. Tom, C. M. Mate, X. D. Zhu, J. E. Crowell, T. F. Heinz, G. A. Somorjai, and Y. R. Shen, Phys. Rev. Lett. 52, 348 (1984).
6. Th. Rasing, Y. R. Shen, M. W. Kim, P. Valint, Jr., and J. Bock, Phys. Rev. A 31, 537 (1985).
7. G. F. Graines, Jr., Insoluble Monolayers at Liquid-Gas Interfaces (Wiley, New York, 1966).

Second-Harmonic Generation:
A Probe of Symmetry and Order in Crystalline Surfaces

T.F. Heinz, M.M.T. Loy, and W.A. Thompson

IBM Thomas J. Watson Research Center, Yorktown Heights, NY 10598, USA

Introduction

Since optical second-harmonic generation (SHG) is (electric-dipole) forbidden within a centrosymmetric medium, the process exhibits a high degree of sensitivity to the character of the medium's surface, the region where the inversion symmetry of the bulk is broken. This intrinsic surface sensitivity, combined with the high spectral and temporal resolution and wide range of applicability afforded by optical techniques, makes SHG an attractive tool for the study of surfaces and interfaces. To date, the SHG technique has been exploited primarily in investigations of adsorbate-covered surfaces and interfaces [1]. In this paper, we describe the extension of the SHG technique to studies of clean, ordered surfaces, showing that the polarization dependences of the SH signal reflect the symmetry and ordering of the surface atomic structure [2,3]. Results are presented here for Si(111)-2x1 and 7x7 reconstructed surfaces and for Si(111)-7x7 surfaces modified by the room-temperature deposition of Si atoms [4]. Note that all of the materials under study consist of pure Si, both at the surface and in the bulk. We can, nonetheless, readily examine their surface properties because of the intrinsic surface sensitivity provided by the SHG process.

Symmetry Analysis

In this section we describe the process of SHG in centrosymmetric media in general terms and outline the method for predicting SH polarization dependences based on symmetry considerations. Formulas are given for the case of pump excitation at normal incidence to the surface and explicit relations are derived for (111) faces of cubic materials.

We start with the following expression for the nonlinear source polarization of a centrosymmetric medium with a surface defined by $z = 0$:

$$\vec{P}^{NLS}(2\omega) = \vec{P}_s^{NLS}(2\omega)\delta(z) + \vec{P}_b^{NLS}(2\omega), \tag{1a}$$

with

$$\vec{P}_s^{NLS}(2\omega) = \chi_s^{(2)}:\vec{E}(\omega)\vec{E}(\omega) \tag{1b}$$

$$\vec{P}_b^{NLS}(2\omega) = \chi_b^{(2)}:\vec{E}(\omega)\nabla\vec{E}(\omega). \tag{1c}$$

Here $\vec{E}(\omega)$ is the pump electric field oscillating at frequency ω. The bulk nonlinear polarization, $\vec{P}_b^{NLS}$, characterized by the fourth-rank tensor $\chi_b^{(2)}$, arises from magnetic-

311

dipole and electric-quadrupole (non-local) contributions. This bulk polarization is, typically, several orders of magnitude weaker than the corresponding dipole-allowed polarization expected in a non-centrosymmetric medium. In a centrosymmetric medium, however, the bulk polarization may still be significant because stronger dipole-allowed terms are only present in a much smaller volume in the surface layer. We represent the surface contribution phenomenologically by a sheet of polarization. This idealization is justified since the surface layer in which the bulk description of the medium breaks down is only on the order of Angstroms in thickness, much less than the wavelength of light [5]. Consequently, as far as macroscopic electrodynamics is concerned, we can treat the induced nonlinear currents as a dipole layer. This macroscopic model does not necessarily imply that the surface nonlinear response should be considered as arising solely from dipolar terms in the usual calculation in second-order perturbation theory: higher-order (non-local) contributions may also be important because of the rapid change in the normal component of the electric field occurring at the surface [6].

For our present purposes, the key feature of (1) is that the nonlinear response is described by the third- and fourth-rank tensors $\chi_s^{(2)}$ and $\chi_b^{(2)}$ reflecting the symmetry of the surface and bulk regions, respectively. The general solution for the radiated SH fields can be calculated for the given source polarization from Maxwell's Equations. We shall restrict ourselves here to the special case of exciting the medium with normally incident pump radiation. We then find that the intensity of the reflected SH wave has a polarization dependence given by

$$I(2\omega) \propto | \ \hat{e}(2\omega) \cdot \chi_{s,\,eff}^{(2)} : \hat{e}(\omega)\hat{e}(\omega) \ |^2,$$ (2a)

where

$$[\chi_{s,\,eff}^{(2)}]_{ijk} = [\chi_s^{(2)}]_{ijk} - \frac{1}{2}[1 + n^{-1}(\omega)n(2\omega)]^{-1}[\chi_b^{(2)}]_{ijzk}.$$ (2b)

In this formula, $\hat{e}(\omega)$ and $\hat{e}(2\omega)$ denote the polarization vectors of the pump and SH radiation, respectively, and n represents the refractive index of the nonlinear medium (assumed to be independent of polarization) for a wave propagating along the z-axis.

An interesting consequence of the form of (2) is that no SHG is possible in the absence of ordering either in the surface or the bulk. This property makes normal incidence excitation particularly well suited for studies of ordering and symmetry. It should also be noted that no SHG from the surface or bulk can occur when the input and output polarization vectors both lie perpendicular to a mirror plane in the corresponding region of the material.

Let us now apply this analysis to the (111) face of Si. The bulk contribution to $\chi_{s,\,eff}^{(2)}$, governed by the m3m symmetry of the Si crystal, will exhibit 3m symmetry. An ideal, unreconstructed Si(111) surface would also display 3m symmetry with one of the three mirror planes lying perpendicular to the [01$\bar{1}$] direction. The two independent components of the SH reflection then have intensities

$$I_x(2\omega) = A \ | \ [\chi_{2,\,eff}^{(2)}]_{xxx} \ |^2\cos^2 2\theta$$ (3a)

$$I_y(2\omega) = A \ | \ [\chi_{2,\,eff}^{(2)}]_{xxx} \ |^2\sin^2 2\theta,$$ (3b)

where x and y correspond, respectively, to the [2$\bar{1}\bar{1}$] and [01$\bar{1}$] directions in the surface plane; pump polarization angle θ is measured with respect to the x-axis; and A is a constant determined by the frequency, the linear dielectric constants, and the intensity of the pump laser.

When surface reconstruction occurs, the resulting surface may exhibit a lower symmetry than the ideal surface. This change in symmetry will in general modify the SH polarization dependences. For example, if the three-fold symmetry is broken and only a single plane of mirror symmetry remains, the relations analogous to (3) are

$$I_x(2\omega) = A \mid [\chi^{(2)}_{s,\,eff}]_{xxx}\cos^2\theta + [\chi^{(2)}_{s,\,eff}]_{xyy}\sin^2\theta \mid^2 \tag{4a}$$

$$I_y(2\omega) = A \mid [\chi^{(2)}_{s,\,eff}]_{yxy} \mid^2 \sin^2 2\theta. \tag{4b}$$

Here the mirror plane is assumed to lie perpendicular to the $[01\bar{1}]$ direction. The lower symmetry of the surface in this case gives rise to additional independent tensor elements of the nonlinear susceptibility and, consequently, more complex SH polarization dependences.

Experimental

The Si(111)-2x1 and 7x7 surfaces studied in this work were prepared and maintained under ultrahigh vacuum conditions [2,3]. Deposition of Si atoms was accomplished using a resistively heated evaporative source. The flux of atoms on the sample surface was calibrated with a quartz microbalance. The pump radiation for the SH measurements was provided by a Q-switched Nd:YAlG laser operating at $1.06\mu m$ with 8-ns pulses and was directed onto the sample at normal incidence. Typical pulse energies were $\sim$10 mJ, which, when focused to a $\sim$1-mm spot, gave rise to $> 10^3$ SH photons in the reflected beam.

Results and Discussion

We have measured the intensity of the SH radiation polarized along the $[2\bar{1}\bar{1}]$ and $[01\bar{1}]$ directions as a function of the pump polarization for both the Si(111)-2x1 and 7x7 reconstructed surfaces. We consider first the case of the Si(111)-7x7 surface. Figure 1 displays the SH intensity for two fixed output polarizations as a function of the pump polarization. The dotted curves are experimental values and the solid lines represent theoretical predictions for 3m symmetry given by (3). The good agreement with theory is obtained simply by matching the overall scale, the shape of the curves being completely specified.

While the analysis of the data in Fig. 1 indicates 3m symmetry, this result is perhaps not intuitively obvious from the form of the curves. The symmetry properties of the sample are more readily apparent if we transform into a reference frame in which the output polarization tracks the pump polarization. For normally incident excitation, this is equivalent to collecting data with a rotating sample and fixed input and output polarizations. Such transformed data [7], presented in Fig. 2, display clearly the expected three-fold symmetry [8]. The three distinct planes of mirror symmetry can be identified, as discussed above, from the null points in the SH response for parallel polarizations.

The SH polarization dependences for the Si(111)-2x1 surface are shown in Fig. 3. These data, corresponding to those of Fig. 1 for the Si(111)-7x7 surface, clearly reflect the lower symmetry of 2x1 surface. Comparing Figs. 1 and 3, we see immediately the importance of the surface contribution to the SH signal, as only this contribution can differ in the two cases. In fact, by oxidizing the clean Si(111) surfaces, we have determined that the surface con-

313

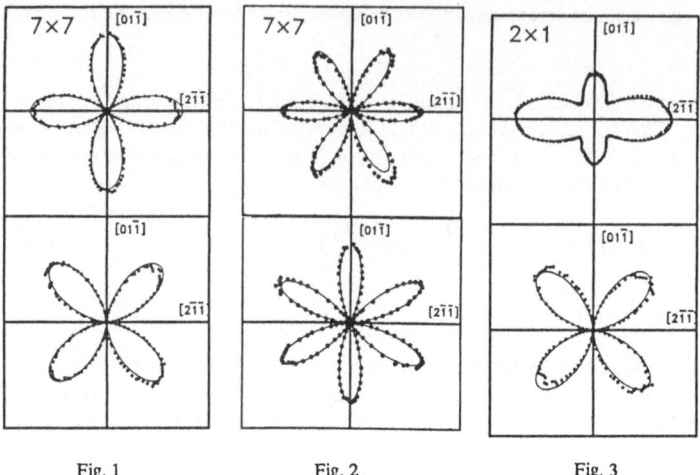

Fig. 1   Fig. 2   Fig. 3

Figs. 1-3. SH intensity for Si(111) surfaces as a function of pump polarization. Figs. 1 and 2 refer to the 7x7 reconstructed surfaces; Fig. 3 to the 2x1 reconstruction. The upper and lower panels in Figs. 1 and 3 display the SH signal polarized along the [211] and [011] directions, respectively. In Fig. 2, the upper and lower panels correspond to the SH signal polarized parallel and perpendicular to the pump polarization, respectively. In all figures, the solid lines are theory.

tribution exceeds that of the bulk by a factor of > 100 [3]. We believe that the large relative contribution of the surface nonlinearity compared with the bulk is associated with strong, nearly resonant transitions between surface states occurring at the fundamental and SH frequencies [9].

The data of Fig. 3 can be reproduced well by the symmetry analysis for a surface with a single mirror plane perpendicular to the [01$\bar{1}$] direction [Eqn. (4)]. From the fit to experiment, we can deduce the magnitudes of the three independent tensor elements of $\chi^{(2)}_{s, \text{eff}} \simeq \chi^{(2)}_{s}$ for the Si(111)-2x1 surface. They are found to be roughly comparable to one another and to the non-vanishing elements of $\chi^{(2)}_{s, \text{eff}}$ for the Si(111)-7x7 surface.

We have examined the presence of the planes of mirror symmetry in the Si(111)-2x1 and 7x7 surfaces in a direct manner by searching for the null in the SH radiation predicted for input and output polarizations perpendicular to any mirror plane. In this manner, we have deduced an upper bound of $< 3\text{x}10^{-3}$ for the symmetry-forbidden signals relative to symmetry-allowed ones. This result is of special interest for the Si(111)-2x1 surface. For the 2x1 surface, the atomic structure is now generally believed to be that of a π-bonded chain [10]. This structure, in its idealized form, is consistent with the observed m symmetry. There remains, however, the question of whether the chains of surface atoms dimerize, forming alternately longer and shorter bonds, as occurs for the organic π-bonded chains in polyacetylene. Such a dimerization would break the mirror symmetry perpendicular to the direction of the chains ([01$\bar{1}$]). Our measurements suggest that any such dimerization must be very weak, since the surface electronic structure, as reflected in the SHG measurement, still responds to high accuracy as if a mirror plane were present [11].

314

Finally, we discuss a real-time measurement of the change in the surface structure during the deposition of Si atoms on the Si(111)-7x7 surface. Similar real-time measurements of surface structure have been performed for the thermally driven Si(111)-2x1 → Si(111)-7x7 surface phase transformation [2] and for surface disordering of the Si(111)-7x7 surface induced by ion bombardment [3].

Our SH data were collected in this case with parallel input and output polarizations aligned along the [2$\overline{1}\overline{1}$] axis. Since this direction does not lie perpendicular to a mirror plane for the Si(111)-7x7 surface, a strong SH signal is observed. Upon deposition of Si atoms, the SH intensity (Fig. 4) falls off sharply and reaches a low value after a film of roughly monolayer thickness has been formed. The decay in SHG corresponds to the disordered, non-epitaxial nature of the adlayer created by deposition on a Si surface held at room temperature.

We can model the SH response by assuming that each incoming Si atom striking a fresh piece of the sample causes the surface nonlinear susceptibility to drop to a lower value over an area A. Taking the deposited atoms to occupy random sites, we then expect $\chi_S^{(2)}$ to decay exponentially as $\exp(-N_S A)$, where N_S denotes the fluence of deposited Si atoms. A small residual surface nonlinearity remains, presumably associated with sub-surface ordering. As shown in Fig. 4, the experimental data can be reproduced quite accurately by this simple model. The average effective area disordered by an incoming Si atom is found to be $A \sim 15 \text{Å}^2$. This value is comparable to the area per surface atom of the ideal Si(111) surface and is consistent with the notion that deposited Si atoms do not strongly perturb the existing Si(111)-7x7 surface structure, but rather cover it by an amorphous overlayer [12]. Further studies of annealing of disordered Si adlayers and of direct epitaxial growth are currently underway.

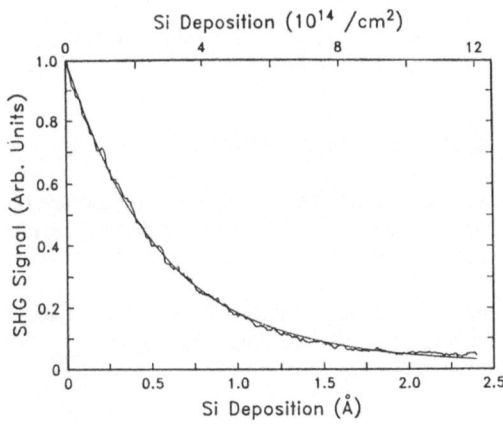

Fig. 4. SH intensity during Si deposition on a Si(111)-7x7 surface at 300K. The heavy line is theory.

Conclusion

The surface-specific process of SHG has been demonstrated to be sensitive to the symmetry and order of the surface layer of a crystal. The symmetries of the Si(111)-2x1 and 7x7 surfaces were determined from SH polarization dependences and surface mirror planes were identified with particular accuracy by a nulling scheme. The method holds promise for a variety of time-resolved measurements of surface structural change, as was illustrated here by monitoring the disordering of the Si(111)-7x7 surface upon deposition of atomic Si at room temperature.

This work was supported in part by the U. S. Office of Naval Research.

[1]   See, for example, H.W.K. Tom et al., Phys. Rev. Lett. 52, 348 (1984).

[2]   T.F. Heinz, M.M.T. Loy, and W.A. Thompson, Phys. Rev. Lett. 54, 63 (1985).

[3]   T.F. Heinz, M.M.T. Loy, and W.A. Thompson, to be published in J. Vac. Sci. Techn. B.

[4]   For SHG measurements of oxidized Si samples, see H.W.K. Tom, T.F. Heinz, and Y.R. Shen, Phys. Rev. Lett. 51, 1983 (1983).

[5]   The bulk treatment of the medium cannot be applied where either the material properties are changing or strong electric-field gradients are present.

[6]   Earlier work on SHG from centrosymmetric media generally treated the surface response as a quadrupolar effect. See N. Bloembergen, R.K. Chang, S.S. Jha, and C.H. Lee, Phys. Rev. 174, 813 (1968).

[7]   The procedure for transforming the data is described in [2].

[8]   The surface symmetry cannot be six-fold, since under this higher symmetry no surface SHG is allowed for normally incident excitation.

[9]   For Si(111)-2x1 surfaces, see, for example, P. Chiaradia, A. Cricenti, S. Selci, and G. Chiarotti, Phys. Rev. Lett. 52, 1145 (1984); M.A. Olmstead and N.M. Amer, Phys. Rev. Lett. 52, 1148 (1984).

[10]   K.C. Pandey, Phys. Rev. Lett. 47, 1913 (1981); 49, 223 (1982).

[11]   Rigorously speaking, we can only state that the averaged character of the surface on the macroscopic scale probed by optical fields dispalys mirror symmetry. In the presence of microscopic domains, the local symmetry of each domain might be lower.

[12]   This behavior is suggested by ion scattering studies for Ge on Si(111)-7x7 surfaces. See H.-J. Gossmann, L.C. Feldman, and W.M. Gibson, Phys. Rev. Lett. 53, 294 (1984).

Advances in Laser Spectroscopy
of Refractory Cluster Beams

R.E. Smalley

Rice Quantum Institute and Department of Chemistry, Rice University,
Houston, TX 77251, USA

Developments in the techniques of generating supersonic cluster beams of
refractory elements have recently begun to open an entirely new approach to
the study of surfaces [1]. By using laser vaporization of appropriate
targets in a supersonic pulsed nozzle, it is now possible to routinely
generate cold beams of virtually any element in the periodic table --
including such high boiling elements as tungsten and silicon [2]. Small
clusters of these elements are expected to be tightly bound, and their
surfaces are expected to exhibit many of the interesting properties of the
bulk surface,while still preserving the virtue of being microscopic,
molecular species small enough for detailed treatment at a high level of
theory.

Three key developments in this new field have resulted from our recent
work at Rice University. The first may be of considerable interest even to
researchers with little interest in metal clusters -- it is the development
of a general beam source for cold cluster ions [3]; a technique which
should work quite well even for more ordinary molecules. It involves the
use of an ArF excimer laser directed into the throat of a pulsed supersonic
nozzle operating with a helium carrier gas at very high densities.
Photoionization of the metal clusters by this excimer laser as they exit
the nozzle produces a dense cloud of ions and electons which are cooled
together as they pass down the supersonic free jet expansion. In effect,
the high density of helium carrier gas prevents the loss of the
photoelectrons, and produces instead a cold plasma. Since this plasma is
neutral overall, the cluster ions are cooled in collisions with the
expanding helium without being effected by space charge. Futhermore, the
cold plasma density is sufficiently high that the Debye screening length is
extremely short (less than 0.01 cm). As a result of this short screening
length, the cold plasma beam is easily able to travel long distances in the
apparatus without being affected by small stray electric or magnetic
fields.

This cold cluster ion source also is capable of producing rather
intense beams of cold negative cluster ions as well. In this case the
excimer laser is intentionally directed to strike the (aluminum) metal
surface at the end of the supersonic nozzle. Photoelectrons from this
surface are then entrained in the expanding helium, and are found to
efficiently attach to metal clusters possessing more than a few atoms.
Detailed spectral studies are currently in progress using
photodissociation to monitor the absorption event in mass-selected cold
cluster ions pulse-extracted from these cold plasma beams. Initial
spectral results indicate the ions have been cooled to below a rotational
temperature of 15K.

A second key development is the successful marriage of supersonic
cluster beam technology to the technique of Fourier transform ion cyclotron

resonance (FT-ICR) [4]. Here the key problem to overcome was the necessity of injecting the cold, mass-selected cluster ions through the fringing field of a 6.0 Tesla superconducting magnet, such that they could be efficiently trapped and analyzed at high mass resolution in the rectangular Penning-like cell of a FT-ICR spectrometer. As we have illustrated with cluster ions of iron and niobium, this problem has now been solved by a electrostatic lens/deceleration system that directs the ions smoothly down the axis of the magnetic field, such that the ion trajectories are never far from parallel to the lines of force of the magnet. For these metal clusters an overall trapping efficiency of %40 was obtained, with a FT-ICR detected mass resolution of better than 20000.

A third area of development in this new field of refractory cluster beam spectroscopy is the generation of clusters of semiconducting elements such as silicon and germanium, as well as III-V semiconductors such as Ga_xAs_y [5]. Using a newly-developed rotating disc source, it is now routinely possible to prepare clusters by laser-vaporization of any material obtainable in disc form. Previous versions of these laser-vaporization nozzles required the target material be in the form of a rod which was rotated and translated during the vaporization process. The new disc source is vital to such materials as III-V semiconductors where thin wafers are effectively the only available form. A survey of the photoionization, fragmentation, and radiationless transition aspects of small semiconductor clusters produced by these sources indicates they are, in fact, drastically different in almost all respects from that of equivalent metal clusters. For example, semiconductor clusters are found to fragment by a fission mechanism [6], whereas metal clusters generally decay by the loss of single atoms [7]. Furthermore, unlike metal clusters in which radiationless decay of excited electronic states is generally found to be extremely rapid, semiconductor clusters -- particularly those of silicon and germanium -- display efficient resonant two-photon ionization spectra with rather long-lived excited states [6].

Acknowledgement

The research outlined here involved the work of a number of excellent students, postdoctoral associates, and colleagues as documented in the cited references. Research on bare metal clusters in this laboratory is funded by the U.S. Department of Energy, Division of Chemical Sciences, while study of non-metal adducts on the surfaces of these clusters is funded by the National Science Foundation. Our research into the properties of semiconductor clusters is funded by the U.S. Army Research Office. Support from the Robert A. Welch Foundation, the Exxon Education Foundation, and the Petroleum Research Fund is also gratefully acknowledged.

References

1. R. E. Smalley: in Comparison of ab initio Theory with Experiment
   R. J. Bartlett, ed. (D. Reidel, Holland) 1985.

2. a. T. G. Dietz, M. A. Duncan, D. E. Powers, and R. E. Smalley:
      J. Chem. Phys. 74 6511 (1981).
   b. D. E. Powers, S. G. Hansen, M. E. Geusic, A. C. Puiu, J. B. Hopkins,
      T. G. Dietz, M. A. Duncan, P. R. R. Langridge-Smith, and
      R. E. Smalley: J. Phys. Chem. 86, 2556 (1982).

c. J. B. Hopkins, P. R. R. Langridge-Smith, M. D. Morse, and
R. E. Smalley: J. Chem. Phys. $\underline{78}$ 1627 (1983).

3. L-. S. Zheng, P. J. Brucat, C. L. Pettiette, S. Yang, and R. E. Smalley:
J. Chem. Phys. (submitted).

4. J. M. Alford, P. E. Williams, D. J. Trevor, and R. E. Smalley:
Int. J. Ion. Spectrom. and Ion Phys. (in press).

5. S. C. O'Brien, Yuan Liu, Qing Ling Zhang, J. R. Heath, R. F. Curl,
R. E. Smalley, and F. K. Tittle: J. Phys. Chem. (submitted).

6. J. R. Heath, Yuan Liu, S. C. O'Brien, Qing Ling Zhang, R. F. Curl,
R. E. Smalley, and F. K. Tittle: J. Chem. Phys. (submitted).

7. P. J. Brucat, L-. S. Zheng, C. L. Pettiette, S. Yang, C. Karner, and
R. E. Smalley: J. Chem. Phys. (submitted).

Negative and Positive Clusters of Semiconductor Ions

L.A. Bloomfield, M.E. Geusic, R.R. Freeman, and W.L. Brown

AT & T Bell Laboratories, Murray Hill, NJ 07974, USA

We have developed an apparatus that produces and mass resolves either positive or negative clusters of semiconductor ions. This apparatus does not make use of a secondary ionizing agent (such as an electron beam or a laser) to ionize neutral clusters, but produces the clusters directly in the laser-vaporization source[1]. The apparatus is shown in Fig.1: the ions from the source enter the pulsed plate region where they are accelerated to a constant potential and mass dispersed along the time-of-flight region. When the pulsed mass selector is activated, it allows only one cluster size to enter the second set of plates. There the chosen cluster is slowed down, and exposed to a high intensity laser. Either photofragmentation spectroscopy (positive ions) or photodetachment spectroscopy (negative ions) is performed in this region, and the products are mass-analyzed in the second time-of-flight portion of the apparatus.

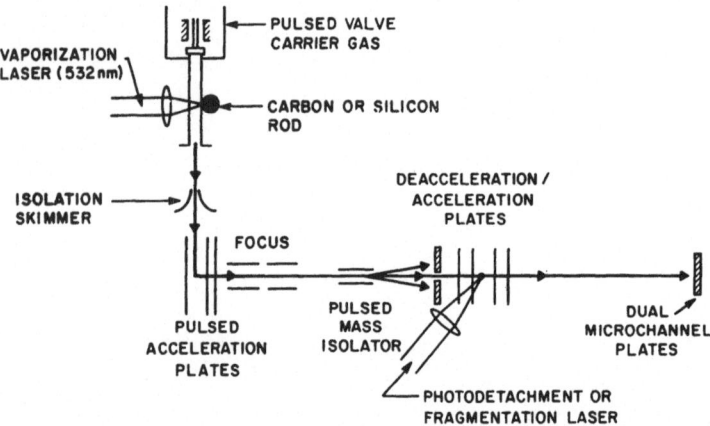

Fig. 1 Apparatus used to produce negative and positive cluster ions

Figure 2 shows the spectrum of positive silicon cluster ions obtained, and the inset shows what happens when Si_{12}^+ is isolated by the plates and fragmented by an intensed pulse of 266 nm laser radiation. Analysis of data like this shows that Si_6^+ is a particularly stable cluster ion, a so-called "magic number"[2].

Figure 3 shows the spectrum of both C_n^- (a) and C_n^+ produced directly from the source. Our *estimate* of the number of clusters of each mass arriving at the detector is between 1-10 thousand per shot. Figure 3 shows that there are clearly different

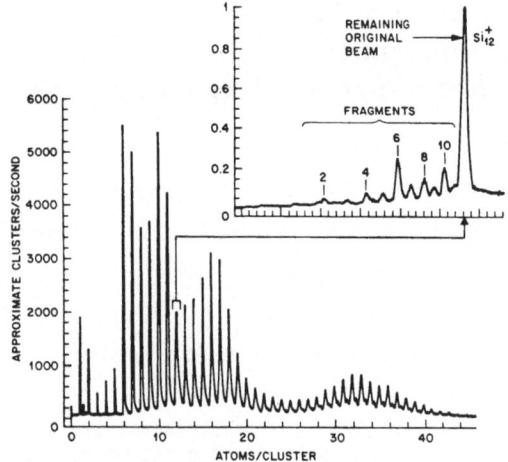

Fig. 2 Si_n^+ spectrum showing how a single cluster ion mass can be selected and fragmented

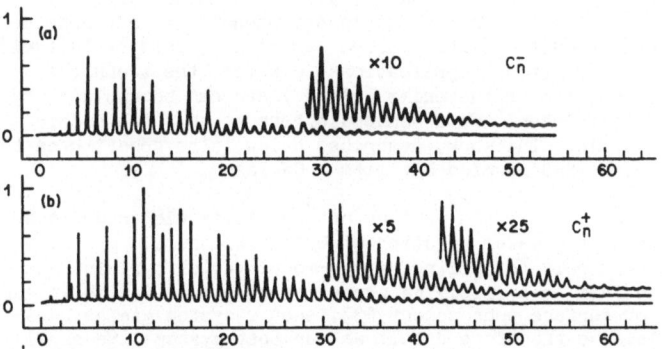

Fig. 3 Spectra of negative (a) and positive (b) clusters of carbon

"magic numbers" in carbon than in silicon, and that the structure of C_n^+ is most likely quite different than that of C_n^-.

REFERENCES

1. D.E. Powers *et.al*, J. Phys. Chem. *86*, 2556 (1982)

2. L. A. Bloomfield *et.al*, Phys. Rev. Lett. *54*, 2246 (1985)

Second-Harmonic Generation from Sub-Monolayer Molecular Adsorbates Using a CW Diode Laser - Maui Surface Experiment

G.T. Boyd and Y.R. Shen

Department of Physics, University of California, Berkeley, CA 94720, and

Materials and Molecular Research Division, Lawrence Berkeley Laboratory, Berkeley, CA 94720, USA

T.W. Hänsch
Department of Physics, Stanford University, Palo Alto, CA 94305, USA

Optical second-harmonic generation (SHG) can be an extremely sensitive tool for surface studies. The technique is capable of probing adsorbed molecules at various interfaces [1]. It is based on the idea that SHG is forbidden in a medium with inversion symmetry, but necessarily allowed at a surface. To see such a surface nonlinear optical effect, high laser intensity is often needed. Thus, in the experiments reported so far, pulsed lasers were used exclusively. From the consideration for practical applications, however, the technique would look much more attractive if the bulky pulsed laser can be replaced by a simple inexpensive CW diode laser. We demonstrate here at this conference that this is indeed possible. The work, described below, also constitutes the first experiment of surface SHG carried out with a CW laser.

The system we choose to probe is the surface of the silver electrode in an electrolytic cell containing a water solution of 0.1 M KCl and 0.1 M pyridine. Following appropriate electrolytic cycling, a sub-monolayer of pyridine can be adsorbed on or desorbed from Ag by applying proper biasing. It was shown earlier [2] that because of surface enhancement [3], such a system yields unusually strong SH signals. We therefore use it as our test system. We shall show, however, that with future improvement on the diode laser and the detection system, the surface enhancement will not be necessary.

The CW diode laser was purchased from Sharp (LT024MD). It emits 20 mW at 784 nm at room temperature. The laser beam is focused to a $\sim 100~\mu m^2$ spot on the Ag electrode and the diffuse second-harmonic light generated is collected by a photomultiplier with appropriate filtering. The laser is modulated at 500 Hz so that the signal can be processed by a lock-in amplifier.

Figure 1 shows the second-harmonic intensity and the electrolytic current as a function of time. We began with an SH signal from a monolayer of pyridine adsorbed on Ag with a negative bias potential, -0.78V, on Ag (relative to a standard calomel electrode). The signal corresponds to ~ 4000 photons/second at the photodetector. At t = 0, the bias potential is switched from -.78 to +0.08V, and the SH signal drops abruptly as pyridine is desorbed from Ag. The residual signal at t > 0 arises from AgCl layers formed on Ag [2]. Since SHG is only sensitive to the first one or two surface AgCl layers, the signal does not vary appreciably with the multilayer formation or reduction of AgCl on Ag. At t = 2 min, the bias potential is switched from +0.08 to -0.01V, and the AgCl layers start being reduced. The reduction is complete at t $\sim$ 5 min when the bias potential drops to -0.78V and the monolayer of pyridine again gets adsorbed on Ag. The process is readily repeatable. The signal is confirmed to be from SHG by its quadratic power dependence and spectral purity.

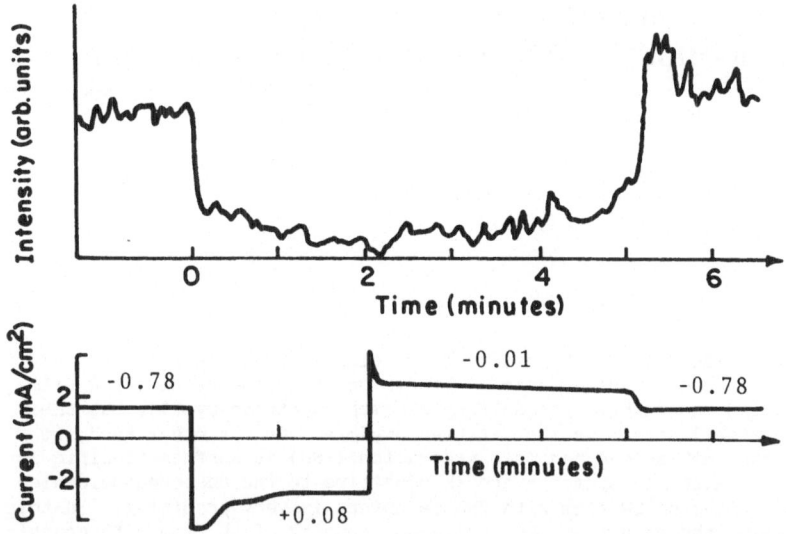

Fig. 1 SH intensity (top) from the Ag electrode and cell current (bottom) vs
time. Ag bias potentials are relative to a standard calomel electrode

The dominant noise source in the present case is the dark current in the photodetector.

The above results indicate that one can indeed monitor the adsorption of a
monolayer of molecules on an electrode in an electrolytic cell by SHG using a
single diode laser. If 100-mW diode lasers become available and focusing of
the laser beam to 10 μm^2 is possible, then a signal of ~ 4 × 10^6 photons/sec
from pyridine on a Ag electrode could be obtained. With such a setup, even
monolayer adsorbates on smooth substrates (without surface enhancement) would
be easily detectable. Further improvement could be made by using a diode array and a photon-counting detection system.

In conclusion, we have demonstrated that surface SHG using a CW diode laser
to study surfaces and interfaces is feasible. Compactness of the setup could
make this system an extremely attractive and useful surface tool. Among the
many applications of such a system, we only mention here the possibility of an
optical second-harmonic surface microscope, capable of displaying surface
structure and morphology with sub-micron resolution.

This experiment was conceived at the Maui Surf Hotel, Kaanapali Beach, Maui,
Hawaii. We wish to thank the Hotel for its hospitality in accomodating the
preparation of this experiment in its Queen Suite and the subsequent demonstration in its conference hall. GTB and YRS acknowledge partial support from
the U.S. Department of Energy under Contract No. DE-ACO3-76SF00098, and TWH
acknowledges partial support from the U.S. Office of Naval Research under Contract No. ONR N00014-C-78-0304.

References
1. T. F. Heinz, H. W. K. Tom, and Y. R. Shen, Laser Focus, May 1983.
2. C. K. Chen, T. F. Heinz, D. Ricard, and Y. R. Shen, Phys. Rev. Lett. 46,
    1010 (1981).
3. G. T. Boyd, Th. Rasing, J. R. R. Leite, and Y. R. Shen, Phys. Rev. B 30,
    519 (1984).

Surface-Enhanced SHG Study
on Langmuir-Blodgett Mono-Molecular Layers

Zhan Chen, Wei Chen, Jia-biao Zheng, Wen-cheng Wang,
and Zhi-ming Zhang

Department of Physics, Fudan University,
Shanghai, People's Republic of China

1. Introduction

There is growing interest in the study of the classic Langmuir-Blodgett (LB) mono-molecular layers in recent years. Linear optical properties of LB film have been studied thoroughly with surface plasmon spectroscopy [1], but their nonlinear characteristics have not yet been explored. On the other hand, it is already known that second harmonic generation (SHG) is surface-specific for the medium with inversion symmetry and is sensitive to the adsorbed molecules [2], hence the study of LB film with SHG technique is very promising. In this paper, we present the result on the nonlinear property of LB film with arachidic acid [$CH_3(CH_2)_{18}COOH$] as our example. We excite the surface plasmons to enhance the SHG so that we can use a rather low-power incident laser beam to prevent the LB film from being damaged when a reasonable amount of SHG signal can be detected.

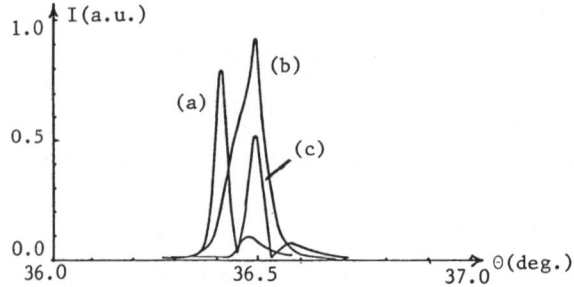

Fig. 1 Theoretical calculation for SHG from (a) bare silver film, (b) with a single molecular layer, (c) with two molecular layers

2. Theory

The contributions to SHG from the metal film is from the quadrupolar term of the bulk polarization and two dipole sheets at the glass-metal interface and metal-LB film interface. We use the hydrodynamics theory of the electron gas [3] to calculate these contributions. For LB mono-molecular film we made some approximations to calculate the SHG signal. For a single molecular layer, SHG will mainly come from the polarization along the direction perpendicular to the film, i.e., only d_{zz} contributes. The reason for this is that the LB film is formed by closely-packed organic molecules arranged parallel in their long-chain direction but randomly-oriented in the plane of the film. They can be treated as isotropic in the x-y plane,so that the nonlinear polarization in this plane does not contribute. For the LB film with an even number of mono-molecular layers, we can deduce from the symmetry that the dipole contribution vanishes. For simplicity, we ignored the higher-order contributions and approximate the two-layer LB film as an isotropic dielectric slab. The theoretical curves are shown in Fig. 1. Explicit expressions can be found elsewhere [4].

3. Experimental Result

A Q-switched YAG laser with output energy limited below 6 mJ/pulse is used as the incident beam. The signal is detected by a photo-detector and averaged by a boxcar. Attenuated total reflection (ATR) spectra is exploited to obtain the thickness of the silver film and the dielectric constants of both the silver film and arachidic acid at 1.06μ and 0.53μ for theoretical calculation. We scan the incident angle around the peak of surface plasmons excitation. The experimental curves are shown in Fig. 2. They are in good agreement with the theoretical calculations shown in Fig. 1, not only for the peak positions, but also the relative intensities. Except for the two-layer LB film, the experimental curve is slightly higher than the theoretical prediction. This discrepancy could be expected if we considered the higher order of polarization due to the interaction between the first molecular layer with the silver substrate and the exponential decaying nature of the evanescent wave. Thus, our result did show us that the nonlinear property of a two-layer LB film is quite different from that of a single-layer LB system. Furthermore, by fitting the experimental curves of SHG from the bare silver film and the film covered with one monolayer with the theoretical formula, we deduced the nonlinear polarizability coefficient along the long chain for a single arachidic acid molecule in LB film is $d_{zz} = 0.32 \times 10^{-29} \pm 0.01 \times 10^{-29}$ esu.

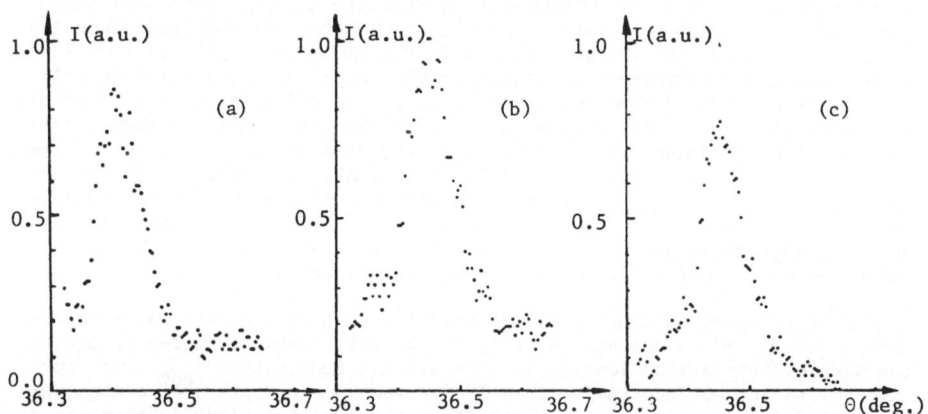

Fig. 2 Experimental curves for SHG from: (a) bare silver film, (b) covered with one single molecular layer, (c) covered with two molecular layers

References

1. I. Pockrand, J. D. Swalen, J. G. Gordon II, and M. R. Philpott, Surf. Sci. 74, 237 (1974).
2. Y. R. Shen, The Principles of Nonlinear Optics, Ch.25 (J. Wiley, New York, 1984).
3. E. Sipe, V. C. Y. So, M. Fakui, and G. I. Stegeman, Phys. Rev. B 21, 4389 (1980).
4. Z. Chen et al., to be published in Acta Optica Sinica 5, (1985).

Laser Investigation of the Dynamics of Molecule-Surface-Interaction

J. Häger, B. Simon, H. Vach, and H. Walther

Max-Planck-Institut für Quantenoptik,
Sektion Physik der Universität München,
D-8046 Garching, Fed. Rep. of Germany

Combining molecule surface scattering with laser induced fluorescence and laser resonance ionization opens the possibility to investigate the internal energy distribution as well as the state and angle resolved velocity distributions of surface scattered molecules.

In the present experiment, a supersonic beam of NO molecules is scattered from a cleaved pyrographite surface. The incoming or scattered molecules are excited by the light of a frequency tunable dye laser and the fluorescence intensity is recorded as a function of the laser wavelength. From the line intensities of the resulting spectra, the population density of the rotational and vibrational levels can be deduced [1]. In addition, it is possible to obtain information on the angle and state resolved velocity distributions of the scattered molecules: a cylindrical cage of metal wire mesh is set in front of the sample with the focussed laser beam propagating along the cage axis. Molecules scattered through the centre of the cage are ionized from a specific rotational state by a resonant two-photon transition. The ions produced have the same velocity as the parent molecules. Their velocity can therefore be analysed by time-of-flight measurements. Changing the position of the surface relative to the cage leads to a detection of molecules with different scattering angles [2].

The incoming NO molecules can be characterized by a rotational temperature of 30 K and an average velocity of 750 m/s. They are heated up during the surface interaction leading to a rotational temperature T_{rot} which corresponds to the surface temperature T_s for low T_s and approaches a constant value of 250 K for high T_s [3]. Increasing the incoming kinetic energy from 80 meV to about 200 meV (NO seeded in He) leads to a rotational temperature being about 25% higher; this indicates that the kinetic energy of the incoming beam has only a minor influence on the final rotational temperature. This is supported by the velocity distribution measurements. If the rotational excitation results from a direct transfer of translational into rotational energy the molecules in different rotational states should show different velocities. The measurements, however, yield time-of-flight spectra nearly independent of the rotational state [2].

The time-of-flight spectra are composed of a high peak of very fast molecules and a longer tail resulting from particles with slower velocities. This is characteristic for molecules scattered into specular direction. Molecules scattered under a smaller angle exhibit, for low surface temperatures, a velocity distribution with a smaller average velocity corresponding to a diffusive part in the angular distributions [2]. This diffusive part with very slow molecules disappears when the particles are scattered from a sufficiently hot surface. In this case, the scattered molecules show a clean lobular angular distribution where the peak velocity increases with increasing surface temperature and is dependent on both the incidence and

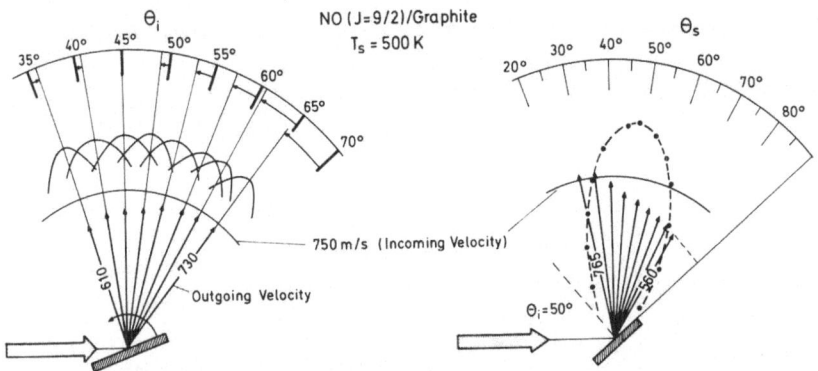

Fig.1: Average velocity of scattered NO molecules (arrows) in dependence of the incidence angle θ_i (left) and the scattering angle θ_s (right). The maximum of the scattering lobe is shifted to smaller scattering angles for $\theta_i > 45°$ and to larger angles for $\theta_i < 45°$ (left).

the scattering angle. This is demonstrated in Fig. 1. Increasing the incidence angle leads to an increasing shift of the angular distribution maximum to lower scattering angles. Simultaneously the average velocity in the direction of the angular maximum increases. However, the velocities inside a scattering lobe are not constant: they increase with decreasing scattering angle. This effect can be so substantial, that a part of the scattered molecules shows a larger velocity than the incoming ones, especially for high surface temperature, large incidence and small scattering angles.

These results reflect the interaction of the NO molecules with the surface phonons in normal and tangential direction. Molecules at low scattering angles show a gain and those at large angles show a loss of normal velocity. In addition, the decreasing outgoing velocity with decreasing incidence angle indicates a loss of normal velocity due to an increasing recoil into the surface. In tangential direction, the particles experience a loss of tangential velocity, where an increasing loss leads to a decreasing scattering angle. This molecule surface interaction which allows the molecules to transfer a net average momentum along the surface is contrary to the assumptions of the hard cube model [4]. This model, on the other hand, is very helpful for the understanding of the scattering results obtained in normal direction.

References

1. F. Frenkel, J. Häger, W. Krieger, H. Walther, C.T. Campbell, G. Ertl, H. Kuipers, and J. Segner: Phys. Rev. Lett. 46, 152 (1981)
2. J. Häger, Y.R. Shen, and H. Walther: Phys. Rev. A 31, 1962 (1985)
3. F. Frenkel, J. Häger, W. Krieger, H. Walther, G. Ertl, J. Segner, and W. Vielhaber: Chem. Phys. Lett. 90, 225 (1982)
4. W.L. Nichols, and J.H. Weare: J. Chem. Phys. 63, 379 (1975).

Surface Analysis with Second Harmonic Generation

G. Marowsky, B. Dick, and A. Gierulski

Max-Planck-Institut für Biophysikalische Chemie, Abteilung Laserphysik
D-3400 Göttingen, Fed. Rep. of Germany

G.A. Reider and A.J. Schmidt

IAEE, Abteilung für Quantenelektronik und Lasertechnik,
Technische Universität, A-1040 Vienna, Austria

This paper reviews recent results obtained by SH-generation of
rhodamine 6G-layers adsorbed at quartz-surfaces. A complete
analytical description of the $\chi^{(2)}$-tensor of this adsorbate is
presented, which has been obtained by polarization-sensitive
measurements both in transmission and reflection. The first
part of this contribution is devoted to the theoretical model,
the second section describes recent experimental results together with an interesting application of SHG: frequency doubling of 308 nm excimer laser radiation and measurement of picosecond duration pulses.

The surface-layer is treated as a thin nonlinear planparal-
lel slab between two linear, centrosymmetric media, air and
fused silica glass. Maxwell s equations for this model have
been solved by BLOEMBERGEN and PERSHAN [1]. Each beam (cf.
Fig.1) is characterized by its angle with respect to the surface normal and a refractive index: The fundamental as incident beam (δ_I, n_I) in the first linear medium, and as the source
wave (δ_S, n_S) in the surface layer. Four waves at the second
harmonic frequency are generated: One in reflection (n_R, δ_R),
one in transmission (δ_T, n_T) and two travelling inside the nonlinear medium (δ_M, n_M).
All angles and refrac-
tive indices are re-
lated by:

$$n_x \sin\delta_x = n_I \sin\delta_I \qquad (1)$$
$$x = S, R, T, M$$

The generated field com-
ponents parallel and per-
pendicular to the plane
of incidence can be writ-
ten as a product of the
nonlinear polarization
and nonlinear Fresnel
factors $\tilde{f}$ [2]:

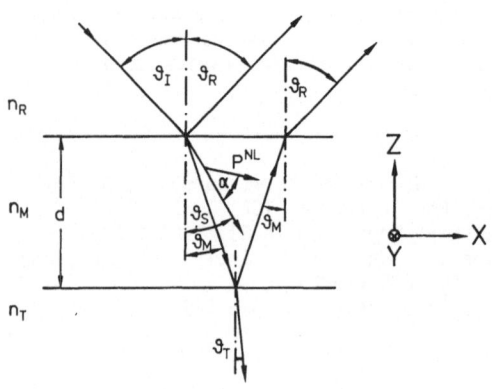

$$\varepsilon_\perp = \tilde{f}_y P_y^{NL};$$

$$\varepsilon_\parallel = \tilde{f}_x P_x^{NL} + \tilde{f}_z P_z^{NL} \qquad (2)$$

Fig.1. Schematic of three-layer-model
used in derivation of eq. (1) - (4)

The nonlinear Fresnel factor for the reflected wave is:

$$\tilde{f}^R = \frac{4\pi i d(2\omega)\sin\delta_T}{cn_R\sin(\delta_R+\delta_T)} \cdot \begin{bmatrix} -\cos\delta_T/\cos(\delta_T-\delta_R) \\ 1 \\ (n_T/n_M)^2\sin\delta_T/\cos(\delta_T-\delta_R) \end{bmatrix} \qquad (3)$$

The corresponding expression for the transmitted wave has the indices T and R interchanged in the bracket and the sign for the x-component reserved.

The nonlinear polarization is given by:

$$P_x^{NL} = 2\chi_{xxz}\cdot\epsilon_x\epsilon_z$$

$$P_y^{NL} = 2\chi_{xxz}\cdot\epsilon_y\epsilon_z \qquad (4)$$

$$P_z^{NL} = \chi_{zxx}\cdot(\epsilon_x^2 + \epsilon_y^2) + \chi_{zzz}\epsilon_z^2$$

The field ϵ in eq. (4) is the fundamental <u>after</u> refraction into the surface layer.

Measurements of the second harmonic intensities as a function of the polarization and angle of incidence of the fundamental can yield the relative magnitude of the three tensor components χ_{xxz}, χ_{zxx}/n_M, and χ_{zzz}/n_M without any prior assumption about dominant molecular contributions or adsorbate orientation. In addition the unknown refractive index n_S can be determined from the angle of incidence dependence of the transmitted second harmonic signal. We interpret this in terms of the glass-character of the surface layer, which turns out to be approximately 75 %. The same glass-character can be used to interpolate n_M between n_T and n_R. The results obtained in this way for rhodamine 6G on fused silica are as follows:

%-glass	χ_{zzx}	χ_{zxx}	χ_{zzz}
50	1.000	−0.181	0.795
75	1.000	−0.158	0.692
100	1.000	−0.190	0.830

As novel application of surface-SHG Fig.2 shows experimental results obtained by frequency doubling of 308 nm excimer laser

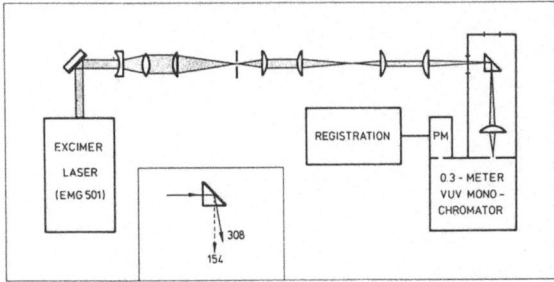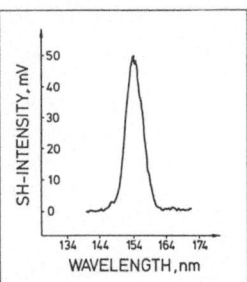

Fig.2. Experimental setup for SHG of 308 nm. Spectrum shows VUV-monochromator-scan in the vicinity of 154 nm

radiation (details cf. Ref. [3]), using CaF_2-optics and coverages of the dye p-NBA (p-nitro-benzoic-acid).

An example for use of surface-SHG for picosecond pulse duration measurements is shown in Fig.3 (cf. Ref. [4]). With coverages of the dye Nile Blue A the autocorrelation trace depicted in the right part of the plate has been recorded. Assuming a form factor of 1.5 good agreement was obtained with simultaneous single shot streak camera measurements. In the meantime this technique was applied to fs-duration measurements and into spectral ranges where frequency doubling crystals are unavailable.

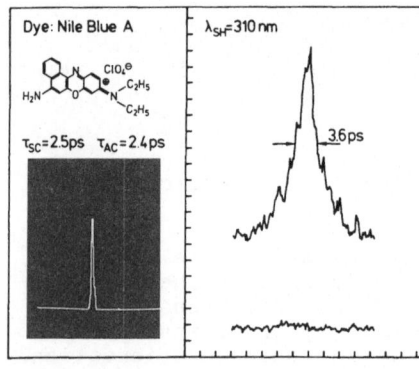

Fig.3. *Right:* Collinear autocorrelation trace taken with ps-laser operating at 620 nm and Nile Blue A dye layer. *Left:* For comparison, streak camera measurement; displayed pulse width τ_{SC} = 2.5 ps as compared to 2.4 ps, the result of autocorrelation measurement

References

1. N. Bloembergen, P. S. Pershan: Phys. Rev. 128, 606 (1962)
2. B. Dick, A. Gierulski, G. Marowsky, G. A. Reider: Appl. Phys. B, in press
3. A. Gierulski, D. Epperlein, N. Vorob'ev, G. Marowsky: Opt. Lett., submitted
4. A. Gierulski, G. Marowsky, B. Nikolaus, N. Vorob'ev: Appl. Phys. B 36, 133 (1985)

Laser Raman Spectroscopy:
The Adsorption of CO on Ni Single Crystal

H.A. Marzouk and E.B. Bradley

Department of Electrical Engineering, University of Kentucky, Lexington, KY 40506, USA

For the first time we have succeeded in detecting Raman signals from a monolayer of CO molecules adsorbed on the surfaces of a single crystal nickel at UHV pressures ($<10^{-8}$ Torr). Normal Unenhanced Raman Spectroscopy (NURS) is employed in these studies [1, 2, 3]. The confirmation of the observed Raman bands in the previously-mentioned studies was done employing two techniques: (a) reproducibility, and (b) taking advantage of the fact that the Raman effect is independent of the wavelength of the exciting line.

In this letter we report for this system the independent confirmation of the previously observed CO frequencies [1] by the use of isotopic substitution. The relation between ^{13}C-O stretch frequency (ν) and that of ^{12}C-O is $\nu(^{13}CO)/\nu(^{12}CO) = \sqrt{\mu(^{12}CO)/\mu(^{13}CO)}$ where μ is the reduced mass of the species in question. Since the ^{13}CO used is of 99% purity and since ^{12}CO is present in the residual gas we expect to observe the previously ^{12}C-O stretches in addition to the isotopically-shifted ^{13}C-O and Ni-^{13}C bands. Figure 1(a and b) shows the spectral region 1900 - 2100 cm^{-1} (an illustrative portion of a complete reproducible spectrum).

In these two figures we display two sets of bands, a set corresponding to ^{12}C-O stretches [entry A] and the second set [entry B] corresponds to their equivalent ^{13}C-O stretches. The high resolution of Raman spec-

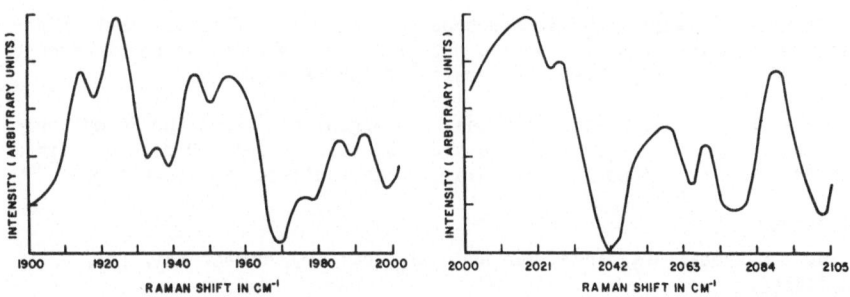

Figure 1[a, b]. A partial Raman spectrum in the region 1900-2100 cm after a 10^6L ^{13}CO exposure on Ni(111). T = -165°C. The laser power = 2 W, 4880 A, and the count time is 20s/step. The slit widths = 300 μ.

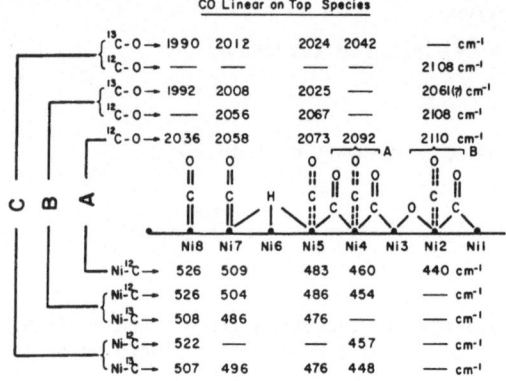

Fig.2. Tentative models proposed to explain the observed Raman bands in the CO & NiC regions for the system described in the above figure's caption.

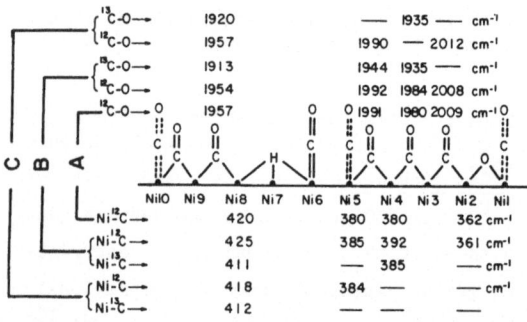

troscopy of 1-2 cm^{-1} (one of the advantages it holds over EELS technique) led to the observance of such rich spectra. Figure [2] depicts proposed models to explain the results.

These models propose that CO molecules are coadsorbed in their different configurations as well as with H and O atoms. The number of electrons backdonated from the metal ($1\pi + 5\sigma$) orbitals into the $2\pi^*$ antibonding orbitals of the CO species control the strength of these different CO bonds and hence their stretching frequencies (see reference [1]).

A unique advantage Raman spectroscopy holds over EELS is that high-pressure studies can be carried out with the use of Raman spectroscopy. Entry "C" on Fig. [2] shows the results of such a study.

Preliminary theoretical studies [4] suggest that when the CO molecule in the linear configuration is attached to a 2- fold CO molecule via a surface atom, the linear CO molecule must be inclined with some angle.

References

1. H.A. Marzouk, K.A. Arunkumar and E.B. Bradley, Surface Sci. 147, 477 (1984).
2. H.A. Marzouk, E.B. Bradley and K.A. Arunkumar, Spec. Letters 18(3), 189(1985).
3. K.A. Arunkumar, H.A. Marzouk and E.B. Bradley, Rev. Sci. Instr. 55, 905 (1984).
4. A. Companion, private communications [on-going research work].

State-to-State Molecular Beam-Surface Scattering via Laser Spectroscopic Techniques

J. Misewich, H. Zacharias, and M.M.T. Loy

IBM Thomas J. Watson Research Center, Yorktown Heights, NY 10598, USA

Laser spectroscopic techniques are finding increasing applications in surface science experiments. In our laboratory, the state-specificity and time resolution capabilities of laser techniques are applied to molecular beam-surface scattering experiments in order to perform state-to-state studies in which the angular, velocity and internal energy distributions are determined for the scattering of a laser excited incident molecular beam in a well-defined vibrationally excited single quantum state.

Experimentally, two laser beams intersect a supersonic molecular beam of NO. First, tunable infrared radiation excites the incident beam to a single vibrational-rotational state, $(v=1, J=3/2, \Omega=1/2)$. Then tunable ultraviolet radiation state-selectively probes the molecular beam both before and after interaction with a well-defined crystal surface in a UHV scattering chamber. Angular distributions for specific final states are obtained by moving the probe laser beam in the scattering plane. Also, because of the short duration of the exciting infrared laser pulse, the beam of vibrationally excited molecules is well-defined in time as well as energy. The short duration of the probing ultraviolet laser pulse then allows the direct determination of velocity distributions for specific final scattering channels. Surfaces used in these experiments were Ag(111) (prepared in the standard way with order and cleanliness checked by LEED and Auger) and freshly cleaved LiF(100).

Extensive characterization of the dynamics of vibrationally elastic scattering from LiF(100) has recently been completed in our laboratory.[1] Angular and velocity distributions have been obtained for vibrationally elastic scattering into final rotational states between $J=1.5$ and $J=20.5$ for both spin-orbit components which indicate a direct inelastic type of scattering and only a small dependence on final rotational energy. Integration over the electronic, rotational, angular and translational distributions allowed us to determine the

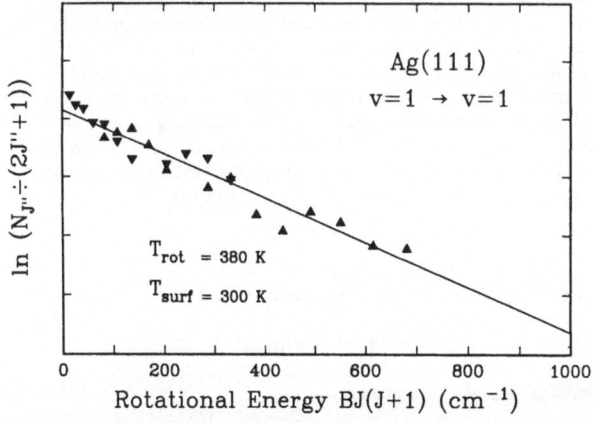

Figure 1: Boltzman plot of rotational populations for vibrationally elastic scattering of $v=1$ NO from a Ag(111) surface.

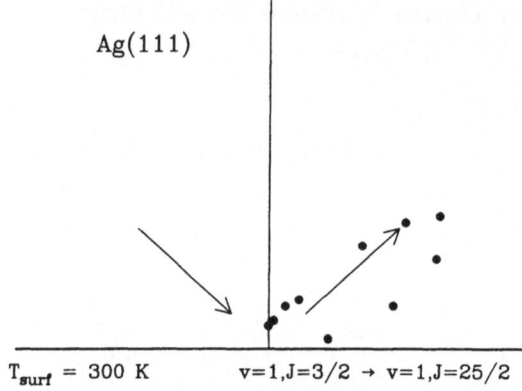

$T_{surf} = 300$ K $v=1,J=3/2 \rightarrow v=1,J=25/2$

Figure 2: State-specific angular distribution for scattering of $NO(v=1,J=3/2,\Omega=1/2)$ -->
$(v=1,J=25/2,\Omega=1/2)$ from a Ag(111) surface.

vibrational energy survival probability to be 0.9 for an incident beam translational energy of
~310 meV (the incident angle, θ_i, was 45 degrees, so the normal energy was ~155 meV).

More recently, we have been studying the scattering from a Ag(111) surface. Analysis of
UV multiphoton ionization spectra of scattered NO gives the rotational distribution for
vibrationally elastic scattering (i.e. $v=1$ --> $v=1$ scattering) presented in figure 1. Rotational
distributions taken at our incident beam kinetic energy of ~310 meV ($\theta_i = 45$ degrees) can be
reasonably described by a Boltzman distribution at 380 K similar to $v=0$ --> $v=0$ rotationally
inelastic scattering. Figure 2 shows the angular distribution we obtained for scattering from
$NO(v=1,J=3/2,\Omega=1/2)$ --> $(v=1,J=25/2,\Omega=1/2)$, indicating a direct-inelastic scattering
mechanism. Analysis of time-of-flight data will allow us to complete our investigation of the
scattering of vibrationally excited $NO(v=1,J=3/2,\Omega=1/2)$ from Ag(111) and determine the
vibrational energy survival probability.

We note that similar angular and rotational distributions have been observed for the
scattering of vibrationally excited NO from LiF(100) and Ag(111) surfaces. Since the
attractive well in the NO-surface interaction potential for both the LiF(100) and Ag(111)
surfaces is shallow, it is quite reasonable that both surfaces demonstrate direct-inelastic
scattering for NO. It will be of interest to compare the vibrational energy survival probabilities for scattering from LiF and Ag. Because of the existence of conduction electrons in the
Ag surface, a new deactivation channel through excitation of electron-hole pairs is possible.
On the other hand, since NO undergoes direct-inelastic scattering and spends only a very small
time near the surface, the electron-hole pair mechanism in metal surface scattering experiments
might contribute only a small amount to the deactivation.[2,3] We plan to answer the question of
the role of electron-hole pair excitation in direct-inelastic scattering of vibrationally excited
NO from Ag(111) by comparing its survival probability with that from LiF(100).

Acknowledgement

This work was supported in part by the Office of Naval Research.

1. J. Misewich, H. Zacharias and M. M. T. Loy, submitted for publication.

2. B. N. J. Persson and M. Persson, Surf. Sci. *97*, 609 (1980).

3. A. Leibsch, Phys. Rev. Lett. *54*, 67 (1985).

Part XI

Ultrashort Pulses
and Applications to Solids

Coherent Time Domain Far-Infrared Spectroscopy

D.H. Auston and K.P. Cheung

AT & T Bell Laboratories, Murray Hill, NJ 07974, USA

A new approach to far-infrared spectroscopy is described which uses extremely short far-infrared pulses to measure the dielectric properties of materials. These pulses are generated by rectification of femtosecond optical pulses in electro-optic materials, and produce far-infrared radiation of approximately one cycle in duration in the THz spectral range. Phase sensitive detection of the electric field of these pulses provides a capability for measuring precise changes in the shape of the waveform following reflection or transmission. This method, which is equivalent to having a tunable laser in the spectral range from 0.1 to 5 THz, is illustrated by measurements of the dielectric response of solid-state materials.

1. THE ELECTRO-OPTIC CHERENKOV EFFECT

A substantial amount of work has been directed at the use of nonlinear optical techniques for the generation of coherent far-infrared radiation[1]. Picosecond pulses are often considered as potential sources due to their broad spectral content and high intensities. A key difficulty, however, has been the large difference in the far-infrared and optical indices of refraction of most suitable nonlinear materials. A novel approach to this phase-matching problem was recently demonstrated which makes use of a Cherenkov analogue to produce a conical waveform of extremely short duration[2]. This is accomplished by focussing femtosecond pulses into an electro-optical material to produce a moving dipole moment which has a spatial extent that is less than the coherence length for a phase-matched interaction. This dipole moment radiates a pulse which has a duration comparable to the excitation pulse. Its shape is approximately a single cycle of radiation in the far-infrared. When femtosecond optical pulses are used for excitation, this technique makes possible the generation of coherent continuum pulses whose spectra span many teraHertz ($1\text{THz} = 33 \ cm^{-1}$).

In addition, we have also developed a phase-sensitive coherent detection technique which permits accurate measurements of the electric field of these pulses. This method is based on an extension of the electro-optic sampling technique developed by Valdmanis and co-workers[3] for the detection of short electrical transients. It utilizes the electro-optic effect to probe the local electric field of the far-infrared wave by measuring the small birefringence induced in a second femtosecond optical pulse. This is accomplished in perfect synchronism with the the generation process by propagating the probe pulse parallel to the pump pulse with a small separation between them. A differential delay between these pulses enables the waveform to be measured by the stroboscopic method.

For our experiments femtosecond optical pulses were obtained from a colliding pulse mode-locked ring dye laser[4]. The pulse duration was 50 fs (FWHM) at 625 nm. A pulse energy of only 10^{-10} *J* at a repetition rate of 125 MHz was sufficient for the experiments described in this paper.

2. COHERENT TIME DOMAIN SPECTROSCOPY

The specific geometry we have used to measure the far-infrared properties of materials by coherent time-domain reflection spectroscopy is illustrated in figure 1. The Cherenkov cone of far-infrared radiation emanating from the pump pulse is measured by the probe pulse before and after reflection from the interface between the material of interest and the host electro-optic crystal. Since the detection method is phase sensitive and measures the electric field(not the intensity) of the waveforms, it is possible to determine the complex reflectivity spectrum of the material by simple numerical Fourier analysis. From the Fresnel reflectivity law, the full dielectric response of the material can then be determined. This gives us both the real and imaginary parts of the dielectric response over the entire spectral range spanned by the far-infrared pulse.

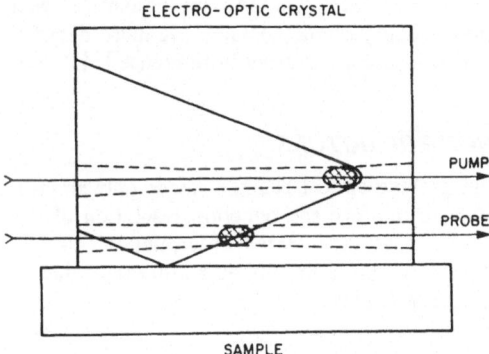

Figure 1

The method is thus fully equivalent to having a tunable monochromatic coherent source in the far-infrared. The full power of this procedure is apparent in its ability to obtain both the real and imaginary parts of the dielectric response over a broad range of frequencies from a single waveform. Incoherent methods measure only the magnitude of the reflectivity and must resort to Kramers-Kronig analyses to determine the complex dielectric function.

An example of some incident and reflected waveforms is shown in figure 2. The reflected waveform from a gold film is used to establish a timing reference. Notice that it has a shape which is nearly identical to the incident waveform, except that it is inverted(the attenuation is due to far-infrared absorption in the lithium tantalate electro-optic crystal). The waveform reflected from a germanium sample, however, shows a dramatically distorted shape. This is due to the dipsersion arising in this sample from a free electron plasma resonance which was centered close to 1 THz. From a Fourier analysis of these waveforms it is possible to determine the electron density and elastic scattering times of the free electron plasma.

We have applied this technique to other semiconductors, including, a gallium arsenide/gallium aluminum arsenide multi-quantum well. For example, we have determined the electron momentum relaxation time in our multi-quantum well sample

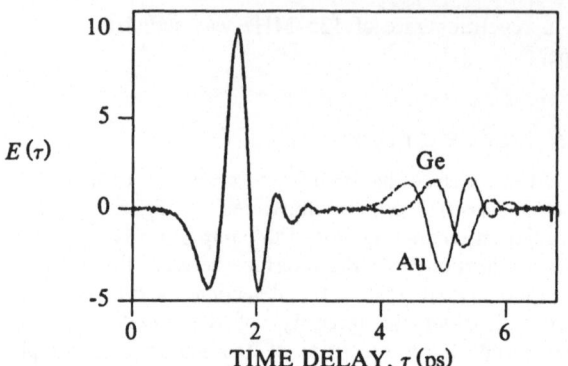

Figure 2

to be 190 fs at room temperature. This figure is in good agreement with estimates based on Hall mobility measurements. This method gives an independent measure of electron mobility which does not require electrical contacts on the sample. This is an important advantage for materials such as the GaAs/GaAlAs multiquantum well sample to which it is difficult to make good quality ohmic contacts. A more detailed description of this approach to far-infrared spectroscopy is given in reference [5].

3. DIRECT OBSERVATION OF PHONON POLARITONS

If the distance between the generating and probing optical beams is made only slightly greater than the beam waists, it is possible to improve the temporal resolution of the measurement approach by reducing the effects of far-infrared absorption. This has enabled us to directly observe the contribution to the electro-optic nonlinearity due to the TO phonon resonance in lithium tantalate at 6 THz.

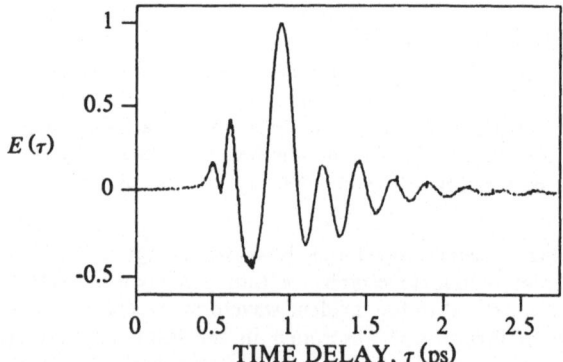

Figure 3

As shown in figure 3, this manifests itself as a ringing on the trailing edge of the waveform. These oscillations are due to the ionic lattice vibrations, and since our detection method is phase-coherent, we can see the individual cycles of the damped resonance. From the decay rate of the ringing we find this damping rate is equal to 190 fs, which is close to estimates based on far-infrared spectral measurements of the resonance lineshape. By varying the spacing between the pump and probe, we can measure the spatial dispersion of this phonon polariton wave as it propagates through the crystal.

REFERENCES

[1] For a review, see Y. R. Shen, Prog. Quant. El., **4**, 207 (1976).

[2] D. H. Auston, K. P. Cheung, J. A. Valdmanis, and D. A. Kleinman, Phys. Rev. Lett., **53**, 1555 (1984).

[3] J. A. Valdmanis, G. A. Mourou, and C. W. Gabel, IEEE J. Quant. Electr., **QE-19**, 664 (1983).

[4] R. L. Fork, B. I. Greene, and C. V. Shank, Appl. Phys. Lett. **38**, 671 (1981); J. A. Valdmanis, R. L. Fork, and J. P. Gordon, Opt. Lett., **10**, 11 (1985).

[5] D. H. Auston and K. P. Cheung, J. Opt. Soc. Am. B, **2**, 606 (1985).

Picosecond Time-Resolved Photoemission Spectroscopy of Semiconductor Surfaces

J. Bokor and R. Haight*

AT & T Bell Laboratories, Holmdel, NJ 07733, USA

R.R. Freeman and P.H. Bucksbaum

AT & T Bell Laboratories, Murray Hill, NJ 07974, USA

In angle-resolved ultraviolet photoelectron spectroscopy (ARUPS) a measurement of the kinetic energy and vector momenta of electrons photoemitted from a crystal surface by monochromatic ultraviolet radiation is used to determine the binding energy and crystal momenta of those electrons before the photon absorption event. Energy versus momentum band dispersions can be obtained in this way for normally occupied bands, i.e. those lying below the Fermi level. We have used nonlinear optical techniques to produce picosecond pulses in the vacuum ultraviolet wavelength region suitable for ARUPS experiments. By exploiting the pump and probe technique of picosecond laser spectroscopy, with visible or infrared pump laser pulses and using ARUPS as a probe, we have studied the dynamics of optically excited electrons on semiconductor surfaces.

A schematic diagram of the experimental apparatus is shown in Fig. 1. Probe radiation at 118.2 nm is produced by harmonic generation starting with a Nd:YAG laser system which produces

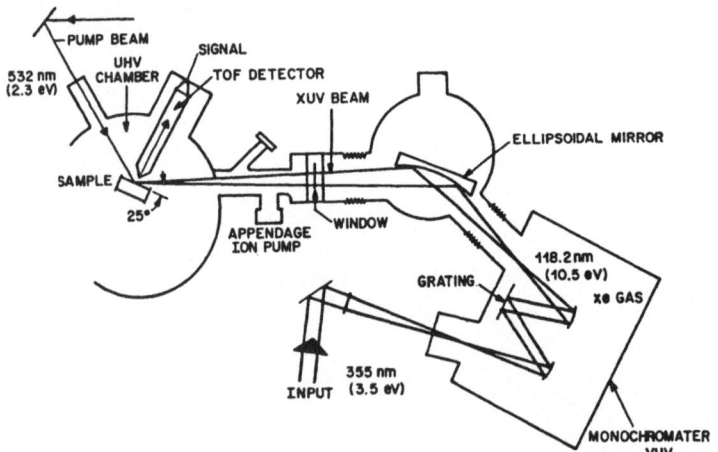

Fig.1: Schematic diagram of the photoemission chamber and optical beamline used to generate and transport the 118 nm probe radiation.

---

* Present address: IBM Thomas J. Watson Research Center, Yorktown Heights, New York 10598.

70 psec pulses at 1.06 μm. The 355 nm third harmonic of the YAG laser is first generated in KDP crystals, and this radiation is then frequency tripled in Xe gas to produce the final output at 118.2 nm. A time-of-flight photoelectron spectrometer capable of processing multiple photoelectron counts per laser shot is used to take the ARUPS spectra. The geometric angular acceptance of the detector is $\pm 2.5^\circ$ and the energy resolution is 105 meV at 6.5 eV electron energy.

This system has been used so far to study the dynamics of intrinsic surface states on both InP(110) and 2x1 reconstructed Si(111) surfaces. In the case of InP(110), a normally unoccupied surface state was observed in the photoemission spectra following excitation of a bulk carrier plasma with 532 nm excitation pulses [1]. Some representative spectra are shown in Fig. 2. The band minimum for this state, labelled C_3, was found to lie 1.47 eV above the valence band maximum at the center of the surface Brillouin zone. This means that it lies in the conduction band (the band gap of InP at room temperature is 1.35 eV), and rapid equilibration between the surface and bulk bands occurs [1] on a time scale of a picosecond or less. The C_3 surface band was found to be dispersive with an anisotropic effective mass m_{eff}: $0.20 \le m_{eff}/m_e \le 0.24$ along

the $\langle \bar{1}10 \rangle$ direction and $0.10 \le m_{eff}/m_e \le 0.15$ along the $\langle 001 \rangle$

direction, where m_e is the free electron mass.

We have also begun an examination of the cleaved Si(111) surface exhibiting the well known 2x1 reconstruction. Optical studies [2] have identified a normally unoccupied surface state lying in the middle of the optical band gap. This empty state is responsible for the strong infrared absorption band found

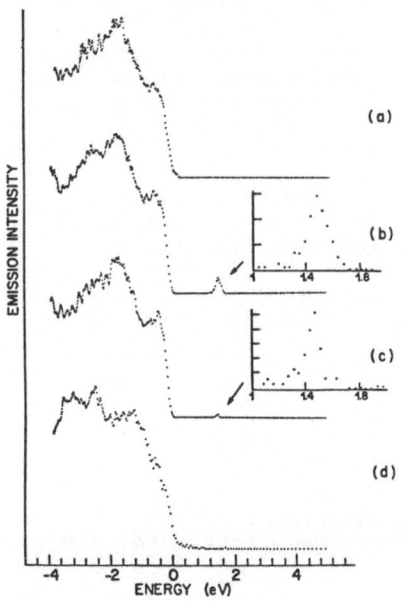

Fig. 2 Pump-probe ARUPS spectra from InP (110) for time delays of: a) t = -133 psec, b) t = 0, c) t = 266 psec, and d) t = 0 where the surface has been exposed to hydrogen.

[2] at 2.8 µm. To produce population in the excited surface
state, we are using 2.8 µm pulsed laser radiation produced by
second Stokes Raman shifting a 1.06 µm laser pulse in high
pressure methane. Representative spectra for this case are
shown in Fig. 3. The energy scale in Fig. 3 is adjusted such
that zero corresponds to the bulk valence band maximum.
Previous photoemission studies of occupied surface states [3]
have identified the prominent peak at -0.15 eV as a highly
dispersive dangling bond surface band. The spectra shown in
Fig. 3 were taken at an emission angle of 45° along the <011>
azimuthal direction. This corresponds to the $\bar{J}$ point in the
surface Brillouin zone where the occupied surface band reaches
its maximum [3], and where the unoccupied surface band minimum
is predicted [4] to lie. We indeed find the unoccupied state to
have its band minimum at $\bar{J}$, at an energy of 0.5+0.05 eV above
the unoccupied state. This is in good agreement with the
optical studies [2]. It is also consistent with the finding
[5] that, on this surface, the Fermi-level is pinned 0.4+0.03
eV above the valence band maximum.

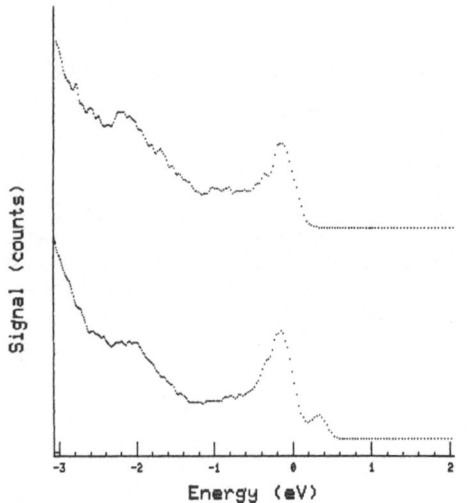

Fig. 3 Upper spectrum:
ARUPS spectrum from 2x1
reconstructed Si (111)
without infrared
excitation. The electron
emission angle is 45° and
the zero of energy is
adjusted to the position
of the bulk valence band
maximum. Lower spectrum:
ARUPS spectrum from Si
(111) 2x1 with
simultaneous excitation
by 2.8µm radiation. The
infrared fluence was
approximately 3mJ/cm².

A study of the time-dependence of the excited surface state
population is currently in progress. In contrast to the InP
case, here the surface state is well isolated from the bulk
bands. In this case, the lifetime of the surface state will be
decoupled from the dynamics of bulk carriers. This should
allow a fundamental understanding of the physics of surface
recombination processes.

References

1. R. Haight, J. Bokor, J. Stark, R. H. Storz, R. R.
   Freeman, and P. H. Bucksbaum, Phys. Rev. Lett. **54**, 1302
   (1985).

2. M. A. Olmstead and N. M. Amer, Phys. Rev. Lett. $\underline{52}$, 1148 (1984); S. Nannarone, P. Chiaradia, F. Ciccacci, R. Memeo, P. Sassaroli, S. Selci, and G. Chiarotti, Solid State Commun. $\underline{33}$, 593 (1980); J. Assman and W. Mönch, Surf. Sci. $\underline{99}$, 34 (1980).

3. R. I. G. Uhrberg, G. V. Hansson, J. M. Nicholls, and S. A. Flodström, Phys. Rev. Lett. $\underline{48}$, 1032 (1982).

4. J. E. Northrup and M. L. Cohen, Phys. Rev. Lett. $\underline{49}$, 1349 (1982).

5. F. J. Himpsel, G. Hollinger, and R. A. Pollak, Phys. Rev. $\underline{B28}$, 7014 (1983).

Study of Excitonic Particles in CuCl by Laser Spectroscopy

A. Mysyrowicz

Groupe de Physique des Solides de l'E.N.S., Université Paris VII, Tour 23 - 2, place Jussieu, F-75251 Paris Cedex 05, France

In many cases, electronically excited states of pure non-metallic crystals can be described in terms of a chemistry taking place within a system of transient quasi-particles similar to hydrogen : photo-generated electrons and holes (playing the role of protons) bind into atomic-like entities called excitons. At increasing densities, excitons further couple into molecules called biexcitons. At still higher densities, excitonic particles dissociate spontaneously into a ionized plasma state (the analog of metallic hydrogen predicted at very high pressure). Because of the very light effective mass of the electrons and holes, quantum statistical effects can become apparent at relatively high temperatures. For bound electron-hole pairs, i.e. excitons, Bose statistics applies, with the prospect of observing Bose-Einstein condensation. On the other hand, Fermi statistics will prevail for the system of unbound e-h pairs in the high density plasma. By using an ultrafast optical pulse to create the plasma, and by following its subsequent relaxation with time-resolved optical detection techniques, as it evolves towards a system of bound (Bose) particles, one has a unique system at hand for exploring the dynamics of transition from one quantum system to the other.

To study such effects experimentally, it may prove advantageous to select simple crystals showing well-resolved excitonic lines, already understood in the linear or weakly non-linear excitation regime. CuCl provides such a canonical excitonic material. This direct gap compound has isotropic, parabolic bands of low symmetry with extrema at the center of the Brillouin zone. Fig. 1 shows the excitation diagram (energy versus K vector) of the n = 1 term of the lowest exciton series as determined experimentally by two-photon and hyper-Raman scattering [1]. The parabolic dependance of the longitudinal and transverse exciton branch (reflecting the particle kinetic energy), extends over a surprisingly large portion of the Brillouin zone. At small K vector ($K \sim K_o$ (photon) = $4.45 \ 10^5$ cm^{-1}), the coupling between the transverse exciton and the radiation field leads to the characteristic polariton dispersion of the mixed photon-exciton mode.

In the same figure, the biexciton excitation diagram given by direct two-photon absorption from the crystal ground state and by two-step absorption via the transverse polariton is also plotted [2]; again the quadratic law illustrates the validity of the particle concept for the biexciton.

Figure 2a) shows a luminescence spectrum of CuCl at 25K, recorded under well controlled homogeneous volume excitation, using two counter-propagating pulsed dye laser beams tuned at the biexciton resonance (K = 0 excitation)[3]. The density of created biexcitons in quasi cw excitation regime (laser pulse intensity $\sim 10^5$ W/cm^2 ; pulse duration = 5.10^{-9} s ; biexciton lifetime $\tau = 10^{-9}$ s) is less than 10^{18} cm^{-3}. The open circles correspond to the calculated spectrum expected from a gas of biexcitons assumed to be thermally

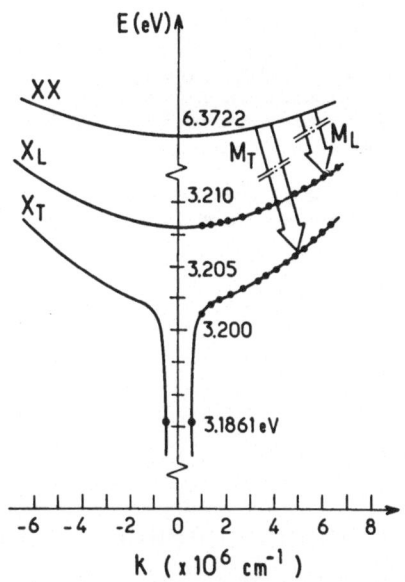

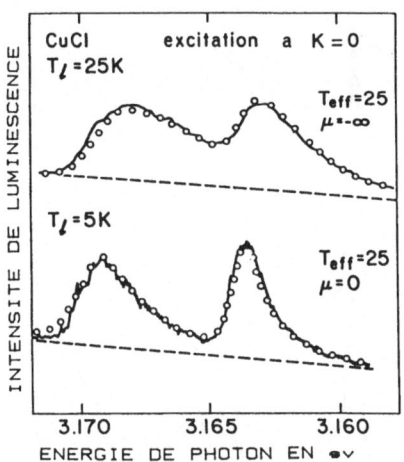

Fig. 1) Excitation diagram of the n = 1 exciton and of the biexciton in CuCl (after Mita et al) [1,2].

Fig. 2) a) (top) luminescence spectrum due to a low density of biexcitons in CuCl (n < 10^{18} cm-3). Initial excitation occurs at K = 0. b) same as in a), except for a higher density of decaying particles (n > 10^{18} cm-3). Lines around 3.163 eV and 3.168 eV correspond to transitions M_L and M_T of fig. 1). Open circles correspond to calculated lineshapes with effective temperature T_{eff} and chemical potential μ as indicated.

distributed according to Maxwell Boltzmann statistics. Except for an over-all scaling factor, the only adjustable parameter is the gas effective temperature (here equal to the lattice temperature);initial and final states of the transition are those shown in fig. 1. Thermalisation of the particles inside their kinetic energy band involves interparticle collisions, with a mean time between collision in the picosecond range (3,4).

If the density of biexcitons is gradually increased, deviations from MB statistics appear below a sample temperature T ∿ 50K (3). The emission spectra are reproduced by introducing Bose distribution functions, with increasing degree of degeneracy as the temperature is lowered and/or the excitation intensity is increased. Particles densities extracted from line-shape analysis (n ∿ 10^{18} - 10^{19} cm-3) agree with those estimated from the experimental parameters. At the lowest crystal temperatures T < 25K, a value μ = 0 of the chemical potential is reached, signaling the onset of BEC (see fig. 2b). Note that for a condensation occurring at K = 0, only thermally excited particles are apparent in the luminescence spectrum : the radiative decay from the condensate itself to the longitudinal exciton is forbidden by selection rules, whereas its decay to the transverse branch is shifted towards higher photon energies, because of the polariton effect, as readily seen in fig. 1.

The luminescence obtained with single beam two-photon resonant excitation is markedly different (see fig. 3). A narrow emission peak develops above a temperature-dependant excitation threshold. This line exhibits a strong spatial anisotropy (a unusual feature for a cubic material). This

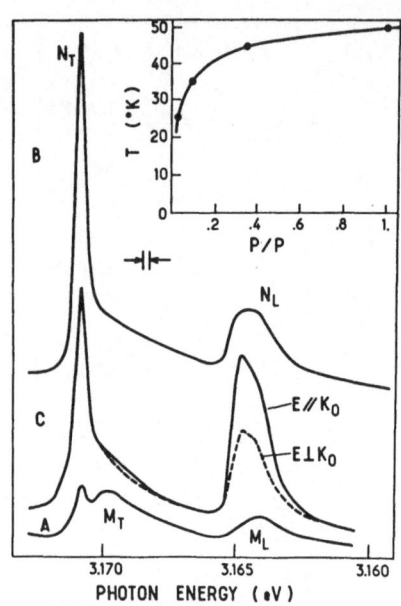

Fig. 3) Biexciton luminescence of CuCl obtained under similar conditions as in fig. 2b, except for initial excitation at $K = 8.9.10^5$ cm^{-1}. The spectra are recorded in forward A), side C) and backward B) detection geometry. Inset shows approximate temperature-dependence for appearance of sharp line N_T.

behaviour is readily explained under the assumption of a condensate forming at a small but finite K value ($K = 8.9.10^5$ cm^{-1}) imposed by the excitation process (2 K_o excitation) (3).

Evidence for plasma formation in CuCl has long proved elusive. Only with ultrashort UV laser pulses ($\lambda = 310$ nm, pulse duration 120 fs, intensity $\sim 10^9 - 10^{10}$ W/cm^2) could the required excitation levels be reached without destroying the sample (5). The formation of a plasma is signaled by the appearance above a threshold excitation of a characteristic broad emission band located on the low-energy side of the biexciton luminescence. Informations upon the parameters describing the plasma (density n, temperature, chemical potential) can be obtained from subpicosecond time-resolved transmission spectroscopy (6). In the presence of free carriers, excitons are screened, leading to an increase of transmission at the corresponding wavelenths. Also, the band gap energy is reduced, by an amount which depends, in first approximation, on the carrier density only. Interband transitions across the reduced gap lead to additional absorption or optical gain, depending on band state filling, (and thus on density and temperature). The value of the plasma chemical potential fixes the crossover from gain to loss and depends both on carrier density and temperature.

Results of such time-resolved transmission experiments have been compared to a calculation, using the many-body formalism of the density dependant dielectric function developed by Haug and coworkers (7). It allows to characterize the plasma during the first picosecond following carrier injection. The initially very hot ($T \sim 2500$ K), superdense ($n \sim 3.10^{20}$ cm^{-3}), non-degenerate plasma relaxes at approximatively constant density to a temperature $T < 200$K within 1 ps. The subsequent evolution of the plasma has been modeled more recently with a simplified treatment (6) known to give accurate results in GaAs. This model reproduces well the main experimental features, including the time and wavelength dependence of induced absorption and the onset of optical gain.

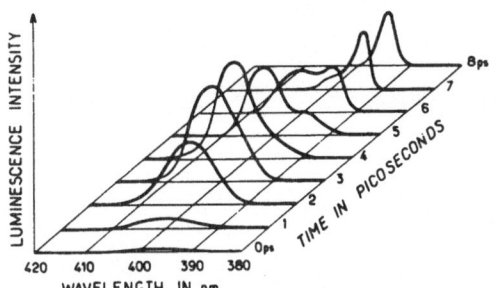

Fig. 4) Time-resolved luminescence spectra of highly excited CuCl (initial electron-hole density n > 10^{20} cm^{-3}), showing the conversion of the plasma into biexciton luminescence after a delay of 6 ps.

As stated above, a strongly excited crystal offers the possibility to explore the dynamics of the reverse Mott transition between the plasma phase of Fermi particles and the dielectric phase of Bose particles, which occurs when the free relaxing plasma reaches the critical conditions. A sequence of time-resolved luminescence spectra displayed in fig. 4, shows directly the kinetics of conversion from one phase to the other. The spectra were recorded through an optically activated Kerr shutter, of picosecond resolution, consisting of a 3 mm benzene cell between crossed polarizers (8).

The conversion from the metallic plasma to the dielectric excitonic fluid occurs very abruptly, in less than 3 ps. At the time of conversion, the plasma is Fermi degenerate, with a e-h density estimated to be $n = 1.10^{19}$ cm^{-3} and a temperature $T \sim 100$ K. Because of the large exciton dissociation energy (B = 0.2 eV), the condition kT/B << 1 is fulfilled during the Mott transition. However, the electronic temperature is too high for a Bose degeneracy with $\mu \sim 0$ to be achieved in the excitonic phase immediatly after conversion. It would be interesting to reduce the initial plasma temperature (by minimizing the excess energy imparted to each electron-hole pair in the creation process), in order to explore the Mott transition between two <u>highly</u> degenerate systems.

[1]   B. Hönerlage, A. Bivas, V.D. Phach : Phys. Rev. Lett. <u>41</u>, 49 (1978).
        T. Mita, K. Sotome, M. Ueta : Solid. State Comm. <u>33</u>, 1135 (1980).

[2]   T. Mita, K. Sotome, M. Ueta : J. Phys. Soc. Jap. <u>48</u>, 496 (1980).

[3]   N. Peyghambarian, L.L. Chase, A. Mysyrowicz : Phys. Rev. B. <u>27</u>, 2325 (1983).

[4]   T. Itoh, T. Kathono, M. Ueta : J. Phys. Soc. Jap. <u>53</u>, 844 (1984).

[5]   D. Hulin, A. Antonetti, L.L. Chase, J.L. Martin, A. Migus,
        A. Mysyrowicz : Optics Comm. <u>42</u>, 260 (1982).

[6]   A. Antonetti, D. Hulin, A. Migus, A. Mysyrowicz, L.L. Chase, to
        appear in JOSAB (1985).

[7]   J.P. Löwenau, S. Schmitt-Rink, H. Haug : Phys. Rev. Lett. <u>49</u>, 1511 (1982).
        D. Hulin et al. : Phys. Rev. Lett. <u>52</u>, 779 (1984).

[8]   D. Hulin, J. Etchepare, A. Antonetti, L.L. Chase, G. Grillon, A. Migus
        A. Mysyrowicz : Appl. Phys. Lett. <u>45</u>, 993 (1984).

Part XII

Spectroscopic Sources and Techniques

Frequency Doubled, Laser Diode Pumped, Miniature, Nd: YAG Oscillator - Progress Toward an All Solid State Sub-Kilohertz Linewidth Coherent Source

R.L. Byer, G.J. Dixon, T.J. Kane, and W. Kozlovsky

Department of Applied Physics, Stanford University,
Stanford, CA 94305, USA

Bingkun Zhou

Department of Radio Electronics, Tsinghua University,
Peking People's Republic of China

1. Introduction

Laser diode pumping of solid state laser sources is a subject with a decade long history.[1-4] However, rapid progress in diode laser performance has revived research efforts in laser diode pumped miniature laser sources.[5] Recent experimental results have demonstrated that laser diode pumped solid state lasers are efficient at converting the divergent, broadband diode radiation into TEM_{00} mode, single axial mode output.[6,7] Furthermore, the solid state laser medium with its long storage time can be Q-switched or mode locked. The high Q, low loss, solid state laser medium also offers the potential of achieving Schawlow/Townes limited linewidths of 1 Hz at 1 mW output power from a monolithic miniature resonator device.[6] This linewidth is six orders of magnitude less than can be achieved directly in diode laser sources.

The solid state laser medium improves both the spatial and spectral mode characteristics compared to the direct laser diode radiation. This brightness amplification in turn leads to the possibility of efficient nonlinear conversion of the solid state laser source. Direct nonlinear frequency conversion of the laser diode radiation is much more difficult to achieve.

2. Laser Diode Pumped Miniature Nd:YAG Oscillator

Figure 1 shows the schematic of the laser diode pumped miniature monolithic Nd:YAG oscillator. The laser diode output is focussed with a gradient index lens collinearly with the fundamental transverse mode of the monolithic Nd:YAG oscillator. The Nd:YAG oscillator consists of a 2 mm diameter rod onto which curved surfaces have been polished and coated for reflectivity at 1064 nm. Early oscillators were 5 mm in length with 18 mm curvature reflecting surfaces polished onto the crystal. The dielectric coatings transmitted the 809 nm laser diode pump radiation and were highly reflecting and 0.3% transmitting at 1064 nm. The measured threshold was 2.2 mW and the slope efficiency was 25%. Recent work with high power multi-stripe laser diode pump sources has yielded improved slope efficiency and overall electrical efficiency.[8]

The simple standing wave miniature oscillators operate single axial mode at power levels up to 1 mW. However, spatial hole burning and residual stress in the Nd:YAG medium causes these oscillators to operate in a few axial modes at power levels as low as 100 μW. To obtain single axial mode operation, Kane and Byer [7] introduced the out-of-plane ring

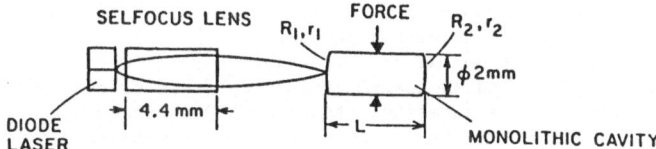

SELFOCUS LENS R_1, r_1 FORCE R_2, r_2

ϕ2mm

DIODE
LASER $\leftarrow$ 4.4 mm $\rightarrow$ $\leftarrow$ L $\rightarrow$ MONOLITHIC CAVITY

Fig. 1--Schematic diagram of the diode laser pumped monolithic Nd:YAG laser.

resonator with a built-in optical diode. Recent results have yielded uni-
direction, single axial mode output of 10 mW with 100 mW of laser diode
pumping.

The monolithic nature of the crystal oscillators coupled with the very
stable output power of the laser diode pump source, gives the potential
for very narrow linewidth operation. The linewidth of two independent
oscillators was investigated by a simple beat frequency experiment.[6]
The experiment, conducted without shielding from laboratory noise and
vibrations, gave a frequency jitter of 10 kHz over a 0.3 sec. period and
a linewidth of less than 300 Hz over a 0.03 sec. period. The Schawlow/
Townes linewidth for this miniature oscillator is predicted to be 1 Hz at
1 mW output power.[6] Experiments are planned to investigate the linewidth limits of these miniature laser diode pumped oscillators.

The temperature tuning rate of a monolithic Nd:YAG oscillator is
3 GHz/K. Current oven technology allows temperature control to less than
a micro-degree level so that it is possible to control the frequency of
these oscillators to less than the kilohertz level. Other crystal or
Nd:Glass media offer improved temperature tuning rates. In preliminary
work we have pumped Nd:YLF and have achieved thresholds below the 1 mW
level.[10] We have also achieved oscillation on the 1321 nm transition
of Nd:YAG with a threshold of 6 - 8 mW. Nd:GGG and Nd:Glass monolithic
oscillators have been fabricated for laser diode pumping.

3. Amplification

For many applications, including high resolution Raman spectroscopy, the
source must be amplified to kilowatt power levels. We have successfully
demonstrated a 60 dB gain, multipass, slab geometry, flashlamp pumped,
Nd:YAG amplifier.[11] The amplifier takes advantage of angular multi-
plexing in the slab geometry to multipass the gain medium three or four
times. The amplifier operates at an equivalent noise input level that is
only ten times the quantum noise limit or 0.5 μW of input power. In line
with the amplifier is an acousto- optic switch to control the pulse width.
The high gain, high fidelity, unsaturated amplifier allows the experimenter
to select the pulse width and through the Fourier relations the linewidth
of the amplified radiation. To date we have achieved a 10 kW output power
level in a 3 μsec pulse at a pulse repetition rate of 10 Hz with the flashlamp pumped amplifier.

4. Nonlinear Frequency Conversion

The intrinsically low loss solid state laser media yield high standing
wave power internal to the miniature laser resonator. This high circulating
power allows the possibility of efficient harmonic generation even for
milliwatt cw power levels.

Three approaches are possible for efficient harmonic generation :
internal SHG; resonant external SHG; and self-doubling in a nonlinear

laser medium. To date we have experimentally explored internal SHG using MgO:LiNbO3 and KTP and self-doubling using Nd:MgO:LiNbO3. The internal SHG experiment consisted of a 5 mm long Nd:YAG crystal adjacent to a 10 mm long MgO:LiNbO3 crystal. Dielectrically coated curved end-faces of the two crystals formed the resonator. The harmonic output was 500 μW in each direction when pumped by 215 mW of dye radiation at 590 nm. The threshold of the composite resonator was 6.9 mW which should allow laser diode pumping. The MgO:LiNbO3 crystal phasematched at 108 C to generate the green output at 532 nm.

The self-doubling experiment used a 1 cm anti-reflection coated Nd:MgO: LiNbO3 crystal within an external mirror resonator. The output power in the green vs the dye laser input power is shown in Fig. 2.[12,13] Again the threshold is below 10 mW so that laser diode pumping is feasible. The advantages of the self-doubling approach are simplicity and the ability to Q-switch or mode lock the source by using the intrinsic electro-optic effect of MgO:LiNbO3.

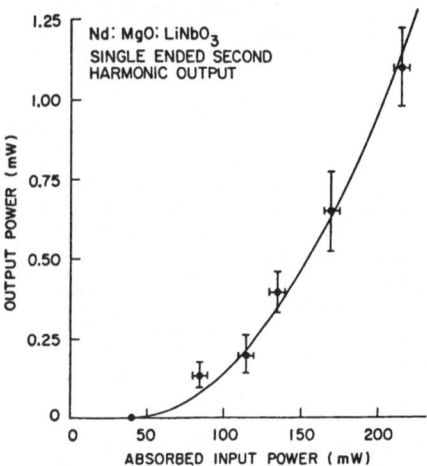

Fig. 2--Observed second harmonic data. Solid line is best fit quadratic.

These preliminary results have demonstrated the feasibility of harmonic generation of the mW power level cw sources. The key to efficient frequency conversion is the availability of low loss nonlinear crystals of sufficient optical quality to allow high standing wave power levels. The generation of 10 mW output power at 532 nm has recently been reported.[14] We are continuing to explore methods of generating green from these miniature laser sources.

5. Conclusion

In this paper we have discussed the potential of laser diode pumped miniature solid state laser sources. The improvement in laser diode capability should allow high power levels and efficient harmonic conversion of these narrow linewidth sources. The preliminary harmonic generation results suggest that it may be possible to frequency stabilize the miniature source at 532 nm using saturated absorption in I_2 vapor. The tuning range of Nd:YAG at 532 nm is known to include a number of strong I_2 absorption lines. [15]

A single axial mode, narrow linewidth green source is also a potential pump for a cw optical parametric oscillator.[6] The single axial mode nature of the pump source allows low threshold doubly resonant parametric oscillator operation. The threshold for a cw OPO is calculated to be between 30 and 60 mW for a 2.5 cm confocally focussed MgO:LiNbO3 crystal. The advantage of the parametric oscillator for this application is the retention of the coherence of the narrow linewidth pump source. The all solid state, monolithic construction also should lead to excellent linewidth at the tunable output.

The laser diode pump sources are efficient sources of radiation. We have shown that by using the laser diodes as pump sources the advantage of high efficiency is maintained. However, the advantages provided by the solid state laser medium of narrow linewidth, diffraction limited spatial mode output, and the ability to harmonically convert the near infrared radiation into the visible are gained. With the availability of higher power laser diodes or laser diode arrays, the performance level of laser diode pumped solid state sources should improve such that laser spectroscopy using tunable all solid state devices is possible.

Acknowledgements

This work was supported by the Army Research Office and the National Aeronautics and Space Administration. We want to acknowledge cooperation with Spectra Physics during early phases of this work. William Kozlovsky gratefully acknowledges support by the Fannie and John Hertz Foundation.

References

1. Monte Ross, Proc. of the IEEE, 56, 196 (1968).
2. L.J. Rosenkrantz, J. Appl. Phys. 43, 4603 (1972).
3. J. Stone and C.A. Burrus, Applied Optics, 13, 1256 (1974).
4. L.C. Conant and C.W. Reno, Applied Optics, 13, 2457 (1974).
5. L.B. Allen Jr., R.R. Rice, H.G. Koenig and D.D. Meyer, SPIE, 247 100 (1980).
6. B.K. Zhou, T.J. Kane, G.J. Dixon and R.L. Byer, Optics Letters, 10, 62 (1985).
7. T.J. Kane and R.L. Byer, Optics Letters, 10, 65 (1985).
8. D.L. Sipes, Appl. Phys. Letts. 47, 74 (1985).
9. T.J. Kane and R.L. Byer, Proc. of the 3rd Topical Meeting on Coherent Laser Radar - Technology & Applications, Great Malvern, Worcs. U.K. 1985.
10. T.Y. Fan, Stanford University, (private communication).
11. T.J. Kane. W.J. Kozlovsky and R.L. Byer, Technical Digest of the Conference on Lasers and Electro-optics (CLEO '85), ThK5, Baltimore, MD. 1985.
12. T.Y. Fan, A. Cordova-Plaza, M.J.F. Digonnet, R.L. Byer and H.J. Shaw, "Nd:MgO:LiNbO3 : Spectroscopy and Laser Devices", to be published.
13. T.Y. Fan, A. Cordova-Plaza, M.J.F. Digonnet, H.J. Shaw and R.L. Byer, "Laser Action and Self-Frequency Doubling in Nd:MgO:LiNbO3", to be published.
14. T. Baer and M.S. Keirstead, Post-deadline Papers of the Conference on Lasers and Electro-optics (CLEO '85), ThZZ1, Baltimore, MD. 1985.
15. R.L. Byer, R.L. Herbst and R.N. Fleming, "A Broadly Tunable IR Source", in Laser Spectroscopy, ed. by S. Haroche, et.al., (Springer-Verlag, Berlin 1975).
16. R.L. Byer, "Optical Parametric Oscillators", in Quantum Electronics, ed. by R. Rabin and C.L. Tang, (Academic Press, New York 1975).

Linewidth Reduction of Semiconductor Diode Lasers [1]

J. Harrison and A. Mooradian

Lincoln Laboratory, Massachusetts Institute of Technology,
Lexington, MA 02173, USA

The semiconductor diode laser (SL) has a number of properties that make it a desirable light source for spectroscopy. Aside from the advantages in size, cost and complexity, the SL can operate over a broad spectral range and may be easily modulated at rates up to several gigahertz. Here we discuss the issue of the ultimate SL linewidth. Included are comments on the prospect of line narrowing through active frequency control,as well as heterodyne measurements between two independently operating external cavity (GaAl)As lasers.

Measurements of the linewidth of (GaAl)As lasers as a function of power [1,2,3] first uncovered two anomalies in the spectral characteristics. The first was the appearance of a significantly broader (~ 25 times) power-dependent Lorentzian linewidth than that indicated by the SCHAWLOW-TOWNES analysis [1]. The major contribution to the power-dependent linewidth comes from the intrinsic coupling of amplitude and phase noise and is the same as that predicted for detuned gas lasers. The theory of this coupling was first presented for SL's by HENRY [4].

The coupling of amplitude and phase noise is manifested in SL characteristics other than the linewidth. For example, sidebands occur in the power spectrum at the relaxation oscillation frequency due to carrier dynamics [5,6,7]. These typically exhibit different intensities,indicating a combination of amplitude and frequency modulation [8,9]. Measurements indicate a strong correlation between amplitude and frequency noise [10, 11].

The second feature of SL spectra is the power-independent linewidth [3]. This appears to present a practical limit to the coherence and is on the order of 2 MHz at room temperature in (GaAl)As lasers [3]. A number of mechanisms have been proposed to explain the effect. Occupation fluctuations around the Fermi level [12] do not appear to account for the magnitude of the effect [13], especially at very low temperatures. Longitudinal mode beating [14] between the primary mode and small (possibly transient) submodes leads to an effect with a strict dependence on the ratio of total power in the primary mode that is not observed experimentally in "single-mode" lasers. A phenomenological model [3] suggests a relationship with electron number fluctuations. This model considers the variation in the Fabry-Perot cavity frequency with the active region electron density (due to the dependence of the refractive index on carrier concentration). Good agreement exists between this model and data taken at a number of temperatures down to 1.7 K [3,13]. While injection-rate variations have been shown to be channeled by gain saturation into the spectral sidebands [6], other types of carrier noise may directly impact the cen-

---

[1]This work was sponsored by the Department of the Air Force.

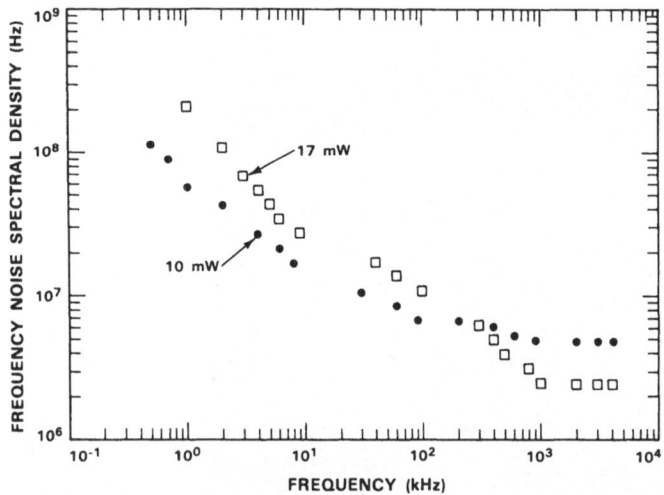

Fig. 1. Frequency noise spectral density of a single-frequency (GaAl)As
diode laser at room temperature

tral Lorentzian linewidth. An example is carrier trapping. A rigorous
treatment of such phenomena by ZEIGER [15] indicates that trapping may
significantly contribute to the power-independent linewidth. Further evidence of this relationship is the 1/f noise that is observed in the frequency (and amplitude) noise spectra of SL's [10,11,16-17] (Fig. 1). This
noise component represents fluctuations in the laser center frequency.
The 1/f cutoff frequency is about 1 MHz at room temperature.

The nature of the power-independent linewidth suggests that the linewidth at high powers may be reduced by adding to the laser drive current a
frequency-dependent component to compensate for the intrinsic center frequency fluctuations. This requires an optical discriminator in a feedback
loop,the bandwidth of which is at least equal to the linewidth. Results
have been obtained by SAITO et al. [18] in which the heterodyne signal between a narrow line master SL and a broader slave SL was reduced from
20 MHz to 1 MHz by applying feedback to the slave laser.

Optical feedback can be used to achieve dramatic linewidth reduction.
By placing a SL in an external cavity the linewidth may be reduced by as
much as the square of the increase in the photon lifetime (i.e., several
orders of magnitude). The external cavity SL is broadly tunable when a
dispersive element is included in the cavity. By employing compact,
stable cavity structures and carefully isolating the laser from acoustic
sources, one can construct a very narrow line, tunable source that can
deliver tens of milliwatts. High power pulsed SL's can also be used in
external cavities to produce much higher peak powers in a single longitudinal mode for efficient frequency mixing.

We have constructed two compact external cavity lasers (Fig. 2) using
Hitachi HLP-1400 (GaAl)As lasers. AR coatings were applied to the internal facets with residual reflectivities of about 1%. The lasers were
imaged onto 5% output couplers using commercial microscope objectives that
transmitted 62% of the incident light at 8300 Å. Line coincidence was

Fig. 2. Compact external cavity diode laser

achieved by current tuning,and no effort was made to thermally stabilize
the lasers. The heterodyne signal was derived from a Ge avalanche PIN
photodiode and observed on an HP 8566B spectrum analyzer (Fig. 3) on a
linear scale. The apparent heterodyne linewidth is about 30 kHz during
the 30 ms sweep. This is very close to the resolution halfwidth of the
analyzer with a 10 kHz resolution bandwidth. The heterodyne frequency
typically varied less than 25 MHz in ten minutes and is attributed to slow
thermal drift.

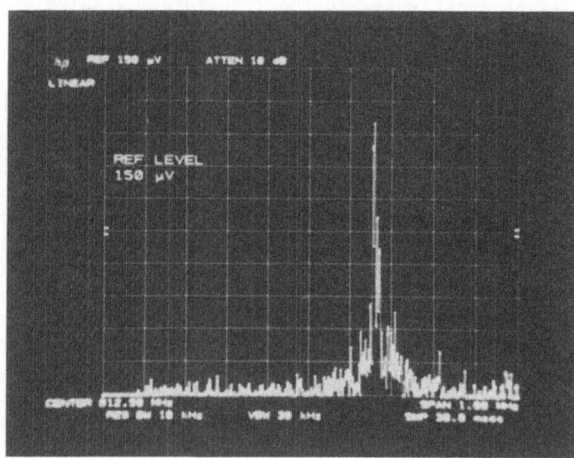

Fig. 3. Heterodyne beat spectrum between two free-running external cavity (GaAl)As diode lasers

The linewidth of the external cavity laser is given by

$$\Gamma = \Gamma_{SL} \left[(\ell/L)(\alpha\ell - \ln R)/(\alpha\ell - \ln \sqrt{RR_0}) \right]^2 \quad ,$$

where Γ and Γ_{SL} are the Lorentzian FWHM linewidths of the external cavity and solitary SL, respectively; ℓ is the SL length; L is the total external cavity length; α is the total internal loss coefficient; R is the facet reflectivity; and R_0 is the lumped reflectivity of the combined objective and output mirror. The external cavity lasers were operated at an output power of 0.44 mW corresponding to an internal mode power of 14 mW. At this level, the measured linewidth of the solitary device (before AR coating) was 15 MHz. Given ℓ = 300 μm, L = 9 cm, R = 0.35 and using a measured value of α = 25 cm^{-1} [19] we calculate that the ultimate observable heterodyne linewidth was 2Γ = 340 Hz. The dominant broadening mechanism in the experiment appeared to be coupling of acoustic noise to the cavity. External cavities promise to produce tunable light at subkilohertz linewidths.

References:
1. M. W. Fleming and A. Mooradian, Appl. Phys. Lett. 38, 511 (1981).
2. D. Welford and A. Mooradian, Appl. Phys. Lett. 40, 865 (1982).
3. D. Welford and A. Mooradian, Appl. Phys. Lett. 40, 560 (1982).
4. C. H. Henry, IEEE J. Quantum Electron. QE-18, 259 (1982).
5. C. H. Henry, IEEE J. Quantum Electron. QE-19, 1391 (1983).
6. K. Vahala and A. Yariv, IEEE J. Quantum Electron. QE-19, 1102 (1983).
7. B. Daino, P. Spano, M. Tamburrini and S. Piazzolla, IEEE J. Quantum Electron. QE-19, 266 (1983).
8. K. Vahala, C. Harder and A. Yariv, Appl. Phys. Lett. 42, 211 (1983).
9. S. Kobayashi, Y. Yamamoto, M. Ito and T. Kimura, IEEE J. Quantum Electron. QE-18, 582 (1982).
10. K. Kikuchi and T. Okoshi, Electron. Lett. 19, 812 (1983).
11. A. Dandridge and H. F. Taylor, IEEE J. Quantum Electron. QE-18, 1738 (1982).
12. K. Vahala and A. Yariv, Appl. Phys. Lett. 43, 140 (1983).
13. J. Harrison and A. Mooradian, Appl. Phys. Lett. 45, 318 (1984).
14. W. Elsasser and E. O. Gobel, IEEE J. Quantum Electron. QE-21, 687 (1985).
15. H. J. Zeiger, to be published in Appl. Phys. Lett.
16. F. G. Walther and J. E. Kaufman, Paper TUJ5 at the Topical Meeting on Optical Fiber Communication, March 1983.
17. G. Tenchio, Electron. Lett. 12, 562 (1976).
18. S. Saito, O. Nilsson and Y. Yamamoto, Appl. Phys. Lett. 46, 3 (1985).
19. B. C. Johnson, Ph.D. Thesis, MIT (1985).

Laser Frequency Division and Stabilization

R.G. DeVoe, C. Fabre*, and R.G. Brewer

IBM Research Laboratory, 5600 Cottle Road, K01/281,
San Jose, CA 95193, USA

The current method for measuring an optical frequency relative to the primary time standard, the cesium beam standard at ~9.2 GHz, utilizes a complex frequency synthesis chain involving harmonics of laser and klystron sources. The method has been extended recently to the visible region [1], to the 633 nm He-Ne laser locked to a molecular iodine line, with an impressive accuracy of 1.6 parts in 10^{10}. With the new definition of the meter, the distance traversed by light in vacuum during the fraction 1/299 792 458 of a second, the speed of light is now fixed and both time and length measurements can be realized with the same accuracy as an optical frequency measurement. In view of the complexity of optical frequency synthesis, these developments set the stage for originating complementary techniques for stabilizing and measuring laser frequencies which are more convenient.

This paper reports a sensitive optical interferometric technique *dual frequency modulation* (DFM) for measuring and stabilizing a laser frequency by comparison, in a single step, to a radio frequency rf standard. Conversely, a low-noise rf source can be stabilized by a laser frequency reference. A prototype [2] has demonstrated a resolution of 2 parts in 10^{10}, but devices currently under development should have a resolution of 10^{-12} and an absolute accuracy of $\sim 10^{-10}$. The method may be competitive with the optical frequency synthesis chain in accuracy and its simplicity suggests its convenient use in metrology, high-precision optical spectroscopy, and gravity wave detection.

The principle of the technique rests on phase-locking the mode spacing c/2L of an optical cavity to a radio frequency standard and simultaneously phase-locking a laser to the n-th order of the same cavity. When these two conditions are satisfied, the optical frequency ω_0 and the radio frequency ω_1 are simply related,

$$\omega_0 = n\omega_1 , \tag{1}$$

neglecting for the moment diffraction and mirror phase-shift corrections. The idea of locking a laser to a cavity is of course a well-established subject [3,4], but the concept of phase-locking an optical cavity to a radio frequency source is new. Interferometric rf-optical frequency comparisons of lower sensitivity have previously been performed by Bay *et al.* [5] using a related idea based on amplitude modulation (AM) rather than frequency modulation.

To introduce the DFM technique, first consider a single frequency modulation scheme. An electrooptic phase modulator driven at ω_1 generates a comb of optical frequencies $\omega_0 \pm m\omega_1$ which are compared to cavity modes of frequency $n\sigma$, where σ is the cavity free spectral range, m=0,1,2,... and n is a large integer $\sim 10^6$. The cavity response perturbs the balanced phase relationships between the sidebands and transforms frequency

---

*On leave from the Laboratoire de Spectroscopie Hertzienne de l'ENS, Paris

modulation into intensity modulation at ω_1. A photodetector, viewing the cavity either in reflection or transmission, then generates an error signal at the heterodyne beat frequency ω_1. This signal yields a null when the laser frequency ω_0 equals $n\sigma$ and the radio frequency ω_1 matches the mode spacing σ. In this circumstance, the comb of optical frequencies all resonate with their corresponding cavity modes. Although simple in principle, this single FM technique generates a complex error signal which depends not only on the rf detuning $\delta=\omega_1-\sigma$ but also on the optical detuning $\Delta=\omega_0-n\sigma$ and is therefore unsuitable for locking.

The DFM technique overcomes this problem by using two phase modulators, driven at frequencies ω_1 and ω_2 respectively, where $\omega_2/2\pi$ is nonresonant with c/2L. Dual frequency modulation creates, in lowest order, sidebands at $\omega_0\pm\omega_1$, $\omega_0\pm\omega_2$, and $\omega_0\pm\omega_1\pm\omega_2$. A photodetector views the cavity in reflection and two error signals are derived, one at ω_2 and the other at the *intermodulation* frequency $\omega_1\pm\omega_2$. The first signal at ω_2 allows locking the laser to the reference cavity as described elsewhere[3,4] and is independent of ω_1 tuning. The second signal at $\omega_1\pm\omega_2$ allows locking the cavity to the rf reference. This signal varies directly with the rf detuning $\delta=\omega_1-\sigma$ and provides the desired null at the rf resonance condition $\omega_1=\sigma$, while being independent of laser detuning Δ.

As has been shown elsewhere [2], this intermodulated DFM signal is given by

$$I(\omega_1\pm\omega_2) = I_0 \cdot C \cdot \sin\omega_2 t \cdot \{\text{Im}[g(\Delta + \delta) - g(\Delta - \delta)]\cos\omega_1 t\}$$

where I is the light intensity on the photodiode, g is the complex cavity lineshape function, and $C=4J_0(\beta_1)J_0(\beta_2) J_1(\beta_1)J_1(\beta_2)$, where β_1 and β_2 are the modulation indices of the phase modulators. Assuming a Lorentzian lineshape function g

$$g(\Delta) = \frac{\Delta(\Delta - i\Gamma)}{(\Delta^2 + \Gamma^2)} ,$$

the error signal becomes, for $\delta<\Gamma$

$$I(\omega_1\pm\omega_2) = I_0 \cdot C \cdot \frac{\delta}{\Gamma} \cdot [\sin(\omega_1 + \omega_2)t + \sin(\omega_1 - \omega_2)t]$$

By detecting the beat either at $\omega_1+\omega_2$ or $\omega_1-\omega_2$ in a double balanced mixer, an error signal proportional to the rf detuning δ can be derived for locking an optical cavity to an rf standard, or conversely an rf source to a cavity. Second, the error signal is independent of laser detuning Δ and thus optical frequency jitter. Third, there is no background signal. Fourth, the DFM signal has excellent signal-to-noise ratio since $C=4J_0(\beta_1)J_1(\beta_1)J_0(\beta_2)J_1(\beta_2)=.45$ at $\beta_1=\beta_2=1$.

An examination of the above equations shows that the intermodulated DFM technique realizes an *FM Differential Interferometer* which produces a locking signal similar to conventional FM laser locking, but in which the optical tuning parameter $\Delta=\omega_0-2\pi\cdot n\cdot c/2L$ is replaced by the radio frequency (differential) tuning parameter $\delta=\omega_1-2\pi\cdot c/2L$. The DFM sideband structure creates an optical subtraction in the photodiode of the (n+1)-th cavity resonance curve from the (n-1)-th curve and thus permits accurate, low noise measurements of the cavity mode spacing.

A prototype DFM standard (Fig. 1) [2] was constructed to verify the principle and study resolution and systematic errors. A home-made cw dye ring laser contains an intracavity ADP crystal for laser phase-locking to an external cavity at 6000Å. DFM sidebands are applied by two phase modulators driven at $\omega_1/2\pi=c/2L\sim300$ MHz and $\omega_2/2\pi=19.4$ MHz. The laser beam is then mode-matched onto the reference cavity through an optical circulator which both isolates the laser from the cavity and directs the

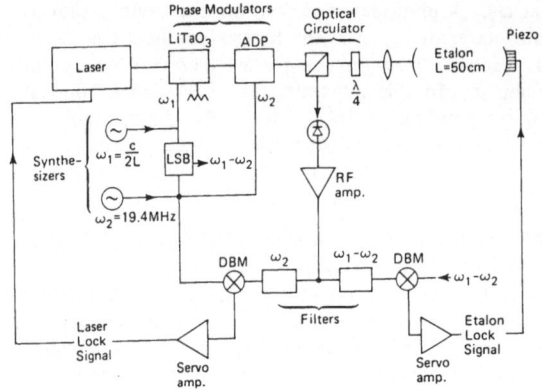

Figure 1. Block diagram of an optical frequency divider showing two servo loops where the laser is locked to a reference cavity and the cavity to a radio frequency standard. The LiTaO$_3$ modulator is driven at ω_1 and the ADP modulator at ω_2. LSB denotes a mixer and filter which generates the difference frequency $\omega_1-\omega_2$ for the double balanced mixer (DBM) in the cavity-rf servo.

reflected light, the signal, to a high-speed photodiode. The detector photocurrent contains rf beats at ω_2 and $\omega_1\pm\omega_2$ which are amplified and then separately filtered. The error signal at ω_2 is sent to an FM sideband servo which locks the dye laser to the n-th cavity fringe so that $\omega_0=n\sigma$ with a short-term error $\Delta/2\pi\sim300$ Hz RMS or less [4]. The DFM signal at $\omega_1-\omega_2$ is coherently detected in a double-balanced mixer. This signal can control the cavity length L via a piezo so that the resonance condition $\sigma=\omega_1$ is satisfied. In initial tests the piezo control loop was opened and noise levels of 60 millihertz RMS, corresponding to a fractional frequency deviation σ_y (Allan variance) of 2×10^{-10} in 0.1 sec integration time, were measured. The measurement of divider noise of 60 millihertz in the presence of 300 Hz of optical jitter verifies the differential behavior of the DFM standard and indicates that laser jitter is cancelled at the 10^{-3} to 10^{-4} level.

Four improvements have been made to the prototype to increase the resolution to 10^{-12} and to make possible absolute frequency comparisons at the 10^{-10} level. First, the interferometer itself has been placed in an acoustically isolated vacuum chamber and constructed from a 5 cm diameter by 47.5 cm long Zero-Dur spacer. Super-polished ring laser gyro quality mirrors have been used to achieve a finesse of >20,000 or a linewidth $\Gamma/2\pi<10$ kHz. Secondly, $\omega_1/2\pi$ has been raised to 4095 MHz to resonate with two cavity modes separated by 13 times the free spectral range of 315 MHz. Although this does not change the locking error δ, it reduces the *fractional* error and noise δ/ω_1 by a factor of 13. A simple but highly efficient resonant cavity microwave phase modulator has also been developed which will be described elsewhere. Third, two I$_2$-stabilized He-Ne lasers of an NBS design have been constructed to serve as optical frequency standards. Fourth, a low noise microwave source has been built whose frequency is known to $\sim10^{-11}$ by radio comparison to WWVB and Loran C.

These improvements described above have increased the effective Q of the system by a factor of >200, and thus resolution of 10^{-12} is expected. The theoretical shot-noise limited Allan variance of the new system is $\Gamma/(\omega_1\sqrt{N\tau})$ where N is the number of photoelectrons/sec and τ is the integration time. For $N=10^{15}$ sec^{-1} and $\tau=0.1$ sec $\sigma_y=2\times10^{-13}$.

New techniques have also been developed to measure systematic errors common to precision interferometry. Equation (1) must be modified by corrections due to phase shifts in the cavity arising from (a) the multilayer dielectric coatings on the mirrors and (b) diffraction effects. Previous workers [6] have corrected for (a) by using two different cavity lengths, which require disassembly and realignment of the interferometer. The DFM standard, on the other hand, is designed to reach 10^{-10} accuracy with mirrors which can be quickly moved in vacuum, without disturbing alignment, by only 4 cm. In addition, the diffraction phase shift (b) can also be determined by measuring the higher order transverse mode spacing.

The DFM technique, therefore, should provide an independent and accurate measurement of an optical frequency that can be compared to the results of the optical frequency synthesis chain [1].

We are indebted to H. P. Layer for providing us with a detailed design of the NBS He-Ne iodine stabilized laser, to J. L. Hall for informing us of his design of an acoustically isolated interferometer chamber, to K. L. Foster for assistance in all phases of their fabrication, and to F. Walls for help with ultra-stable crystal oscillators. This work is supported in part by the U.S. Office of Naval Research.

References

[1]   D. A. Jennings, C. R. Pollack, F. R. Peterson, R. E. Drullinger, K. M. Evenson, and J. S. Wells: Optics Letters $\underline{8}$, 136 (1983).
[2]   R. G. DeVoe and R. G. Brewer: Phys. Rev. $\underline{A30}$, 2827 (1984).
[3]   R. W. P. Drever, J. L. Hall, E. V. Kowalski, H. Hough, G. M. Ford, A. J. Munley, and H. Ward: Appl. Phys. $\underline{B31}$, 97 (1983).
[4]   R. G. DeVoe and R. G. Brewer: Phys. Rev. Lett. $\underline{50}$, 1269 (1983).
[5]   Z. Bay, G. G. Luther, and J. A. White: Phys. Rev. Lett. $\underline{29}$, 189 (1972).
[6]   H. P. Layer, R. D. DesLattes, and W. G. Schweitzer: Appl. Optics $\underline{15}$, 734 (1976).

FM Dye Lasers for Use in Optical Metrology

D.M. Kane, S.R. Bramwell, and A.I. Ferguson

Department of Physics, University of Southampton,
Southampton, S09 5NH, England

We have been exploring the use of frequency modulated dye lasers as a
source of broadband laser radiation which will provide a comb of modes
for precision comparison of laser frequencies. An ideal FM laser output
consists of a laser beam which is constant in amplitude but sinusoidally
varying in frequency. This provides us with a source of many laser modes
which are equally spaced by the frequency at which an intracavity phase
modulator is driven.

In application to spectroscopy and metrology, a laser of unknown
frequency can be compared with a standard laser frequency by heterodyning
both lasers and FM laser on a nonlinear detector. Large laser frequency
intervals can be compared by measuring three low-frequency beats. These
are the driving frequency on the phase modulator, a beat between the
known laser and one of the FM modes and thirdly a beat between the
unknown laser and another of the FM modes. The frequency difference
between the lasers can then be accurately measured without systematic
errors caused by phase-shifts on coatings, diffraction or alignment effects
which are characteristic of interferometry.

HARRIS and TARG (1) first demonstrated the FM laser using a 633nm HeNe
laser. The linear theory of the FM laser developed by HARRIS and McDUFF
(2) shows that by coupling FM sidebands closely to the longitudinal mode
spacing of a laser $\Delta\Omega$, an enhancement of the modulation index Γ is
obtained and is given by $\Gamma = (\delta/\pi)(\Delta\Omega/\Delta\nu)$ where δ is the single pass
phase retardation, $\Delta\Omega$ is the laser mode spacing and $\Delta\nu$ is the detuning
of the RF modulation frequency ν_m from the cavity mode spacing, given by
$\Delta\nu = \nu_m - \Delta\Omega$. As $\Delta\nu$ is decreased, amplitude modulation increases and
the output is mode-locked for a small range of detunings about $\Delta\nu = 0$.
AMMAN et al (3) showed that the detailed behaviour of the FM HeNe laser
fitted very well with that predicted by theory. KUIZENGA and SIEGMAN (4)
showed good agreement with this theory in the homogeneously broadened
Nd:YAG laser. They obtained FM spectra with a bandwidth of up to 50GHz
which appeared to be near ideal Bessel intensity spectra.

The dye laser we have used is a single frequency CR699-21 run as a
standing wave laser of extended length (~1.2m). The phase modulator is
located close to the output mirror which is mounted on a translation stage.
A single Brewster angled crystal of 45°y-cut ADP with dimensions of 4mm x
6mm x 30mm, electroded on the 6mm x 30mm faces, is used as the phase
modulator. This is resonated with a lumped inductance at 126-128 MHz and
matched into a 50Ω cable. Single pass phase retardations of up to 0.6
radians are obtained using up to 2W of RF power.

Initially, the laser was operated using the standard intracavity assembly
(ICA) containing etalons of 10GHz and 225GHz free spectral range, to ensure

single frequency operation. Intracavity phase modulation of **this** laser gives FM bandwidths up to 3GHz. A sample spectrum for a small value of δ is shown in fig. 1. A computer-generated Bessel intensity spectrum is also shown for comparison. The values of δ calculated from the values of Γ obtained by fitting a series of Bessel intensity spectra to the single frequency FM dye laser spectra, using the known detunings, is the same as that measured by extracavity phase modulation of a HeNe laser, within the experimental error (10%).

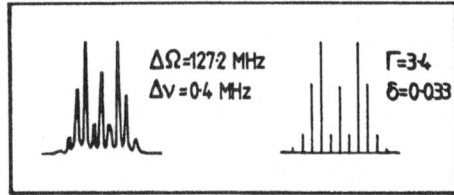

Fig. 1 FM spectrum obtained by intracavity phase modulation of a single frequency dye laser as compared to a Bessel function spectrum.

For small values of Δν mode-locked behaviour is observed. Pulses with repetition rates of c/2L and 2(c/2L) are seen. The latter occurs when both of the possible mode-locked pulse states oscillate simultaneously, and occurs over most of the mode-locked region. Pulses at the driving frequency occur for a smaller range of detunings as δ is increased.

When the laser is operating as a near ideal FM dye laser, the RF beat spectrum of the output shows beats at the first, second and third harmonic of the RF driving frequency (with a resolution of -80dBm). Many more RF beats (up to 10 are observed in the mode-locked region and these are at least 20dBm higher in power than those observed in the FM region.

To increase the FM bandwidth a broader bandwidth free running dye laser is modulated. The ICA is replaced with a single solid etalon with a free spectral range (FSR) of 225GHz, 400GHz or 900GHz leading to free running linewidths of 3GHz, 7GHz and 16GHz respectively. Modulated bandwidths of up to 12GHz, 20GHz and > 30GHz respectively are obtained from these free running dye lasers. With no etalons in the dye laser cavity and just the three plate birefringent filter free running linewidth of 40GHz is observed and this leads to a modulated bandwidth of up to 175GHz. The bandwidth and time-resolved behaviour of the dye laser operated with a single 225GHz FSR etalon is shown in Figure 2. The oscillatory time behaviour for large detunings is due to a passive interaction between the thin etalon and the phase modulator. Again, a large enhancement in the RF beat amplitudes is observed in the mode-locked region. The optical frequency spectrum of the FM laser appears like an ideal FM spectrum (Fig. 2 Δν = 200KHz) but the RF beat spectrum is very similar to that for the free running laser. It will be necessary to generate single frequency UV by sum frequency mixing, as has been proposed (4,5) or to use the FM laser for spectroscopy to test the FM status of the output.

The FM laser spectrum is free to drift in frequency as the laser cavity length changes due to fluctuations in the dye jet, thermal expansion and vibration. For spectroscopic applications,it is necessary to stabilise the carrier frequency. We have locked the FM output to an interferometer with a FSR equal to the RF driving frequency as has been done for mode-locked lasers (6). Also, we have frequency offset-locked the FM dye laser to a single frequency dye laser using a scheme we developed for

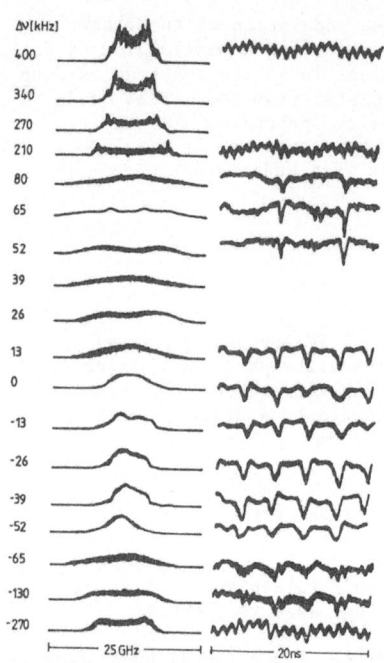

$\Delta \nu$[kHz]

400
340
270
210
80
65
52
39
26
13
0
-13
-26
-39
-52
-65
-130
-270

|← 25 GHz →| |← 20ns →|

Fig. 2. The bandwidth and time-resolved behaviour of a phase-modulated dye laser ($\nu_m = 127$MHz) with a free running linewidth of 3GHz. Note that occurrence of pulsing at c/2L and 2(c/2L) and the complex series of changes in the mode envelope in the mode-locked region.

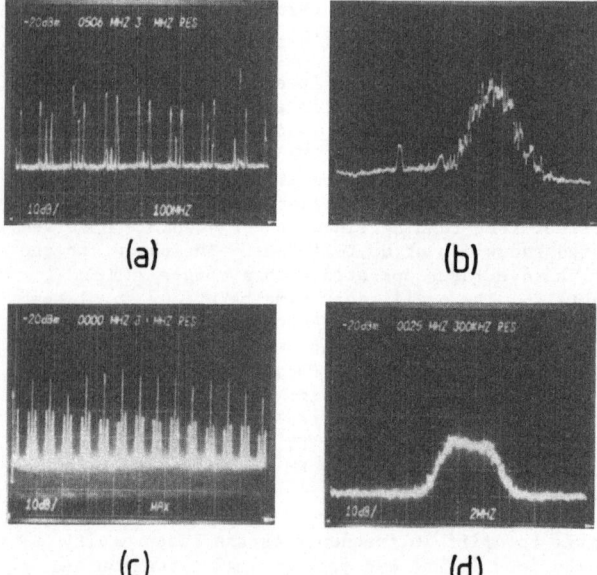

(a) (b)

(c) (d)

Fig. 3 The RF beat spectrum obtained by frequency offset-locking a FM (a), a FM mode-locked laser (c), to a single frequency dye laser. In (b) and (d) an expanded view of the lowest frequency stabilised beat is shown for the FM and FM mode-locked lasers respectively.

frequency stabilising synchronously pumped mode-locked lasers (7). This scheme works well for the FM and FM mode-locked laser, as is shown in Fig. 3. The RF beats occur at multiples of 127MHz and are flanked by pairs of heterodyne beats between the two lasers. Mode stability of less than 3MHz is observed for the FM laser, and of less than 4MHz for the FM mode-locked laser. Note that Figs. 3a and 3c also show the difference in the RF beat spectrum for the FM and FM mode-locked lasers respectively.

Acknowledgements

This work was supported by the National Physical Laboratory and the Science and Engineering Research Council.

References

1. S E Harris and R Targ, Appl. Phys. Lett. 5, 202 (1964)
2. S E Harris and O P McDuff, IEEE J.Quant.Electron.QE-1, 245 (1965)
3. E O Amman, B J McMurtry and M K Oshman, IEEE J.Quant. Electron. QE-1, 263 (1965)
4. D J Kuizenga and A E Siegman, IEEE J.Quant. Electron. QE-6, 673 (1970)
5. H P Weber and E Mathieu, IEEE J.Quant. Electron. QE-3, 376 (1967)
6. A I Ferguson and R Taylor, Opt. Commun, 41, 271 (1982)
7. S R Bramwell, D M Kane and A I Ferguson, submitted to Opt. Comm.

Tunable Far Infrared Spectroscopy

K.M. Evenson, D.A. Jennings, K.R. Leopold, and L.R. Zink

Time and Frequency Division, National Bureau of Standards,
Boulder, CO 80303, USA

Introduction

In this paper we describe the generation of far infrared (FIR) radiation
with the metal-insulator-metal (MIM) diode and the operation of a
spectrometer employing this diode. This technique is an extension of the
use of the MIM diode from its use in the measurement of frequencies to
the generation of far infrared radiation between 0.3 and 6.0 THz. The MIM
diode has previously been used in frequency measurements yielding a
definitive value for the speed of light and in the measurement of the
frequency of visible radiation.

The MIM diode is the non-linear device used in the direct frequency
measurement of the 88 THz (3.39 micron) helium-neon laser stabilized on
methane. This frequency when multiplied by the wavelength of that laser,
gave a value of the speed of light which was a hundred times more accurate
than the previous values [1].

We have used the MIM diode to measure frequencies up to 200 THz (1.5
microns), which is about the upper frequency limit of the MIM. To reach
the visible, bulk doublers were used, and two accurate values of molecular
iodine absorptions were measured [2, 3]. These led the way to the
redefinition of the meter: "The meter is the length of the path travelled
by the light in vacuum during the time interval 1/299 792 458 of a second"
[4]. This definition fixes the value of c at exactly 299 792 458 meters
per second and permits the use of the laser in realizing the meter.

The MIM Diode

In the FIR we couple radiation in or out of the MIM diode by using long
wire antenna coupling. When the wavelength of the radiation becomes
comparable to the diameter of the whisker (in the IR) we use conical
antenna coupling [5].

In the measurement of frequencies, the MIM diode is used as a harmonic-generator and mixer, and a radio frequency heterodyne difference (a
beat note) is generated. In the measurement of methane, the radio
frequency signal is generated from 3 times the frequency of CO_2 plus the
microwave heterodyning with the methane stabilized He-Ne radiation [6].
The diode is used in this case to actually generate the radio frequency
difference.

Contribution of the National Bureau of Standards, not subject to
copyright.
Supported in part by NASA grant no. W-15, 047,and The Chemical Manufactur-
ers Association

Generation of FIR Radiation

We have observed very large signals when we mix the radiation from two CO_2 lasers a few MHz apart. These signals led us to speculate that FIR radiation might radiate from the diode if we selected CO_2 laser lines differing by a far infrared frequency. Using about 500 milliwatts of CO_2 radiation i.e., 250 milliwatts from each laser, we were able to radiate 0.2 microwatt of far infrared radiation [7]; we have now increased that to 0.7 microwatt from 100 mW from each CO_2 laser. We have generated radiation from 0.3 to 6.3 THz. The tunability of the FIR radiation results from the use of a high-pressure wave guide CO_2 laser tunable by about 150 megahertz. By using three isotopes of CO_2 most of the far infrared can be covered using various pairs of CO_2 lasers.

To demonstrate the fact that we had produced tunable far infrared radiation, we took spectra of a rotational line of CO. Figure 1a shows one of our first traces.

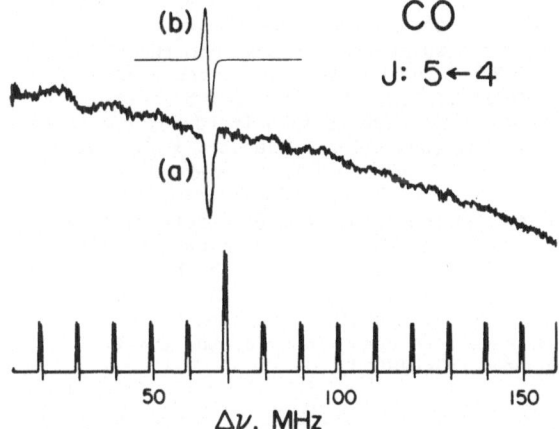

Fig. 1. CO ground state absorption at 576 GHz. Trace (a) is the first trace made using CO_2 difference radiation, and trace (b) is a recent trace using the spectrometer shown in figure 2.

FIR Spectroscopy

The FIR region had been a fairly inactive region because there were neither good sources of radiation nor good detectors. That situation is now changed: first, there are over a thousand fixed frequency FIR laser lines; second, very good detectors have become available which are limited by room temperature blackbody radiation. With these advances, radio astronomy expanded into the FIR, and measurements made on our atmosphere from balloons or high-flying airplanes are also made in the FIR region.

Two types of transitions lie in this region: fine structure transitions in atoms and molecules (these are magnetic dipole transitions), and rotational transitions in lighter molecules. The latter are very strong transitions and yield a high sensitivity. For example, FIR transitions are some 10^5 times stronger than microwave transitions. They are about an order of magnitude stronger than infrared transitions (vibrational), but they are about a hundred times weaker than electronic transitions in the visible. (These relative strengths are based on the absorption per molecule at a pressure of 133 Pa.)

There are several other techniques of generating FIR radiation. Some backward wave oscillators operate up to a little above 1 THz [8]. The oldest technique is that of generating harmonics from a klystron or backward wave oscillator [9, 10]. A more efficient technique is that of generating a far infrared laser sideband [11,12]. Difference frequency radiation between CO_2 lasers can be generated using either non-linear crystals [13] or the MIM diode [7]. Although the MIM diode does not provide as much power as the other techniques, it is convenient and by using CO_2 stabilized lasers, it provides an accuracy of 35 kHz and a spectral purity of about 10 kHz.

The Spectrometer

The complete spectrometer (which we call a TuFIR spectrometer) is shown in fig. 2. The two drive CO_2 lasers (CO_2 laser number I and the CO_2 waveguide laser) are combined on the beam splitter and then are focused on the MIM diode where the far infrared radiation is generated. Laser I is frequency-modulated and the derivative of the absorption signal is observed following lock-in detection. Laser II is used to control the frequency of the waveguide laser with the radio frequency sweep. Opto-acoustic modulators are used to isolate the lasers from the MIM diode and to provide an additional 90 MHz of tunability. The improvements in the spectrometer are indicated in Fig.1b in which the same transition as in Fig. 1a was recorded. The signal to noise is much improved; the noise level now corresponds to a fractional absorption of about 1×10^{-4} with a 1 s time-constant.

Seven CO lines have been observed and the rotational constants calculated. These are used to predict the rotational frequencies with uncertainties from about 30 kHz for low J lines to about a 100 kHz on the high J lines. These frequencies are very useful as calibration standards and will be published elsewhere. The first four rotational frequencies of HF which are about 12 times those of CO, have also been measured, and their rotational constants have also been calculated.

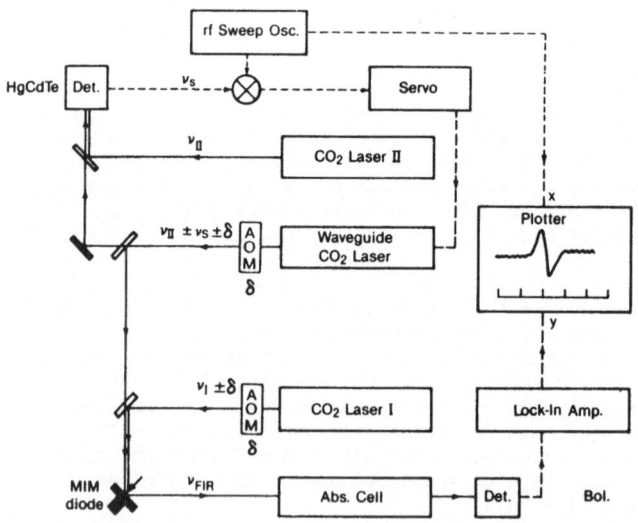

Fig. 2. Tunable Far Infrared spectrometer using a MIM diode to generate the difference frequency from two CO_2 lasers.

The high sensitivity of the technique is demonstrated by the observation of the spectra of the OH free radical. The rotational constants have been calculated and then used to predict the rotational frequencies to an accuracy of about ± 100 kHz. The result is an improvement in the accuracy of these lines by about two orders of magnitude. These OH transition frequencies can now be used by radio astronomers and by aeronomers to detect and study this species important in our atmosphere and in interstellar space. Another important application of this spectroscopy will be in line broadening studies of OH so that atmospheric spectroscopic data can be used to yield the concentrations of OH in our atmosphere as a function of altitude.

Two recently observed transient molecules are NaH and MgH. We have observed six rotational transitions of NaH in the $v = 0$ manifold and six transitions in the $v = 1$ manifold. The data are being analyzed and will be published elsewhere. MgH is a very important astronomical molecule and the observation of more lines will allow the prediction of its frequencies for astronomical searches.

The exciting uses of FIR spectroscopy have been indicated, and we have shown some of the interesting molecules which are available for future studies.

Future

How efficient is the MIM diode? The I-V curve of tungsten-nickel is nearly a straight line; but its small non-linearity allows the MIM diode to detect and to generate difference frequencies. Because of its small non-linearity, it is not a very efficient device. By comparison, the Schottky diode (which does not operate well above 1 THz) has a very sharp bend in its I-V curve. Are there other more efficient devices that will also work at this high speed? Klaus Siemsen from the Canadian National Research Council tested a tin telluride diode in our NBS laboratory. It is about five times more efficient than nickel; however, it is less stable, and we still prefer nickel as a base.

The I-V curve of tungsten-cobalt is nearly symmetrical (tungsten-nickel's curve is asymmetrical). This diode does not detect well and does not produce difference radiation as predicted by its symmetrical I-V curve; however, it does generate a 3rd order signal very well. This permits the generation of the difference between two CO_2 lasers plus and minus microwave radiation. This is very exciting, because we now obtain nearly 100 times more tunability. For example, in 2nd order (CO_2 difference), a tunability of ± 300 MHz is possible, but in 3rd order with fixed CO_2 lasers and a microwave sweeper, the tunability is ± 20,000 MHz. Thus we can now cover the entire far infrared by using 150 pairs of CO_2 differences plus and minus microwave radiation.

The MIM diode is more and more perplexing. The mechanism responsible for the operation of the diode seems to be tunneling; however, we are not even sure of that [5].

Does the diode also radiate the sum frequency? It probably does, but this is in the region where blackbody room-temperature radiation swamp the potential IR radiation from the diode; we will need a heterodyne or interferometric technique to detect it.

Thin film metal-insulator metal diodes would be much more reliable. Some initial research has been done along this line [14], but more is desirable. Communication's applications with terahertz bandwidth is a very exciting possibility with such devices.

The use of other metals may extend the operation of the MIM diode beyond 1.5 microns to the visible. Third-order diodes would be very useful in the visible so that one could directly heterodyne various pairs of visible oscillators with IR oscillators to obtain highly accurate frequency measurements across the visible.

Future Spectroscopic Measurements

The Rydberg will certainly be directly frequency-measured in the near future. The time standard itself presently is cesium in the microwave region: it will be improved by optically pumping cesium with a laser [15]. Further in the future, a stored ion standard in the visible [16] may become the frequency standard, then frequency multiplication from the microwave to the visible will be absolutely essential.

In closing, the future of this field of frequency measurement and synthesis looks very bright indeed!

The authors would like to acknowledge the assistance in bolometer design and construction provided by I. G. Nolt, J. V. Radostitz, S. Predko, and P.A.R. Ade; to E.C.C. Vasconcellos for testing various MIM diodes; to Klaus Siemsen for bringing the tin telluride base to us; to A. Hinz for his measurements on the spectrometer; to M. Mizushima for numerous discussions on diatonic molecules; and to J. M. Brown for his experimental and theoretical expertise.

1. K. M. Evenson, J. S. Wells, F. R. Petersen, B. L. Danielson, G. W. Day R. L. Barger, and J. L. Hall, Phys. Rev. Lett., 29, 1346 (1972).
2. C. R. Pollock, D. A. Jennings, F. R. Petersen, J. S. Wells, R. E. Drullinger, E. C. Beaty, and K. M. Evenson, Opt. Lett., 8, 133 (1983).
3. D. A. Jennings, C. R. Pollock, F. R. Petersen, R. E. Drullinger, K. M. Evenson, and J. S. Wells, Opt. Lett. 8, 136 (1983).
4. "Documents Concerning the New Definition of the Metre", Metrologia 19, pp. 163-177 (Springer-Verlag, 1984).
5. K. M. Evenson, M. Inguscio, and D. A. Jennings, J. Appl. Phys. 57, 956 (1985).
6. K. M. Evenson, J. S. Wells, F. R. Petersen, B. L. Danielson, and G. W. Day, Appl. Phys. Lett. 22, 192 (1973).
7. K. M. Evenson, D. A. Jennings, and F. R. Petersen, Appl. Phys. Lett. 44 576 (1984).
8. A. F. Krupnov, "Modern Aspect of Microwave Spectroscopy", ed. G. W. Chantry (Academic Press, New York, 1979), p. 217.
9. P. Helminger, J. K. Messer, and F. C. DeLucia, Appl. Phys. Lett. 42, 309 (1983).
10. E. A. Cohen and H. M. Pickett, J. Mol. Spectrosc. 93, 83 (1982).
11. D. D. Bícaníc, B. F. J. Zuidberg, and A. Dyamanus, Appl. Phys. Lett. 32, 367 (1978).
12. H. R. Fetterman, P. E. Tannenwald, B. J. Clifton, W. D. Fitzgerald, and N. R. Erickson, Appl. Phys. Lett. 33, 151 (1978).
13. R. L. Aggarwal, B. Lax, H. R. Fetterman, P. E. Tannenwald, and B. J. Clifton, J. Appl. Phys. 45, 3972 (1974).
14. K. M. Evenson, M. Inguscio, and D. A. Jennings, J. Appl. Phys. 57, 960 (1985).
15. R. E. Drullinger, private communication.
16. J. C. Bergquist, D. J. Wineland, W. M. Itano, H. Hemmati, H.-U. Daniel, and G. Leuchs, "Two-Photon Optical Spectroscopy of Trapped HgII" (chapter in this volume).

Photothermal Spectroscopy Using a Pyroelectric Calorimeter

H. Coufal

IBM Research Laboratory, 5600 Cottle Road, K34/281,
San Jose, CA 95193, USA

1. Introduction

In photothermal spectroscopy, the sample under investigation is excited with a modulated or pulsed light source [1]. Via radiationless decay part of the absorbed light is released in the sample as heat. With the incident energy being either modulated or pulsed, the heat generation will also show a corresponding time dependence. Thermal waves are, therefore, generated and, due to thermal expansion, acoustic waves are induced. These waves can be detected with suitable transducers, such as pyro- or piezoelectric transducers. With a pyroelectric thin film calorimeter recently, a sensitivity of nanojoules was obtained at a time resolution of tens of nanoseconds [2]. Only the *absorbed* and via radiationless decay into heat converted fraction of the incident light energy contributes towards the photothermal signal. The fact that the absorbed energy is determined directly and not as the difference of incident and reflected or transmitted light energy makes photothermal techniques particularly suitable for weakly absorbing samples, strongly scattering or opaque samples. Combined with laser excitation, this detection method, therefore, offers a unique combination of advantages, such as high sensitivity, high spectral resolution, high time resolution and instrumental simplicity.

2. Pyroelectric Thin Film Calorimeter

Assuming a temperature independent pyroelectric coefficient and homogeneous pyro- and dielectric properties, the charge generated in a pyroelectric element is proportional to the change of the heat content of this element and independent of the temperature distribution. If heat losses are negligible, as for example in a thin, self-supporting pyroelectric film, the element is effectively a calorimeter. For a sample deposited onto such calorimeter, the heat flow from the film into the calorimeter is proportional to the temperature averaged over the interface. Assuming the validity of classical heat conduction processes, a metal film with a thickness of 30 nm, as used here as electrode, introduces, due to its finite thermal diffusivity, a time delay of several picoseconds between heat generation at the irradiated surface of the film and the arrival of the heat pulse at the interface between film and calorimeter.

In this study, ferroelectric polyvinylidene fluoride (PVDF) foils (Pennwalt KYNAR®) were used as the active pyroelectric element. The PVDF foils were coated on both sides with 30 nm thick electrodes. The coated foil was stretched over a 11 mm diameter stainless steel supporting ring and held in place by a retaining ring. These rings served as heat sinks and as electrical contacts and mechanical support. The complete assembly is contained in a standard Inficon® microbalance housing for rf-shielding. With the described setup rise times as fast as 20 ns have been obtained; the minimum detectable amount of energy is of the order of 1 nJ.

3. Applications

By now a number of applications has been explored with this detector [3]. Vibrational *spectroscopy* of adsorbates demonstrated a sensitivity sufficient to detect few thousands of a monolayer under rigorous UHV conditions [4]. Experiments with Langmuir-Blodgett films underline the compatibility of the detector with most media. In these experiments, the distance of a dye layer from the silver electrode of the calorimeter was varied by a increasing number of sublayers. The photothermal signal as a function of distance shows the expected electric field distribution and the quenching of the dye fluorescence close to the metal electrode [5]. In another series of experiments, a periodic surface profile was generated via a holographic exposure of a photoresist. When overcoated by a silver film, light from a CW-Laser can be coupled into surface plasmon polaritons of the silver film [6]. The shift of this resonance due to adsorbates or dielectric thin films allowed for a direct observation of the dispersion relation.

This new device enables, in addition, highly time-resolved studies of *transient laser induced heating processes* at surfaces and in thin films. Exciting 30 nm thick Te-films with pulsed lasers melting, boiling and recrystallization of the film was observed [7]. These effects were found to be independent from the exciting wavelength. In another experiment the surface temperature rise during laser-induced thermal desorption of Xenon atoms from a copper surface was studied [8]. For the first time, actual surface temperatures are correlated with translational temperatures of desorbed particles as determined by time-of-flight mass spectrometry. In another type of application, heat pulses, generated with a suitable laser, were used to probe the thermal diffusivity of thin films [9] and to obtain nondestructive depth profiles of samples [10]. As a by-product of this technique, a thermal wave phaseshifter was developed [11]. With this technique, the transmitted energy, as well as the absorbed and into heat converted energy, can be determined independently. The quantum yield for radiationless deexcitation can be determined from these data directly.

4. Conclusion

In conclusion, the viability and versatility of pyroelectric thin film calorimeters, as fast and sensitive detectors for laser spectroscopy and laser induced thermal processes, has been demonstrated

References

[1] See for example C. K. N. Patel and A. C. Tam: Rev. Mod. Phys. 53, 517 (1981).
[2] H. Coufal: Appl. Phys. Lett. 44, 59 (1984).
[3] Detailed accounts will be presented at the 4th Int. Topical Meeting on Photoacoustics, Photothermal and Related Sciences, Montreal, August 1985, and are included in the Technical Digest of this meeting.
[4] H. Coufal, F. Träger, T. Chuang and A. Tam: Surf. Sci. 145, L504 (1984).
[5] W. Knoll and H. Coufal: Appl. Phys. A., to be published.
[6] R. Grygier, W. Knoll and H. Coufal [3].
[7] H. Coufal and W. Lee, Springer Series in Chemical Physics, 39, 25 (1984).
[8] I. Hussla, H. Coufal, F. Träger and T. Chuang [3].
[9] P. Hefferle and H. Coufal [3].
[10] H. Coufal: J. Photoacous. 1, 413 (1984).
[11] H. Coufal: Appl. Phys. Lett. 45, 516 (1984).

Interferometric Frequency Measurement of ^{130}Te$_2$ Transitions in the Region of the Balmer β Line in Hydrogen and Deuterium

J.M. Girkin, J.R.M. Barr, and A.I. Ferguson

Department of Physics, Southampton University, U.K.

G.P. Barwood, P. Gill, W.R.c. Rowley, and R.C. Thompson

The National Physical Laboratory, Teddington, U.K.

Atomic hydrogen has always been used to test and refine atomic theory and to measure atomic constants. Of particular interest, in recent years, has been the 1S-2S transition. This transition can be investigated using two-photon Doppler-free spectroscopy at 243nm and has a natural linewidth of a few Hertz. Previous experiments have compared this transition with Balmer β to measure the 1S Lamb shift, but they have not provided an absolute energy calibration (1). We have calibrated two Doppler-free lines

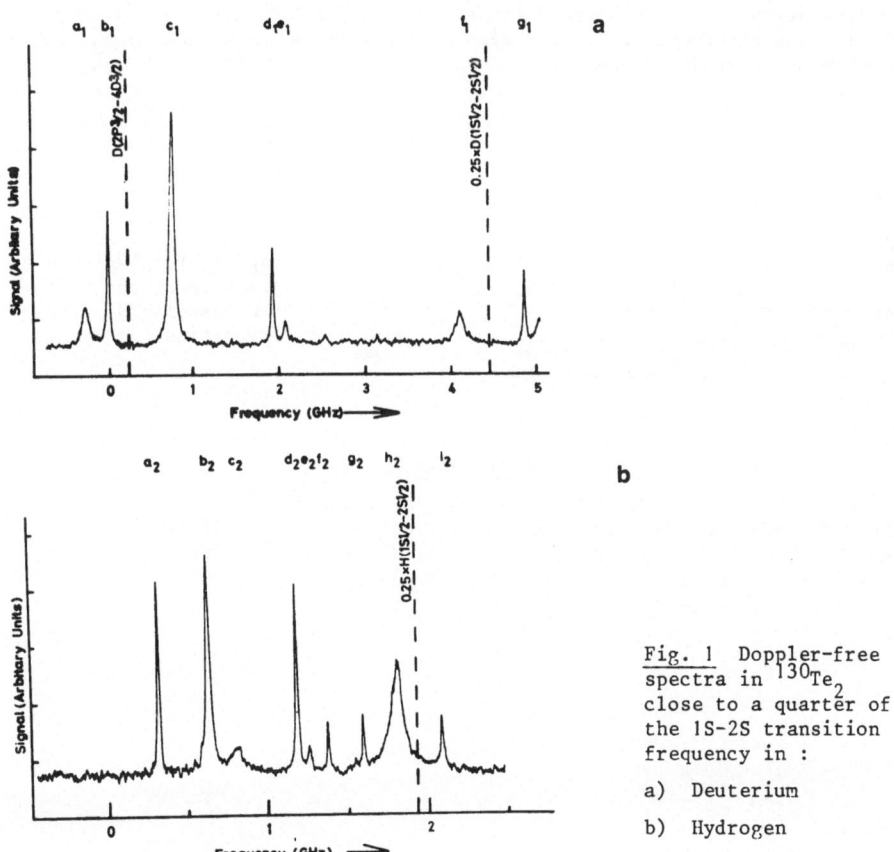

Fig. 1 Doppler-free spectra in ^{130}Te$_2$ close to a quarter of the 1S-2S transition frequency in :

a) Deuterium

b) Hydrogen

in Te$_2$ vapour for use as a frequency standard for hydrogen and deuterium spectroscopy at 486nm.

The system consists of ^{130}Te$_2$ vapour enclosed in a quartz cell heated to 630°C. A continuous wave dye laser was then locked to the selected Doppler-free components in a standard saturation spectrometer using a first derivative locking technique. After the preliminary wavemeter measurement (to 30MHz) to determine the fringe order number, the laser light was sent into a 150MHz Fabry-Perot interferometer at the N.P.L. (2). During these precision measurements,the pressure shifts due to cell temperature changes were monitored and an empirical formula derived. The two spectra are shown in fig. 1(a) and 1(b). The frequencies of the two calibrated lines were found to be (3).

$$b_1 = 616\ 678\ 857.5 \pm 0.25 \text{ MHz}$$
$$b_2 = 616\ 513\ 896.3 \pm 0.25 \text{ MHz}$$

At Southampton we are now using these lines in measurements on the 1S-2S, and Balmer β transitions in hydrogen and deuterium. A C.W. dye laser beam is split, part sent into the Te$_2$ spectrometer and part into a four-stage travelling wave amplifier pumped by a frequency tripled NdYAG laser. Then frequency doubled in Urea. The resulting 243nm radiation is used to carry out a Doppler-free two-photon resonantly enhanced three-photon ionisation experiment on the 1S-2S transition in atomic hydrogen. This spectrum is shown in fig. 2. These measurements will be used in precision determination of the Rydberg constant, the 1S Lamb shift, and the isotope shift. The reference lines will also be used with a metastable atomic beam experiment to study the Balmer β transition.

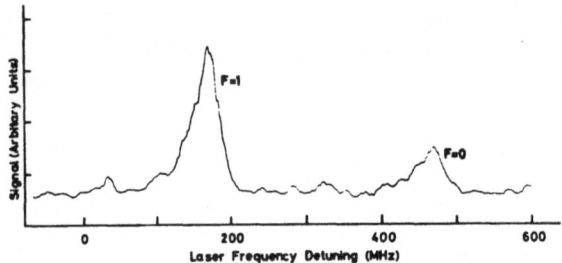

Fig. 2 Doppler-free spectrum of the hydrogen 1S-2S transition

References

1. C Weiman, T W Hansch. Phys. Rev. A. 22, 192 (1980)
2. P T Woods, K C Shotton, W R C Rowley.Appl. Opt. 17 1048 (1978)
3. J R M Barr, J M Girkin, A I Ferguson, G P Barwood, P Gill, W R C Rowley, R C Thompson. Opt. Commun; to be published.

A Computer-Controlled Two-Dye-Laser Heterodyne Spectrometer of High Precision

S. Grafström, U. Harbarth, J. Kowalski, R. Neumann, S. Noehte, and G. zu Putlitz

Physikalisches Institut der Universität Heidelberg, Philosophenweg 12, D-6900 Heidelberg, Fed. Rep. of Germany

A computer-controlled heterodyne spectrometer [1,2], consisting of two self-built free-jet ring dye lasers (output power 600 mW each), was constructed for optical measurements of high precision and excellent signal-to-noise ratio. It substitutes

1. Fabry-Perot interferometers of exactly known and constant free spectral range for the precise calibration of laser frequency scans;

2. expensive microwave systems, including different sets of microwave sources, amplifiers, waveguides, tunable resonant cavities, etc. in order to cover large frequency ranges, e.g. from 1 to 150 GHz.

The present device (see Fig.1) makes it possible to measure optical-line splitting frequencies e.g. via saturation dip or Doppler-free two-photon absorption technique with a precision in general only obtained by microwave spectroscopy. This is achieved in the following way:

One laser is locked to a line of I_2 (10^{-10} long-term stability) at approximately equal frequency separations from both optical lines under investigation, while the second laser is tuned in a precisely controlled way over the line profiles. The beat frequency of the lasers is produced by a fast avalanche diode and counted with a calibrated high-frequency counter. The computer realizes a series of beat frequency values by comparing each desired frequency with the counter reading (actual value) and changing the tunable laser frequency via a digital-analog converter until counter reading and desired value coincide. In order to guarantee exact laser tuning the computer works with a closed-loop control system, using learning-mode and error analysis operations, and on-line monitoring various spectrometer parameters, such as frequency jitter and drifts, intensity instabilities etc. Further,

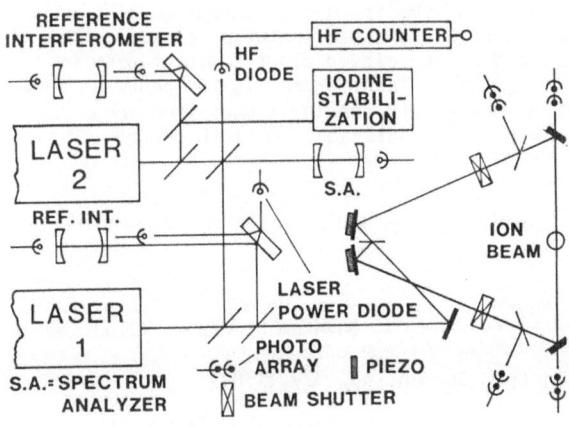

Fig.1 Experimental setup of two-dye-laser heterodyne spectrometer, with angle-controlled counterpropagating laser beams crossing an ion beam at right angles

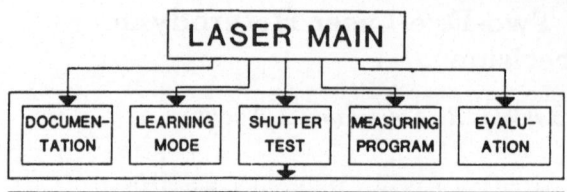

Fig.2 Schematic of overlay structure, illustrating administration of overlay programs and data exchange by main program, and independent administration of reentrant I/O modules by overlay programs

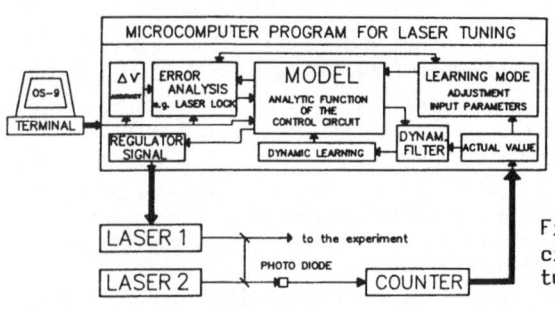

Fig.3 Software design of control circuit for closed-loop laser tuning

all measured line profiles are shown on a graphic display, and related values concerning Lamb dip position, signal shape, laser power, laser beams geometry, etc. are registered continuously. A specific data acquisition technique eliminates the fluorescence signal Doppler background. Fig.2 shows the complete package of the involved computer programs. The software design of the control circuit for closed-loop laser tuning, as a part of the measuring program, is displayed in Fig.3.

A shift of the Lamb dip position with respect to the 'natural' transition frequency arises in particle beam spectroscopy when the angle between the counter-propagating laser light beams deviates from 180°. But such a deviation negligibly affects the measurement of the frequency separation of two lines if the angle remains constant during the whole measuring time. In this setup the angle is actively controlled not to exceed about 2 microradian by continuously monitoring the position of both laser beams via two photo arrays. Additional mixing of microwave frequencies with the laser beat frequency was also realized, and further enlarged the operation bandwidth of the heterodyne spectrometer.

Acknowledgement. This work was sponsored by the Deutsche Forschungsgemeinschaft.

References:
[1] M. Englert, J. Kowalski, F. Mayer, R. Neumann, S. Noehte, R. Schwarzwald, H. Suhr, K. Winkler, G. zu Putlitz: Sov. J. Quantum Electron. 12(5),664 (1982)
[2] J.C. Bergquist, H.-U. Daniel: Opt. Commun. 48, 327 (1984)

Kinetic Spectroscopy Using a Color Center Laser

J.L. Hall, D. Zeitz, J.W. Stephens, R.F. Curl, Jr., J.V.V. Kasper, and F.K. Tittel

Departments of Chemistry and of Electrical and Computer Engineering, Rice Quantum Institute, Rice University, Houston, TX 77251 USA

The study of both the spectra and chemical kinetics of gaseous free radicals has been given a significant boost in the past few years by the development of infrared kinetic spectroscopy employing narrow-bandwidth tunable IR laser probes and pulsed excimer lasers for photolysis. The infrared region is particularly advantageous since almost all molecules have absorption features there, and simple calculations indicate that typical absorptions for species formed in this way will be on the order of a few percent using commercially available excimer lasers [1].

The short (14 nsec) excimer laser photolysis pulse coupled with fast IR detectors (<1 μsec rise time) permits observation of short-lived species, measurement of nascent distributions, and study of chemical kinetics on the microsecond time scale. Such investigations are currently under way in our laboratory, where a color center laser (2.3 – 3.3 μm) is used as an IR probe and a Lambda Physik EMG-101 excimer laser as the photolysis source. In initial studies of the Br, OH and NH_2 radicals, we have explored various schemes for monitoring such species, depending on their lifetime and possible precursor absorption interferences [2].

Spectral features corresponding to absorptions of 1% have been measured with a signal-to-noise ratio of 100 in this work. The vibrational fundamental of $^1\Delta$ NH has been detected for the first time [3]. Photolysis of the HN_3 precursor using 193 nm radiation was observed to produce rotationally hot (~10000 K) NH radicals having a lifetime of 6 microseconds [3].

Currently we are investigating the kinetic scheme of the reaction between NH_2, formed by photolysis of NH_3, and NO with the aim of identifying the primary products and subsequent reactions. This reaction is thought to play a primary role in the conversion of NO to N_2 in the industrial "DeNOX" process, where NH_3 is added to the flue gases of power plants to effect this conversion. Computer modeling of the reactions involved have been severely hampered by presently unknown branching ratios into what are thought to be the major product channels of the NH_2 + NO reaction. A recent experimental study [4] has proposed the following distributions:

$$NH_2 + NO \rightarrow N_2 + H_2O \quad : \; > 29\%$$
$$\rightarrow N_2O + H_2 \quad : \; < 1\%$$
$$\rightarrow N_2OH + H \quad : \; < 5\%$$
$$\rightarrow N_2H + OH \quad : \; > 65\%$$

Our preliminary investigations have confirmed that 193 nm photolysis of NH_3 produces hot NH_2 radicals which can be effectively cooled by added He buffer gas. The growth of the OH signal, however, appears to be

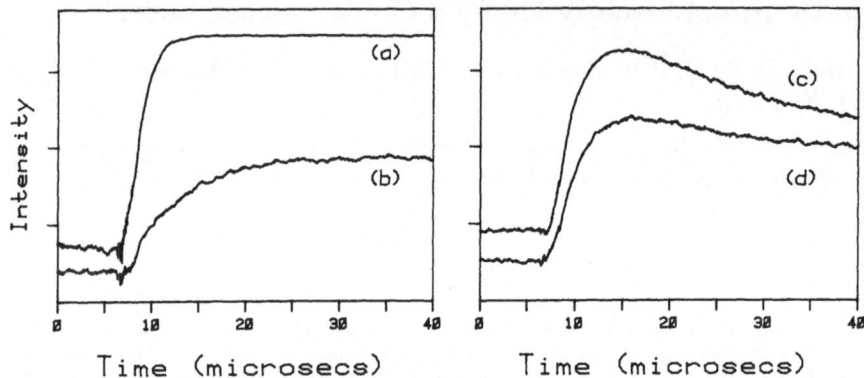

Figure 1   Kinetic behavior of the NH_2 and OH radicals.   Ground state NH_2 absorption is shown with [(a)] and without [(b)]   He buffer gas.   The corresponding traces for a ground state OH line are shown in (c) and (d). (The excimer flash occurs at $6.4\ \mu secs$.)

independent of temperature (see Fig. 1).   Our measurement of the rate constant for this reaction is in agreement with the previously determined value of $1 \times 10^{-11}\,cm^3\,s^{-1}$ [4].

Of more interest is the appearance of the H_2O signal after an induction period of 6 μsec in contrast with the prompt appearance of OH absorption. Although the H_2O being formed is clearly hot, this induction period is still observed in the presence of large pressures (10 Torr) of He buffer gas ,and is independent of the rotational transition monitored.   A similar delayed disappearance of NH_3 was also observed ,and appears to correlate with this appearance of H_2O but we have been so far unable to propose a reaction scheme to account for both observations.   Our work is strongly suggestive that H_2O is not, as has been thought, an original product of the reaction between NH_2 and NO.   Although preliminary, this work is illustrative of the power of the method.   We have been able to follow the time-dependence of absorption lines arising from NH_3, NH_2, OH and H_2O all in the same system.   Future experiments will monitor lines of NO and hopefully the as yet unidentified HN_2 species.

References

1.   Jeffrey L. Hall, Horst Adams, L.A. Russell, J.V.V. Kasper, F.K. Tittel and R.F. Curl, Proceedings of the International Conference on Lasers '83, R.C. Powell, ed. (STS, McLean, VA, 1985)

2.   Horst Adams, Jeffrey L. Hall, L.A. Russell, J.V.V. Kasper, F.K. Tittel and R.F. Curl, J. Opt. Soc. Am. B 2, 776 (1985)

3.   Jeffrey L. Hall, Horst Adams, J.V.V. Kasper, R.F. Curl and F.K. Tittel, J. Opt. Soc. Am. B 2, 781 (1985)

4.   P. Anderson, A. Jacobs, C. Kleinermann and J. Wolfrum, Nineteenth Symposium (Int.) on Combustion, The Combustion Institute, p.11 (1983)

Very High Resolution Doppler-Free Spectroscopy in a Hollow-Cathode Discharge

P. Hannaford and D.S. Gough

CSIRO Division of Chemical Physics, P.O. Box 160,
Clayton, Victoria, 3168 Australia

The phenomenon of cathodic sputtering in a hollow-cathode discharge (HCD) or glow discharge has found increasing application in recent years as a means of generating vapours of ground-state and metastable atoms for almost any solid element in the Periodic Table. The method has been used in Doppler-free saturation spectroscopy by a number of workers [1-6], but the resolution attained to date has been limited to linewidths of 25-130 MHz, and the narrow, Doppler-free components have normally been accompanied by broad, Doppler-limited pedestals which originate from velocity-changing collisions with rare-gas atoms in the discharge. The extraneous line-broadening has been attributed to perturbations due to charged-particle and neutral-particle collisions and to electric fields within the HCD.

In this paper we report the results of a detailed investigation of laser saturation spectra in a HCD for a wide range of atomic systems. Figure 1 shows the main features of a saturated absorption spectrum recorded for the 613.5 nm transition in Zr I using a 5mm bore HCD operated at 0.4 Torr of Ar and 9 mA. The width of the individual isotopic components is 7 MHz FWHM and reduces to 6 MHz at lower laser powers. This width is found to vary with Ar pressure at the rate of 8 MHz/Torr and to approach a value of 4 MHz at zero pressure. The zero-pressure width can be accounted for totally in terms of the following contributions: natural broadening of the upper level (0.3 MHz), laser bandwidth (2 MHz), residual saturation broadening ($\approx$1 MHz), residual Doppler broadening ($\approx$0.5 MHz), and unresolved Zeeman splitting due to stray magnetic fields ($\approx$0.5 MHz). Thus the actual homogeneous linewidth in the HCD at 0.4 Torr of Ar is estimated to be only 3 MHz. The above results also imply that any extraneous line-broadening due to effects of the discharge is negligibly small (< 1 MHz). This view is further supported by measurements on the Sm 570.7 nm transition: the observed linewidths for a given number density

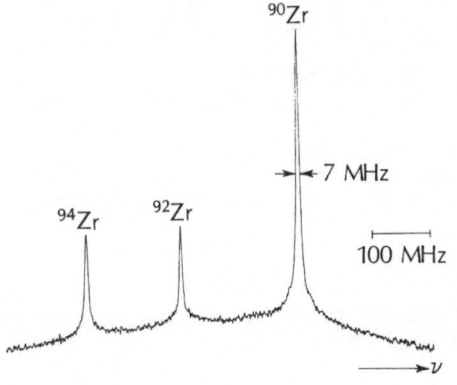

Figure 1. Part of the saturated absorption spectrum for the 613.5 nm (a^3F_2-$z^3F_2^o$) transition in Zr I. Hollow-cathode discharge: 0.4 Torr Ar, 9 mA. Laser beams: 10 mW cm^{-2}

of argon atoms were the same whether the samarium vapour was generated in a HCD or in a thermal vapour cell in the absence of a discharge.

Similar measurements on other transitions in Zr (614.3, 612.7 and 595.5 nm), Y (619.2 and 602.3 nm), Mo (603.1 nm). Ta (578.1 nm), V (621.6 nm) and Sm (570.7 nm) yield linewidths which are similar to that of Zr 613.5 nm for all except Mo 603.1 nm. This latter transition, which is one of the Mo $a^5D-z^5p^0$ group for which linewidths of 25-30 MHz have previously been reported [2], has a relatively large contribution from natural broadening (8 MHz). Likewise, substantial contributions from natural broadening, and also pressure and saturation broadening in some cases, appear to be responsible for the large linewidths previously observed in HCDs for other systems [1-6].

We note that the saturated absorption spectrum in Fig. 1 is almost free from Doppler-limited pedestals. The ratio of the height of the Doppler-free peak to that of the pedestal (R), however, is strongly dependent on the current in the HCD; for example, at 0.7 Torr of Ar, R varies from 0.4 to 10 over the current range 0-50 mA. A factor of about four of this increase in R can be accounted for by the increase in temperature in the HCD (300-900 K), while the remaining factor of six we attribute to a shortening of the lifetime of the lower level by electron impact collisions in the HCD and the corresponding reduction in the number of velocity-changing collisions that can occur during this lifetime [2]. Similar suppression of the pedestals by the HCD is found to occur for all other transitions we have investigated.

The high resolution attainable in HCDs is currently being applied in studies of isotope shifts and hyperfine structures in a number of transitions in the refractory elements Zr, Y, Ta and V. The quality of the spectra, in terms of linewidth and freedom from pedestals, is comparable with that recently obtained in the collision-free, field-free environment of an atomic beam [7].

We conclude that the atomic vapour produced by cathodic sputtering in a HCD provides a suitable environment for performing very high resolution laser saturation spectroscopy on a wide range of atomic systems, including highly refractory metal atoms. In particular, any effects of the HCD on the measured linewidths appear to be negligibly small, and the high current density in a HCD is effective in suppressing the broad, Doppler-limited pedestals normally associated with laser saturation spectra obtained in a rare gas discharge.

References

1. D.C. Gerstenberger, E.L. Latush and G.J. Collins: Opt. Comm. 31, 28 (1979)
2. A. Siegel, J.E. Lawler, B. Couillaud and T.W Hansch: Phys. Rev. A 23 2457 (1981); J. Appl. Phys. 52, 4375 (1981)
3. Ph. Dabkiewicz and T.W. Hansch: Opt. Comm. 38, 351 (1981)
4. N.Beverini, M. Galli, M. Inguscio, F. Strumia and G. Bionducci: Opt. Comm. 43, 261 (1982)
5. M. Inguscio: J. Physique Suppl. C7 44, 217 (1983)
6. D.M. Kane and M.H. Dunn: Opt. Comm. 50, 219 (1984)
7. D.W. Duquette, D.K. Doughty and J.E. Lawler: Phys. Lett. 99A, 307 (1983)

Harmonic Mixing Properties of Tungsten-Nickel Point-Contacts in the 30-120 THz Frequency Region

*H.H. Klingenberg**

DFVLR, Institut für Technische Physik,
D-7000 Stuttgart, Fed. Rep. of Germany

Tungsten-nickel point-contact (W-Ni) diodes are useful means as rectifiers and harmonic-generator mixers in the frequency region above 4 THz up to about 200 THz. In this paper, an investigation of six harmonic mixing processes is reported using W-Ni diodes as mixers /1/. 4th to 7th order mixing processes have been studied (Table). Processes of the same mixing order shown in the table are different with respect to the mixing products. The behaviour of W-Ni point-contacts as rectifiers and harmonic generators in the mid-infrared have been studied recently /2/. A material specific static current-voltage (I-V) characteristic of the diodes was measured by subtracting the Ohmic portion of 300 Ω /3/. W-Ni diodes showed a nearly parabolic dependence of current on voltage. It was found that the observed beat notes for the 5th order mixing processes exhibited the same bias dependence as the corresponding 5th derivative of the static current-voltage characteristic. These findings initiated an investigation of mixing processes of different mixing orders and products. When the mixing properties

MIXING PROCESS	ORDER	S/N Ratio in 100 kHz BW (dB)
$3 \times f_{R(30)CO_2} - f_{CCL}$	4	25
$3 \times f_{R(30)CO_2} - f_{CCL} \pm f_{MW}$	5	20
$3 \times f_{R(32)CO_2} - f_{He-Ne} - f_{MW}$	5	20
$4 \times f_{P(12)CO_2} - f_{CCL}$	5	20
$4 \times f_{P(12)CO_2} - f_{CCL} + f_{MW}$	6	17
$4 \times f_{P(12)CO_2} - f_{CCL} + 2 \times f_{MW}$	7	10-15

Table   Summary of the investigated mixing processes

* This investigation was carried out while the author was with the Physikalisch-Technische Bundesanstalt, Braunschweig.

in the infrared region were determined by the dc I-V characteristic of the diodes, the beat signal amplitudes measured for a mixing process should follow the corresponding derivative of the dc I-V curve.

The bias dependence of the beat notes for the various mixing processes has been studied. The best signal-to-noise ratio (S/N) for the beat signals obtained in a bandwidth of 100 kHz are summarized in the table. A 4th order mixing process: the 3rd harmonic of a CO_2 laser radiation was generated and mixed with the radiation of a color-center laser (CCL). The CO_2 laser operated at the 9 μm R (30) line, the CCL operated at a wavelength of 3.07 μm, respectively. The measured beat amplitude exhibited a S/N ratio of 25 dB in a bandwidth of 100 kHz provided the diode had a negative bias voltage of -80 mV. The beat notes of this process disappeared when the diode was biased by a voltage of +80 mV. 5th order mixing processes: to the 4th order process, a microwave frequency of a Gunn oscillator, operating at 8.9 GHz was added and in a second experiment the microwave frequency was subtracted. The beat notes obtained for both processes against the bias voltages to the diode showed the same result, evaluating the maximum beat signals around zero bias. This result indicated that only the mixing order is important, no matter whether the microwave frequency was added or subtracted. Furthermore, the measurements are consistent with data obtained for two other 5th order mixing processes reported earlier /2/ which are also listed in the table. 6th order mixing process: the 4th harmonic of the 10 μm P (12) CO_2 laser radiation was mixed with the radiation of a CCL operating at 2.62 μm and of a Gunn oscillator. The best obtained S/N ratio for this mixing process yielded 17 dB in a bandwidth of 100 kHz around zero-bias voltage applied to the diode. 7th order process: same as described for the 6th order process, this time the second harmonic of the microwave radiation was used to measure the beat signals. As a result no considerable bias dependence of the beat signals was found. The S/N ratio obtained was 10-15 dB in 100 kHz bandwidth.

For all the investigated harmonic mixing processes, i.e. 4th, 5th, 6th and 7th order, the corresponding derivatives of the dc I-V characteristic were calculated and compared with the measured bias-dependence of the beat signals. The surprising result was that in most cases the beat signal amplitudes observed for the various mixing processes can be described qualitatively by the static I-V curve of the W-Ni diode. This seems to indicate that the nonlinear I-V curve of the W-Ni point-contact diode is a suitable means to describe harmonic mixing processes up to at least 120 THz.

References

1. H.H. Klingenberg, Appl. Phys. B 37 (1985)
2. H.H. Klingenberg and C.O. Weiss, Appl. Phys. Lett. 43, 361 (1983)
3. G. Kramer, Proceedings of the 2nd Frequency Standard and Metrology Symposium, Copper Mountain, July 5, 1976, p.469.

Thermo-Optical Absorption Spectroscopy Utilising Transient Volume Holograms

D.J. McGraw

Physics Department, University of Utah, Salt Lake City, UT 84112, USA

J.M. Harris

Chemistry Department, University of Utah, Salt Lake City, UT 8412, USA

Passive self-pumped phase conjugation by the photorefractive effect in $BaTiO_3$ is used as a sensitive volume spatial filter in thermal lens spectroscopy, allowing the extension of thermal lens spectroscopy to scattering and optically inhomogeneous samples.

Application of Self-Pumped Phase Conjugation in $BaTiO_3$ to Thermal Lens Spectroscopy

A thermal lens is produced in an absorbing sample by the non-radiative conversion of electronic energy into heat. For liquids where $dn/dT < 0$, and for a Gaussian excitation intensity distribution, a defocusing thermal lens-like element is generated. At steady state, when radial conduction balances the rate of heat being deposited, the lens has an effective focal length inversely proportional to the absorption [1].

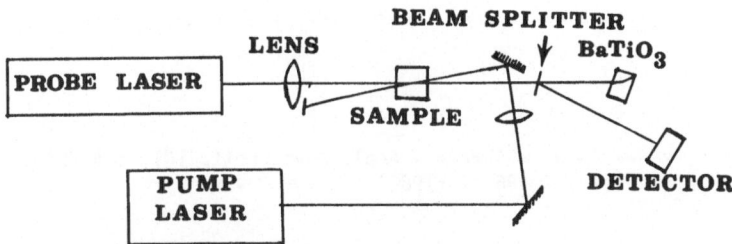

Fig. 1. Dual beam thermal lens setup

In our dual beam thermal lens setup shown in figure 1, a 20 mW c.w. probe beam is focused through the sample and into the $BaTiO_3$, where, over 20 seconds, a steady state phase conjugate reflection develops and is monitored using a beam splitter [2]. A pump beam creates a thermal lens in the sample and the resulting defocusing of the probe causes a decrease in the phase conjugation intensity. Figure 2a shows the fractional drop in the phase conjugate intensity after the formation of the thermal lens. We note that while an absorbance as low as 10^{-5} cm^{-1} is detected, the monitoring of the far field steady state probe center intensity is more sensitive as shown in figure 2c. An increase in $BaTiO_3$ sensitivity to transient wavefront aberrations is found by placing a diffuser in front of the crystal as shown in figure 2b. This suggests the use of $BaTiO_3$ with scattering samples.

Preliminary results demonstrate a sensitivity to absorptions of less than 10^{-4} cm^{-1} in highly scattering samples. In these experiments, the detected

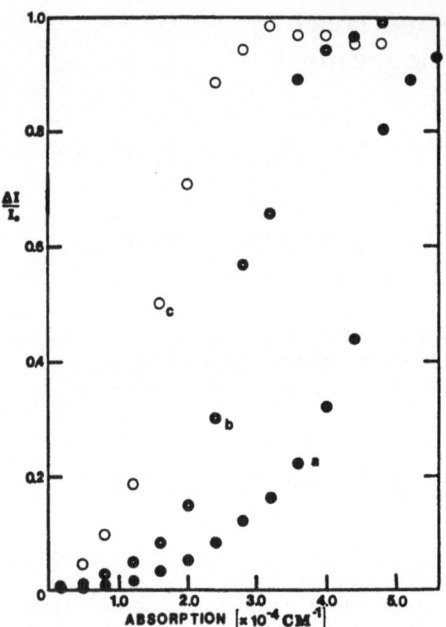

Fig. 2. Fractional decrease in thermal lens signal vs sample absorbance for three detection methods: a) phase conjugation from $BaTiO_3$, b) $BaTiO_3$ with diffuser, c) probe center intensity measurement

phase conjugate signal passes back through the sample, recreating the TEM_{00} beam without any thermal blooming present and without the speckle that makes center intensity measurements of thermal lensing from scattering samples imprecise.

References

1. R.C. Leite, R.S. Moore, J.R. Whinnery: Appl. Phys. Lett. 141, 5 (1964).
2. J. Feinberg: Optics Letters 486, 7 (1982).

High Resolution, Mass Resolved Spectra of Lutetium Isotopes

C.M. Miller, Engleman, Jr., and R.A. Keller

Los Alamos National Laboratory, Los Alamos, NM 87545, USA

A combination of resonance-enhanced photoionization and mass analysis of the resulting ions is an ultrasensitive tool for trace-species analysis. For example, isotope ratios of lutetium were accurately determined in 60 ng samples containing trace amounts of ^{173}Lu and ^{174}Lu and a thousand fold excess of isobaric interfering isotopes ^{173}Yb and ^{174}Yb [1]. ^{173}Lu and ^{174}Lu are rare isotopes with half-lives of 1.4 and 3.3 years. The ratio ^{173}Lu/^{175}Lu was measured to be $(0.44 \pm 0.07) \times 10^{-6}$ on a sample containing only 10^8 atoms of ^{173}Lu. In this work, and in most resonance-ionization mass spectrometry (RIMS), the photoionization step uses a broadband laser (FWHM 1-2 cm^{-1} or greater) and consequentially is element selective but not isotope selective; isotope ratios are measured by dispersion of the ions in a mass spectrometer. In order to increase the accuracy and dynamic range of isotope ratio measurements, it may be necessary to use single frequency lasers and isotopically selective photoionization. To be successful in this task it is necessary to know isotope shifts and hyperfine structure of optical transition(s) used for the isotopes of interest. We are particularly interested in rare and sometimes highly radioactive elements for which this information is not readily available.

We demonstrate that the sensitivity and mass selectivity of RIMS itself can be used to obtain the needed spectral data. High resolution spectra of the $^2D_{3/2}^{\;0} \leftarrow {}^2D_{3/2}$ transition at 22 125 cm^{-1} for ^{173}Lu, ^{174}Lu, ^{175}Lu and ^{176}Lu are shown in the figure. These spectra were taken with 60 mW of single frequency, tunable, cw laser radiation focused to a beam diameter of 0.05 mm, positioned ~ 2 mm above a rhenium filament, heated to 1600 K, and containing 1-2 µg of sample. The total amount of ^{173}Lu and ^{174}Lu in our sample was 2 $\times 10^{-11}$g and 2 x 10^{-12}g. A single color, 1+1 photoionization scheme was utilized. Analysis of the spectra shown in the figure yield the parameters listed in the table. The nuclear spin of ^{174}Lu was uncertain; we confirm that it is unity. Values underlined in the table were determined in this work.

Although isotope shifts are less than the Doppler broadened line widths, the hyperfine splittings of the various isotopes are sufficiently different to allow some isotope selectivity in the photoionization. Investigation of the spectra of other rare isotopes, as well as Doppler-free spectra, is underway. A complete description of this work has been accepted for publication [2].

PARAMETERS FOR THE LUTETIUM, $^2D_{3/2}{}^0 \leftarrow {}^2D_{3/2}$ TRANSITION

ISOTOPE	176	175	174	173
NUCLEAR SPIN	7	7/2	1	7/2
HALF LIFE (y)	3.6×10^{10}	STABLE	3.3	1.4
DOPPLER WIDTH[a]	0.036	0.036	0.036	0.036
LORENTZ WIDTH[a]	0.050	0.050	0.050	0.050
μ(nm)	3.139	2.3799	1.94	2.34
Q (barns)	8.0	5.68	0.1+0.5	5.7+0.6
ISOTOPE SHIFT[a]	0.1	0.00	-0.009	-0.037

[a]Units in wavenumbers.
Lifetime of $^2D_{3/2}{}^0$ state = 43 ns

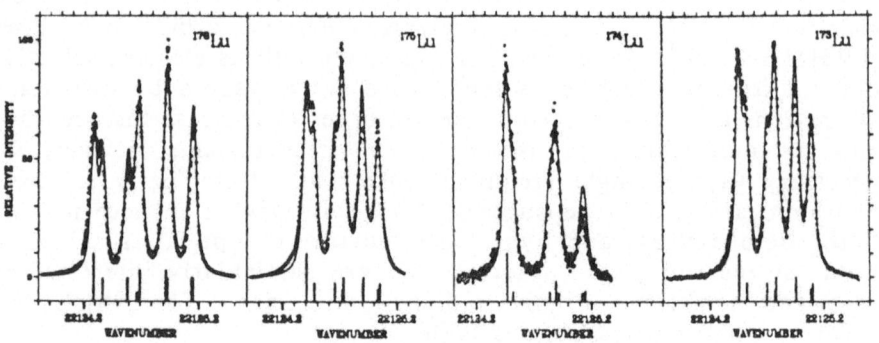

Hyperfine spectra of lutetium isotopes. Dots are data and solid lines are best fit. Bars represent positions and classical intensities of hyperfine components.

References
[1] N. S. Nogar, S. W. Downey, and C. M. Miller: "Analytical Capabilities of RIMS: Absolute Sensitivity and Isotopic Analysis", in Resonance Ionization Spectroscopy 1984, G. S. Hurst and M. G. Payne, eds. (Institute of Physics, Bristol, England, 1984), p.91.
[2] C. M. Miller, R. Engleman, Jr., and R. A. Keller: "Resonance-Ionization Mass Spectrometry for High-resolution, Mass-resolved Spectra of Rare Isotopes", J. Opt. Soc. Am. B, 2, 000 (1985).

Optical Pumping Effect of Four-Level System in Doppler-Free Laser Spectroscopic Techniques

S. Nakayama and S. Tsutsumi

Department of Electronics, Kyoto Institute of Technology, Matsugasaki, Sakyo-ku, Kyoto 606, Japan

Theoretical analysis of Doppler-free signals in saturation spectroscopy, velocity-selective optical pumping (VSOP) or polarization intermodulated excitation (POLINEX) spectroscopy, and polarization spectroscopy is presented by using the optical pumping theory in a four-level system in order to determine the signal magnitude and sign [1,2]. For the application to arbitrary polarization combinations of the pump and probe beams, both circular and linear optical anisotropy are considered [3]. The analysis of the signal magnitude and sign in atomic and crossover resonances is important to investigate how atoms interact with laser fields.

A three-level approximation used commonly is inadequate to interpret the signals such as D lines of alkali atoms (see Fig. 1), because four-level crossover resonances occur in the same order as three-level crossover resonances. The four-level approximation is proposed as a

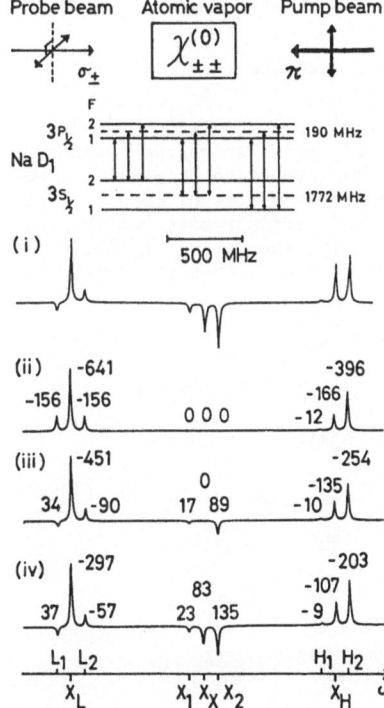

Fig. 1 Saturation spectroscopy of the Na D_1 line. The pump and probe beams are $\pi-$ and σ_+-polarized, respectively. The quantization axis is chosen in the direction of the linear polarization of the pump beam. (i) Saturated absorption signal observed by NEUKAMMER, RINNEBERG, and ZHANG [6] (the power density of the pump beam Is: a few $\mu W/mm^2$, that of the probe beam Ip: below 1 $\mu W/mm^2$, and the vapor temperature T: 403 K), and theoretical curves (ii) in three-level system without the spontaneous emission, (iii) in the three-level system with the spontaneous emission, and (iv) in the four-level system with the spontaneous emission (The Doppler width $\Delta\nu_D$ (FWHM):1519 MHz and the line width $\Delta\nu_L$ (FWHM):16 MHz)

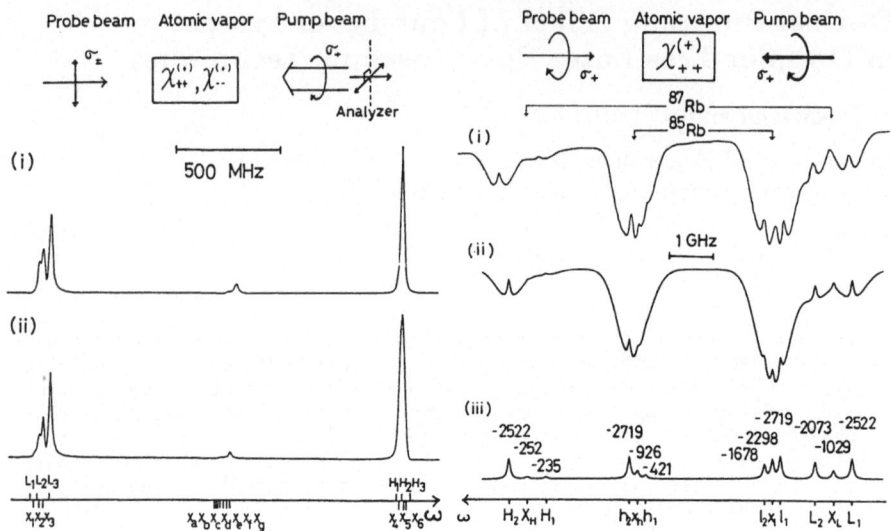

Fig. 2 Polarization spectroscopy of the Na D_2 line. (i) Experimental curve observed by GAWLIK [7] (Is=25 μW/mm², Ip=0.5 μW/mm², T=423 K), and (ii) theoretical curve calculated using the four-level optical pumping theory ($\Delta\nu_D$=1558 MHz, $\Delta\nu_L$=16 MHz)

Fig. 3 Saturation spectroscopy of the Rb D_1 line. (i) Experimental curve observed by YAMAGUCHI [8] (Is=370 μW/mm², Ip=57 μW/mm², T=298 K), (ii) theoretical curve calculated using the four-level optical pumping theory ($\Delta\nu_D$=664 MHz, $\Delta\nu_L$=80 MHz), and (iii) the theoretical Lamb dips

method of calculating the optical anisotropy by selecting two-level principal resonances and three- and four-level crossover resonances among many Zeeman sublevels between lower and upper hyperfine states. We derive the general formula of the magnitude and sign of the Doppler-free resonances due to circular and linear optical anisotropies which are induced in those spectroscopic techniques. The theoretical analysis is in good agreement with the experiments [4-6] of the D lines of Na and Rb as shown in Figs. 1, 2 and 3.

References

1. S. Nakayama: J. Phys. Soc. Jpn. **50**, 609 (1981)
2. S. Nakayama: J. Phys. Soc. Jpn. **53**, 3351 (1984)
3. S. Nakayama: Jpn. J. Appl. Phys. **24**, 1 (1985)
4. J. Neukammer, H. Rinneberg, and Z.X. Zhang: Opt. Commun. **38**, 361 (1981)
5. W. Gawlik: private communication, July 1984
6. S. Yamaguchi: private communication, April 1985

Ultra-Sensitive Laser Isotope Analysis of Krypton in an Ion Storage Ring*

J.J. Snyder, T.B. Lucatorto, P.H. Debenham, R.E. Bonanno, and C.W. Clark

National Bureau of Standards, Gaithersburg, MD 20899, USA

We are developing a novel instrument [1] for ultra-sensitive isotope analysis of krypton based on multi-stage separation of ions stored in a small ring. This instrument combines magnetic mass selection, resonant charge-exchange and laser photoionization to achieve isotopic abundance sensitivities we expect will approach 10^{-13}, several orders of magnitude better than is possible using conventional mass spectrometers.

Our proposed instrument consists of a small, racetrack-shaped ion storage ring as shown schematically in Fig. 1. Ions from the sample are continuously injected into the ring, stored for several orbits while the undesired species are removed, and then directed into a particle counter. During each orbit the ions are mass-separated by two 180° bending magnets, neutralized by resonant charge-exchange, and re-ionized by laser photoionization. The use of multiple orbits offers very high rejection of adjacent isotopes, as well as reducing contamination due to low-angle scattering. In addition, the resonant charge-exchange and photoionization processes virtually eliminate isobaric and molecular interferences.

In order to prevent back contamination of the highly purified ions in the ring, successive ion orbits follow different trajectories. Orbit separation is accomplished by decelerating the ion beam as it enters the 75 volt potential of the electrically isolated charge exchange heat pipe. Part of the resulting neutral beam is re-ionized by the laser after it leaves the heat

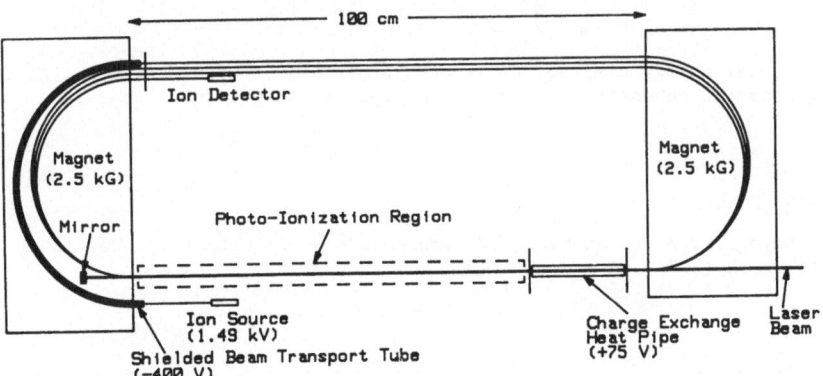

Fig. 1. Schematic of the ion storage ring configured for detection of ^{81}Kr

---

* Work supported in part by DOE Contract No. DE-A-101-85ER60302

pipe, but before it enters the magnet on the left. Ions which are neutralized within the heat pipe and photoionized outside it will not be re-
accelerated, and will therefore follow a new orbit with a diameter about 1 cm
smaller than before. The result is that each atom of the desired species
follows a decaying spiral path until it reaches the detector.

The charge-exchange gas is rubidium, which has a large resonant cross-
section for transfer of an electron into the metastable first excited state
of krypton. Because of the internal energy of the metastable krypton beam
emerging from the charge-exchange heat pipe, the beam can be easily
photoionized by two visible photons via a resonant intermediate p state to
an ns or nd autoionizing level, as shown in Fig. 2.

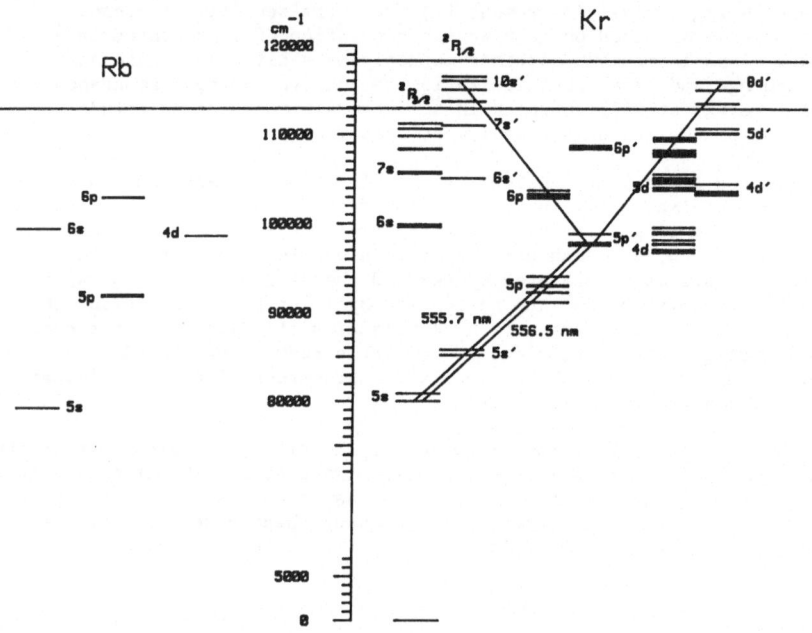

Fig. 2. Partial energy diagrams of krypton and rubidium, drawn relative to a
common ionization potential. Two possible photoionizing transitions are
shown

References

1. J.J. Snyder, T.B. Lucatorto, P.H. Debenham and S. Geltman, J. Opt. Soc.
Am. B, in press (1985).

A Wide Scan Range Single-Mode Dye Laser

K. Uehara and N. Ukaji

Department of Physics, Faculty of Science and Technology,
Keio University, Hiyoshi, Kohoku-ku, Yokohama 223, Japan

T. Kasuya

The Institute of Physical and Chemical Research,
Wako, Saitama 351, Japan

1. Introduction

Dye lasers are useful light sources in visible spectroscopy. Their broad
tunability is especially important for studies of a wide-spread spectrum such
as an entire band system of a molecule. However, the continuous scan range of
single-mode dye lasers developed so far is less than 1 cm^{-1} (30 GHz) with the
exception of a few models [1-3]. In this report we describe a dye laser which
is scannable over 200 cm^{-1} with a very simple control system, while retaining a
narrow spectral width of 15 MHz.

2. Method of Wide Scanning

Figure 1 shows the block diagram of the optical layout and the control circuits of the present dye laser. The three-mirror laser cavity of 45-cm optical length is supported by an Invar rod of 5 cm in diameter. The wide scanning is accomplished by repeated linear sweeps of both the cavity and an
internal etalon, along with a continuous tuning of a birefringent filter. The
etalon of 2.5-cm^{-1} FSR and the output mirror are driven synchronously by a
piezoelectric translator and an electromagnetic translator, respectively. The
extension of each of the repeated sweeps is set to a little wider than twice
the FSR of the etalon, so that there will be a frequency overlap rather than a
gap between sweeps. The birefringent filter is rotated by a dc motor. The
center frequency of the filter is a nonlinear function of its rotational
angle. Therefore, for a given fixed rate of the filter rotation, there is a
limit to the extension of a jumpless scan. A numerical calculation shows that
the limit is 225 cm^{-1} at 17000 cm^{-1}.

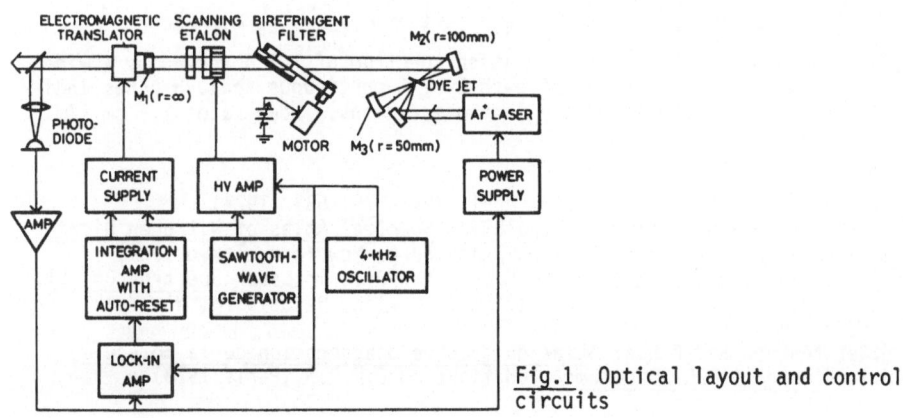

Fig.1 Optical layout and control circuits

3. New Servo Amplifier

To prevent mode jumps caused by the mismatch of the scan rate between the internal etalon and the cavity, the latter is locked to the former by a servo loop. The error signal is provided by 4-kHz modulation of the etalon transmission,followed by lock-in detection of the modulated output power. A general problem in a servo system is that it becomes inactive when the servo amplifier is saturated by a large error signal. A novel type of servo amplifier with an auto-reset function was devised for the present servo system. This servo amplifier consists of a resetable integrator, voltage comparators, and a monostable multi-vibrator which sends a reset pulse to the integrator. When the integrated error signal grows to a preset voltage, +V or -V, it is automatically reset to zero quickly. Thus the servo loop is free from the dead time,and the mode jumps are eliminated throughout each continuous scan. Another servo loop stabilizes the dye laser output power by controlling the excitation power from an argon-ion laser.

4. Results

With a rhodamine 6G/ethylene glycol solution,a stable output power of a few milliwatts was obtained at any wavelength between 569 and 605 nm under the excitation by no more than 1 W of the 514.5-nm line. Power fluctuation during a scan was less than 3% when measured with a time-constant of 10 ms. The spectral width was measured with a spectrum analyzer to be less than 15 MHz.

To evaluate the scanning capabilities, the absorption spectrum of I_2 and the Stark modulation spectrum of IC1 (Fig.2) were observed with the present dye laser. A published spectral atlas of I_2 [4] was used to calibrate the wavelength and check the continuity of the scan. So far, by rotation of the birefringent filter at a fixed speed, a jumpless scan range of up to 160 cm^{-1} has been achieved,as compared with the theoretical limit, 225 cm^{-1}. The scan range is extendable beyond this limit by varying the rotational rate of the filter during a scan according to the theoretical formula.

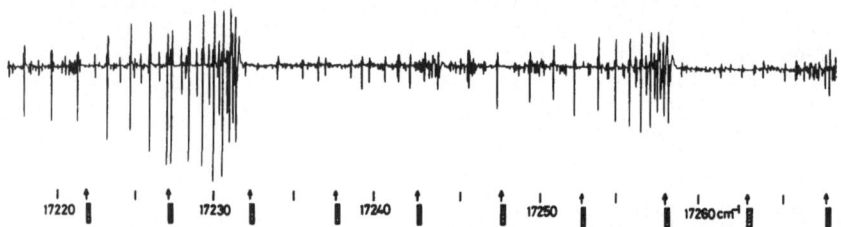

Fig.2 A portion of the Stark modulation spectrum of IC1 observed by a wide scan. Arrows indicate the end of each 5-cm^{-1} continuous scan. Hatches indicate overlapping of the frequency in those regions. Scan rate: 1.2 cm^{-1}/min.

References

1. K.Uehara and K.Shimoda: Jpn. J. Appl. Phys. 16, 633 (1977)
2. K.Uehara and H.Sasada: High Resolution Spectral Atlas of Nitrogen Dioxide 559 to 597 nm (Springer, Berlin, Heidelberg, New York, Tokyo 1985)
3. G.H.Williams, J.L.Hobart, and T.F.Johnston,Jr., in Laser Spectroscopy VI, ed. by H.P.Weber and W.Lüthy (Springer, Berlin, Heidelberg, New York, Tokyo 1983), p.422
4. S.Gerstenkorn and P.Luc: Atlas du spectre d'absorption de la molécule d'iode entre 14800-20000 cm^{-1} (Editions du C.N.R.S.,Paris 1978)

Servo Control of Amplitude Modulation in FM Spectroscopy: Shot-Noise Limited Measurement of Water Vapor Pressure-Broadening

*N.C. Wong and J.L. Hall**

Joint Institute for Laboratory Astrophysics, University of Colorado and National Bureau of Standards, Boulder, CO 80309, USA

Phase or frequency modulation (FM) spectroscopy is a powerful technique which offers the advantage of zero-background phase-sensitive heterodyne detection, and has been used in various types of spectroscopic investigation and in laser frequency locking. However, spurious amplitude modulation (AM) causes fluctuation in the baseline. This reduces the detection sensitivity and causes error in laser frequency locking. Here we have developed and demonstrated an active servo system[1] to suppress the AM noise of a phase-modulated laser beam to achieve shot-noise limited detection in a linear absorption experiment.

AM noise in FM spectroscopy is generally caused by temperature-dependent birefringence variation of the phase modulator crystal, scattering and etalon effects, spatial inhomogeneity of the modulation electric field in the crystal, and intensity fluctuation of the laser. The AM heterodyne power I_{AM} detected at the modulation frequency is given by

$$I_{AM} \propto c\, I_0\, J_1(M)\, \sin(\Delta\phi + \Delta\phi^{dc}) \quad ,$$

where c is a constant involving the polarizer and analyzer angles relative to the crystal's principal axes, I_0 is the laser intensity, M is the modulation index, J_1 is the first-order Bessel function, $\Delta\phi$ is the natural birefringence, and $\Delta\phi^{dc}$ is the induced birefringence due to an applied dc voltage.

We monitor the AM in the phase-modulated beam and apply a feedback dc voltage to the modulator crystal to compensate for its fluctuating birefringence $\Delta\phi$. Thus, the polarization of the light leaving the crystal is actively controlled and aligned such that the AM noise is suppressed, i.e., $I_{AM} = 0$. To remove the spatial inhomogeneity of the AM, a polarization-preserving optical fiber is used as a spatial filter.

The performance of the servo system is evaluated for various modulation frequencies up to 1 GHz in a linear absorption experiment in iodine. The AM noise is consistently suppressed to within 1 dB of the shot-noise level. An absorption of ~3 × 10^{-6} gives an experimental signal-to-noise ratio of unity (with only 100 µW of laser power, at a modest modulation index of 0.15, with an avalanche photodiode of low quantum efficiency of 0.2, and a 1 Hz detection bandwidth). The servo system does not require periodic adjustment or any additional modulation, thus producing a purely phase-modulated light at all times. In addition, it is not necessary to match photodiodes or to adjust incident optical powers, making it a relatively simple system to use.

---

*Staff Member, Quantum Physics Division, National Bureau of Standards.

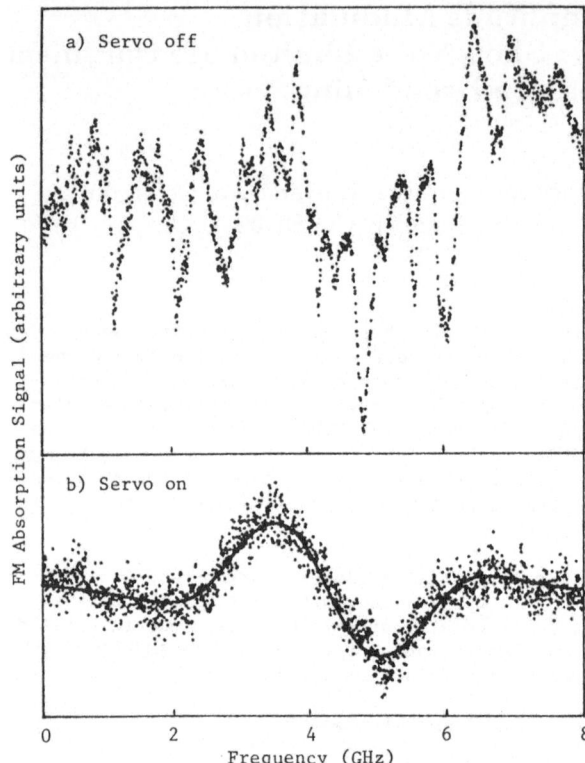

Fig. 1. Water vapor FM absorption signal in a single sweep with (a) the servo off, and (b) the servo on, illustrating an AM noise reduction of more than 20 dB with the servo system. The solid line in (b) is a least-squares fit using a Lorentzian lineshape yielding a linewidth (HWHM) of 1.0 GHz. Modulation frequency is 1.5772 GHz and detection bandwidth is 50 Hz.

We have applied the servo-FM technique to observe a weak absorption signal in water vapor at λ_{vac} = 5903.1 Å. The vibrational and rotational assignments for the upper and lower states of the observed transition in the electronic ground state of water vapor are v'(upper) = 401, $J'K'_a K'_c$ = 202, and v"(lower) = 000, $J"K"_a K"_c$ = 101. Figure 1(a) shows a single sweep of FM signal when the servo system is turned off, where the absorption signal is masked by the large AM noise. When the servo is on, the water vapor absorption signal is clearly revealed in a single sweep in Fig. 1(b). The solid line is a least-squares fit to the data using a Lorentzian lineshape yielding a linewidth (HWHM) of 1.0 GHz. The peak absorption in Fig. 1(b) corresponds to an absorption coefficient of ~8 × 10^{-7}/cm in a 50 cm single pass cell. We have also measured air pressure broadening up to 1 atmosphere. Preliminary analysis shows a broadening coefficient of ~6 MHz (HWHM)/Torr.

This work has been supported in part by the Office of Naval Research and the National Science Foundation, and in part by the National Bureau of Standards under its program of precision measurement for application in the basic standards area.

1. N. C. Wong and J. L. Hall, in the special issue of J. Opt. Soc. Am. B on Ultrasensitive Laser Spectroscopy (Sept. 1985).

Part XIII

Miscellaneous Applications

Laser Induced Fluorescence Studies on Ultraviolet Laser Ablation of Polymers

R. Srinivasan and R. W. Dreyfus

IBM Thomas J. Watson Research Center, Yorktown Heights, NY 10598, USA

In organic molecules, the absorption of photons of ultraviolet wavelength gives rise to electronic excitation [1]. At wavelengths > 200nm the resulting transitions are from from bonding to anti-bonding levels of the valence electrons. At shorter wavelengths, Rydberg transitions are also important. Polymeric organic materials undergo similar excitation processes except that a single molecule, which can consist of 10,000 atoms, may contain a thousand atomic groups called chromophores which can each absorb a UV photon.

When intense UV laser radiation became readily available with the invention of the excimer laser, it was observed [2] that a pulse of radiation (15-30nsec FWHM) from the laser is capable of spontaneously etching the surface of an organic polymer. Typically, at 248nm a pulse of 25nsec half-width and of 0.2 J/cm^2 will remove 0.4 μm of PMMA [3]. Nearly all organic solids including biological tissue can be etched by this method although the sensitivities to a given wavelength and fluence depend upon the structure and composition. The features of this dry-etching process which are of technological [4] and medical importance are the precision with which the geometry (especially the depth) of the cut can be defined and the absence of thermal damage in the substrate to a microscopic level. In Fig.1 the etch response of a typical polymer to two different wavelengths are shown. All polymers have a threshold fluence (0.02 − 0.08 J/cm^2) below which the etch depth/pulse is negligible. At the other ex-

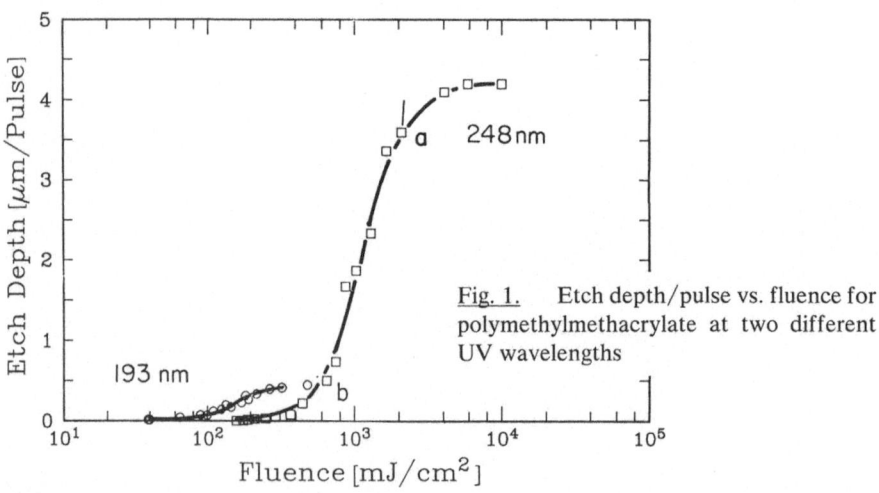

Fig. 1. Etch depth/pulse vs. fluence for polymethylmethacrylate at two different UV wavelengths

treme, when the fluence exceeds $2 - 3 \ J/cm^2$, thermal damage to the substrate becomes evident.

On the basis of studies at 193nm (= 6.4 eV), it has been proposed [6] that at this wavelength, electronic excitation of the polymer leads to an efficient breaking of the atomic bonds in the excited state itself. The resulting fragments would still contain ~3 eV of the photon energy in translational, vibrational, and rotational degrees of freedom. The fragments also have a larger (30%) specific volume than the polymer molecules that they replace. The resulting temperature and pressure rise is considered to be the propelling force for the expulsion of the products at supersonic velocities. This photochemical mechanism which is well supported by the chemical analysis of the products in several instances has been given the name "Ablative Photodecomposition" [7]. At longer wavelengths, (248nm or 308nm), there is experimental evidence that the temperature of the ablated products is several hundred degrees above room temperature even though the substrate may show as little thermal damage as with 193nm radiation. The mechanism at these longer wavelengths may still be purely photochemical if either thermally activated photochemistry or two-photon (per chromophore) processes are involved. Electronic excitation energy can also be internally converted to vibrational excitation in the ground electronic state. This is the photothermal pathway for ablation. It has not been clearly established under what conditions of wavelength and absorption cross-section, photothermal processes compete successfully with "ablative photodecomposition".

The chemical and physical complexities of the UV laser ablation reaction has required the use of numerous experimental methods to establish some understanding of the mechanism in a given polymer. We shall describe here the use of laser-induced fluorescence to track the ejection of two diatomic species (CN and C_2) from two polymers. Obviously, these data cannot be expected to give a complete picture in themselves.

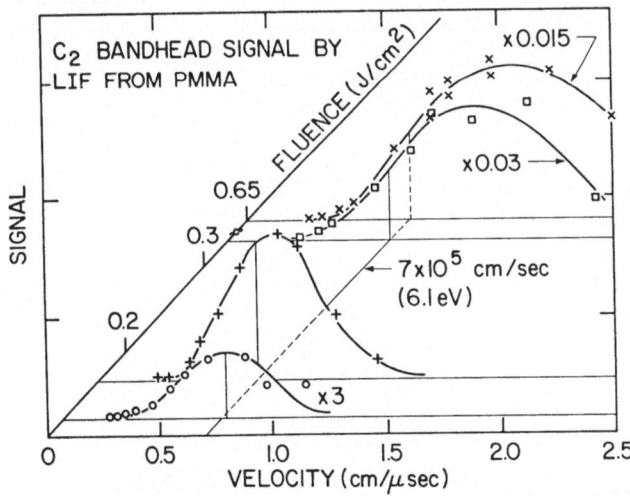

Fig. 2. LIF signal from C_2 radicals indicating the distributions in velocities as determined by time-of-flight. The solid curves are for purposes of visualization

Figure 2 shows the C_2 bandhead signal from PMMA that was obtained when the ablating wavelength was 248nm. The pumping wavelength corresponded to 438.2nm [(2,0) of the $A^3\Pi g \leftarrow X'^3\Pi_3$ transistion] and the detection wavelengths of 471.5nm corresponded to (2,1) in the same transition. The velocities were calculated from the temporal delay between the UV pulse and the fluorescence excitation pulse since the ejected material was sampled at a known distance (1.65 cm) above the surface of the polymer. Note that the intensity of the signal increases by a factor of 30 at the fluence at which significant etching is observed according to the data in Fig. 1. Chemical analysis shows [3] that in region from <u>b</u> to <u>a</u> in Fig. 1, ablation results in the ejection of all of the material down to a certain depth. The velocity calculated for C_2 in this instance is of the same order as that reported for CH from laser pulses at 193nm on PMMA [6]. In this instance, LIF was not used but the ablation fluence (0.3 J/cm^2) was large enough so that CH emission from the $A^2\Delta \rightarrow X^2\Pi$ transition could be monitored. An attempt to fit the velocity distribution observed here at a fluence of 0.16 J/cm^2 to a Maxwell-Boltzmann distribution is shown in Fig. 3. The fit is obviously unsatisfactory. This can be contrasted to reports [7,8] where satisfactory fits were obtained to the Maxwell-Boltzmann distribution of the velocities of copper atoms and Cu(II) ablated from a surface at 248nm at $3 - 5$ J/cm^2 (same pulsewidth as here) and of the velocities of polyatomic organic molecules photothermally desorbed and ablated by a CO_2 laser at fluences of $0.3 - 0.7$ J/cm^2 and a pulsewidth of 200 nsec. UV laser ablation of polymers which we identify to be a photochemical process would not be expected to raise the products to a uniform translational temperature nor would the vibrational temperature match the translational temperature of a given species. The latter point was already made [6] from studies mentioned earlier on PMMA at 193nm. Hydrodynamical effects related to supersonic expansion can also narrow the velocity distribution as observed [9,10].

Results of LIF studies on other polymers are analogous to those on PMMA. Figure 4 shows the velocity distribution in CN, a diatomic product from the irradiation of polyimide by 248nm pulses. The LIF pumping wavelength was at 399.3nm which cor-

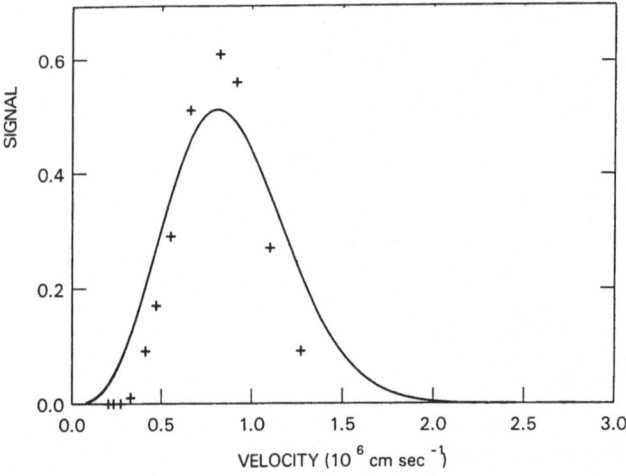

<u>Fig. 3.</u>   LIF signal from C_2 radicals as a function of their velocity. The solid curve represents the free expansion of a Boltzmann distribution

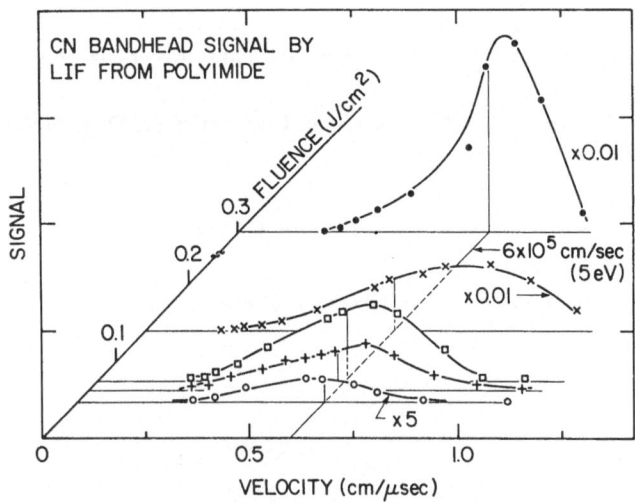

<u>Fig. 4.</u>   LIF signal from CN radical indicating the distributions in velocities as determined by time-of-flight.   The solid curves are for purposes of visulization.

responds to (0,0) in the $B^2\Sigma \leftarrow X^2\Sigma$ transition, while the detecting wavelength was 419.5nm (0,1). The sharp rise in the signal at about 0.1 J/cm^2 corresponds well to the threshold for significant etching [11].

A model of laser ablation of organic polymers has been described recently [9]. The velocities and angular distributions predicted on this model are of the same order as the values reported here and elsewhere by others. But the critical test that is needed is a measurement of velocities and their distribution from organic polymers using laser wavelengths that would lead only to thermal excitation. Such experiments are currently under way.

We wish to acknowledge the cooperation of R. Walkup, J. Jasinski and L. Urbach in obtaining and analyzing the present results.

References

1.   For a review, see F. P. Schäfer: <u>Dye Lasers</u> (Topics in Applied Physics, Vol. I, Springer-Verlag, New York 1973)

2.   (a) R. Srinivasan and V. Mayne-Banton: Appl. Phys. Lett. <u>41</u>, 576 (1982); (b) M. W. Geis, J. N. Randall, T. F. Deutsch. P. D. DeGraff, K. E. Krohn and L. A. Stern: <u>ibid.</u> <u>43</u>, 74 (1983)

3.   B. Braren and R. Srinivasan: (to be published)

4.   R. Srinivasan: <u>Photophysics and Photochemistry above 6eV</u> ed. F. Lahmani, Elsevier, Amsterdam 1985, p. 595

5. R. Srinivasan: J. Vac. Sci. Tech. B1, 1, 923 (1983)

6. G. M. Davis, M. C. Gower, C. Fotakis, T. Efthimiopoulos and P. Argyrakis: Appl. Phys. A36, 27 (1985)

7. R. Viswanathan and I. Hussla: Laser Processing and Diagnostics ed. D. Bauerle (Chemical Physics, Springer-Verlag 1984) p. 148

8. M. Buck, B. Schafer and P. Hess: Surf. Sci. (to be published)

9. B. J. Garrison and R. Srinivasan: J. Appl. Phys. 57, 2909 (1985)

10. Y. S. Lou: J. Appl. Phys. 42, 536 (1971)

11. R. Srinivasan and B. Braren: J. Polymer Sci. (Chem), 22, 2601 (1984)

Tumour Localization by Means of Laser-Induced Fluorescence in Hematoporphyrin Derivative (HPD) – Bearing Tissue

S. Andersson[1], *J. Ankerst*[2], *E. Kjellén*[3], *S. Montán*[1], *E. Sjöholm*[1],
K. Svanberg[2], *and S. Svanberg*[1]

[1]Department of Physics, Lund Institute of Technology, P.O. Box 118,
S-221 00 Lund, Sweden
[2]Wallenberg Laboratory and Department of Internal Medicine,
Lund University Hospital, S-221 85 Lund, Sweden
[3]Department of Oncology, Lund University Hospital, S-221 85 Lund, Sweden

1. Introduction

Following systemic injection, hematoporphyrin derivative (HPD) is known to be selectively retained in malignant tissue. This property can be used in a two-fold way: a) for localizing tumours by observing the characteristic dual-peaked laser-induced fluorescence from HPD, and b) for photodynamic therapy (HPD-PDT) using a localized HPD-assisted singlet oxygen release induced by irradiation of 630 nm laser light. This rapidly developing field has recently been reviewed by DOUGHERTY [1].

We are involved in spectroscopic studies of tissue fluorescence aimed at the development of powerful techniques for early cancer tumour detection. We have performed extensive studies on normal and malignant tissues of rats that had been injected with HPD [2,3,4]. Studies of the natural fluorescence from excised human tumours have also been performed. The measurements have been performed using a nitrogen laser in conjunction with optical multi-channel detection. Spectroscopic contrast enhancement, allowing the monitoring of weak HPD fluorescence in the presence of strong natural fluorescence, has been emphasized both in point measurements and in recent imaging measurements [5]. These techniques are of particular interest in endoscopic lung, bladder and gastro-intestinal applications.

2. Spectroscopic Point Measurements

A set-up for spectroscopic point measurements on tissue is shown in Fig. 1. The beam of a nitrogen laser, emitting light at 337 nm was directed onto the tissue sample and the fluorescence was detected using a Tracor Northern TN-1223-4IG intensified diode array detector in the focal plane of a Jobin-Yvon UFS-200 spectrograph. A spectrum from an induced rat tumour is shown in Fig. 2 with HPD peaks (A and C) above a smoothly decreasing blue background. It was noted that, for tumours, the HPD fluorescence (e.g. signal A) increases at the same time as the blue fluorescence (B) decreases. Contrast enhancement is thus accomplished by studying the ratio A/B. Such a dimensionless quantity also has the advantage of being immune to temporal and spatial illumination fluctuations, to surface

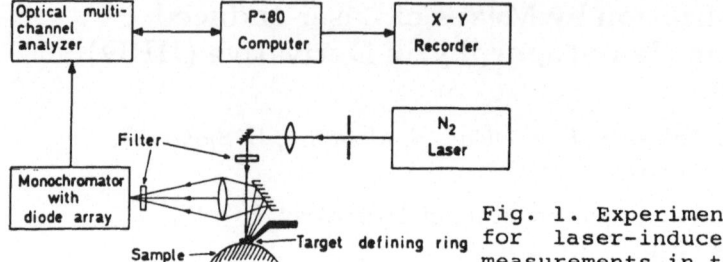

Fig. 1. Experimental arrangement for laser-induced fluorescence measurements in tissue (Ref. 2).

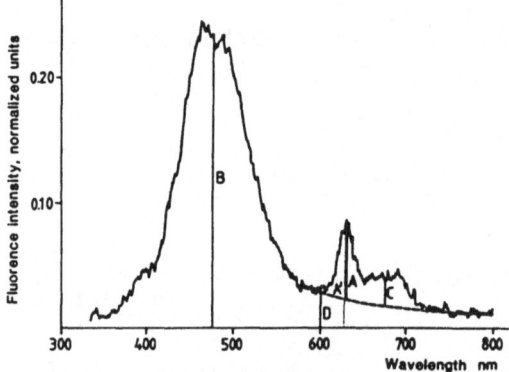

Fig. 2. Fluorescence spectrum from a tumour in a rat that had been injected (5 mg/kg of bodyweight) 8 hours before the investigation (Ref. 3).

topography and to distance variations. Results for different rat organs are shown in Fig. 3 at varying time delays after HPD injection. The strong contrast between tumour and surrounding muscle is evident. High A/B ratios are also obtained for other organs. Another way of discriminating cancerous from healthy tissue is to monitor the C/A ratio, which is again dimensionless. As shown in Fig. 4 this ratio has a high value for tumours even after several days, while for most other tissues the ratio decreases. For the rat tissues investigated it thus seems possible to use the A/B and the C/A ratios together with chosen discriminator levels c_1 and c_2 for identifying tumour tissue using the combined criteria $A/B > c_1$; $C/A > c_2$ as illustrated in Fig. 5 ("automatic tumour detection"). By increasing the discriminator levels, the chances of false positive indications are reduced, but at the same time some tumours may be overlooked. It will be very interesting to investigate to what extent similar concepts could be used in human applications.

3. Multi-Colour Fluorescence Imaging

Clearly, it would be of great diagnostic interest to extend the procedures outlined above for point measurements to the case of spatial resolution. The ultimate goal for such work would be equipment with video read-out showing only malignant tumours, well discriminated from the surrounding tissue by selecting a proper combination of discrimination levels c_i . Obviously, there is a long way to go, but as a first step we have demonstrated multi-colour fluorescence imaging along a line [5]. In

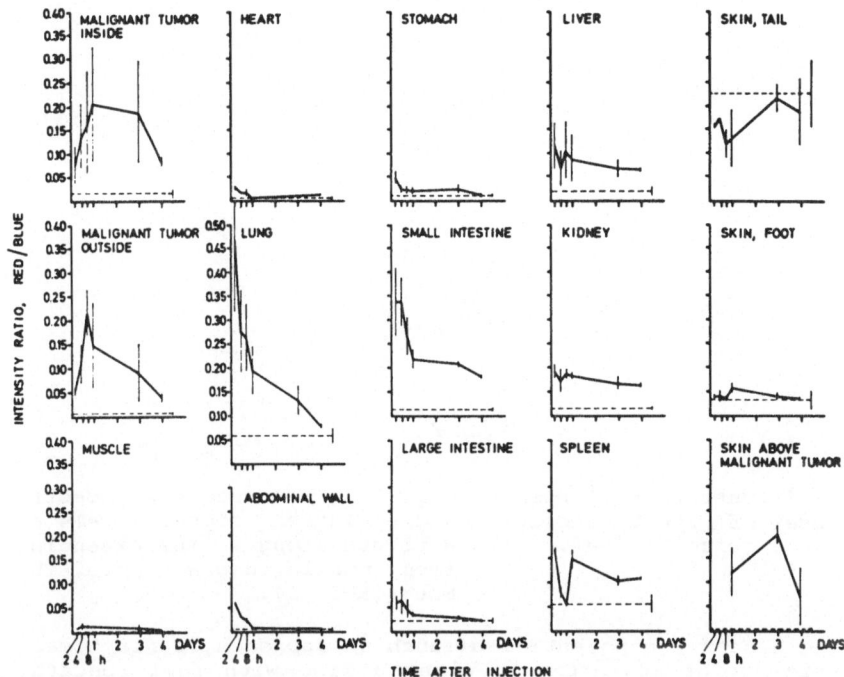

Fig. 3. Ratio of signal intensities A and B for different rat tissues, investigated at different time delays after HPD injection. The background level for non-injected rats is indicated by a broken line (Ref. 3).

C/A

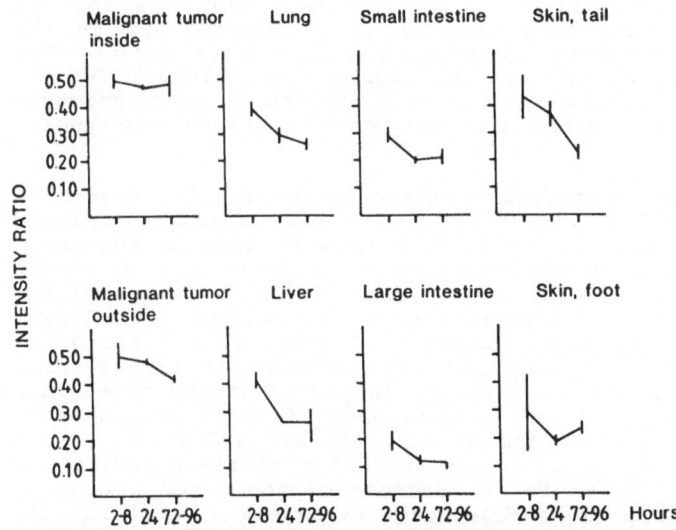

Fig. 4. Ratio of signal intensities C and A for different rat tissues (Ref. 3).

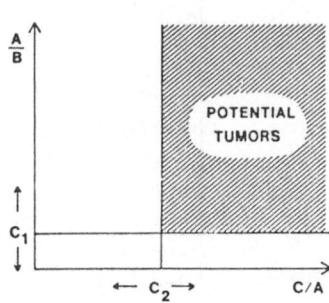

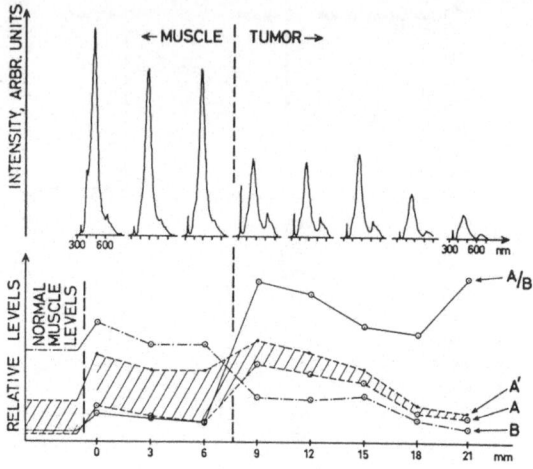

Fig. 5. Illustration of possible use of spectroscopic
acceptance criteria (Ref. 3).

Fig. 6. Fluorescence spectra
and evaluated signal levels at
8 points along a line extending
from muscle into a cancer tumour (Ref. 3).

order to assess the potential of such an approach,we first made
successive point measurements along a line with full spectral
resolution at each point. The results of such a scan starting
in normal muscle and extending into a tumour are shown in Fig.
6. The importance of monitoring background-free HPD peaks and
using the A/B ratio for contrast enhancement is evident. The
result of a similar experiment is shown in Fig. 7 [6], in which
a specially constructed fibre-optical bronchoscopic fluorescence monitoring system [7] was used in a scan of a rat
tumour. This system utilized a mercury lamp (405 nm) for the
excitation, and fluorescence monitoring is switched between the
red and blue regions by a rotating filter arrangement. The
fluorescence monitoring is interrupted by visual optical
inspection utilizing the visible part of the Hg lamp spectrum
in a 17 Hz cycle. During the inspection periods the photo-
electric detection system is blocked.

The lay-out of a laser-based three-colour imaging system is
shown in Fig. 8. Using cylindrical lenses, a nitrogen laser
beam is shaped into a 20 mm long, 1 mm wide line at the position of the object. A three-mirror arrangement and a common
achromatic lens are used to form three individual images of the
streak of laser-induced fluorescence by the object. By adjusting the mirrors, the three images (demagnified by a factor 4)
are arranged side by side, and the intensified linear-array
detector, that was discussed above, is placed at the three-fold
image line position. An interference filter arrangement,
selecting 5-10 nm wide bands at 630, 600 and 488 nm is placed
in front of the array, allowing spatially resolved LIF to be
detected in these three bands. Because of the gated action of
the image intensifier in the detector arrangement, measurements
could be made in full ambient illumination. By subtracting the
600 nm image line from that obtained for 630 nm and dividing by

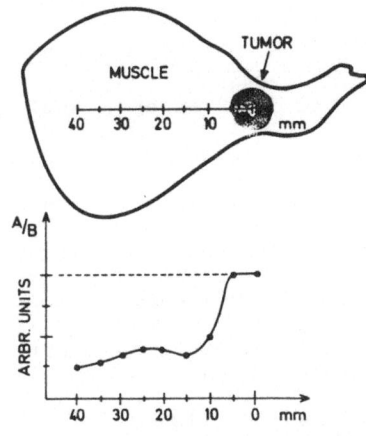

Fig. 7. A/B curve obtained in a scan of a rat leg with a cancer tumour in the muscle. A fibre-optical bronchoscope with a Hg lamp was used in this test (Refs. 6,7).

the blue data, the boundaries of tumours clearly appear, even for cases when e.g. superficial tumour blood staining leads to strongly erroneous results if only red fluorescence is monitored. Clearly, this type of line imaging can be extended to two-dimensional imaging using area illumination and a matrix or vidicon detector [8,4].

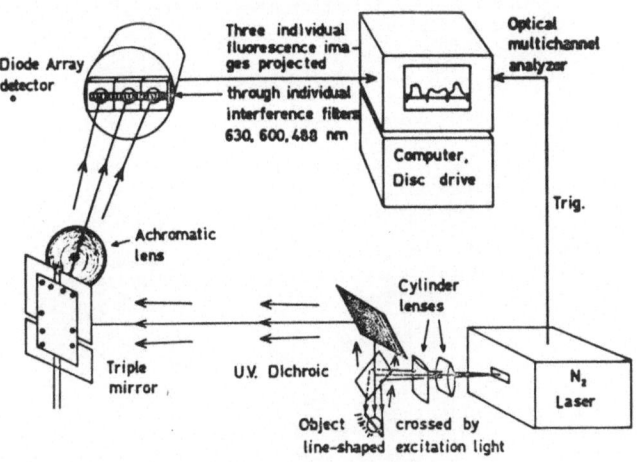

Fig. 8. Experimental arrangement for spatially resolved three-colour fluorescence measurements (Ref. 5).

Acknowledgements

Support by the Lund HPD group, in particular Professors A. Gustafson, D. Killander and U. Stenram is gratefully acknowledged. This work was supported by the Swedish Cancer Foundation (RMC).

References

1. T.J. Dougherty: <u>CRC Critical Reviews in Oncology/Hematology</u>, S. Davis, Ed. (CRC Press, Florida 1984).
2. J. Ankerst, S. Montán, K. Svanberg and S. Svanberg: Appl. Spectr. <u>38</u>, 890 (1984).
3. K. Svanberg, E. Kjellén, J. Ankerst, S. Montán, E. Sjöholm and S. Svanberg: to be published.
4. J. Ankerst, S. Montán, E. Sjöholm, K. Svanberg and S. Svanberg: L.I.A. ICALEO <u>43</u>, 52 (1984).
5. S. Montán, K. Svanberg and S. Svanberg: Opt. Letters <u>10</u>, 56 (1984).
6. U.A.S. Andersson, S.-E. Karlsson and S. Svanberg: Unpublished results.
7. U.A.S. Andersson and T. Persson, Diploma paper, Lund Institute of Technology, Lund Reports on Atomic Physics LRAP-39 (1984).
8. S. Montán and S. Svanberg: Unpublished report (1984).

IR-Laser Spectroscopy for Measurement Applications in the Industrial Environment

H. Ahlberg and S. Lundqvist

Department of Electrical Measurements, Chalmers University of Technology, S-412 96 Gothenburg, Sweden

The increasing availability of suitable and compact tunable lasers in the infrared wavelength range has initiated the development of a new generation of laser spectroscopic instrumentation for measurement applications in the industrial environment. The laser sources most often used in these instruments are the carbon-dioxide and the tunable diode lasers. A number of pollutant and hazardous gases have absorption bands in the infrared spectral region where these lasers operate. This has led to a number of industrial applications such as measurement of the total mass flow of diffusely leaking hydrocarbons from a petrochemical industry [1], air quality management in the working environment [2], and on-line process monitoring and quality control [3].

We will present the results of a five year program to evaluate the use of CO_2-laser based instrumentation for industrial measurement applications. Real time monitoring of several gaseous species in different types of workshops have been made. Another application for this kind of instrument is to trace hydrocarbon leaks in chemical plants. Results from plume profile measurements and measurements of the total mass flow from a petrochemical plant will be presented. The broad and weak spectroscopic features of several solvent vapours at atmospheric pressure have necessitated a series of modifications of the instrument.

The increasing use of very high voltage power transmission systems has created a large interest in new measurement techniques for analyzing the electrical field distribution around critical components. Insulating gases such as SF_6 are used to facilitate a compact design of substation high-voltage components. Since the photon energy is sufficiently low, IR-laser spectroscopy can be an alternative technique for analyzing the field distribution in such components without affecting the sparking potential. It would be possible to probe electrical field-induced absorption lines in the SF_6 gas, but the integrated absorption coefficient is proportional to the square of the electrical field and about 10^6 times weaker than the ordinary linear dipole absorption [4]. The electrical breakdown characteristics of SF_6 will not be affected by the introduction of a small amount ($\sim 0.1\%$) of a trace gas molecule such as NO [5].

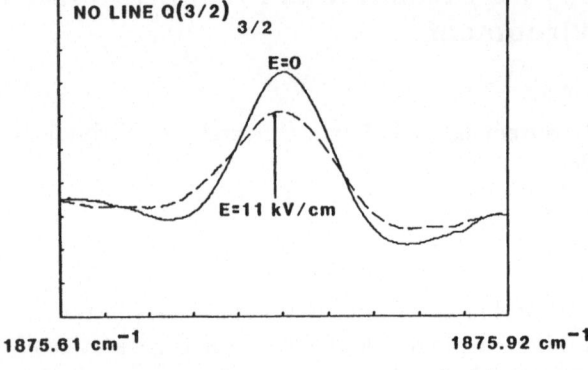

Fig. 1. Second derivative of the $Q(3/2)_{3/2}$ NO line with and without an electric field

In this presentation we will show that the linear Stark effect in this molecule can be utilized as a sensitive field probe in gas-insulated high voltage components. A tunable semiconductor laser was locked onto the line center of the $Q(3/2)_{3/2}$ transition in NO. The laser beam was then directed through an absorption cell equipped with a pair of uniform field electrodes. Figure 1 shows the influence on the second derivative signal when a field strength of 11 kV/cm is applied. The amplitude of the second derivative of the absorption line will approximately vary linearly with the applied electric field as energy is shifted away from the line center. The collision broadening will reduce the measurement signal at low electric fields as shown in Fig. 2. However, high SF_6-gas pressures are only used when the electrical field strength is proportionally higher, so this effect will be no serious limitation to this measurement method.

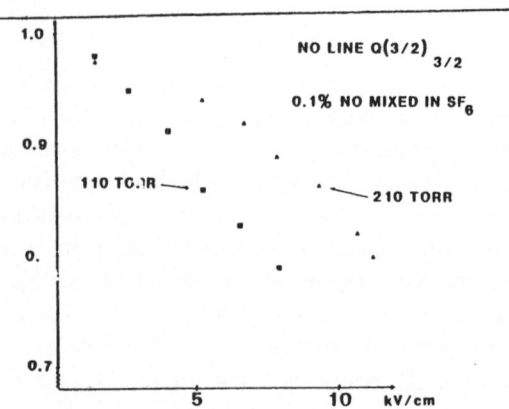

Fig. 2. Second derivative amplitude as a function of electric field strength for two different SF_6 pressures

References
1 U. Persson, J. Johansson, B. Marthinsson, S.T. Eng:
  Appl. Opt. **21**, 4417 (1982)
2 U. Persson, S. Lundqvist, B. Marthinsson, S.T.Eng:
  Appl. Opt. **23**, 998 (1984)
3 D.L. Wall, E.F. Pearson, A.W. Mantz: Proceedings of the 4th
  International IFAC Conference, Ghent, Belgium (1980) p.295
4 E.U. Gordon: Phys. Rev. **41** (1932)
5 J.Dutton, F.M. Harris, D.B. Hughes: Proc. IEE, **121** (1974)

Two-Step Saturated Fluorescence Detection of Atomic Hydrogen in Flames

J.E.M. Goldsmith

Combustion Research Facility, Sandia National Laboratories, Livermore, CA 94550, USA

This paper describes some recent measurements of atomic hydrogen in a hydrogen-air diffusion flame made with a new two-step saturated fluorescence technique [1]. This method provides new capabilities for studying atomic hydrogen and other species, and has been used for the first demonstration of imaging of hydrogen atoms in flames [2].

Details of the excitation-detection scheme used for two-step saturated fluorescence detection of atomic hydrogen appear in the literature [1]. Briefly, a 243-nm laser beam focused at the point to be probed in the atmospheric-pressure hydrogen-air diffusion flame raises some atoms to the 2S state by two-photon excitation of the 1S-2S transition. The focal point of the 243-nm beam is along the centerline of the uniform, unfocused beam of a second laser tuned to 656 nm, which saturates the 2S-3P transition. The 656-nm fluorescence emitted by subsequent 3P-2S decay is imaged onto the entrance slit of a spectrometer, detected by a photomultiplier, and monitored by a gated integrator with a 1-nsec gate set at the peak of the pulse. This scheme provides 656-nm excitation that is uniform in time and in space, eliminating the beam-edge effects encountered in conventional saturated fluorescence.

In addition to saturating the 2S-3P transition (described in detail in [1]), the 656-nm beam shifts and splits the atomic energy levels due to the optical Stark effect. The uniform excitation intensity of the 656-nm beam is very useful for observing this effect. Figure 1 shows a series

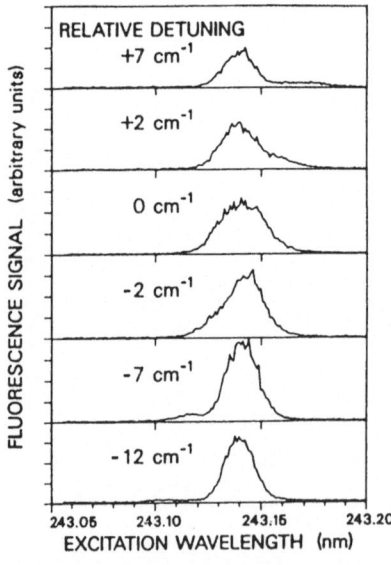

Fig. 1 Two-step saturated fluorescence spectra of atomic hydrogen recorded by scanning the wavelength of the 243-nm beam, with the 656-nm wavelength fixed with various detunings from the 2S-3P resonance. The value for "relative detuning" in the figure refers to the wavelength of the 656-nm beam relative to that which produced a symmetric profile (marked 0 cm⁻¹), with decreasing detuning representing increasing wavelength of the 656-nm laser. All curves are plotted to the same scale, with nothing changed between the recordings of the spectra except the detuning of the 656-nm laser.

of excitation scans recorded with an intensity of ~4 MW/cm² at 656 nm that illustrate the influence of the relative detuning of this wavelength from the 2S-3P resonance on the two-step fluorescence signal. As this detuning is varied, the dominant peak remains essentially fixed in wavelength, but the smaller peak shifts from longer excitation wavelength for positive detuning to shorter wavelength for negative detuning. A theory has been developed that describes this behavior [3]. It can be explained physically by ascribing the larger peak to two-step excitation, where the observed excitation peak of the 1S-2S transition is independent of the wavelength of the second excitation step (with the energy defect made up by the linewidth of the lasers and the transitions), and ascribing the smaller peak to direct three-photon excitation (where the sum of the excitation energies must correspond to the energy of the 1S-3P transition).

Figure 2 shows a series of excitation scans recorded at the indicated intensities with fixed detuning of the 656-nm laser from the 2S-3P transition. The saturation of the two-step signal is evident from the nearly constant amplitude of the larger peak for an eightfold variation in the 656-nm excitation intensity. The amplitude of the smaller peak grows with increasing excitation intensity, with a small but observable increase in separation between the two peaks with increasing intensity.

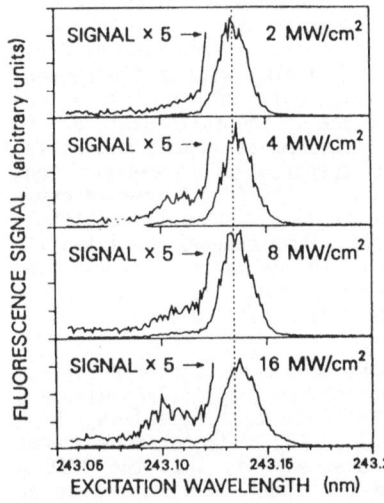

Fig. 2 Two-step saturated fluorescence spectra of atomic hydrogen recorded by scanning the wavelength of the 243-nm beam at the indicated 656-nm excitation intensities, with the detuning of the latter wavelength fixed at the value shown as -7 cm⁻¹ in Fig. 1. All curves are plotted to the same scale, with nothing changed between the recordings of the spectra except the intensity of the 656-nm laser beam.

We plan to repeat these experiments in a low-pressure discharge flow system, where the decreased collisional and Doppler broadening should make the shifts and splittings more evident, and where the theory can be tested in an environment very different from that of an atmospheric-pressure flame. Applications of two-step excitation schemes for studying other species are also being pursued. This research is supported by the U.S. Department of Energy, Office of Basic Energy Sciences.

1. J. E. M. Goldsmith, "Two-Step Saturated Fluorescence Detection of Atomic Hydrogen in Flames," Opt. Lett. **10**, 116 (1985).
2. J. E. M. Goldsmith and R. J. M. Anderson, "Imaging of Atomic Hydrogen in Flames with Two-Step Saturated Fluorescence Detection," Appl. Opt. **24**, 607 (1985).
3. A. M. F. Lau, "Theory of Dynamical Stark Splitting with Application to Laser Combustion Experiments," Sandia Report No. 85-8697, 1985.

Application of Fluorescence to Measurement of Surface Temperature*

L.P. Goss and A.A. Smith

Systems Research Laboratories, Inc., 2800 Indian Ripple Road, Dayton, OH 45440, USA

Diagnosing condensed-phase combustion of energetic materials in intermediate (3-20 MPa) and high (> 300 MPa) pressure regions is extremely challenging. Previous techniques for measuring the surface temperature of these materials are of limited utility. Infrared emissivity measurements are limited to atmospheric-pressure combustion by interference from plume emission. Thermocouples suffer from slow response and poor spatial resolution. An optical diagnostic technique, laser-induced fluorescence, holds the most promise of meeting the criteria for an effective diagnostic for an energetic material undergoing combustion. The absence of naturally occurring fluorescent species in propellants, however, requires doping with a fluorescent material exhibiting temperature-dependent emission over the range of interest. Rare-earth ions which are known to fluoresce in the solid state and be quite sensitive to temperature changes were selected as dopant materials. Temperature-induced changes of the emission spectrum of rare-earth ions include line broadening due to quenching which results from increased crystal lattice vibration, frequency shifting due to the reduced crystal field resulting from thermal expansion of the lattice, and changes in intensity distribution due to Boltzmann thermalization of energy levels. Of the many temperature effects demonstrated by the rare-earth ions, thermalization and lifetimes are the most sensitive in the temperature range 300-1300 K.

Four rare-earth ion crystals were shown to be good candidates for surface temperature measurements by laser-induced fluorescence. Dy^{+3}:CaF_3 displays a thermalization process with an energy gap of 1070 cm^{-1}, resulting in a fluorescence intensity change of $\sim$ 200 over a 700 K temperature - range (Fig. 1). An added feature of Dy^{+3}:LaF_3 fluorescence is that the 4768 Å line displays little change with temperature and, thus, can be used as an internal standard for calibration of the temperature-sensitive transitions. This is important when working in adverse environments where an absolute-intensity measurement of a single line is impossible but a relative measurement to a nearby line is practical. The remaining three crystals display strong multiphonon quenching of the laser-induced fluorescence. The prime example of this process is Cr^{+3}:Al_2O_3 (ruby) which displays an extreme temperature sensitivity to the lifetime of the R-fluorescence lines [see Fig. 2(a)]. These lifetimes change by a factor of $\sim$ 230 over a 500 K temperature range. This extreme sensitivity of the fluorescence lifetime is due to the large energy gap of the R-lines from their next lowest energy levels, which requires a high-order multiphonon process for quenching. Er^{+3}:LaF_3 and Ho^{+3}:LaF_3 are less temperature sensitive than ruby due to the closer spacing of the excited fluorescent lines to low energy levels (lower phonon number for quenching) but, as a result, cover a wider temperature-range [Fig. 2(b)].

---

*Supported by USAF Office of Scientific Research, Contract F49620-83-C-1038.

Presently, measurements of signal-to-noise ratio as a function of crystal size are being performed to determine the lower size limit which will, in turn, determine the temporal response of the dopant to temperature fluctuations on surfaces. The ultimate goal is to imbed these crystals into a propellant-like material to determine the surface temperature during combustion or heating from a laser source.

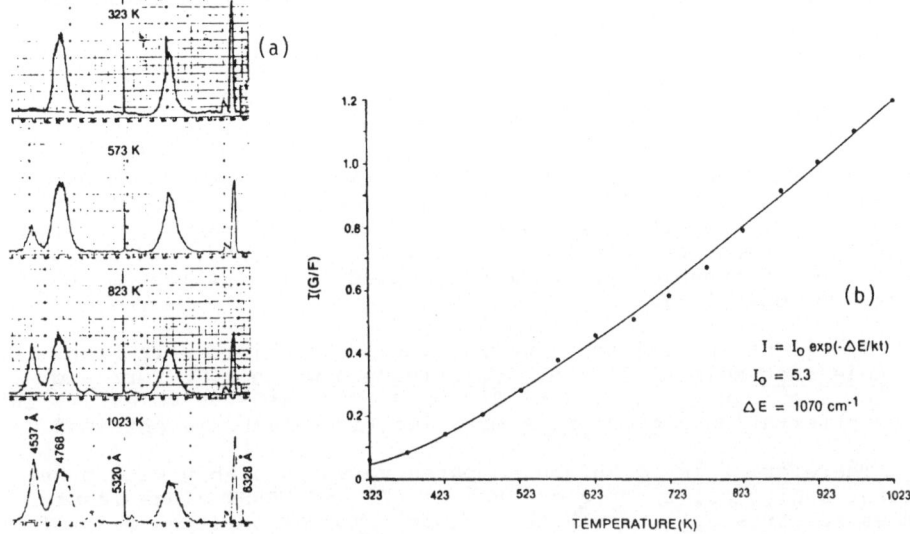

Figure 1. (a) Spectral variation and (b) ratio of G to F fluorescence levels of Dy^{+3}:LaF_3 with temperature

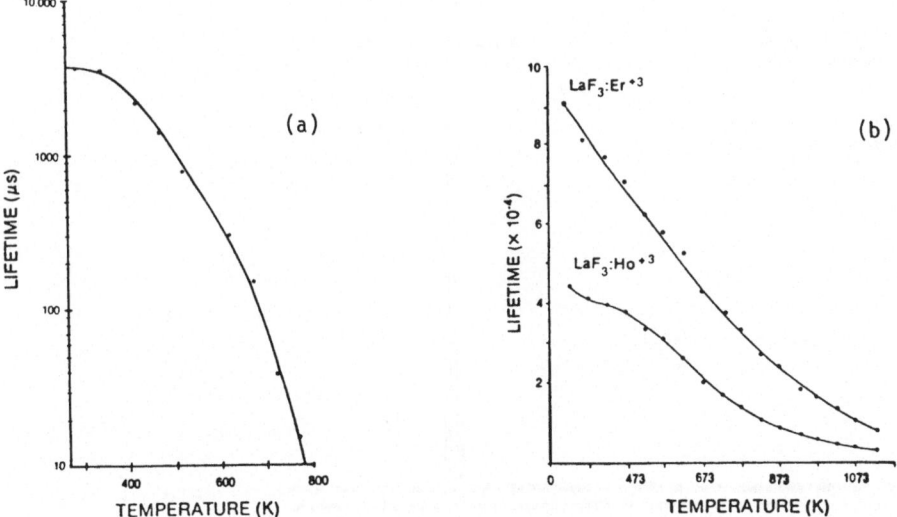

Figure 2. Lifetime variation of (a) Cr^{+3}:Al_2O_3 and (b) Er^{+3}:LaF_3 and Ho^{+3}:LaF_3 with temperature

413

Laser Mass Spectroscopy of Organometallic Compounds and Applications for Laser CVD

M. Stuke and R. Fantoni*

Max-Planck-Institut für Biophysikalische Chemie, Abteilung Laserphysik,
P.O.B. 2841, D-3400 Göttingen, Fed. Rep. of Germany

Fast, sensitive, and selective fingerprint detection of organo-metallic precursors of III-V and II-VI semiconductors, with emphasis on CH_3TeCH_3, $CH_3TeTeCH_3$, $C_2H_5TeC_2H_5$, CH_3SeCH_3, $(CH_3)_3Ga$, $(CH_3)_3In$, AsH_3, and others is achieved, using short and ultra-short pulse tunable dye laser controlled time-of-flight mass spectroscopy [1-3].

Excellent spatial and time resolutions combined with high single shot selectivity and sensitivity down to 10^{-7} Torr and below (corresponding to 3×10^{19} cm^{-3} or about 3000 molecules in the observation region) plus mass identification are obtained.

Therefore this technique compares very favourably with other laser analytical techniques such as LIF and CARS, since laser mass spectroscopy gives both - rough information (i.e. mass of the species) - and in addition spectroscopic information, which makes this method ideally suited for identification and charac-terization of MOCVD (Metal Organic Chemical Vapor Deposition) and LCVD (Laser Chemical Vapor Deposition) processes. Especially

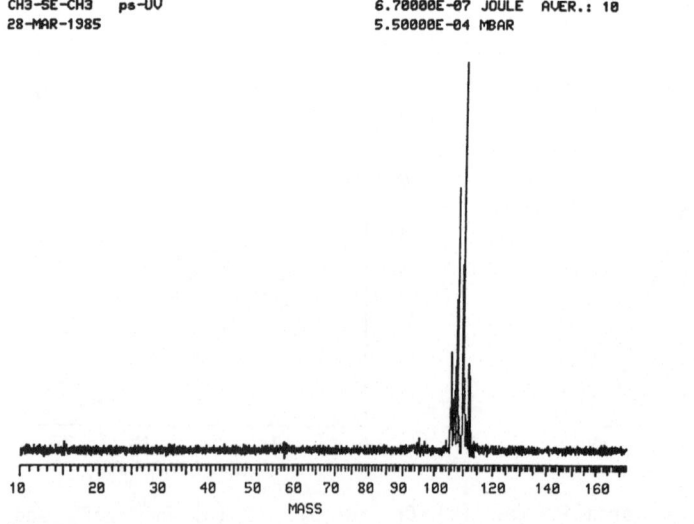

CH3-SE-CH3   ps-UV
28-MAR-1985

6.70000E-07 JOULE AVER.: 10
5.50000E-04 MBAR

MASS

Fig. 1

---

*Permanent address: Laboratorio Spettroscopia Molecolare ENEA CRE Frascati, Frascati (Roma), Italy

in cases where spectroscopic information is lacking, the observation of the parent molecular ion is <u>essential</u> for quick species
identification. Often, however, especially in the case of organometallics, the parent ion has very low abundance. This problem
can be overcome - if necessary - using picosecond laser excitation, as shown in Fig.1, where for CH_3SeCH_3 only the parent
molecular ion is observed. Since this enhancement of the parent
molecular ion abundance is typical for virtually all of the
molecules listed above, laser mass spectroscopy is ideally suited
for the detection and identification of even labile molecules
like organometallics.

References

1. M. Stuke: Appl. Phys. Lett. <u>45</u>, 1175 (1984
2. R. Fantoni, M. Stuke: Appl. Phys. B (1985)
3. Technical details: SUMOTEC GmbH, P. O. B. 3311,
   D-3400 Göttingen

Index of Contributors

Laser Spectroscopy VI

Proceedings of the Sixth International Conference, Interlaken, Switzerland, June 27–July 1, 1983
Editors: **H.P.Weber, W.Lüthy**

1983. 258 figures. XVII, 442 pages. (Springer Series in Optical Sciences, Volume 40). ISBN 3-540-12957-X

Contents: Photons in Spectroscopy. – Spectroscopy of Elementary Systems. – Coherent Processes. – Novel Spectroscopy. – High Selectivity Spectroscopy. – High Resolution Spectroscopy. – Cooling and Trapping. – Collisions and Thermal Effects on Spectroscopy. – Atomic Spectroscopy. – Rydberg-State Spectroscopy. – Molecular Spectroscopy. – Transient Spectroscopy. – Surface Spectroscopy. – NL-Spectroscopy. – Raman and CARS. – Double Resonance and Multiphoton Processes. – XUV – VUV Generation. – New Laser Sources and Detectors. – Index of Contributors.

Laser Spectroscopy V

Proceedings of the Fifth International Conference, Jasper Park Lodge, Alberta, Canada, June 29–July 3, 1981

Editors: **A.R.W.McKellar, T.Oka, B.P.Stoicheff**

1981. 319 figures. XI, 495 pages. (Springer Series in Optical Sciences, Volume 30). ISBN 3-540-10914-5

Laser Spectroscopy IV

Proceedings of the Fourth International Conference, Rottach-Egern, Federal Republic of Germany, June 11–15, 1979

Editors: **H.Walther, K.W.Rothe**

1979. 411 figures, 19 tables. XIII, 652 pages (Springer Series in Optical Sciences, Volume 21) ISBN 3-540-09766-X

Springer-Verlag
Berlin
Heidelberg
New York
Tokyo

Laser Spectroscopy III

Proceedings of the Third International Conference, Jackson Lake Lodge, Wyoming, USA, July 4–8, 1977

Editors: **J.L.Hall, J.L.Carlsten**

1977. 296 figures. XI, 468 pages. (Springer Series in Optical Sciences, Volume 7). ISBN 3-540-08543-2

Picosecond Phenomena II

Proceedings of the Second International Conference on
Picosecond Phenomena, Cape Cod, Massachusetts, USA,
June 18–20, 1980

Editors: **R. Hochstrasser, W. Kaiser, C. V. Shank**

1980. 252 figures, 17 tables. XII, 382 pages. (Springer Series
in Chemical Physics, Volume 14). ISBN 3-540-10403-8

Picosecond Phenomena III

Proceedings of the Third International Conference on Picosecond Phenomena, Garmisch-Partenkirchen, Federal
Republic of Germany, June 16–18, 1982

Editors: **K. B. Eisenthal, R. M. Hochstrasser, W. Kaiser,
A. Laubereau**

1982. 288 figures. XIII, 401 pages. (Springer Series in
Chemical Physics, Volume 23). ISBN 3-540-11912-4

Ultrafast Phenomena IV

Proceedings of the Fourth International Conference,
Monterey, California, June 11–15, 1984

Editors: **D. H. Auston, K. B. Eisenthal**

1984. 370 figures. XVI, 509 pages. (Springer Series in
Chemical Physics, Volume 38). ISBN 3-540-13834-X

W. Demtröder
Laser Spectroscopy

Basic Concepts and Instrumentation

2nd corrected printing. 1982. 431 figures. XIII, 696 pages
(Springer Series in Chemical Physics, Volume 5)
ISBN 3-540-10343-0

B. Y. Zel'dovich, N. F. Pilipetsky, V. V. Shkunov
Principles of Phase Conjugation

1985. 70 figures. X, 250 pages. (Springer Series in Optical
Sciences, Volume 42). ISBN 3-540-13458-1

Contents: Introduction to Optical Phase Conjugation. –
Physics of Stimulated Scattering. – Properties of Speckle-
Inhomogeneous Fields. – OPC by Backward Stimulated
Scattering. – Specific Features of OPC-SS. – OPC in Four-
Wave Mixing. – Nonlinear Mechanisms for FWM. – Other
Methods of OPC. – References. – Subject Index.

Springer-Verlag
Berlin
Heidelberg
New York
Tokyo